Astronomy 2e

Volume 2/2:
- Chapters 17-30
- Appendices A-M
- Index

SENIOR CONTRIBUTING AUTHORS

Andrew Fraknoi, Foothill College
David Morrison, National Aeronautics and Space Administration
Sidney C. Wolff, National Optical Astronomy Observatory (Emeritus)

ORIGINAL PUBLICATION YEAR 2022

1 2 3 4 5 6 7 8 9 10 RS 21

⊟ CONTENTS

Preface 1

1 **Science and the Universe: A Brief Tour 9**

Introduction 9
1.1 The Nature of Astronomy 11
1.2 The Nature of Science 11
1.3 The Laws of Nature 12
1.4 Numbers in Astronomy 13
1.5 Consequences of Light Travel Time 15
1.6 A Tour of the Universe 16
1.7 The Universe on the Large Scale 21
1.8 The Universe of the Very Small 24
1.9 A Conclusion and a Beginning 26
For Further Exploration 28

2 **Observing the Sky: The Birth of Astronomy 29**

Thinking Ahead 29
2.1 The Sky Above 30
2.2 Ancient Astronomy 39
2.3 Astrology and Astronomy 46
2.4 The Birth of Modern Astronomy 50
Key Terms 57
Summary 57
For Further Exploration 58
Collaborative Group Activities 59
Exercises 60

3 **Orbits and Gravity 63**

Thinking Ahead 63
3.1 The Laws of Planetary Motion 64
3.2 Newton's Great Synthesis 69
3.3 Newton's Universal Law of Gravitation 74
3.4 Orbits in the Solar System 78
3.5 Motions of Satellites and Spacecraft 81
3.6 Gravity with More Than Two Bodies 84
Key Terms 88
Summary 89
For Further Exploration 90
Collaborative Group Activities 91
Exercises 91

4 | **Earth, Moon, and Sky   95**

Thinking Ahead   95
4.1 Earth and Sky   96
4.2 The Seasons   99
4.3 Keeping Time   105
4.4 The Calendar   108
4.5 Phases and Motions of the Moon   111
4.6 Ocean Tides and the Moon   116
4.7 Eclipses of the Sun and Moon   119
Key Terms   126
Summary   126
For Further Exploration   127
Collaborative Group Activities   129
Exercises   130

5 | **Radiation and Spectra   135**

Thinking Ahead   135
5.1 The Behavior of Light   136
5.2 The Electromagnetic Spectrum   142
5.3 Spectroscopy in Astronomy   150
5.4 The Structure of the Atom   154
5.5 Formation of Spectral Lines   159
5.6 The Doppler Effect   163
Key Terms   168
Summary   169
For Further Exploration   170
Collaborative Group Activities   171
Exercises   171

6 | **Astronomical Instruments   175**

Thinking Ahead   175
6.1 Telescopes   176
6.2 Telescopes Today   182
6.3 Visible-Light Detectors and Instruments   192
6.4 Radio Telescopes   195
6.5 Observations outside Earth's Atmosphere   202
6.6 The Future of Large Telescopes   207
Key Terms   210
Summary   210
For Further Exploration   211
Collaborative Group Activities   212
Exercises   213

7 | **Other Worlds: An Introduction to the Solar System   217**

Thinking Ahead   217
7.1 Overview of Our Planetary System   218
7.2 Composition and Structure of Planets   229
7.3 Dating Planetary Surfaces   233
7.4 Origin of the Solar System   236
Key Terms   239
Summary   239
For Further Exploration   240
Collaborative Group Activities   241
Exercises   242

8 | **Earth as a Planet   245**

Thinking Ahead   245
8.1 The Global Perspective   246
8.2 Earth's Crust   250
8.3 Earth's Atmosphere   257
8.4 Life, Chemical Evolution, and Climate Change   261
8.5 Cosmic Influences on the Evolution of Earth   266
Key Terms   272
Summary   273
For Further Exploration   273
Collaborative Group Activities   275
Exercises   276

9 | **Cratered Worlds   279**

Thinking Ahead   279
9.1 General Properties of the Moon   280
9.2 The Lunar Surface   286
9.3 Impact Craters   290
9.4 The Origin of the Moon   296
9.5 Mercury   297
Key Terms   304
Summary   304
For Further Exploration   304
Collaborative Group Activities   306
Exercises   307

10 | **Earthlike Planets: Venus and Mars   311**

Thinking Ahead   311
10.1 The Nearest Planets: An Overview   311
10.2 The Geology of Venus   317
10.3 The Massive Atmosphere of Venus   322

10.4 The Geology of Mars   325
10.5 Water and Life on Mars   334
10.6 Divergent Planetary Evolution   345
Key Terms   347
Summary   347
For Further Exploration   348
Collaborative Group Activities   350
Exercises   350

11   The Giant Planets   353

Thinking Ahead   353
11.1 Exploring the Outer Planets   354
11.2 The Giant Planets   359
11.3 Atmospheres of the Giant Planets   365
Key Terms   375
Summary   375
For Further Exploration   375
Collaborative Group Activities   377
Exercises   378

12   Rings, Moons, and Pluto   381

Thinking Ahead   381
12.1 Ring and Moon Systems Introduced   382
12.2 The Galilean Moons of Jupiter   383
12.3 Titan and Triton   392
12.4 Pluto and Charon   397
12.5 Planetary Rings (and Enceladus)   404
Key Terms   415
Summary   415
For Further Exploration   416
Collaborative Group Activities   418
Exercises   419

13   Comets and Asteroids: Debris of the Solar System   421

Thinking Ahead   421
13.1 Asteroids   422
13.2 Asteroids and Planetary Defense   432
13.3 The "Long-Haired" Comets   436
13.4 The Origin and Fate of Comets and Related Objects   445
Key Terms   453
Summary   453
For Further Exploration   454
Collaborative Group Activities   456

Exercises   457

14 Cosmic Samples and the Origin of the Solar System   459

Thinking Ahead   459
14.1 Meteors   460
14.2 Meteorites: Stones from Heaven   465
14.3 Formation of the Solar System   470
14.4 Comparison with Other Planetary Systems   476
14.5 Planetary Evolution   480
Key Terms   486
Summary   486
For Further Exploration   487
Collaborative Group Activities   489
Exercises   489

15 The Sun: A Garden-Variety Star   493

Thinking Ahead   493
15.1 The Structure and Composition of the Sun   494
15.2 The Solar Cycle   504
15.3 Solar Activity above the Photosphere   509
15.4 Space Weather   513
Key Terms   520
Summary   520
For Further Exploration   521
Collaborative Group Activities   523
Exercises   523

16 The Sun: A Nuclear Powerhouse   527

Thinking Ahead   527
16.1 Sources of Sunshine: Thermal and Gravitational Energy   527
16.2 Mass, Energy, and the Theory of Relativity   530
16.3 The Solar Interior: Theory   539
16.4 The Solar Interior: Observations   545
Key Terms   551
Summary   551
For Further Exploration   552
Collaborative Group Activities   552
Exercises   553

17 Analyzing Starlight   557

Thinking Ahead   557

17.1 The Brightness of Stars   557
17.2 Colors of Stars   561
17.3 The Spectra of Stars (and Brown Dwarfs)   563
17.4 Using Spectra to Measure Stellar Radius, Composition, and Motion   570
Key Terms   579
Summary   579
For Further Exploration   580
Collaborative Group Activities   581
Exercises   581

18 | **The Stars: A Celestial Census   585**

Thinking Ahead   585
18.1 A Stellar Census   585
18.2 Measuring Stellar Masses   589
18.3 Diameters of Stars   596
18.4 The H–R Diagram   601
Key Terms   610
Summary   610
For Further Exploration   611
Collaborative Group Activities   611
Exercises   612

19 | **Celestial Distances   617**

Thinking Ahead   617
19.1 Fundamental Units of Distance   618
19.2 Surveying the Stars   621
19.3 Variable Stars: One Key to Cosmic Distances   630
19.4 The H–R Diagram and Cosmic Distances   636
Key Terms   640
Summary   640
For Further Exploration   641
Collaborative Group Activities   642
Exercises   643

20 | **Between the Stars: Gas and Dust in Space   647**

Thinking Ahead   647
20.1 The Interstellar Medium   648
20.2 Interstellar Gas   651
20.3 Cosmic Dust   659
20.4 Cosmic Rays   666
20.5 The Life Cycle of Cosmic Material   668
20.6 Interstellar Matter around the Sun   670
Key Terms   672

Summary 672

For Further Exploration 673

Collaborative Group Activities 674

Exercises 675

21 The Birth of Stars and the Discovery of Planets outside the Solar System 679

Thinking Ahead 679

21.1 Star Formation 680

21.2 The H–R Diagram and the Study of Stellar Evolution 689

21.3 Evidence That Planets Form around Other Stars 692

21.4 Planets beyond the Solar System: Search and Discovery 695

21.5 Exoplanets Everywhere: What We Are Learning 703

21.6 New Perspectives on Planet Formation 709

Key Terms 713

Summary 713

For Further Exploration 714

Collaborative Group Activities 716

Exercises 717

22 Stars from Adolescence to Old Age 719

Thinking Ahead 719

22.1 Evolution from the Main Sequence to Red Giants 720

22.2 Star Clusters 726

22.3 Checking Out the Theory 729

22.4 Further Evolution of Stars 735

22.5 The Evolution of More Massive Stars 743

Key Terms 747

Summary 747

For Further Exploration 748

Collaborative Group Activities 749

Exercises 750

23 The Death of Stars 753

Thinking Ahead 753

23.1 The Death of Low-Mass Stars 754

23.2 Evolution of Massive Stars: An Explosive Finish 759

23.3 Supernova Observations 766

23.4 Pulsars and the Discovery of Neutron Stars 772

23.5 The Evolution of Binary Star Systems 779

23.6 The Mystery of the Gamma-Ray Bursts 782

Key Terms 791

Summary 791

For Further Exploration   792
Collaborative Group Activities   794
Exercises   795

24 Black Holes and Curved Spacetime   799

Thinking Ahead   799
24.1 Introducing General Relativity   800
24.2 Spacetime and Gravity   805
24.3 Tests of General Relativity   808
24.4 Time in General Relativity   811
24.5 Black Holes   813
24.6 Evidence for Black Holes   820
24.7 Gravitational Wave Astronomy   824
Key Terms   829
Summary   829
For Further Exploration   830
Collaborative Group Activities   832
Exercises   833

25 The Milky Way Galaxy   835

Thinking Ahead   835
25.1 The Architecture of the Galaxy   836
25.2 Spiral Structure   844
25.3 The Mass of the Galaxy   848
25.4 The Center of the Galaxy   851
25.5 Stellar Populations in the Galaxy   857
25.6 The Formation of the Galaxy   859
Key Terms   865
Summary   865
For Further Exploration   866
Collaborative Group Activities   868
Exercises   868

26 Galaxies   871

Thinking Ahead   871
26.1 The Discovery of Galaxies   872
26.2 Types of Galaxies   875
26.3 Properties of Galaxies   880
26.4 The Extragalactic Distance Scale   883
26.5 The Expanding Universe   887
Key Terms   894
Summary   894
For Further Exploration   895

Collaborative Group Activities   896
Exercises   897

27 | **Active Galaxies, Quasars, and Supermassive Black Holes   899**

Thinking Ahead   899
27.1 Quasars   899
27.2 Supermassive Black Holes: What Quasars Really Are   907
27.3 Quasars as Probes of Evolution in the Universe   915
Key Terms   922
Summary   922
For Further Exploration   922
Collaborative Group Activities   924
Exercises   925

28 | **The Evolution and Distribution of Galaxies   927**

Thinking Ahead   927
28.1 Observations of Distant Galaxies   928
28.2 Galaxy Mergers and Active Galactic Nuclei   935
28.3 The Distribution of Galaxies in Space   942
28.4 The Challenge of Dark Matter   955
28.5 The Formation and Evolution of Galaxies and Structure in the Universe   962
Key Terms   969
Summary   969
For Further Exploration   970
Collaborative Group Activities   972
Exercises   974

29 | **The Big Bang   977**

Thinking Ahead   977
29.1 The Age of the Universe   978
29.2 A Model of the Universe   985
29.3 The Beginning of the Universe   993
29.4 The Cosmic Microwave Background   999
29.5 What Is the Universe Really Made Of?   1006
29.6 The Inflationary Universe   1012
29.7 The Anthropic Principle   1016
Key Terms   1019
Summary   1019
For Further Exploration   1021
Collaborative Group Activities   1022
Exercises   1023

30   Life in the Universe   1027

   Thinking Ahead   1027
   30.1 The Cosmic Context for Life   1028
   30.2 Astrobiology   1030
   30.3 Searching for Life beyond Earth   1039
   30.4 The Search for Extraterrestrial Intelligence   1047
   Key Terms   1058
   Summary   1058
   For Further Exploration   1059
   Collaborative Group Activities   1061
   Exercises   1062

A   How to Study for an Introductory Astronomy Class   1065

B   Astronomy Websites, Images, and Apps   1067

C   Scientific Notation   1073

D   Units Used in Science   1077

E   Some Useful Constants for Astronomy   1079

F   Physical and Orbital Data for the Planets   1081

G   Selected Moons of the Planets   1083

H   Future Total Eclipses   1087

I   The Nearest Stars, Brown Dwarfs, and White Dwarfs   1089

J   The Brightest Twenty Stars   1093

K   The Chemical Elements   1095

L   The Constellations   1101

M   Star Chart and Sky Event Resources   1107

Preface

Welcome to *Astronomy 2e*, an OpenStax resource. This textbook was written to increase student access to high-quality learning materials, maintaining highest standards of academic rigor at little to no cost.

About OpenStax

OpenStax is a nonprofit based at Rice University, and it's our mission to improve student access to education. Our first openly licensed college textbook was published in 2012 and our library has since scaled to over 50 books for college and AP® courses used by hundreds of thousands of students. OpenStax Tutor, our low-cost personalized learning tool, is being used in college courses throughout the country. Through our partnerships with philanthropic foundations and our alliance with other educational resource organizations, OpenStax is breaking down the most common barriers to learning and empowering students and instructors to succeed.

About OpenStax resources

Customization

Astronomy 2e is licensed under a Creative Commons Attribution 4.0 International (CC BY) license, which means that you can distribute, remix, and build upon the content, as long as you provide attribution to OpenStax and its content contributors.

Because our books are openly licensed, you are free to use the entire book or pick and choose the sections that are most relevant to the needs of your course. Feel free to remix the content by assigning your students certain chapters and sections in your syllabus, in the order that you prefer. You can even provide a direct link in your syllabus to the sections in the web view of your book.

Instructors also have the option of creating a customized version of their OpenStax book. The custom version can be made available to students in low-cost print or digital form through their campus bookstore. Visit your book page on OpenStax.org for more information.

Errata

All OpenStax textbooks undergo a rigorous review process. However, like any professional-grade textbook, errors sometimes occur. Since our books are web based, we can make updates periodically when deemed pedagogically necessary. If you have a correction to suggest, submit it through the link on your book page on OpenStax.org. Subject-matter experts review all errata suggestions. OpenStax is committed to remaining transparent about all updates, so you will also find a list of past errata changes on your book page on OpenStax.org.

Format

You can access this textbook for free in web view or PDF through OpenStax.org, and for a low cost in print.

About *Astronomy 2e*

Astronomy 2e is written in clear non-technical language, with the occasional touch of humor and a wide range of clarifying illustrations. It has many analogies drawn from everyday life to help non-science majors appreciate, on their own terms, what our modern exploration of the universe is revealing. The book can be used for either a one-semester or two-semester introductory course (bear in mind, you can customize your version and include only those chapters or sections you will be teaching.) It is made available free of charge in electronic form (and low cost in printed form) to students around the world.

Coverage and scope

Astronomy 2e by its three senior authors (see below) and was updated, reviewed, and vetted by a wide range of astronomers and astronomy educators in a strong community effort. It is designed to meet scope and sequence requirements of introductory astronomy courses nationwide.

Chapter 1: Science and the Universe: A Brief Tour

Chapter 2: Observing the Sky: The Birth of Astronomy

Chapter 3: Orbits and Gravity

Chapter 4: Earth, Moon, and Sky

Chapter 5: Radiation and Spectra

Chapter 6: Astronomical Instruments

Chapter 7: Other Worlds: An Introduction to the Solar System

Chapter 8: Earth as a Planet

Chapter 9: Cratered Worlds

Chapter 10: Earthlike Planets: Venus and Mars

Chapter 11: The Giant Planets

Chapter 12: Rings, Moons, and Pluto

Chapter 13: Comets and Asteroids: Debris of the Solar System

Chapter 14: Cosmic Samples and the Origin of the Solar System

Chapter 15: The Sun: A Garden-Variety Star

Chapter 16: The Sun: A Nuclear Powerhouse

Chapter 17: Analyzing Starlight

Chapter 18: The Stars: A Celestial Census

Chapter 19: Celestial Distances

Chapter 20: Between the Stars: Gas and Dust in Space

Chapter 21: The Birth of Stars and the Discovery of Planets outside the Solar System

Chapter 22: Stars from Adolescence to Old Age

Chapter 23: The Death of Stars

Chapter 24: Black Holes and Curved Spacetime

Chapter 25: The Milky Way Galaxy

Chapter 26: Galaxies

Chapter 27: Active Galaxies, Quasars, and Supermassive Black Holes

Chapter 28: The Evolution and Distribution of Galaxies

Chapter 29: The Big Bang

Chapter 30: Life in the Universe

Appendix A: How to Study for Your Introductory Astronomy Course

Appendix B: Astronomy Websites, Pictures, and Apps

Appendix C: Scientific Notation

Appendix D: Units Used in Science

Appendix E: Some Useful Constants for Astronomy

Appendix F: Physical and Orbital Data for the Planets

Appendix G: Selected Moons of the Planets

Appendix H: Upcoming Total Eclipses

Appendix I: The Nearest Stars, Brown Dwarfs, and White Dwarfs

Appendix J: The Brightest Twenty Stars

Appendix K: The Chemical Elements

Appendix L: The Constellations

Appendix M: Star Charts and Sky Event Resources

Currency and accuracy

Astronomy 2e has information and images from the LIGO and VIRGO gravitational-wave detectors, the Perseverance/Ingenuity mission to Mars, the Juno mission to Jupiter, and many other recent projects in astronomy. The discussion of exoplanets has been updated with recent information—indicating not just individual examples, but trends in what sorts of planets seem to be most common. Black holes receive their own chapter, and the role of supermassive black holes in active galaxies and galaxy evolution is clearly explained. Chapters have been reviewed by subject-matter experts for accuracy and currency.

Flexibility

Because there are many different ways to teach introductory astronomy, we have made the text as flexible as we could. Math examples are shown in separate sections throughout, so that you can leave out the math or require it as you deem best. Each section of a chapter treats a different aspect of the topic being covered; a number of sections could be omitted in shorter overview courses and can be included where you need more depth. And, as we have already discussed, you can customize the book in a variety of ways that have never been possible in traditional textbooks.

Student-centered focus

This book is written to help students understand the big picture rather than get lost in random factoids to memorize. The language is accessible and inviting. Helpful diagrams and summary tables review and encapsulate the ideas being covered. Each chapter contains interactive group activities you can assign to help students work in teams and pool their knowledge.

Interactive online resources

Interesting "Links to Learning" are scattered throughout the chapters, which direct students to online animations, short videos, or enrichment readings to enhance their learning. Also, the resources listed at the end of each chapter include links to websites and other useful educational videos.

Feature boxes that help students think outside the box

A variety of feature boxes within the chapters connect astronomy to the students' other subjects and humanize the face of astronomy by highlighting the lives of the men and women who have been key to its progress. Besides the math examples that we've already mentioned, the boxes include:

> **Making Connections.** This feature connects the chapter topic to students' experiences with other fields, from poetry to engineering, popular culture, and natural disasters.
> **Voyagers in Astronomy.** This feature presents brief and engaging biographies of the people behind historically significant discoveries, as well as emerging research.
> **Astronomy Basics.** This feature explains basic science concepts that we often (incorrectly) assume students know from earlier classes.
> **Seeing for Yourself.** This feature provides practical ways that students can make astronomical observations on their own.

End-of-chapter materials to extend students' learning

> **Chapter Summaries.** Summaries give the gist of each section for easy review.
> **For Further Exploration.** This section offers a list of suggested articles, websites, and videos so students can delve into topics of interest, whether for their own learning, for homework, extra credit, or papers.
> **Review Questions.** Review questions allow students to show you (or themselves) how well they understood the chapter.
> **Thought Questions.** Thought questions help students assess their learning by asking for critical reflection on principles or ideas in the chapter.
> **Figuring For Yourself.** Mathematical questions, using only basic algebra and arithmetic, allow students to apply the math principles given in the example boxes throughout the chapter.
> **Collaborative Group Activities.** This section suggests ideas for group discussion, research, or reports.

Beautiful art program

Our comprehensive art program is designed to enhance students' understanding of concepts through clear and effective illustrations, diagrams, and photographs. Here are a few examples.

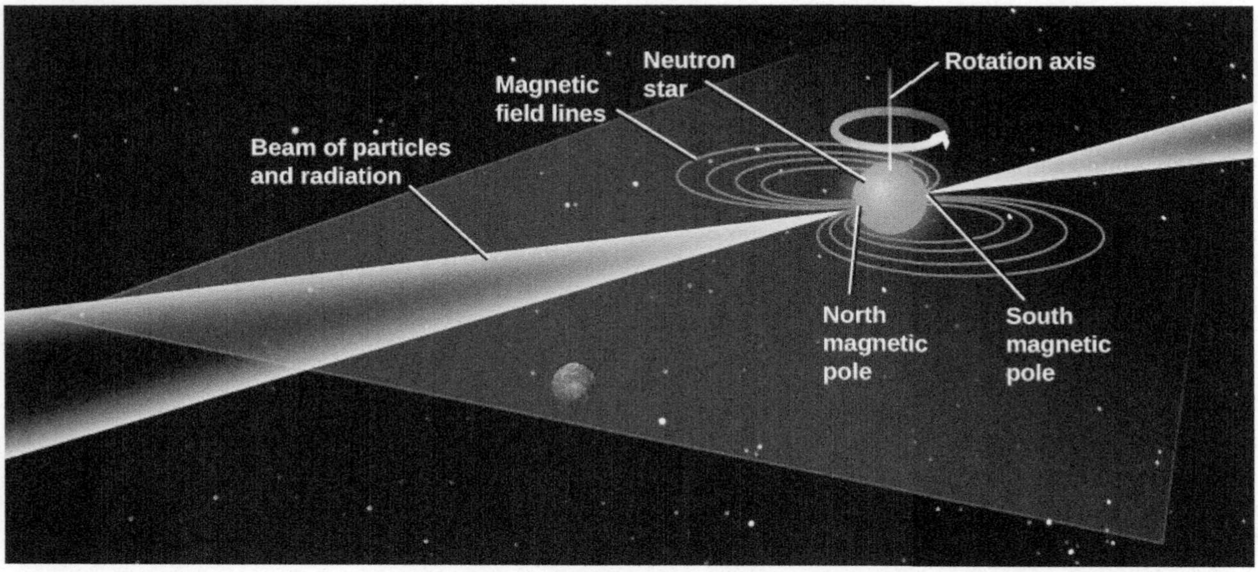

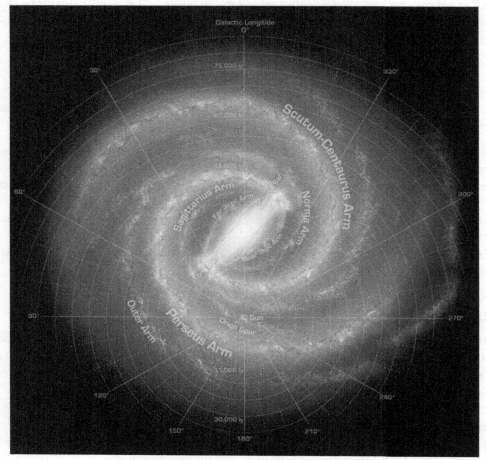

Figure 1 How a Pulsar Beam Sweeps over Earth.

Figure 2 Structure of the Milky Way Galaxy.

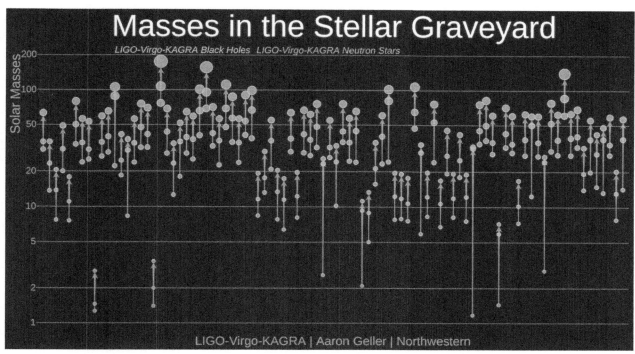

Figure 3 Masses in the Stellar Graveyard.

Figure 4 Pluto Close Up.

Additional resources

Student and instructor resources

We've compiled additional resources for both students and instructors, including Getting Started Guides, PowerPoint slides, and an instructor answer guide. Instructor resources require a verified instructor account, which you can apply for when you log in or create your account on OpenStax.org. Take advantage of these resources to supplement your OpenStax book.

Community Hubs

OpenStax partners with the Institute for the Study of Knowledge Management in Education (ISKME) to offer Community Hubs on OER Commons – a platform for instructors to share community-created resources that support OpenStax books, free of charge. Through our Community Hubs, instructors can upload their own materials or download resources to use in their own courses, including additional ancillaries, teaching material, multimedia, and relevant course content. We encourage instructors to join the hubs for the subjects most relevant to your teaching and research as an opportunity both to enrich your courses and to engage with other faculty.

To reach the Community Hubs, visit https://www.oercommons.org/hubs/OpenStax (https://www.oercommons.org/hubs/OpenStax).

Partner resources

OpenStax Partners are our allies in the mission to make high-quality learning materials affordable and accessible to students and instructors everywhere. Their tools integrate seamlessly with our OpenStax titles at a low cost. To access the partner resources for your text, visit your book page on OpenStax.org.

About the authors

Senior contributing authors

Figure 5 Senior contributing authors: Andrew Fraknoi (left), David Morrison (center), Sidney C. Wolff (right)

Andrew Fraknoi, Foothill College

Andrew Fraknoi is Chair of the Astronomy Department at Foothill College and served as the Executive Director of the Astronomical Society of the Pacific from 1978–1992. His work with the society included editing *Mercury* Magazine, *Universe in the Classroom*, and *Astronomy Beat*. He's taught at San Francisco State University, Canada College, and the University of California Extension. He is editor/co-author of *The Universe at Your Fingertips 2.0*, a collection of teaching activities, and co-author of *Solar Science,* a book for middle-school teachers. He was co-author of a syndicated newspaper column on astronomy, and appears regularly on local and national radio. With Sidney Wolff, he was founder of *Astronomy Education Review*. He serves on the Board of Trustees of the SETI Institute and on the Lick Observatory Council. In addition, he has organized six national symposia on teaching introductory astronomy. He received the Klumpke-Roberts Prize of the ASP, the Gemant Award of the American Institute of Physics, and the Faraday Award of the NSTA.

David Morrison, National Aeronautics and Space Administration

David Morrison is a Senior Scientist at NASA Ames Research Center. He received his PhD in astronomy from Harvard, where he was one of Carl Sagan's graduate students. He is a founder of the field of astrobiology and is known for research on small bodies in the solar system. He spent 17 years at University of Hawaii's Institute for Astronomy and the Department of Physics and Astronomy. He was Director of the IRTF at Maunakea Observatory. Morrison has held senior NASA positions including Chief of the Ames Space Science Division and founding Director of the Lunar Science Institute. He's been on science teams for the Voyager, Galileo, and Kepler missions. Morrison received NASA Outstanding Leadership Medals and the NASA Exceptional Achievement Medal. He was awarded the AAS Carl Sagan medal and the ASP Klumpke-Roberts prize. Committed to the struggle against pseudoscience, he serves as Contributing Editor of *Skeptical Inquirer* and on the Advisory Council of the National Center for Science Education.

Sidney C. Wolff, National Optical Astronomy Observatories (Emeritus)

After receiving her PhD from the UC Berkeley, Dr. Wolff was involved with the astronomical development of Maunakea. In 1984, she became the Director of Kitt Peak National Observatory, and was director of National Optical Astronomy Observatory. Most recently, she led the design and development of the 8.4-meter Large Synoptic Survey Telescope. Dr. Wolff has published over ninety refereed papers on star formation and stellar atmospheres. She has served as President of the AAS and the ASP. Her recently published book, *The Boundless Universe: Astronomy in the New Age of Discovery*, won the 2016 IPPY (Independent Publisher Book Awards) Silver Medal in Science.

All three senior contributing authors have received the Education Prize of the American Astronomical Society and have had an asteroid named after them by the International Astronomical Union. They have worked together on a series of astronomy textbooks over the past two decades.

Contributing authors

John Beck, Stanford University
Susan D. Benecchi, Planetary Science Institute
John Bochanski, Rider University
Howard Bond, Pennsylvania State University, Emeritus, Space Telescope Science Institute
Jennifer Carson, Occidental College
Bryan Dunne, University of Illinois at Urbana-Champaign
Martin Elvis, Harvard-Smithsonian Center for Astrophysics
Debra Fischer, Yale University
Heidi Hammel, Association of Universities for Research in Astronomy
Tori Hoehler, NASA Ames Research Center
Douglas Ingram, Texas Christian University
Steven Kawaler, Iowa State University
Lloyd Knox, University of California, Davis
Mark Krumholz, Australian National University
James Lowenthal, Smith College
Geoff Mathews, Foothill College
Siobahn Morgan, University of Northern Iowa
Daniel Perley, California Institute of Technology
Claire Raftery, National Solar Observatory
Deborah Scherrer, retired, Stanford University
Phillip Scherrer, Stanford University
Sanjoy Som, Blue Marble Space Institute of Science, NASA Ames Research Center
Wes Tobin, Indiana University East
William H. Waller, retired, Tufts University, Rockport (MA) Public Schools
Todd Young, Wayne State College

Reviewers

Elisabeth R. Adams, Planetary Science Institute
Alfred N. Alaniz, San Antonio College
Charles Allison, Texas A&M University–Kingsville
Douglas Arion, Carthage College
Timothy Barker, Wheaton College
Marshall Bartlett, The Hockaday School
Charles Benesh, Wesleyan College
Gerald B. Cleaver, Baylor University
Kristi Concannon, King's College
Anthony Crider, Elon University
Scott Engle, Villanova University
Matthew Fillingim, University of California, Berkeley
Robert Fisher, University of Massachusetts, Dartmouth
Carrie Fitzgerald, Montgomery College
Christopher Fuse, Rollins College
Shila Garg, Emeritus, The College of Wooster
Richard Gelderman, Western Kentucky University
Lee Hartman, University of Michigan
Beth Hufnagel, Anne Arundel Community College
Francine Jackson, Brown University
Joseph Jensen, Utah Valley University
John Kielkopf, University of Louisville
James C. Lombardi, Jr., Allegheny College
Amy Lovell, Agnes Scott College
Charles Niederriter, Gustavus Adolphus College
Richard Olenick, University of Dallas
Matthew Olmstead, King's College
Zoran Pazameta, Eastern Connecticut State University
David Quesada, Saint Thomas University
Valerie A. Rapson, Dudley Observatory
Joseph Ribaudo, Utica College
Dean Richardson, Xavier University of Louisiana
Andrew Rivers, Northwestern University
Marc Sher, College of William & Mary
Christopher Sirola, University of Southern Mississippi
Ran Sivron, Baker University
J. Allyn Smith, Austin Peay State University
Jason Smolinski, Calvin College
Michele Thornley, Bucknell University
Richard Webb, Union College
Terry Willis, Chesapeake College
David Wood, San Antonio College
Jeremy Wood, Hazard Community and Technical College
Jared Workman, Colorado Mesa University
Kaisa E. Young, Nicholls State University

Analyzing Starlight

Figure 17.1 Star Colors. This long time exposure shows the colors of the stars. The circular motion of the stars across the image is provided by Earth's rotation. The various colors of the stars are caused by their different temperatures. (credit: modification of work by ESO/A.Santerne)

Chapter Outline

17.1 The Brightness of Stars
17.2 Colors of Stars
17.3 The Spectra of Stars (and Brown Dwarfs)
17.4 Using Spectra to Measure Stellar Radius, Composition, and Motion

 Thinking Ahead

Everything we know about stars—how they are born, what they are made of, how far away they are, how long they live, and how they will die—we learn by decoding the messages contained in the light and radiation that reaches Earth. What questions should we ask, and how do we find the answers?

We can begin our voyage to the stars by looking at the night sky. It is obvious that stars do not all appear equally bright, nor are they all the same color. To understand the stars, we must first determine their basic properties, such as what their temperatures are, how much material they contain (their masses), and how much energy they produce. Since our Sun is a star, of course the same techniques, including spectroscopy, used to study the Sun can be used to find out what stars are like. As we learn more about the stars, we will use these characteristics to begin assembling clues to the main problems we are interested in solving: How do stars form? How long do they survive? What is their ultimate fate?

17.1 The Brightness of Stars

Learning Objectives

By the end of this section, you will be able to:

> Explain the difference between luminosity and apparent brightness
> Understand how astronomers specify brightness with magnitudes

Luminosity

Perhaps the most important characteristic of a star is its **luminosity**—the total amount of energy at all wavelengths that it emits per second. Earlier, we saw that the Sun puts out a tremendous amount of energy every second. (And there are stars far more luminous than the Sun out there.) To make the comparison among stars easy, astronomers express the luminosity of other stars in terms of the Sun's luminosity. For example, the luminosity of Sirius is about 25 times that of the Sun. We use the symbol L_{Sun} to denote the Sun's luminosity; hence, that of Sirius can be written as 25 L_{Sun}. In a later chapter, we will see that if we can measure how much energy a star emits and we also know its mass, then we can calculate how long it can continue to shine before it exhausts its nuclear energy and begins to die.

Apparent Brightness

Astronomers are careful to distinguish between the luminosity of the star (the total energy output) and the amount of energy that happens to reach our eyes or a telescope on Earth. Stars are democratic in how they produce radiation; they emit the same amount of energy in every direction in space. Consequently, only a minuscule fraction of the energy given off by a star actually reaches an observer on Earth. We call the amount of a star's energy that reaches a given area (say, one square meter) each second here on Earth its **apparent brightness**. If you look at the night sky, you see a wide range of apparent brightnesses among the stars. Most stars, in fact, are so dim that you need a telescope to detect them.

If all stars were the same luminosity—if they were like standard bulbs with the same light output—we could use the difference in their apparent brightnesses to tell us something we very much want to know: how far away they are. Imagine you are in a big concert hall or ballroom that is dark except for a few dozen 25-watt bulbs placed in fixtures around the walls. Since they are all 25-watt bulbs, their luminosity (energy output) is the same. But from where you are standing in one corner, they do *not* have the same apparent brightness. Those close to you appear brighter (more of their light reaches your eye), whereas those far away appear dimmer (their light has spread out more before reaching you). In this way, you can tell which bulbs are closest to you. In the same way, if all the stars had the same luminosity, we could immediately infer that the brightest-appearing stars were close by and the dimmest-appearing ones were far away.

To pin down this idea more precisely, recall from the Radiation and Spectra chapter that we know exactly how light fades with increasing distance. The energy we receive is inversely proportional to the square of the distance. If, for example, we have two stars of the same luminosity and one is twice as far away as the other, it will look four times dimmer than the closer one. If it is three times farther away, it will look nine (three squared) times dimmer, and so forth.

Alas, the stars do not all have the same luminosity. (Actually, we are pretty glad about that because having many different types of stars makes the universe a much more interesting place.) But this means that if a star looks dim in the sky, we cannot tell whether it appears dim because it has a low luminosity but is relatively nearby, or because it has a high luminosity but is very far away. To measure the luminosities of stars, we must first compensate for the dimming effects of distance on light, and to do that, we must know how far away they are. Distance is among the most difficult of all astronomical measurements. We will return to how it is determined after we have learned more about the stars. For now, we will describe how astronomers specify the apparent brightness of stars.

The Magnitude Scale

The process of measuring the apparent brightness of stars is called *photometry* (from the Greek *photo* meaning "light" and *–metry* meaning "to measure"). As we saw Observing the Sky: The Birth of Astronomy, astronomical photometry began with Hipparchus. Around 150 B.C.E., he erected an observatory on the island of Rhodes in the Mediterranean. There he prepared a catalog of nearly 1000 stars that included not only their positions but also estimates of their apparent brightnesses.

Hipparchus did not have a telescope or any instrument that could measure apparent brightness accurately, so he simply made estimates with his eyes. He sorted the stars into six brightness categories, each of which he called a **magnitude**. He referred to the brightest stars in his catalog as first-magnitudes stars, whereas those so faint he could barely see them were sixth-magnitude stars. During the nineteenth century, astronomers attempted to make the scale more precise by establishing exactly how much the apparent brightness of a sixth-magnitude star differs from that of a first-magnitude star. Measurements showed that we receive about 100 times more light from a first-magnitude star than from a sixth-magnitude star. Based on this measurement, astronomers then defined an accurate magnitude system in which a difference of five magnitudes corresponds exactly to a brightness ratio of 100:1. In addition, the magnitudes of stars are decimalized; for example, a star isn't just a "second-magnitude star," it has a magnitude of 2.0 (or 2.1, 2.3, and so forth). So what number is it that, when multiplied together five times, gives you this factor of 100? Play on your calculator and see if you can get it. The answer turns out to be about 2.5, which is the fifth root of 100. This means that a magnitude 1.0 star and a magnitude 2.0 star differ in brightness by a factor of about 2.5. Likewise, we receive about 2.5 times as much light from a magnitude 2.0 star as from a magnitude 3.0 star. What about the difference between a magnitude 1.0 star and a magnitude 3.0 star? Since the difference is 2.5 times for each "step" of magnitude, the total difference in brightness is 2.5 × 2.5 = 6.25 times.

Here are a few rules of thumb that might help those new to this system. If two stars differ by 0.75 magnitudes, they differ by a factor of about 2 in brightness. If they are 2.5 magnitudes apart, they differ in brightness by a factor of 10, and a 4-magnitude difference corresponds to a difference in brightness of a factor of 40.You might be saying to yourself at this point, "Why do astronomers continue to use this complicated system from more than 2000 years ago?" That's an excellent question and, as we shall discuss, astronomers today can use other ways of expressing how bright a star looks. But because this system is still used in many books, star charts, and computer apps, we felt we had to introduce students to it (even though we were very tempted to leave it out.)

The brightest stars, those that were traditionally referred to as first-magnitude stars, actually turned out (when measured accurately) not to be identical in brightness. For example, the brightest star in the sky, Sirius, sends us about 10 times as much light as the average first-magnitude star. On the modern magnitude scale, Sirius, the star with the brightest apparent magnitude, has been assigned a magnitude of –1.5. Other objects in the sky can appear even brighter. Venus at its brightest is of magnitude –4.4, while the Sun has a magnitude of –26.8. Figure 17.2 shows the range of observed magnitudes from the brightest to the faintest, along with the actual magnitudes of several well-known objects. The important fact to remember when using magnitude is that the system goes backward: the *larger* the magnitude, the *fainter* the object you are observing.

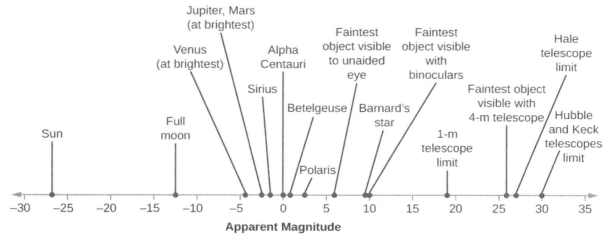

Figure 17.2 Apparent Magnitudes of Well-Known Objects. The faintest magnitudes that can be detected by the unaided eye, binoculars, and large telescopes are also shown.

EXAMPLE 17.1

The Magnitude Equation

Even scientists can't calculate fifth roots in their heads, so astronomers have summarized the above discussion in an equation to help calculate the difference in brightness for stars with different magnitudes. If m_1 and m_2 are the magnitudes of two stars, then we can calculate the ratio of their brightness $\left(\frac{b_2}{b_1}\right)$ using this equation:

$$m_1 - m_2 = 2.5 \log\left(\frac{b_2}{b_1}\right) \qquad \text{or} \qquad \frac{b_2}{b_1} = 2.5^{m_1-m_2}$$

Here is another way to write this equation:

$$\frac{b_2}{b_1} = \left(100^{0.2}\right)^{m_1-m_2}$$

Let's do a real example, just to show how this works. Imagine that an astronomer has discovered something special about a dim star (magnitude 8.5), and she wants to tell her students how much dimmer the star is than Sirius. Star 1 in the equation will be our dim star and star 2 will be Sirius.

Solution

Remember, Sirius has a magnitude of –1.5. In that case:

$$\frac{b_2}{b_1} = \left(100^{0.2}\right)^{8.5-(-1.5)} = \left(100^{0.2}\right)^{10}$$
$$= (100)^2 = 100 \times 100 = 10{,}000$$

Check Your Learning

It is a common misconception that Polaris (magnitude 2.0) is the brightest star in the sky, but, as we saw, that distinction actually belongs to Sirius (magnitude –1.5). How does Sirius' apparent brightness compare to that of Polaris?

Answer:

$$\frac{b_{\text{Sirius}}}{b_{\text{Polaris}}} = \left(100^{0.2}\right)^{2.0-(-1.5)} = \left(100^{0.2}\right)^{3.5} = 100^{0.7} = 25$$

(Hint: If you only have a basic calculator, you may wonder how to take 100 to the 0.7th power. But this is something you can ask Google to do. Google now accepts mathematical questions and will answer them. So try it for yourself. Ask Google, "What is 100 to the 0.7th power?")

Our calculation shows that Sirius' apparent brightness is 25 times greater than Polaris' apparent brightness.

Other Units of Brightness

Although the magnitude scale is still used for visual astronomy, it is not used at all in newer branches of the field. In radio astronomy, for example, no equivalent of the magnitude system has been defined. Rather, radio astronomers measure the amount of energy being collected each second by each square meter of a radio telescope and express the brightness of each source in terms of, for example, watts per square meter.

Similarly, most researchers in the fields of infrared, X-ray, and gamma-ray astronomy use energy per area per second rather than magnitudes to express the results of their measurements. Nevertheless, astronomers in all fields are careful to distinguish between the *luminosity* of the source (even when that luminosity is all in X-rays) and the amount of energy that happens to reach us on Earth. After all, the luminosity is a really important characteristic that tells us a lot about the object in question, whereas the energy that reaches Earth is an accident of cosmic geography.

To make the comparison among stars easy, in this text, we avoid the use of magnitudes as much as possible and will express the luminosity of other stars in terms of the Sun's luminosity. For example, the luminosity of Sirius is 25 times that of the Sun. We use the symbol L_{Sun} to denote the Sun's luminosity; hence, that of Sirius can be written as 25 L_{Sun}.

17.2 | Colors of Stars

Learning Objectives

By the end of this section, you will be able to:

> Compare the relative temperatures of stars based on their colors
> Understand how astronomers use color indexes to measure the temperatures of stars

Look at the beautiful picture of the stars in the Sagittarius Star Cloud shown in Figure 17.3. The stars show a multitude of colors, including red, orange, yellow, white, and blue. As we have seen, stars are not all the same color because they do not all have identical temperatures. To define *color* precisely, astronomers have devised quantitative methods for characterizing the color of a star and then using those colors to determine stellar temperatures. In the chapters that follow, we will provide the temperature of the stars we are describing, and this section tells you how those temperatures are determined from the colors of light the stars give off.

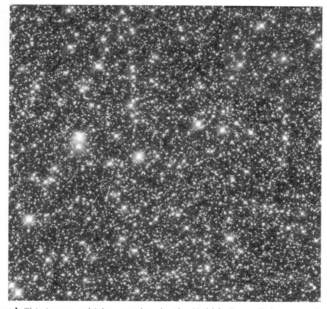

Figure 17.3 Sagittarius Star Cloud. This image, which was taken by the Hubble Space Telescope, shows stars in the direction toward the center of the Milky Way Galaxy. The bright stars glitter like colored jewels on a black velvet background. The color of a star indicates its temperature. Blue-white stars are much hotter than the Sun, whereas red stars are cooler. On average, the stars in this field are at a distance of about 25,000 light-years (which means it takes light 25,000 years to traverse the distance from them to us) and the width of the field is about 13.3 light-years. (credit: Hubble Heritage Team (AURA/STScI/NASA))

Color and Temperature

As we learned in The Electromagnetic Spectrum section, Wien's law relates stellar color to stellar temperature. Blue colors dominate the visible light output of very hot stars (with much additional radiation in the ultraviolet). On the other hand, cool stars emit most of their visible light energy at red wavelengths (with more radiation coming off in the infrared) (Table 17.1). The color of a star therefore provides a measure of its intrinsic or true surface temperature (apart from the effects of reddening by interstellar dust, which will be discussed in Between the Stars: Gas and Dust in Space). Color does not depend on the distance to the object. This should be familiar to you from everyday experience. The color of a traffic signal, for example, appears the same no matter how far away it is. If we could somehow take a star, observe it, and then move it much farther away, its apparent brightness (magnitude) would change. But this change in brightness is the same for all wavelengths, and so its color would remain the same.

Example Star Colors and Corresponding Approximate Temperatures

Star Color	Approximate Temperature	Example
Blue	25,000 K	Spica
White	10,000 K	Vega
Yellow	6000 K	Sun
Orange	4000 K	Aldebaran
Red	3000 K	Betelgeuse

Table 17.1

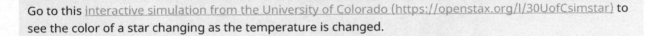

LINK TO LEARNING

Go to this interactive simulation from the University of Colorado (https://openstax.org/l/30UofCsimstar) to see the color of a star changing as the temperature is changed.

The hottest stars have temperatures of over 40,000 K, and the coolest stars have temperatures of about 2000 K. Our Sun's surface temperature is about 6000 K; its peak wavelength color is a slightly greenish-yellow. In space, the Sun would look white, shining with about equal amounts of reddish and bluish wavelengths of light. It looks somewhat yellow as seen from Earth's surface because our planet's nitrogen molecules scatter some of the shorter (i.e., blue) wavelengths out of the beams of sunlight that reach us, leaving more long wavelength light behind. This also explains why the sky is blue: the blue sky is sunlight scattered by Earth's atmosphere.

Color Indices

In order to specify the exact color of a star, astronomers normally measure a star's apparent brightness through filters, each of which transmits only the light from a particular narrow band of wavelengths (colors). A crude example of a filter in everyday life is a green-colored, plastic, soft drink bottle, which, when held in front of your eyes, lets only the green colors of light through.

One commonly used set of filters in astronomy measures stellar brightness at three wavelengths corresponding to ultraviolet, blue, and yellow light. The filters are named: U (ultraviolet), B (blue), and V (visual, for yellow). These filters transmit light near the wavelengths of 360 nanometers (nm), 420 nm, and 540 nm, respectively. The brightness measured through each filter is usually expressed in magnitudes. The difference between any two of these magnitudes—say, between the blue and the visual magnitudes (B–V)—is called a **color index**.

By agreement among astronomers, the ultraviolet, blue, and visual magnitudes of the UBV system are adjusted to give a color index of 0 to a star with a surface temperature of about 10,000 K, such as Vega. The

B–V color indexes of stars range from –0.4 for the bluest stars, with temperatures of about 40,000 K, to +2.0 for the reddest stars, with temperatures of about 2000 K. The B–V index for the Sun is about +0.65. Note that, by convention, the B–V index is always the "bluer" minus the "redder" color.

Why use a color index if it ultimately implies temperature? Because the brightness of a star through a filter is what astronomers actually measure, and we are always more comfortable when our statements have to do with measurable quantities.

17.3 | The Spectra of Stars (and Brown Dwarfs)

Learning Objectives

By the end of this section, you will be able to:

> Describe how astronomers use spectral classes to characterize stars
> Explain the difference between a star and a brown dwarf

Measuring colors is only one way of analyzing starlight. Another way is to use a spectrograph to spread out the light into a spectrum (see the Radiation and Spectra and the Astronomical Instruments chapters). In 1814, the German physicist Joseph Fraunhofer observed that the spectrum of the Sun shows dark lines crossing a continuous band of colors. In the 1860s, English astronomers Sir William Huggins and Lady Margaret Huggins (Figure 17.4) succeeded in identifying some of the lines in stellar spectra as those of known elements on Earth, showing that the same chemical elements found in the Sun and planets exist in the stars. Since then, astronomers have worked hard to perfect experimental techniques for obtaining and measuring spectra, and they have developed a theoretical understanding of what can be learned from spectra. Today, spectroscopic analysis is one of the cornerstones of astronomical research.

Figure 17.4 William Huggins (1824–1910) and Margaret Huggins (1848–1915). William and Margaret Huggins were the first to identify the lines in the spectrum of a star other than the Sun; they also took the first spectrogram, or photograph of a stellar spectrum.

Formation of Stellar Spectra

When the spectra of different stars were first observed, astronomers found that they were not all identical. Since the dark lines are produced by the chemical elements present in the stars, astronomers first thought that the spectra differ from one another because stars are not all made of the same chemical elements. This hypothesis turned out to be wrong. *The primary reason that stellar spectra look different is because the stars have different temperatures.* Most stars have nearly the same composition as the Sun, with only a few exceptions.

Hydrogen, for example, is by far the most abundant element in most stars. However, lines of hydrogen are not seen in the spectra of the hottest and the coolest stars. In the atmospheres of the hottest stars, hydrogen atoms are completely ionized. Because the electron and the proton are separated, ionized hydrogen cannot produce absorption lines. (Recall from the Formation of Spectral Lines section, the lines are the result of electrons in orbit around a nucleus changing energy levels.)

In the atmospheres of the coolest stars, hydrogen atoms have their electrons attached and can switch energy levels to produce lines. However, practically all of the hydrogen atoms are in the lowest energy state (unexcited) in these stars and thus can absorb only those photons able to lift an electron from that first energy level to a higher level. Photons with enough energy to do this lie in the ultraviolet part of the electromagnetic spectrum, and there are very few ultraviolet photons in the radiation from a cool star. What this means is that if you observe the spectrum of a very hot or very cool star with a typical telescope on the surface of Earth, the most common element in that star, hydrogen, will show very weak spectral lines or none at all.

The hydrogen lines in the visible part of the spectrum (called *Balmer lines*) are strongest in stars with intermediate temperatures—not too hot and not too cold. Calculations show that the optimum temperature for producing visible hydrogen lines is about 10,000 K. At this temperature, an appreciable number of hydrogen atoms are excited to the second energy level. They can then absorb additional photons, rise to still-higher levels of excitation, and produce a dark absorption line. Similarly, every other chemical element, in each of its possible stages of ionization, has a characteristic temperature at which it is most effective in producing absorption lines in any particular part of the spectrum.

Classification of Stellar Spectra

Astronomers use the patterns of lines observed in stellar spectra to sort stars into a **spectral class**. Because a star's temperature determines which absorption lines are present in its spectrum, these spectral classes are a measure of its surface temperature. There are seven standard spectral classes. From hottest to coldest, these seven spectral classes are designated O, B, A, F, G, K, and M. Recently, astronomers have added three additional classes for even cooler objects—L, T, and Y.

At this point, you may be looking at these letters with wonder and asking yourself why astronomers didn't call the spectral types A, B, C, and so on. You will see, as we tell you the history, that it's an instance where tradition won out over common sense.

In the 1880s, Williamina Fleming devised a system to classify stars based on the strength of hydrogen absorption lines. Spectra with the strongest lines were classified as "A" stars, the next strongest "B," and so on down the alphabet to "O" stars, in which the hydrogen lines were very weak. But we saw above that hydrogen lines alone are not a good indicator for classifying stars, since their lines disappear from the visible light spectrum when the stars get too hot or too cold.

In the 1890s, Annie Jump Cannon revised this classification system, focusing on just a few letters from the original system: A, B, F, G, K, M, and O. Instead of starting over, Cannon also rearranged the existing classes—in order of decreasing temperature—into the sequence we have learned: O, B, A, F, G, K, M. As you can read in the feature on Annie Cannon: Classifier of the Stars in this chapter, she classified around 500,000 stars over her lifetime, classifying up to three stars per minute by looking at the stellar spectra.

LINK TO LEARNING

For a deep dive into spectral types, explore the interactive project at the Sloan Digital Sky Survey (https://openstax.org/l/30sloandigsky) in which you can practice classifying stars yourself.

To help astronomers remember this crazy order of letters, Cannon created a mnemonic, "Oh Be A Fine Girl, Kiss Me." (If you prefer, you can easily substitute "Guy" for "Girl.") Other mnemonics, which we hope will not be relevant for you, include "Oh Brother, Astronomers Frequently Give Killer Midterms" and "Oh Boy, An F Grade Kills Me!" With the new L, T, and Y spectral classes, the mnemonic might be expanded to "Oh Be A Fine Girl (Guy), Kiss Me Like That, Yo!"

Each of these spectral classes, except possibly for the Y class which is still being defined, is further subdivided into 10 subclasses designated by the numbers 0 through 9. A B0 star is the hottest type of B star; a B9 star is the coolest type of B star and is only slightly hotter than an A0 star.

And just one more item of vocabulary: for historical reasons, astronomers call all the elements heavier than helium *metals*, even though most of them do not show metallic properties. (If you are getting annoyed at the peculiar jargon that astronomers use, just bear in mind that every field of human activity tends to develop its own specialized vocabulary. Just try reading a credit card or social media agreement form these days without training in law!)

Let's take a look at some of the details of how the spectra of the stars change with temperature. (It is these details that allowed Annie Cannon to identify the spectral types of stars as quickly as three per minute!) As Figure 17.5 shows, in the hottest O stars (those with temperatures over 28,000 K), only lines of ionized helium and highly ionized atoms of other elements are conspicuous. Hydrogen lines are strongest in A stars with atmospheric temperatures of about 10,000 K. Ionized metals provide the most conspicuous lines in stars with temperatures from 6000 to 7500 K (spectral type F). In the coolest M stars (below 3500 K), absorption bands of titanium oxide and other molecules are very strong. By the way, the spectral class assigned to the Sun is G2. The sequence of spectral classes is summarized in Table 17.2.

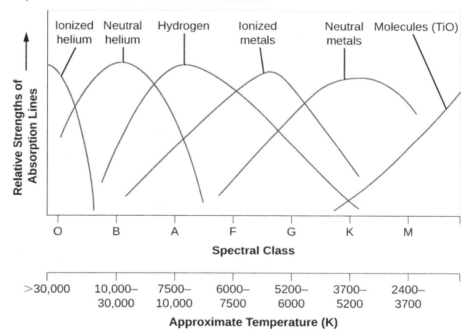

Figure 17.5 Absorption Lines in Stars of Different Temperatures. This graph shows the strengths of absorption lines of different chemical species (atoms, ions, molecules) as we move from hot (left) to cool (right) stars. The sequence of spectral types is also shown.

LINK TO LEARNING

Use the Spectrum Explorer (https://openstax.org/l/30spectexpl) to see how each element's absorption lines change with temperature.

Spectral Classes for Stars

Spectral Class	Color	Approximate Temperature (K)	Principal Features	Examples
O	Blue	> 30,000	Neutral and ionized helium lines, weak hydrogen lines	10 Lacertae
B	Blue-white	10,000–30,000	Neutral helium lines, strong hydrogen lines	Rigel, Spica
A	White	7500–10,000	Strongest hydrogen lines, weak ionized calcium lines, weak ionized metal (e.g., iron, magnesium) lines	Sirius, Vega
F	Yellow-white	6000–7500	Strong hydrogen lines, strong ionized calcium lines, weak sodium lines, many ionized metal lines	Canopus, Procyon
G	Yellow	5200–6000	Weaker hydrogen lines, strong ionized calcium lines, strong sodium lines, many lines of ionized and neutral metals	Sun, Capella
K	Orange	3700–5200	Very weak hydrogen lines, strong ionized calcium lines, strong sodium lines, many lines of neutral metals	Arcturus, Aldebaran
M	Red	2400–3700	Strong lines of neutral metals and molecular bands of titanium oxide dominate	Betelgeuse, Antares
L	Red	1300–2400	Metal hydride lines, alkali metal lines (e.g., sodium, potassium, rubidium)	Teide 1
T	Magenta	700–1300	Methane lines	Gliese 229B
Y	Infrared[1]	< 700	Ammonia lines	WISE 1828+2650

Table 17.2

To see how spectral classification works, let's use Figure 17.5. Suppose you have a spectrum in which the hydrogen lines are about half as strong as those seen in an A star. Looking at the lines in our figure, you see that the star could be either a B star or a G star. But if the spectrum also contains helium lines, then it is a B star, whereas if it contains lines of ionized iron and other metals, it must be a G star.

If you look at Figure 17.6, you can see that you, too, could assign a spectral class to a star whose type was not already known. All you have to do is match the pattern of spectral lines to a standard star (like the ones shown in the figure) whose type has already been determined.

---

1 Absorption by sodium and potassium atoms makes Y dwarfs appear a bit less red than L dwarfs.

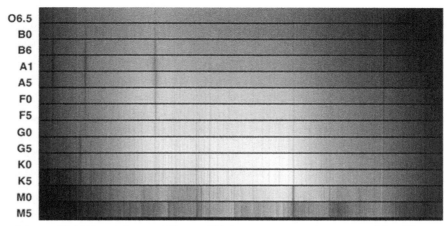

Figure 17.6 Spectra of Stars with Different Spectral Classes. This image compares the spectra of the different spectral classes. The spectral class assigned to each of these stellar spectra is listed at the left of the picture. The strongest four lines seen at spectral type A1 (one in the red, one in the blue-green, and two in the blue) are Balmer lines of hydrogen. Note how these lines weaken at both higher and lower temperatures, as Figure 17.5 also indicates. The strong pair of closely spaced lines in the yellow in the cool stars is due to neutral sodium (one of the neutral metals in Figure 17.5). (Credit: modification of work by NOAO/AURA/NSF)

Both colors and spectral classes can be used to estimate the temperature of a star. Spectra are harder to measure because the light has to be bright enough to be spread out into all colors of the rainbow, and detectors must be sensitive enough to respond to individual wavelengths. In order to measure colors, the detectors need only respond to the many wavelengths that pass simultaneously through the colored filters that have been chosen—that is, to *all* the blue light or *all* the yellow-green light.

VOYAGERS IN ASTRONOMY

Annie Cannon: Classifier of the Stars

Annie Jump Cannon was born in Delaware in 1863 (Figure 17.7). In 1880, she went to Wellesley College, one of the new breed of US colleges opening up to educate young women. Wellesley, only 5 years old at the time, had the second student physics lab in the country and provided excellent training in basic science. After college, Cannon spent a decade with her parents but was very dissatisfied, longing to do scientific work. After her mother's death in 1893, she returned to Wellesley as a teaching assistant and also to take courses at Radcliffe, the women's college associated with Harvard.

Figure 17.7 Annie Jump Cannon (1863–1941). Cannon is well-known for her classifications of stellar spectra. (credit: modification of work by Smithsonian Institution)

In the late 1800s, the director of the Harvard Observatory, Edward C. Pickering, needed lots of help with his ambitious program of classifying stellar spectra. The basis for these studies was a monumental collection of nearly a million photographic spectra of stars, obtained from many years of observations made at Harvard College Observatory in Massachusetts as well as at its remote observing stations in South America and South Africa. Pickering quickly discovered that educated young women could be hired as assistants for one-third or one-fourth the salary paid to men, and they would often put up with working conditions and repetitive tasks that men with the same education would not tolerate. These women became known as the

Harvard Computers. (We should emphasize that astronomers were not alone in reaching such conclusions about the relatively new idea of upper-class, educated women working outside the home: women were exploited and undervalued in many fields. This is a legacy from which our society is just beginning to emerge.)

Cannon was hired by Pickering as one of the "computers" to help with the classification of spectra. She became so good at it that she could visually examine and determine the spectral types of several hundred stars per hour (dictating her conclusions to an assistant). She made many discoveries while investigating the Harvard photographic plates, including 300 variable stars (stars whose luminosity changes periodically). But her main legacy is a marvelous catalog of spectral types for hundreds of thousands of stars, which served as a foundation for much of twentieth-century astronomy.

In 1911, a visiting committee of astronomers reported that "she is the one person in the world who can do this work quickly and accurately" and urged Harvard to give Cannon an official appointment in keeping with her skill and renown. Not until 1938, however, did Harvard appoint her an astronomer at the university; she was then 75 years old.

Cannon received the first honorary degree Oxford awarded to a woman, and she became the first woman to be elected an officer of the American Astronomical Society, the main professional organization of astronomers in the US. She generously donated the money from one of the major prizes she had won to found a special award for women in astronomy, now known as the Annie Jump Cannon Prize. True to form, she continued classifying stellar spectra almost to the very end of her life in 1941.

Spectral Classes L, T, and Y

The scheme devised by Cannon worked well until 1988, when astronomers began to discover objects even cooler than M9-type stars. We use the word *object* because many of the new discoveries are not true stars. A star is defined as an object that during some part of its lifetime derives 100% of its energy from the same process that makes the Sun shine—the fusion of hydrogen nuclei (protons) into helium. Objects with masses less than about 7.5% of the mass of our Sun (about 0.075 M_{Sun}) do not become hot enough for hydrogen fusion to take place. Even before the first such "failed star" was found, this class of objects, with masses intermediate between stars and planets, was given the name **brown dwarfs**.

Brown dwarfs are very difficult to observe because they are extremely faint and cool, and they put out most of their light in the infrared part of the spectrum. It was only after the construction of very large telescopes, like the Keck telescopes in Hawaii, and the development of very sensitive infrared detectors, that the search for brown dwarfs succeeded. The first brown dwarf was discovered in 1988, and, as of the summer of 2015, there are more than 2200 known brown dwarfs.

Initially, brown dwarfs were given spectral classes like $M10^+$ or "much cooler than M9," but so many are now known that it is possible to begin assigning spectral types. The hottest brown dwarfs are given types L0–L9 (temperatures in the range 2400–1300 K), whereas still cooler (1300–700 K) objects are given types T0–T9 (see Figure 17.8). In class L brown dwarfs, the lines of titanium oxide, which are strong in M stars, have disappeared. This is because the L dwarfs are so cool that atoms and molecules can gather together into dust particles in their atmospheres; the titanium is locked up in the dust grains rather than being available to form molecules of titanium oxide. Lines of steam (hot water vapor) are present, along with lines of carbon monoxide and neutral sodium, potassium, cesium, and rubidium. Methane (CH_4) lines are strong in class-T brown dwarfs, as methane exists in the atmosphere of the giant planets in our own solar system.

In 2009, astronomers discovered ultra-cool brown dwarfs with temperatures of 500–600 K. These objects exhibited absorption lines due to ammonia (NH_3), which are not seen in T dwarfs. A new spectral class, Y, was created for these objects. As of 2021, over 50 brown dwarfs belonging to spectral class Y have been discovered, some with temperatures comparable to that of the human body (about 300 K).[2]

Figure 17.8 Brown Dwarfs. This illustration shows the sizes and surface temperatures of brown dwarfs Teide 1, Gliese 229B, and WISE1828 in relation to the Sun, a red dwarf star (Gliese 229A), and Jupiter. (credit: modification of work by MPIA/V. Joergens)

Most brown dwarfs start out with atmospheric temperatures and spectra like those of true stars with spectral classes of M6.5 and later, even though the brown dwarfs are not hot and dense enough in their interiors to fuse hydrogen. In fact, the spectra of brown dwarfs and true stars are so similar from spectral types late M through L that it is not possible to distinguish the two types of objects based on spectra alone. An independent measure of mass is required to determine whether a specific object is a brown dwarf or a very low mass star. Since brown dwarfs cool steadily throughout their lifetimes, the spectral type of a given brown dwarf changes with time over a billion years or more from late M through L, T, and Y spectral types.

Low-Mass Brown Dwarfs vs. High-Mass Planets

An interesting property of brown dwarfs is that they are all about the same radius as Jupiter, regardless of their masses. Amazingly, this covers a range of masses from about 13 to 80 times the mass of Jupiter (M_J). This can make distinguishing a low-mass brown dwarf from a high-mass planet very difficult.

So, what is the difference between a low-mass brown dwarf and a high-mass planet? The International Astronomical Union considers the distinctive feature to be *deuterium fusion*. Although brown dwarfs do not sustain regular (proton-proton) hydrogen fusion, they are capable of fusing deuterium (a rare form of hydrogen with one proton and one neutron in its nucleus). The fusion of deuterium can happen at a lower temperature than the fusion of hydrogen. If an object has enough mass to fuse deuterium (about 13 M_J or

2   The record-holder for the coldest Type Y dwarf is WISE 0855-0714, at around 250 K.

0.012 M_{Sun}), it is a brown dwarf. Objects with less than 13 M_J do not fuse deuterium and are usually considered planets.

17.4 | Using Spectra to Measure Stellar Radius, Composition, and Motion

Learning Objectives

By the end of this section, you will be able to:

> Understand how astronomers can learn about a star's radius and composition by studying its spectrum
> Explain how astronomers can measure the motion and rotation of a star using the Doppler effect
> Describe the proper motion of a star and how it relates to a star's space velocity

Analyzing the spectrum of a star can teach us all kinds of things in addition to its temperature. We can measure its detailed chemical composition as well as the pressure in its atmosphere. From the pressure, we get clues about its size. We can also measure its motion toward or away from us and estimate its rotation.

Clues to the Size of a Star

As we shall see in The Stars: A Celestial Census, stars come in a wide variety of sizes. At some periods in their lives, stars can expand to enormous dimensions. Stars of such exaggerated size are called **giants**. Luckily for the astronomer, stellar spectra can be used to distinguish giants from run-of-the-mill stars (such as our Sun).

Suppose you want to determine whether a star is a giant. A giant star has a large, extended photosphere. Because it is so large, a giant star's atoms are spread over a great volume, which means that the density of particles in the star's photosphere is low. As a result, the pressure in a giant star's photosphere is also low. This low pressure affects the spectrum in two ways. First, a star with a lower-pressure photosphere shows narrower spectral lines than a star of the same temperature with a higher-pressure photosphere (Figure 17.9). The difference is large enough that careful study of spectra can tell which of two stars at the same temperature has a higher pressure (and is thus more compressed) and which has a lower pressure (and thus must be extended). This effect is due to collisions between particles in the star's photosphere—more collisions lead to broader spectral lines. Collisions will, of course, be more frequent in a higher-density environment. Think about it like traffic—collisions are much more likely during rush hour, when the density of cars is high.

Second, more atoms are ionized in a giant star than in a star like the Sun with the same temperature. The ionization of atoms in a star's outer layers is caused mainly by photons, and the amount of energy carried by photons is determined by temperature. But how long atoms *stay* ionized depends in part on pressure. Compared with what happens in the Sun (with its relatively dense photosphere), ionized atoms in a giant star's photosphere are less likely to pass close enough to electrons to interact and combine with one or more of them, thereby becoming neutral again. Ionized atoms, as we discussed earlier, have different spectra from atoms that are neutral.

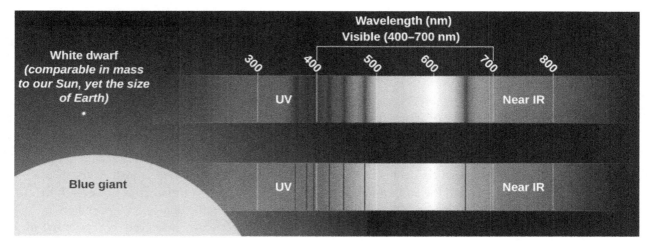

Figure 17.9 Spectral Lines. This figure illustrates one difference in the spectral lines from stars of the same temperature but different pressures. A giant star with a very-low-pressure photosphere shows very narrow spectral lines (bottom), whereas a smaller star with a higher-pressure photosphere shows much broader spectral lines (top). (credit: modification of work by NASA, ESA, A. Field, and J. Kalirai (STScI))

Abundances of the Elements

Absorption lines of a majority of the known chemical elements have now been identified in the spectra of the Sun and stars. If we see lines of iron in a star's spectrum, for example, then we know immediately that the star must contain iron.

Note that the *absence* of an element's spectral lines does not necessarily mean that the element itself is absent. As we saw, the temperature and pressure in a star's atmosphere will determine what types of atoms are able to produce absorption lines. Only if the physical conditions in a star's photosphere are such that lines of an element *should* (according to calculations) be there can we conclude that the absence of observable spectral lines implies low abundance of the element.

Suppose two stars have identical temperatures and pressures, but the lines of, say, sodium are stronger in one than in the other. Stronger lines mean that there are more atoms in the stellar photosphere absorbing light. Therefore, we know immediately that the star with stronger sodium lines contains more sodium. Complex calculations are required to determine exactly how much more, but those calculations can be done for any element observed in any star with any temperature and pressure.

Of course, astronomy textbooks such as ours always make these things sound a bit easier than they really are. If you look at the stellar spectra such as those in Figure 17.6, you may get some feeling for how hard it is to decode all of the information contained in the thousands of absorption lines. First of all, it has taken many years of careful laboratory work on Earth to determine the precise wavelengths at which hot gases of each element have their spectral lines. Long books and computer databases have been compiled to show the lines of each element that can be seen at each temperature. Second, stellar spectra usually have many lines from a number of elements, and we must be careful to sort them out correctly. Sometimes nature is unhelpful, and lines of different elements have identical wavelengths, thereby adding to the confusion. And third, as we saw in the chapter on Radiation and Spectra, the motion of the star can change the observed wavelength of each of the lines. So, the observed wavelengths may not match laboratory measurements exactly. In practice, analyzing stellar spectra is a demanding, sometimes frustrating task that requires both training and skill.

Studies of stellar spectra have shown that hydrogen makes up about three-quarters of the mass of most stars. Helium is the second-most abundant element, making up almost a quarter of a star's mass. Together, hydrogen and helium make up from 96 to 99% of the mass; in some stars, they amount to more than 99.9%. Among the 4% or less of "heavy elements," oxygen, carbon, neon, iron, nitrogen, silicon, magnesium, and sulfur are among the most abundant. Generally, but not invariably, the elements of lower atomic weight are more abundant than those of higher atomic weight.

Take a careful look at the list of elements in the preceding paragraph. Two of the most abundant are hydrogen and oxygen (which make up water); add carbon and nitrogen and you are starting to write the prescription for the chemistry of an astronomy student. We are made of elements that are common in the universe—just mixed together in a far more sophisticated form (and a much cooler environment) than in a star.

As we mentioned in The Spectra of Stars (and Brown Dwarfs) section, astronomers use the term "metals" to refer to all elements heavier than hydrogen and helium. The fraction of a star's mass that is composed of these elements is referred to as the star's *metallicity*. The metallicity of the Sun, for example, is 0.02, since 2% of the Sun's mass is made of elements heavier than helium.

Appendix K lists how common each element is in the universe (compared to hydrogen); these estimates are based primarily on investigation of the Sun, which is a typical star. Some very rare elements, however, have not been detected in the Sun. Estimates of the amounts of these elements in the universe are based on laboratory measurements of their abundance in primitive meteorites, which are considered representative of unaltered material condensed from the solar nebula (see the Cosmic Samples and the Origin of the Solar System chapter).

Radial Velocity

When we measure the spectrum of a star, we determine the wavelength of each of its lines. If the star is not moving with respect to the Sun, then the wavelength corresponding to each element will be the same as those we measure in a laboratory here on Earth. But if stars are moving toward or away from us, we must consider the *Doppler effect* (see The **Doppler** Effect section). We should see all the spectral lines of moving stars shifted toward the red end of the spectrum if the star is moving away from us, or toward the blue (violet) end if it is moving toward us (Figure 17.10). The greater the shift, the faster the star is moving. Such motion, along the line of sight between the star and the observer, is called **radial velocity** and is usually measured in kilometers per second.

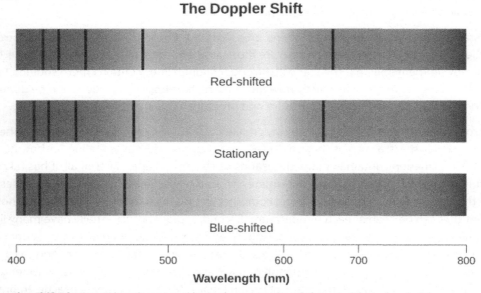

Figure 17.10 Doppler-Shifted Stars. When the spectral lines of a moving star shift toward the red end of the spectrum, we know that the star is moving away from us. If they shift toward the blue end, the star is moving toward us.

William Huggins, pioneering yet again, in 1868 made the first radial velocity determination of a star. He observed the Doppler shift in one of the hydrogen lines in the spectrum of Sirius and found that this star is moving toward the solar system. (Better measurements now indicate that Sirius is actually moving away, but Huggins is still celebrated as a pioneer in making such measurements.) Today, radial velocity can be measured for any star bright enough for its spectrum to be observed. As we will see in The Stars: A Celestial Census, radial velocity measurements of double stars are crucial in deriving stellar masses.

In the Spectrum Constructor (https://openstax.org/l/30spectconst), you can add absorption lines to the spectrum. Then adjust the velocity; negative indicates the star is moving toward the observer, and positive indicates it is moving away. Explore how the absorption lines are seen to shift to shorter and longer wavelengths.

Proper Motion

There is another type of motion stars can have that cannot be detected with stellar spectra. Unlike radial motion, which is along our line of sight (i.e., toward or away from Earth), this motion, called **proper motion**, is *transverse*: that is, across our line of sight. We see it as a change in the relative positions of the stars on the celestial sphere (Figure 17.11). These changes are very slow. Even the star with the largest proper motion takes 200 years to change its position in the sky by an amount equal to the width of the full Moon, and the motions of other stars are smaller yet.

(a)   (b)   (c)

Figure 17.11 Large Proper Motion. Three photographs of Barnard's star, the star with the largest known proper motion, show how this faint star has moved over a period of 20 years. (modification of work by Steve Quirk)

For this reason, with our naked eyes, we do not notice any change in the positions of the bright stars during the course of a human lifetime. If we could live long enough, however, the changes would become obvious. For example, some 50,000 years from now, terrestrial observers will find the handle of the Big Dipper unmistakably more bent than it is now (Figure 17.12).

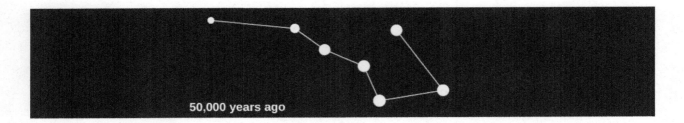

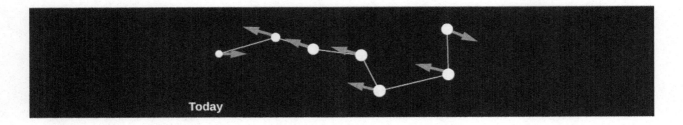

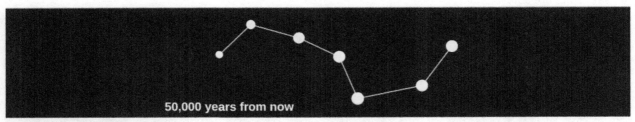

Figure 17.12 Changes in the Big Dipper. This figure shows changes in the appearance of the Big Dipper due to proper motion of the stars over 100,000 years.

We measure the proper motion of a star in arcseconds (1/3600 of a degree) per year. That is, the measurement of proper motion tells us only by how much of an angle a star has changed its position on the celestial sphere. If two stars at different distances are moving at the same velocity perpendicular to our line of sight, the closer one will show a larger shift in its position on the celestial sphere in a year's time. As an analogy, imagine you are standing at the side of a freeway. Cars will appear to whiz past you. If you then watch the traffic from a vantage point half a mile away, the cars will move much more slowly across your field of vision. In order to convert this angular motion to a velocity, we need to know how far away the star is.

To know the true **space velocity** of a star—that is, its total speed and the direction in which it is moving through space relative to the Sun—we must know its radial velocity, proper motion, and distance (Figure 17.13). A star's space velocity can also, over time, cause its distance from the Sun to change significantly. Over several hundred thousand years, these changes can be large enough to affect the apparent brightnesses of nearby stars. Today, Sirius, in the constellation Canis Major (the Big Dog) is the brightest star in the sky, but 100,000 years ago, the star Canopus in the constellation Carina (the Keel) was the brightest one. A little over 200,000 years from now, Sirius will have moved away and faded somewhat, and Vega, the bright blue star in Lyra, will take over its place of honor as the brightest star in Earth's skies.

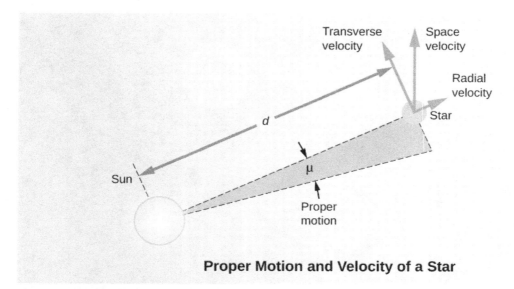

Proper Motion and Velocity of a Star

Figure 17.13 Space Velocity and Proper Motion. This figure shows the true space velocity of a star. The radial velocity is the component of the space velocity projected along the line of sight from the Sun to a star. The transverse velocity is a component of the space velocity projected on the sky. What astronomers measure is proper motion (μ), which is the change in the apparent direction on the sky measured in fractions of a degree. To convert this change in direction to a speed in, say, kilometers per second, it is necessary to also know the distance (d) from the Sun to the star.

Rotation

We can also use the Doppler effect to measure how fast a star rotates. If an object is rotating, then one of its sides is approaching us while the other is receding (unless its axis of rotation happens to be pointed exactly toward us). This is clearly the case for the Sun or a planet; we can observe the light from either the approaching or receding edge of these nearby objects and directly measure the Doppler shifts that arise from the rotation.

Stars, however, are so far away that they all appear as unresolved points. The best we can do is to analyze the light from the entire star at once. Due to the Doppler effect, the lines in the light that come from the side of the star rotating toward us are shifted to shorter wavelengths and the lines in the light from the opposite edge of the star are shifted to longer wavelengths. You can think of each spectral line that we observe as the sum or composite of spectral lines originating from different speeds with respect to us. Each point on the star has its own Doppler shift, so the absorption line we see from the whole star is actually much wider than it would be if the star were not rotating. If a star is rotating rapidly, there will be a greater spread of Doppler shifts and all its spectral lines should be quite broad. In fact, astronomers call this effect *line broadening*, and the amount of broadening can tell us the speed at which the star rotates (Figure 17.14).

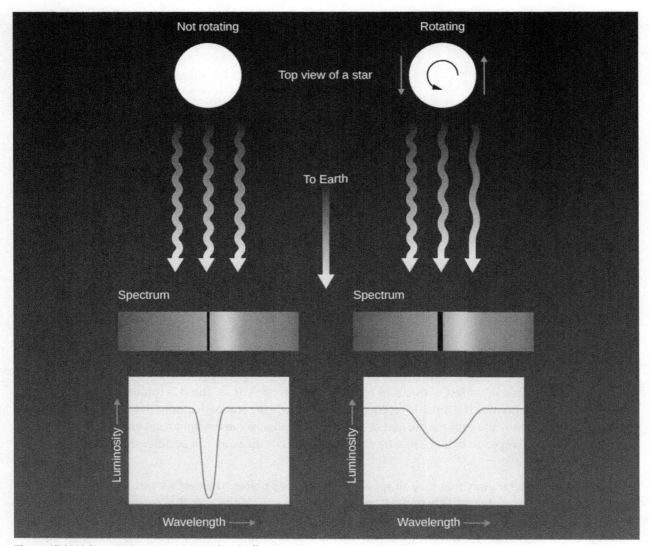

Figure 17.14 Using a Spectrum to Determine Stellar Rotation. A rotating star will show broader spectral lines than a nonrotating star.

Measurements of the widths of spectral lines show that many stars rotate faster than the Sun, some with periods of less than a day! These rapid rotators spin so fast that their shapes are "flattened" into what we call *oblate spheroids*. An example of this is the star Vega, which rotates once every 12.5 hours. Vega's rotation flattens its shape so much that its diameter at the equator is 23% wider than its diameter at the poles (Figure 17.15). The Sun, with its rotation period of about a month, rotates rather slowly. Studies have shown that stars decrease their rotational speed as they age. Young stars rotate very quickly, with rotational periods of days or less. Very old stars can have rotation periods of several months.

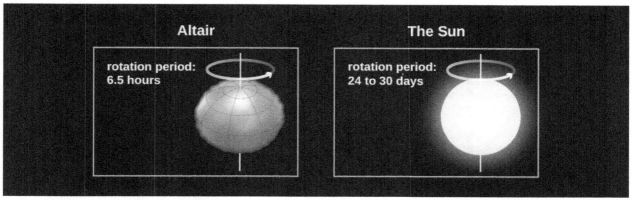

Figure 17.15 Comparison of Rotating Stars. This illustration compares the more rapidly rotating star Altair to the slower rotating Sun.

As you can see, spectroscopy is an extremely powerful technique that helps us learn all kinds of information about stars that we simply could not gather any other way. We will see in later chapters that these same techniques can also teach us about galaxies, which are the most distant objects that can we observe. Without spectroscopy, we would know next to nothing about the universe beyond the solar system.

MAKING CONNECTIONS

Astronomy and Philanthropy

Throughout the history of astronomy, contributions from wealthy patrons of the science have made an enormous difference in building new instruments and carrying out long-term research projects. Edward Pickering's stellar classification project, which was to stretch over several decades, was made possible by major donations from Anna Draper. She was the widow of Henry Draper, a physician who was one of the most accomplished amateur astronomers of the nineteenth century and the first person to successfully photograph the spectrum of a star. Anna Draper gave several hundred thousand dollars to Harvard Observatory. As a result, the great spectroscopic survey is still known as the Henry Draper Memorial, and many stars are still referred to by their "HD" numbers in that catalog (such as HD 209458).

In the 1870s, the eccentric piano builder and real estate magnate James Lick (Figure 17.16) decided to leave some of his fortune to build the world's largest telescope. When, in 1887, the pier to house the telescope was finished, Lick's body was entombed in it. Atop the foundation rose a 36-inch refractor, which for many years was the main instrument at the Lick Observatory near San Jose.

(a)   (b)

Figure 17.16 Henry Draper (1837–1882) and James Lick (1796–1876). (a) Draper stands next to a telescope used for photography. After his death, his widow funded further astronomy work in his name. (b) Lick was a philanthropist who provided funds to build a 36-inch refractor not only as a memorial to himself but also to aid in further astronomical research.

The Lick telescope remained the largest in the world until 1897, when George Ellery Hale persuaded railroad millionaire Charles Yerkes to finance the construction of a 40-inch telescope near Chicago. More recently, Howard Keck, whose family made its fortune in the oil industry, gave $70 million from his family foundation to the California Institute of Technology to help build the world's largest telescope atop the 14,000-foot peak of Maunakea in Hawaii (see the chapter on Astronomical Instruments to learn more about these telescopes). The Keck Foundation was so pleased with what is now called the Keck telescope that they gave $74 million more to build Keck II, another 10-meter reflector on the same volcanic peak.

Now, if any of you become millionaires or billionaires, and astronomy has sparked your interest, do keep an astronomical instrument or project in mind as you plan your estate. But frankly, private philanthropy could not possibly support the full enterprise of scientific research in astronomy. Much of our exploration of the universe is financed by federal agencies such as the National Science Foundation and NASA in the United States, and by similar government agencies in the other countries. In this way, all of us, through a very small share of our tax dollars, are philanthropists for astronomy.

Key Terms

apparent brightness   a measure of the amount of light received by Earth from a star or other object—that is, how bright an object appears in the sky, as contrasted with its luminosity

brown dwarf   an object intermediate in size between a planet and a star; the approximate mass range is from about 1/100 of the mass of the Sun up to the lower mass limit for self-sustaining nuclear reactions, which is about 0.075 the mass of the Sun; brown dwarfs are capable of deuterium fusion, but not hydrogen fusion

color index   difference between the magnitudes of a star or other object measured in light of two different spectral regions—for example, blue minus visual (B–V) magnitudes

giant   a star of exaggerated size with a large, extended photosphere

luminosity   the rate at which a star or other object emits electromagnetic energy into space; the total power output of an object

magnitude   an older system of measuring the amount of light we receive from a star or other luminous object; the larger the magnitude, the less radiation we receive from the object

proper motion   the angular change per year in the direction of a star as seen from the Sun

radial velocity   motion toward or away from the observer; the component of relative velocity that lies in the line of sight

space velocity   the total (three-dimensional) speed and direction with which an object is moving through space relative to the Sun

spectral class   (or spectral type) the classification of stars according to their temperatures using the characteristics of their spectra; the types are O, B, A, F, G, K, and M with L, T, and Y added recently for cooler star-like objects that recent survey have revealed

Summary

17.1 The Brightness of Stars

The total energy emitted per second by a star is called its luminosity. How bright a star looks from the perspective of Earth is its apparent brightness. The apparent brightness of a star depends on both its luminosity and its distance from Earth. Thus, the determination of apparent brightness and measurement of the distance to a star provide enough information to calculate its luminosity. The apparent brightnesses of stars are often expressed in terms of magnitudes, which is an old system based on how human vision interprets relative light intensity.

17.2 Colors of Stars

Stars have different colors, which are indicators of temperature. The hottest stars tend to appear blue or blue-white, whereas the coolest stars are red. A color index of a star is the difference in the magnitudes measured at any two wavelengths and is one way that astronomers measure and express the temperature of stars.

17.3 The Spectra of Stars (and Brown Dwarfs)

The differences in the spectra of stars are principally due to differences in temperature, not composition. The spectra of stars are described in terms of spectral classes. In order of decreasing temperature, these spectral classes are O, B, A, F, G, K, M, L, T, and Y. These are further divided into subclasses numbered from 0 to 9. The classes L, T, and Y have been added recently to describe newly discovered star-like objects—mainly brown dwarfs—that are cooler than M9. Our Sun has spectral type G2.

17.4 Using Spectra to Measure Stellar Radius, Composition, and Motion

Spectra of stars of the same temperature but different atmospheric pressures have subtle differences, so spectra can be used to determine whether a star has a large radius and low atmospheric pressure (a giant star) or a small radius and high atmospheric pressure. Stellar spectra can also be used to determine the chemical composition of stars; hydrogen and helium make up most of the mass of all stars. Measurements of line shifts produced by the Doppler effect indicate the radial velocity of a star. Broadening of spectral lines by the Doppler effect is a measure of rotational velocity. A star can also show proper motion, due to the component of a star's space velocity across the line of sight.

For Further Exploration

Articles

Berman, B. "Magnitude Cum Laude." *Astronomy* (December 1998): 92. How we measure the apparent brightnesses of stars is discussed.

Dvorak, J. "The Women Who Created Modern Astronomy [including Annie Cannon]." *Sky & Telescope* (August 2013): 28.

Hearnshaw, J. "Origins of the Stellar Magnitude Scale." *Sky & Telescope* (November 1992): 494. A good history of how we have come to have this cumbersome system is discussed.

Hirshfeld, A. "The Absolute Magnitude of Stars." *Sky & Telescope* (September 1994): 35.

Kaler, J. "Stars in the Cellar: Classes Lost and Found." *Sky & Telescope* (September 2000): 39. An introduction is provided for spectral types and the new classes L and T.

Kaler, J. "Origins of the Spectral Sequence." *Sky & Telescope* (February 1986): 129.

Skrutskie, M. "2MASS: Unveiling the Infrared Universe." *Sky & Telescope* (July 2001): 34. This article focuses on an all-sky survey at 2 microns.

Sneden, C. "Reading the Colors of the Stars." *Astronomy* (April 1989): 36. This article includes a discussion of what we learn from spectroscopy.

Steffey, P. "The Truth about Star Colors." *Sky & Telescope* (September 1992): 266. The color index and how the eye and film "see" colors are discussed.

Tomkins, J. "Once and Future Celestial Kings." *Sky & Telescope* (April 1989): 59. Calculating the motion of stars and determining which stars were, are, and will be brightest in the sky are discussed.

Websites

Discovery of Brown Dwarfs: http://w.astro.berkeley.edu/~basri/bdwarfs/SciAm-book.pdf (http://w.astro.berkeley.edu/~basri/bdwarfs/SciAm-book.pdf).

Listing of Nearby Brown Dwarfs: http://www.solstation.com/stars/pc10bd.htm (http://www.solstation.com/stars/pc10bd.htm).

Spectral Types of Stars: http://www.skyandtelescope.com/astronomy-equipment/the-spectral-types-of-stars/ (http://www.skyandtelescope.com/astronomy-equipment/the-spectral-types-of-stars/).

Stellar Velocities https://www.e-education.psu.edu/astro801/content/l4_p7.html (https://www.e-education.psu.edu/astro801/content/l4_p7.html).

The Lick Observatory: https://www.ucolick.org/main/index.html (https://www.ucolick.org/main/index.html)

Unheard Voices! The Contributions of Women to Astronomy: A Resource Guide: http://bit.ly/astronomywomen (http://bit.ly/astronomywomen).

Videos

When You Are Just Too Small to be a Star: https://www.youtube.com/watch?v=zXCDsb4n4KU (https://www.youtube.com/watch?v=zXCDsb4n4KU). 2013 Public Talk on Brown Dwarfs and Planets by Dr.

Gibor Basri of the University of California–Berkeley (1:32:52).

Collaborative Group Activities

A. The Voyagers in Astronomy feature on Annie Cannon: Classifier of the Stars discusses some of the difficulties women who wanted to do astronomy faced in the first half of the twentieth century. What does your group think about the situation for women today? Do men and women have an equal chance to become scientists? Discuss with your group whether, in your experience, boys and girls were equally encouraged to do science and math where you went to school.

B. In the section on magnitudes in The Brightness of Stars, we discussed how this old system of classifying how bright different stars appear to the eye first developed. Your authors complained about the fact that this old system still has to be taught to every generation of new students. Can your group think of any other traditional systems of doing things in science and measurement where tradition rules even though common sense says a better system could certainly be found. Explain. (Hint: Try Daylight Savings Time, or metric versus English units.)

C. Suppose you could observe a star that has only one spectral line. Could you tell what element that spectral line comes from? Make a list of reasons with your group about why you answered yes or no.

D. A wealthy alumnus of your college decides to give $50 million to the astronomy department to build a world-class observatory for learning more about the characteristics of stars. Have your group discuss what kind of equipment they would put in the observatory. Where should this observatory be located? Justify your answers. (You may want to refer back to the Astronomical Instruments chapter and to revisit this question as you learn more about the stars and equipment for observing them in future chapters.)

E. For some astronomers, introducing a new spectral type for the stars (like the types L, T, and Y discussed in the text) is similar to introducing a new area code for telephone calls. No one likes to disrupt the old system, but sometimes it is simply necessary. Have your group make a list of steps an astronomer would have to go through to persuade colleagues that a new spectral class is needed.

Exercises

Review Questions

1. What two factors determine how bright a star appears to be in the sky?

2. Explain why color is a measure of a star's temperature.

3. What is the main reason that the spectra of all stars are not identical? Explain.

4. What elements are stars mostly made of? How do we know this?

5. What did Annie Cannon contribute to the understanding of stellar spectra?

6. Name five characteristics of a star that can be determined by measuring its spectrum. Explain how you would use a spectrum to determine these characteristics.

7. How do objects of spectral types L, T, and Y differ from those of the other spectral types?

8. Do stars that look brighter in the sky have larger or smaller magnitudes than fainter stars?

9. The star Antares has an apparent magnitude of 1.0, whereas the star Procyon has an apparent magnitude of 0.4. Which star appears brighter in the sky?

10. Based on their colors, which of the following stars is hottest? Which is coolest? Archenar (blue), Betelgeuse (red), Capella (yellow).

11. Order the seven basic spectral types from hottest to coldest.

12. What is the defining difference between a brown dwarf and a true star?

Thought Questions

13. If the star Sirius emits 23 times more energy than the Sun, why does the Sun appear brighter in the sky?

14. How would two stars of equal luminosity—one blue and the other red—appear in an image taken through a filter that passes mainly blue light? How would their appearance change in an image taken through a filter that transmits mainly red light?

15. Table 17.2 lists the temperature ranges that correspond to the different spectral types. What part of the star do these temperatures refer to? Why?

16. Suppose you are given the task of measuring the colors of the brightest stars, listed in Appendix J, through three filters: the first transmits blue light, the second transmits yellow light, and the third transmits red light. If you observe the star Vega, it will appear equally bright through each of the three filters. Which stars will appear brighter through the blue filter than through the red filter? Which stars will appear brighter through the red filter? Which star is likely to have colors most nearly like those of Vega?

17. Star X has lines of ionized helium in its spectrum, and star Y has bands of titanium oxide. Which is hotter? Why? The spectrum of star Z shows lines of ionized helium and also molecular bands of titanium oxide. What is strange about this spectrum? Can you suggest an explanation?

18. The spectrum of the Sun has hundreds of strong lines of nonionized iron but only a few, very weak lines of helium. A star of spectral type B has very strong lines of helium but very weak iron lines. Do these differences mean that the Sun contains more iron and less helium than the B star? Explain.

19. What are the approximate spectral classes of stars with the following characteristics?
A. Balmer lines of hydrogen are very strong; some lines of ionized metals are present.
B. The strongest lines are those of ionized helium.
C. Lines of ionized calcium are the strongest in the spectrum; hydrogen lines show only moderate strength; lines of neutral and metals are present.
D. The strongest lines are those of neutral metals and bands of titanium oxide.

20. Look at the chemical elements in Appendix K. Can you identify any relationship between the abundance of an element and its atomic weight? Are there any obvious exceptions to this relationship?

21. Appendix I lists some of the nearest stars. Are most of these stars hotter or cooler than the Sun? Do any of them emit more energy than the Sun? If so, which ones?

22. Appendix J lists the stars that appear brightest in our sky. Are most of these hotter or cooler than the Sun? Can you suggest a reason for the difference between this answer and the answer to the previous question? (Hint: Look at the luminosities.) Is there any tendency for a correlation between temperature and luminosity? Are there exceptions to the correlation?

23. What star appears the brightest in the sky (other than the Sun)? The second brightest? What color is Betelgeuse? Use Appendix J to find the answers.

24. Suppose hominids one million years ago had left behind maps of the night sky. Would these maps represent accurately the sky that we see today? Why or why not?

25. Why can only a lower limit to the rate of stellar rotation be determined from line broadening rather than the actual rotation rate? (Refer to Figure 17.14.)

26. Why do you think astronomers have suggested three different spectral types (L, T, and Y) for the brown dwarfs instead of M? Why was one not enough?

27. Sam, a college student, just bought a new car. Sam's friend Adam, a graduate student in astronomy, asks Sam for a ride. In the car, Adam remarks that the colors on the temperature control are wrong. Why did he say that?

Figure 17.17 (credit: modification of work by Michael Sheehan)

28. Would a red star have a smaller or larger magnitude in a red filter than in a blue filter?

29. Two stars have proper motions of one arcsecond per year. Star A is 20 light-years from Earth, and Star B is 10 light-years away from Earth. Which one has the faster velocity in space?

30. Suppose there are three stars in space, each moving at 100 km/s. Star A is moving across (i.e., perpendicular to) our line of sight, Star B is moving directly away from Earth, and Star C is moving away from Earth, but at a 30° angle to the line of sight. From which star will you observe the greatest Doppler shift? From which star will you observe the smallest Doppler shift?

31. What would you say to a friend who made this statement, "The visible-light spectrum of the Sun shows weak hydrogen lines and strong calcium lines. The Sun must therefore contain more calcium than hydrogen."?

Figuring for Yourself

32. In Appendix J, how much more luminous is the most luminous of the stars than the least luminous?

For Exercise 17.33 through Exercise 17.38, use the equations relating magnitude and apparent brightness given in the section on the magnitude scale in The Brightness of Stars and Example 17.1.

33. Verify that if two stars have a difference of five magnitudes, this corresponds to a factor of 100 in the ratio $\left(\frac{b_2}{b_1} \right)$; that 2.5 magnitudes corresponds to a factor of 10; and that 0.75 magnitudes corresponds to a factor of 2.

34. As seen from Earth, the Sun has an apparent magnitude of about –26.7. What is the apparent magnitude of the Sun as seen from Saturn, about 10 AU away? (Remember that one AU is the distance from Earth to the Sun and that the brightness decreases as the inverse square of the distance.) Would the Sun still be the brightest star in the sky?

35. An astronomer is investigating a faint star that has recently been discovered in very sensitive surveys of the sky. The star has a magnitude of 16. How much less bright is it than Antares, a star with magnitude roughly equal to 1?

36. The center of a faint but active galaxy has magnitude 26. How much less bright does it look than the very faintest star that our eyes can see, roughly magnitude 6?

37. You have enough information from this chapter to estimate the distance to Alpha Centauri, the second nearest star, which has an apparent magnitude of 0. Since it is a G2 star, like the Sun, assume it has the same luminosity as the Sun and the difference in magnitudes is a result only of the difference in distance. Estimate how far away Alpha Centauri is. Describe the necessary steps in words and then do the calculation. (As we will learn in the Celestial Distances chapter, this method—namely, assuming that stars with identical spectral types emit the same amount of energy—is actually used to estimate distances to stars.) If you assume the distance to the Sun is in AU, your answer will come out in AU.

38. Do the previous problem again, this time using the information that the Sun is 150,000,000 km away. You will get a very large number of km as your answer. To get a better feeling for how the distances compare, try calculating the time it takes light at a speed of 299,338 km/s to travel from the Sun to Earth and from Alpha Centauri to Earth. For Alpha Centauri, figure out how long the trip will take in years as well as in seconds.

39. Star A and Star B have different apparent brightnesses but identical luminosities. If Star A is 20 light-years away from Earth and Star B is 40 light-years away from Earth, which star appears brighter and by what factor?

40. Star A and Star B have different apparent brightnesses but identical luminosities. Star A is 10 light-years away from Earth and appears 36 times brighter than Star B. How far away is Star B?

41. The star Sirius A has an apparent magnitude of –1.5. Sirius A has a dim companion, Sirius B, which is 10,000 times less bright than Sirius A. What is the apparent magnitude of Sirius B? Can Sirius B be seen with the naked eye?

42. Our Sun, a type G star, has a surface temperature of 5800 K. We know, therefore, that it is cooler than a type O star and hotter than a type M star. Given what you learned about the temperature ranges of these types of stars, how many times hotter than our Sun is the hottest type O star? How many times cooler than our Sun is the coolest type M star?

18 The Stars: A Celestial Census

Figure 18.1 Variety of Stars. Stars come in a variety of sizes, masses, temperatures, and luminosities. This image shows part of a cluster of stars in the Small Magellanic Cloud (catalog number NGC 290). Located about 200,000 light-years away, NGC 290 is about 65 light-years across. Because the stars in this cluster are all at about the same distance from us, the differences in apparent brightness correspond to differences in luminosity; differences in temperature account for the differences in color. The various colors and luminosities of these stars provide clues about their life stories. (credit: modification of work by E. Olszewski (University of Arizona), European Space Agency, NASA)

Chapter Outline

18.1 A Stellar Census
18.2 Measuring Stellar Masses
18.3 Diameters of Stars
18.4 The H–R Diagram

Thinking Ahead

How do stars form? How long do they live? And how do they die? Stop and think how hard it is to answer these questions.

Stars live such a long time that nothing much can be gained from staring at one for a human lifetime. To discover how stars evolve from birth to death, it was necessary to measure the characteristics of many stars (to take a celestial census, in effect) and then determine which characteristics help us understand the stars' life stories. Astronomers tried a variety of hypotheses about stars until they came up with the right approach to understanding their development. But the key was first making a thorough census of the stars around us.

18.1 A Stellar Census

Learning Objectives

By the end of this section, you will be able to:

> Explain why the stars visible to the unaided eye are not typical
> Describe the distribution of stellar masses found close to the Sun

Before we can make our own survey, we need to agree on a unit of distance appropriate to the objects we are studying. The stars are all so far away that kilometers (and even astronomical units) would be very cumbersome to use; so—as discussed in Science and the Universe: A Brief Tour—astronomers use a much larger "measuring stick" called the *light-year*. A light-year is the distance that light (the fastest signal we know) travels in 1 year. Since light covers an astounding 300,000 kilometers per second, and since there are a lot of seconds in 1 year, a light-year is a very large quantity: 9.5 trillion (9.5×10^{12}) kilometers to be exact. (Bear in mind that the light-year is a unit of *distance* even though the term *year* appears in it.) If you drove at the legal US speed limit without stopping for food or rest, you would not arrive at the end of a light-year in space until roughly 12 million years had passed. And the closest star is more than 4 light-years away.

Notice that we have not yet said much about how such enormous distances can be measured. That is a complicated question, to which we will return in Celestial Distances. For now, let us assume that distances have been measured for stars in our cosmic vicinity so that we can proceed with our census.

Small Is Beautiful—Or at Least More Common

When we do a census of people in the United States, we count the inhabitants by neighborhood. We can try the same approach for our stellar census and begin with our own immediate neighborhood. As we shall see, we run into two problems—just as we do with a census of human beings. First, it is hard to be sure we have counted *all* the inhabitants; second, our local neighborhood may not contain all possible types of people.

Table 18.1 shows an estimate of the number of stars of each spectral type[1] in our own local neighborhood—within 21 light-years of the Sun. (The Milky Way Galaxy, in which we live, is about 100,000 light-years in diameter, so this figure really applies to a *very* local neighborhood, one that contains a *tiny* fraction of all the billions of stars in the Milky Way.) You can see that there are many more low-luminosity (and hence low mass) stars than high-luminosity ones. Only three of the stars in our local neighborhood (one F type and two A types) are significantly more luminous and more massive than the Sun. This is truly a case where small triumphs over large—at least in terms of numbers. The Sun is more massive than the vast majority of stars in our vicinity.

Stars within 21 Light-Years of the Sun

Spectral Type	Number of Stars
A	2
F	1
G	7
K	17
M	94
White dwarfs	8
Brown dwarfs	33

Table 18.1

This table is based on data published through 2015, and it is likely that more faint objects remain to be discovered (see Figure 18.2). Along with the L and T brown dwarfs already observed in our neighborhood,

---

1   The spectral types of stars were defined and discussed in Analyzing Starlight.

astronomers expect to find perhaps hundreds of additional T dwarfs and Y dwarfs. Many of these are likely to be even cooler than the coolest currently known T dwarf. The reason the lowest-mass dwarfs are so hard to find is that they put out very little light—ten thousand to a million times less light than the Sun. Only recently has our technology progressed to the point that we can detect these dim, cool objects.

Figure 18.2 Dwarf Simulation. This computer simulation shows the stars in our neighborhood as they would be seen from a distance of 30 light-years away. The Sun is in the center. All the brown dwarfs are circled; those found earlier are circled in blue, the ones found recently with the WISE infrared telescope in space (whose scientists put this diagram together) are circled in red. The common M stars, which are red and faint, are made to look brighter than they really would be so that you can see them in the simulation. Note that luminous hot stars like our Sun are very rare. (credit: modification of work by NASA/ JPL-Caltech)

To put all this in perspective, we note that even though the stars counted in the table are our closest neighbors, you can't just look up at the night sky and see them without a telescope; stars fainter than the Sun cannot be seen with the unaided eye unless they are *very* nearby. For example, stars with luminosities ranging from 1/100 to 1/10,000 the luminosity of the Sun (L_{Sun}) are very common, but a star with a luminosity of 1/100 L_{Sun} would have to be within 5 light-years to be visible to the naked eye—and only three stars (all in one system) are this close to us. The nearest of these three stars, Proxima Centauri, still cannot be seen without a telescope because it has such a low luminosity.

Astronomers are working hard these days to complete the census of our local neighborhood by finding our faintest neighbors. Recent discoveries of nearby stars have relied heavily upon infrared telescopes that are able to find these many cool, low-mass stars. You should expect the number of known stars within 21 light-years of the Sun to keep increasing as more and better surveys are undertaken.

Bright Does Not Necessarily Mean Close

If we confine our census to the local neighborhood, we will miss many of the most interesting kinds of stars. After all, the neighborhood in which you live does not contain all the types of people—distinguished according to age, education, income, race, and so on—that live in the entire country. For example, a few people do live to be over 100 years old, but there may be no such individual within several miles of where you live. In order to sample the full range of the human population, you would have to extend your census to a much larger area. Similarly, some types of stars simply are not found nearby.

A clue that we are missing something in our stellar census comes from the fact that only six of the 20 stars that appear brightest in our sky—Sirius, Vega, Altair, Alpha Centauri, Fomalhaut, and Procyon—are found within 26 light-years of the Sun (Figure 18.3). Why are we missing most of the brightest stars when we take our census of the local neighborhood?

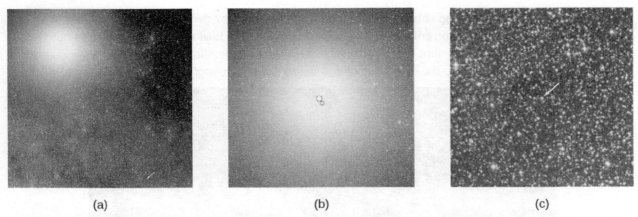

(a)   (b)   (c)

Figure 18.3 The Closest Stars. (a) This image, taken with a wide-angle telescope at the European Southern Observatory in Chile, shows the system of three stars that is our nearest neighbor. (b) Two bright stars that are close to each other (Alpha Centauri A and B) blend their light together. (c) Indicated with an arrow (since you'd hardly notice it otherwise) is the much fainter Proxima Centauri star, which is spectral type M. (credit: modification of work by ESO)

The answer, interestingly enough, is that the stars that appear brightest are *not* the ones closest to us. The brightest stars look the way they do because they emit a very large amount of energy—so much, in fact, that they do not have to be nearby to look brilliant. You can confirm this by looking at Appendix J, which gives distances for the 20 stars that appear brightest from Earth. The most distant of these stars is more than *1000 light-years* from us. In fact, it turns out that most of the stars visible without a telescope are hundreds of light-years away and many times more luminous than the Sun. Among the 9000 stars visible to the unaided eye, only about 50 are intrinsically fainter than the Sun. Note also that several of the stars in Appendix J are spectral type B, a type that is completely missing from Table 18.1.

The most luminous of the bright stars listed in Appendix J emit more than 50,000 times more energy than does the Sun. These highly luminous stars are missing from the solar neighborhood because they are very rare. None of them happens to be in the tiny volume of space immediately surrounding the Sun, and only this small volume was surveyed to get the data shown in Table 18.1.

For example, let's consider the most luminous stars—those 100 or more times as luminous as the Sun. Although such stars are rare, they are visible to the unaided eye, even when hundreds to thousands of light-years away. A star with a luminosity 10,000 times greater than that of the Sun can be seen without a telescope out to a distance of 5000 light-years. The volume of space included within a distance of 5000 light-years, however, is enormous; so even though highly luminous stars are intrinsically rare, many of them are readily visible to our unaided eye.

The contrast between these two samples of stars, those that are close to us and those that can be seen with the unaided eye, is an example of a **selection effect**. When a population of objects (stars in this example) includes a great variety of different types, we must be careful what conclusions we draw from an examination of any particular subgroup. Certainly we would be fooling ourselves if we assumed that the stars visible to the unaided eye are characteristic of the general stellar population; this subgroup is heavily weighted to the most luminous stars. It requires much more effort to assemble a complete data set for the nearest stars, since most are so faint that they can be observed only with a telescope. However, it is only by doing so that astronomers are able to know about the properties of the vast majority of the stars, which are actually much smaller and fainter than our own Sun. In the next section, we will look at how we measure some of these properties.

18.2 | Measuring Stellar Masses

Learning Objectives

By the end of this section, you will be able to:

> Distinguish the different types of binary star systems
> Understand how we can apply Newton's version of Kepler's third law to derive the sum of star masses in a binary star system
> Apply the relationship between stellar mass and stellar luminosity to determine the physical characteristics of a star

The mass of a star—how much material it contains—is one of its most important characteristics. If we know a star's mass, as we shall see, we can estimate how long it will shine and what its ultimate fate will be. Yet the mass of a star is very difficult to measure directly. Somehow, we need to put a star on the cosmic equivalent of a scale.

Luckily, not all stars live like the Sun, in isolation from other stars. About half the stars are **binary stars**—two stars that orbit each other, bound together by gravity. Masses of binary stars can be calculated from measurements of their orbits, just as the mass of the Sun can be derived by measuring the orbits of the planets around it (see Orbits and Gravity).

Binary Stars

Before we discuss in more detail how mass can be measured, we will take a closer look at stars that come in pairs. The first binary star was discovered in 1650, less than half a century after Galileo began to observe the sky with a telescope. John Baptiste Riccioli (1598–1671), an Italian astronomer, noted that the star Mizar, in the middle of the Big Dipper's handle, appeared through his telescope as two stars. Since that discovery, thousands of binary stars have been cataloged. (Astronomers call any pair of stars that appear to be close to each other in the sky *double stars*, but not all of these form a true binary, that is, not all of them are physically associated. Some are just chance alignments of stars that are actually at different distances from us.) Although stars most commonly come in pairs, there are also triple and quadruple systems.

One well-known binary star is Castor, located in the constellation of Gemini. By 1804, astronomer William Herschel, who also discovered the planet Uranus, had noted that the fainter component of Castor had slightly changed its position relative to the brighter component. (We use the term "component" to mean a member of a star system.) Here was evidence that one star was moving around another. It was actually the first evidence that gravitational influences exist outside the solar system. The orbital motion of a binary star is shown in Figure 18.4. A binary star system in which both of the stars can be seen with a telescope is called a **visual binary**.

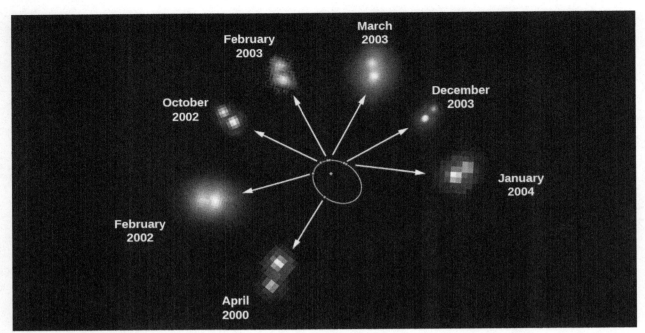

Figure 18.4 Revolution of a Binary Star. This figure shows seven observations of the mutual revolution of two stars, one a brown dwarf and one an ultra-cool L dwarf. Each red dot on the orbit, which is shown by the blue ellipse, corresponds to the position of one of the dwarfs relative to the other. The reason that the pair of stars looks different on the different dates is that some images were taken with the Hubble Space Telescope and others were taken from the ground. The arrows point to the actual observations that correspond to the positions of each red dot. From these observations, an international team of astronomers directly measured the mass of an ultra-cool brown dwarf star for the first time. Barely the size of the planet Jupiter, the dwarf star weighs in at just 8.5% of the mass of our Sun. (credit: modification of work by ESA/NASA and Herve Bouy (Max-Planck-Institut für Extraterrestrische Physik/ ESO, Germany))

Edward C. Pickering (1846–1919), at Harvard, discovered a second class of binary stars in 1889—a class in which only one of the stars is actually seen directly. He was examining the spectrum of Mizar and found that the dark absorption lines in the brighter star's spectrum were usually double. Not only were there two lines where astronomers normally saw only one, but the spacing of the lines was constantly changing. At times, the lines even became single. Pickering correctly deduced that the brighter component of Mizar, called Mizar A, is itself really two stars that revolve about each other in a period of 104 days. A star like Mizar A, which appears as a single star when photographed or observed visually through the telescope, but which spectroscopy shows really to be a double star, is called a **spectroscopic binary**.

Mizar, by the way, is a good example of just how complex such star systems can be. Mizar has been known for centuries to have a faint companion called Alcor, which can be seen without a telescope. Mizar and Alcor form an *optical double*—a pair of stars that appear close together in the sky but do not orbit each other. Through a telescope, as Riccioli discovered in 1650, Mizar can be seen to have another, closer companion that does orbit it; Mizar is thus a visual binary. The two components that make up this visual binary, known as Mizar A and Mizar B, are both spectroscopic binaries. So, Mizar is really a quadruple system of stars.

Strictly speaking, it is not correct to describe the motion of a binary star system by saying that one star orbits the other. Gravity is a *mutual* attraction. Each star exerts a gravitational force on the other, with the result that both stars orbit a point between them called the *center of mass*. Imagine that the two stars are seated at either end of a seesaw. The point at which the fulcrum would have to be located in order for the seesaw to balance is the center of mass, and it is always closer to the more massive star (Figure 18.5).

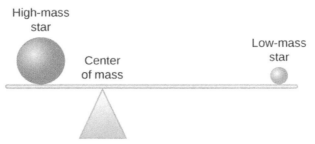

Figure 18.5 Binary Star System. In a binary star system, both stars orbit their center of mass. The image shows the relative positions of two, different-mass stars from their center of mass, similar to how two masses would have to be located on a seesaw in order to keep it level. The star with the higher mass will be found closer to the center of mass, while the star with the lower mass will be farther from it.

Figure 18.6 shows two stars (A and B) moving around their center of mass, along with one line in the spectrum of each star that we observe from the system at different times. When one star is approaching us relative to the center of mass, the other star is receding from us. In the top left illustration, star A is moving toward us, so the line in its spectrum is Doppler-shifted toward the blue end of the spectrum. Star B is moving away from us, so its line shows a redshift. When we observe the composite spectrum of the two stars, the line appears double. When the two stars are both moving across our line of sight (neither away from nor toward us), they both have the same radial velocity (that of the pair's center of mass); hence, the spectral lines of the two stars come together. This is shown in the two bottom illustrations in Figure 18.6.

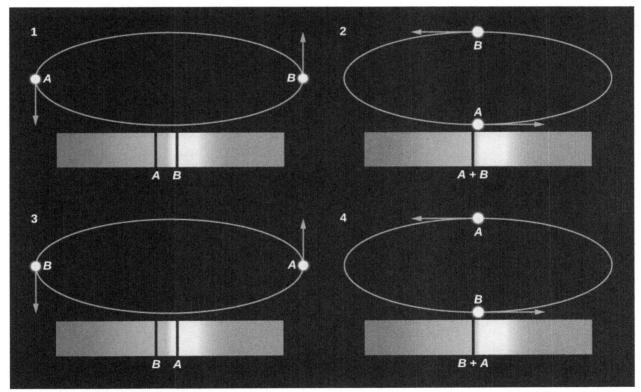

Figure 18.6 Motions of Two Stars Orbiting Each Other and What the Spectrum Shows. We see changes in velocity because when one star is moving toward Earth, the other is moving away; half a cycle later, the situation is reversed. Doppler shifts cause the spectral lines to move back and forth. In diagrams 1 and 3, lines from both stars can be seen well separated from each other. When the two stars are moving perpendicular to our line of sight (that is, they are not moving either toward or away from us), the two lines are exactly superimposed, and so in diagrams 2 and 4, we see only a single spectral line. Note that in the diagrams, the orbit of the star pair is tipped slightly with respect to the viewer (or if the viewer were looking at it in the sky, the orbit would be tilted with respect to the viewer's line of sight). If the orbit were exactly in the plane of the page or screen (or the sky), then it would look nearly circular, but we would see no change in radial velocity (no part of the motion would be toward us or away from us.) If the orbit were perpendicular to the plane of the page or screen, then the stars would appear to move back and forth in a straight line, and we would see the largest-possible radial velocity variations. This diagram has been simplified for clarity. For more realistic situations, see the animation in the following Link to Learning feature box.

A plot showing how the velocities of the stars change with time is called a *radial velocity curve*; the curve for the binary system in Figure 18.6 is shown in Figure 18.7.

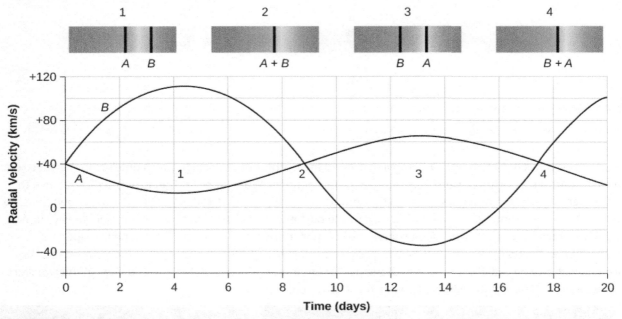

Figure 18.7 Radial Velocities in a Spectroscopic Binary System. These curves plot the radial velocities of two stars in a spectroscopic binary system, showing how the stars alternately approach and recede from Earth. Note that positive velocity means the star is moving away from us, and negative velocity means the star is moving toward us. The center of mass of the system itself is also moving away from us, indicated by the positive velocity of 40 kilometers per second. The positions on the curve corresponding to the illustrations in Figure 18.6 are marked with the diagram number (1–4).

LINK TO LEARNING

This animation (https://openstax.org/l/30binstaranim) lets you follow the orbits of a binary star system in various combinations of the masses of the two stars.

Masses from the Orbits of Binary Stars

We can estimate the masses of binary star systems using Newton's reformulation of Kepler's third law (discussed in Newton's Universal Law of Gravitation). Kepler found that the time a planet takes to go around the Sun is related by a specific mathematical formula to its distance from the Sun. In our binary star situation, if two objects are in mutual revolution, then the period (P) with which they go around each other is related to the semimajor axis (D) of the orbit of one with respect to the other, according to this equation

$$D^3 = (M_1 + M_2)P^2$$

where D is in astronomical units, P is measured in years, and $M_1 + M_2$ is the sum of the masses of the two stars in units of the Sun's mass. This is a very useful formula for astronomers; it says that if we can observe the size of the orbit and the period of mutual revolution of the stars in a binary system, we can calculate the sum of their masses.

Most spectroscopic binaries have periods ranging from a few days to a few months, with separations of usually less than 1 AU between their member stars. Recall that an AU is the distance from Earth to the Sun, so this is a small separation and very hard to see at the distances of stars. This is why many of these systems are known to be double only through careful study of their spectra.

We can analyze a radial velocity curve (such as the one in Figure 18.7) to determine the masses of the stars in a spectroscopic binary. This is complex in practice but not hard in principle. We measure the speeds of the stars from the Doppler effect. We then determine the period—how long the stars take to go through an orbital cycle—from the velocity curve. Knowing how fast the stars are moving and how long they take to go around tells us the circumference of the orbit and, hence, the separation of the stars in kilometers or astronomical units. From Kepler's law, the period and the separation allow us to calculate the sum of the stars' masses.

Of course, knowing the sum of the masses is not as useful as knowing the mass of each star separately. But the relative orbital speeds of the two stars can tell us how much of the total mass each star has. As we saw in our seesaw analogy, the more massive star is closer to the center of mass and therefore has a smaller orbit. Therefore, it moves more slowly to get around in the same time compared to the more distant, lower-mass star. If we sort out the speeds relative to each other, we can sort out the masses relative to each other. In practice, we also need to know how the binary system is oriented in the sky to our line of sight, but if we do, and the just-described steps are carried out carefully, the result is a calculation of the masses of each of the two stars in the system.

To summarize, a good measurement of the motion of two stars around a common center of mass, combined with the laws of gravity, allows us to determine the masses of stars in such systems. These mass measurements are absolutely crucial to developing a theory of how stars evolve. One of the best things about this method is that it is independent of the location of the binary system. It works as well for stars 100 light-years away from us as for those in our immediate neighborhood.

To take a specific example, Sirius is one of the few binary stars in Appendix J for which we have enough information to apply Kepler's third law:

$$D^3 = (M_1 + M_2)P^2$$

In this case, the two stars, the one we usually call Sirius and its very faint companion, are separated by about 20 AU and have an orbital period of about 50 years. If we place these values in the formula we would have

$$(20)^3 = (M_1 + M_2)(50)^2$$
$$8000 = (M_1 + M_2)(2500)$$

This can be solved for the sum of the masses:

$$M_1 + M_2 = \frac{8000}{2500} = 3.2$$

Therefore, the sum of masses of the two stars in the Sirius binary system is 3.2 times the Sun's mass. In order to determine the individual mass of each star, we would need the velocities of the two stars and the orientation of the orbit relative to our line of sight.

The Range of Stellar Masses

How large can the mass of a star be? Stars more massive than the Sun are rare. None of the stars within 30 light-years of the Sun has a mass greater than four times that of the Sun. Searches at large distances from the Sun have led to the discovery of a few stars with masses up to about 100 times that of the Sun, and a handful of stars (a few out of several billion) may have masses as large as 250 solar masses. However, most stars have less mass than the Sun.

According to theoretical calculations, the smallest mass that a true star can have is about 1/12 that of the Sun. By a "true" star, astronomers mean one that becomes hot enough to fuse protons to form helium (as discussed in The Sun: A Nuclear Powerhouse). Objects with masses between roughly 1/100 and 1/12 that of the Sun may produce energy for a brief time by means of nuclear reactions involving deuterium, but they do not become hot enough to fuse protons. Such objects are intermediate in mass between stars and planets and have been given the name **brown dwarfs** (Figure 18.8). Brown dwarfs are similar to Jupiter in radius but have masses from approximately 13 to 80 times larger than the mass of Jupiter.[2]

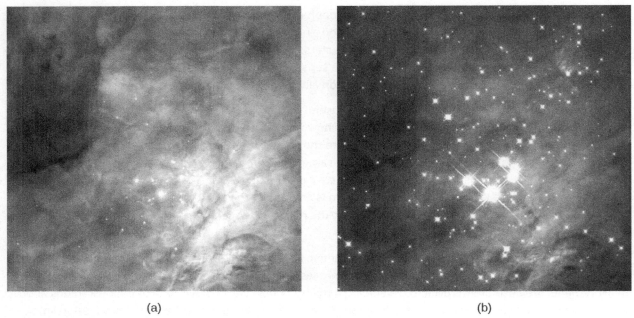

(a)   (b)

Figure 18.8 Brown Dwarfs in Orion. These images, taken with the Hubble Space Telescope, show the region surrounding the Trapezium star cluster inside the star-forming region called the Orion Nebula. (a) No brown dwarfs are seen in the visible light image, both because they put out very little light in the visible and because they are hidden within the clouds of dust in this region. (b) This image was taken in infrared light, which can make its way to us through the dust. The faintest objects in this image are brown dwarfs with masses between 13 and 80 times the mass of Jupiter. (credit a: NASA, C.R. O'Dell and S.K. Wong (Rice University); credit b: NASA; K.L. Luhman (Harvard-Smithsonian Center for Astrophysics) and G. Schneider, E. Young, G. Rieke, A. Cotera, H. Chen, M. Rieke, R. Thompson (Steward Observatory))

Still-smaller objects with masses less than about 1/100 the mass of the Sun (or 10 Jupiter masses) are called planets. They may radiate energy produced by the radioactive elements that they contain, and they may also radiate heat generated by slowly compressing under their own weight (a process called gravitational contraction). However, their interiors will never reach temperatures high enough for any nuclear reactions, to take place. Jupiter, whose mass is about 1/1000 the mass of the Sun, is unquestionably a planet, for example. Until the 1990s, we could only detect planets in our own solar system, but now we have thousands of them elsewhere as well. (We will discuss these exciting observations in The Birth of Stars and the Discovery of Planets outside the Solar System.)

The Mass-Luminosity Relation

Now that we have measurements of the characteristics of many different types of stars, we can search for relationships among the characteristics. For example, we can ask whether the mass and luminosity of a star are related. It turns out that for most stars, they are: The more massive stars are generally also the more luminous. This relationship, known as the **mass-luminosity relation**, is shown graphically in Figure 18.9. Each point represents a star whose mass and luminosity are both known. The horizontal position on the graph shows the star's mass, given in units of the Sun's mass, and the vertical position shows its luminosity in units of the Sun's luminosity.

---

2   Exactly where to put the dividing line between planets and brown dwarfs is a subject of some debate among astronomers as we write this book (as is, in fact, the exact definition of each of these objects). Even those who accept deuterium fusion (see The Birth of Stars and the Discovery of Planets outside the Solar System) as the crucial issue for brown dwarfs concede that, depending on the composition of the star and other factors, the lowest mass for such a dwarf could be anywhere from 11 to 16 Jupiter masses.

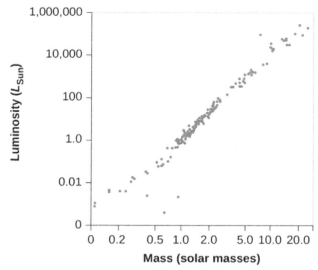

Figure 18.9 Mass-Luminosity Relation. The plotted points show the masses and luminosities of stars. The three points lying below the sequence of points are all white dwarf stars.

We can also say this in mathematical terms.

$$L \sim M^{3.9}$$

It's a reasonably good approximation to say that luminosity (expressed in units of the Sun's luminosity) varies as the fourth power of the mass (in units of the Sun's mass). (The symbol ~ means the two quantities are proportional.) If two stars differ in mass by a factor of 2, then the more massive one will be 2^4, or about 16 times brighter; if one star is 1/3 the mass of another, it will be approximately 81 times less luminous.

EXAMPLE 18.1

Calculating the Mass from the Luminosity of a Star

The mass-luminosity formula can be rewritten so that a value of mass can be determined if the luminosity is known.

Solution

First, we must get our units right by expressing both the mass and the luminosity of a star in units of the Sun's mass and luminosity:

$$L/L_{Sun} = (M/M_{Sun})^4$$

Now we can take the 4th root of both sides, which is equivalent to taking both sides to the 1/4 = 0.25 power. The formula in this case would be:

$$M/M_{Sun} = \left(L/L_{Sun}\right)^{0.25} = \left(L/L_{Sun}\right)^{0.25}$$

Check Your Learning

In the previous section, we determined the sum of the masses of the two stars in the Sirius binary system (Sirius and its faint companion) using Kepler's third law to be 3.2 solar masses. Using the mass-luminosity relationship, calculate the mass of each individual star.

Answer:

In Appendix J, Sirius is listed with a luminosity 23 times that of the Sun. This value can be inserted into the mass-luminosity relationship to get the mass of Sirius:

$$M/M_{Sun} = 23^{0.25} = 2.2$$

The mass of the companion star to Sirius is then 3.2 – 2.2 = 1.0 solar mass.

Notice how good this mass-luminosity relationship is. Most stars (see Figure 18.9) fall along a line running from the lower-left (low mass, low luminosity) corner of the diagram to the upper-right (high mass, high luminosity) corner. About 90% of all stars obey the mass-luminosity relation. Later, we will explore why such a relationship exists and what we can learn from the roughly 10% of stars that "disobey" it.

LINK TO LEARNING

This visualization (https://openstax.org/l/30starmass) illustrates how all the properties of a star are determined by the mass that the star happened to form with and how these properties change when you change the mass.

18.3 | Diameters of Stars

Learning Objectives

By the end of this section, you will be able to:

> Describe the methods used to determine star diameters
> Identify the parts of an eclipsing binary star light curve that correspond to the diameters of the individual components

It is easy to measure the diameter of the Sun. Its angular diameter—that is, its apparent size on the sky—is about 1/2°. If we know the angle the Sun takes up in the sky and how far away it is, we can calculate its true (linear) diameter, which is 1.39 million kilometers, or about 109 times the diameter of Earth.

Unfortunately, the Sun is the only star whose angular diameter is easily measured. All the other stars are so far away that they look like pinpoints of light through even the largest ground-based telescopes. (They often seem to be bigger, but that is merely distortion introduced by turbulence in Earth's atmosphere.) Luckily, there are several techniques that astronomers can use to estimate the sizes of stars.

Stars Blocked by the Moon

One technique, which gives very precise diameters but can be used for only a few stars, is to observe the dimming of light that occurs when the Moon passes in front of a star. What astronomers measure (with great precision) is the time required for the star's brightness to drop to zero as the edge of the Moon moves across the star's disk. Since we know how rapidly the Moon moves in its orbit around Earth, it is possible to calculate the angular diameter of the star. If the distance to the star is also known, we can calculate its diameter in kilometers. This method works only for fairly bright stars that happen to lie along the zodiac, where the Moon (or, much more rarely, a planet) can pass in front of them as seen from Earth.

Eclipsing Binary Stars

Accurate sizes for a large number of stars come from measurements of **eclipsing binary** star systems, and so we must make a brief detour from our main story to examine this type of star system. Some binary stars are lined up in such a way that, when viewed from Earth, each star passes in front of the other during every revolution (Figure 18.10). When one star blocks the light of the other, preventing it from reaching Earth, the brightness of the system decreases, and astronomers say that an eclipse has occurred.

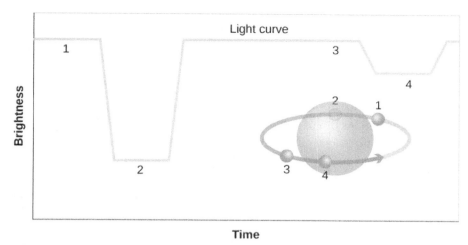

Figure 18.10 Light Curve of an Eclipsing Binary. The light curve of an eclipsing binary star system shows how the combined light from both stars changes due to eclipses over the time span of an orbit. This light curve shows the behavior of a hypothetical eclipsing binary star with total eclipses (one star passes directly in front of and behind the other). The numbers indicate parts of the light curve corresponding to various positions of the smaller star in its orbit. In this diagram, we have assumed that the smaller star is also the hotter one so that it emits more flux (energy per second per square meter) than the larger one. When the smaller, hotter star goes behind the larger one, its light is completely blocked, and so there is a strong dip in the light curve. When the smaller star goes in front of the bigger one, a small amount of light from the bigger star is blocked, so there is a smaller dip in the light curve.

The discovery of the first eclipsing binary helped solve a long-standing puzzle in astronomy. The star Algol, in the constellation of Perseus, changes its brightness in an odd but regular way. Normally, Algol is a fairly bright star, but at intervals of 2 days, 20 hours, 49 minutes, it fades to one-third of its regular brightness. After a few hours, it brightens to normal again. This effect is easily seen, even without a telescope, if you know what to look for.

In 1783, a young English astronomer named John Goodricke (1764–1786) made a careful study of Algol (see the feature on John Goodricke for a discussion of his life and work). Even though Goodricke could neither hear nor speak, he made a number of major discoveries in the 21 years of his brief life. He suggested that Algol's unusual brightness variations might be due to an invisible companion that regularly passes in front of the brighter star and blocks its light. Unfortunately, Goodricke had no way to test this idea, since it was not until about a century later that equipment became good enough to measure Algol's spectrum.

In 1889, the German astronomer Hermann Vogel (1841–1907) demonstrated that, like Mizar, Algol is a spectroscopic binary. The spectral lines of Algol were not observed to be double because the fainter star of the pair gives off too-little light compared with the brighter star for its lines to be conspicuous in the composite spectrum. Nevertheless, the periodic shifting back and forth of the brighter star's lines gave evidence that it was revolving about an unseen companion. (The lines of both components need not be visible for a star to be recognized as a spectroscopic binary.)

The discovery that Algol is a spectroscopic binary verified Goodricke's hypothesis. The plane in which the stars revolve is turned nearly edgewise to our line of sight, and each star passes in front of the other during every revolution. (The eclipse of the fainter star in the Algol system is not very noticeable because the part of it that is covered contributes little to the total light of the system. This second eclipse can, however, be detected by careful measurements.)

Any binary star produces eclipses if viewed from the proper direction, near the plane of its orbit, so that one star passes in front of the other (see Figure 18.10). But from our vantage point on Earth, only a few binary star systems are oriented in this way.

Astronomy and Mythology: Algol the Demon Star and Perseus the Hero

The name Algol comes from the Arabic *Ras al Ghul*, meaning "the demon's head."[3] The word "ghoul" in English has the same derivation. As discussed in Observing the Sky: The Birth of Astronomy, many of the bright stars have Arabic names because during the long dark ages in medieval Europe, it was Arabic astronomers who preserved and expanded the Greek and Roman knowledge of the skies. The reference to the demon is part of the ancient Greek legend of the hero Perseus, who is commemorated by the constellation in which we find Algol and whose adventures involve many of the characters associated with the northern constellations.

Perseus was one of the many half-god heroes fathered by Zeus (Jupiter in the Roman version), the king of the gods in Greek mythology. Zeus had, to put it delicately, a roving eye and was always fathering somebody or other with a human maiden who caught his fancy. (Perseus derives from *Per Zeus*, meaning "fathered by Zeus.") Set adrift with his mother by an (understandably) upset stepfather, Perseus grew up on an island in the Aegean Sea. The king there, taking an interest in Perseus' mother, tried to get rid of the young man by assigning him an extremely difficult task.

In a moment of overarching pride, a beautiful young woman named Medusa had compared her golden hair to that of the goddess Athena (Minerva for the Romans). The Greek gods did not take kindly to being compared to mere mortals, and Athena turned Medusa into a gorgon: a hideous, evil creature with writhing snakes for hair and a face that turned anyone who looked at it into stone. Perseus was given the task of slaying this demon, which seemed like a pretty sure way to get him out of the way forever.

But because Perseus had a god for a father, some of the other gods gave him tools for the job, including Athena's reflective shield and the winged sandals of Hermes (Mercury in the Roman story). By flying over her and looking only at her reflection, Perseus was able to cut off Medusa's head without ever looking at her directly. Taking her head (which, conveniently, could still turn onlookers to stone even without being attached to her body) with him, Perseus continued on to other adventures.

He next came to a rocky seashore, where boasting had gotten another family into serious trouble with the gods. Queen Cassiopeia had dared to compare her own beauty to that of the Nereids, sea nymphs who were daughters of Poseidon (Neptune in Roman mythology), the god of the sea. Poseidon was so offended that he created a sea-monster named Cetus to devastate the kingdom. King Cepheus, Cassiopeia's beleaguered husband, consulted the oracle, who told him that he must sacrifice his beautiful daughter Andromeda to the monster.

When Perseus came along and found Andromeda chained to a rock near the sea, awaiting her fate, he rescued her by turning the monster to stone. (Scholars of mythology actually trace the essence of this story back to far-older legends from ancient Mesopotamia, in which the god-hero Marduk vanquishes a monster named Tiamat. Symbolically, a hero like Perseus or Marduk is usually associated with the Sun, the monster with the power of night, and the beautiful maiden with the fragile beauty of dawn, which the Sun releases after its nightly struggle with darkness.)

Many of the characters in these Greek legends can be found as constellations in the sky, not necessarily resembling their namesakes but serving as reminders of the story. For example, vain Cassiopeia is sentenced to be very close to the celestial pole, rotating perpetually around the sky and hanging upside down every winter. The ancients imagined Andromeda still chained to her rock (it is much easier to see the chain of stars than to recognize the beautiful maiden in this star grouping). Perseus is next to her with the

---

3   Fans of Batman comic books and movies will recognize that this name was given to an archvillain in the series.

head of Medusa swinging from his belt. Algol represents this gorgon head and has long been associated with evil and bad fortune in such tales. Some commentators have speculated that the star's change in brightness (which can be observed with the unaided eye) may have contributed to its unpleasant reputation, with the ancients regarding such a change as a sort of evil "wink."

Diameters of Eclipsing Binary Stars

We now turn back to the main thread of our story to discuss how all this can be used to measure the sizes of stars. The technique involves making a light curve of an eclipsing binary, a graph that plots how the brightness changes with time. Let us consider a hypothetical binary system in which the stars are very different in size, like those illustrated in Figure 18.11. To make life easy, we will assume that the orbit is viewed exactly edge-on.

Even though we cannot see the two stars separately in such a system, the light curve can tell us what is happening. When the smaller star just starts to pass behind the larger star (a point we call *first contact*), the brightness begins to drop. The eclipse becomes total (the smaller star is completely hidden) at the point called *second contact*. At the end of the total eclipse (*third contact*), the smaller star begins to emerge. When the smaller star has reached *last contact*, the eclipse is completely over.

To see how this allows us to measure diameters, look carefully at Figure 18.11. During the time interval between the first and second contacts, the smaller star has moved a distance equal to its own diameter. During the time interval from the first to third contacts, the smaller star has moved a distance equal to the diameter of the larger star. If the spectral lines of both stars are visible in the spectrum of the binary, then the speed of the smaller star with respect to the larger one can be measured from the Doppler shift. But knowing the speed with which the smaller star is moving and how long it took to cover some distance can tell the span of that distance—in this case, the diameters of the stars. The speed multiplied by the time interval from the first to second contact gives the diameter of the smaller star. We multiply the speed by the time between the first and third contacts to get the diameter of the larger star.

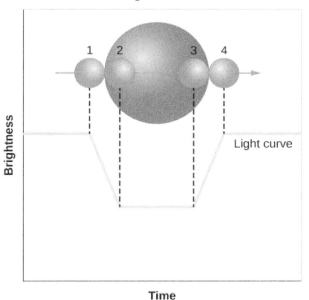

Figure 18.11 Light Curve of an Edge-On Eclipsing Binary. Here we see the light curve of a hypothetical eclipsing binary star whose orbit we view exactly edge-on, in which the two stars fully eclipse each other. From the time intervals between contacts, it is possible to estimate the diameters of the two stars.

In actuality, the situation with eclipsing binaries is often a bit more complicated: orbits are generally not seen exactly edge-on, and the light from each star may be only partially blocked by the other. Furthermore, binary star orbits, just like the orbits of the planets, are ellipses, not circles. However, all these effects can be sorted out from very careful measurements of the light curve.

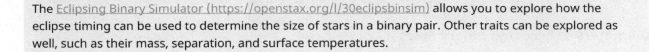

The Eclipsing Binary Simulator (https://openstax.org/l/30eclipsbinsim) allows you to explore how the eclipse timing can be used to determine the size of stars in a binary pair. Other traits can be explored as well, such as their mass, separation, and surface temperatures.

Using the Radiation Law to Get the Diameter

Another method for measuring star diameters makes use of the Stefan-Boltzmann law for the relationship between energy radiated and temperature (see Radiation and Spectra). In this method, the *energy flux* (energy emitted per second per square meter by a blackbody, like the Sun) is given by

$$F = \sigma T^4$$

where σ is a constant and *T* is the temperature. The surface area of a sphere (like a star) is given by

$$A = 4\pi R^2$$

The luminosity (*L*) of a star is then given by its surface area in square meters times the energy flux:

$$L = (A \times F)$$

Previously, we determined the masses of the two stars in the Sirius binary system. Sirius gives off 8200 times more energy than its fainter companion star, although both stars have nearly identical temperatures. The extremely large difference in luminosity is due to the difference in radius, since the temperatures and hence the energy fluxes for the two stars are nearly the same. To determine the relative sizes of the two stars, we take the ratio of the corresponding luminosities:

$$\frac{L_{\text{Sirius}}}{L_{\text{companion}}} = \frac{\left(A_{\text{Sirius}} \times F_{\text{Sirius}}\right)}{\left(A_{\text{companion}} \times F_{\text{companion}}\right)}$$

$$= \frac{A_{\text{Sirius}}}{A_{\text{companion}}} = \frac{4\pi R^2{}_{\text{Sirius}}}{4\pi R^2{}_{\text{companion}}} = \frac{R^2{}_{\text{Sirius}}}{R^2{}_{\text{companion}}}$$

$$\frac{L_{\text{Sirius}}}{L_{\text{companion}}} = 8200 = \frac{R^2{}_{\text{Sirius}}}{R^2{}_{\text{companion}}}$$

Therefore, the relative sizes of the two stars can be found by taking the square root of the relative luminosity. Since $\sqrt{8200} = 91$, the radius of Sirius is 91 times larger than the radius of its faint companion.

The method for determining the radius shown here requires both stars be visible, which is not always the case.

Use the Stellar Luminosity Simulato (https://openstax.org/l/30stellumsim) to explore the relationship between a star's surface temperature, luminosity, and radius. Move the sliders to see what happens. Try to make two stars with the same luminosity but different surface temperatures.

Stellar Diameters

The results of many stellar size measurements over the years have shown that most nearby stars are roughly the size of the Sun, with typical diameters of a million kilometers or so. Faint stars, as we might have expected, are generally smaller than more luminous stars. However, there are some dramatic exceptions to this simple generalization.

A few of the very luminous stars, those that are also red (indicating relatively low surface temperatures), turn out to be truly enormous. These stars are called, appropriately enough, giant stars or supergiant stars. An example is Betelgeuse, the second brightest star in the constellation of Orion and one of the dozen brightest stars in our sky. Its diameter, remarkably, is greater than 10 AU (1.5 *billion* kilometers!), large enough to fill the entire inner solar system almost as far out as Jupiter. In Stars from Adolescence to Old Age, we will look in detail at the evolutionary process that leads to the formation of such giant and supergiant stars.

> **LINK TO LEARNING**
>
> Watch this star size comparison video (https://openstax.org/l/30starsizecomp) for a striking visual that highlights the size of stars versus planets and the range of sizes among stars.

18.4 | The H–R Diagram

Learning Objectives

By the end of this section, you will be able to:

> Identify the physical characteristics of stars that are used to create an H–R diagram, and describe how those characteristics vary among groups of stars

> Discuss the physical properties of most stars found at different locations on the H–R diagram, such as radius, and for main sequence stars, mass

In this chapter and Analyzing Starlight, we described some of the characteristics by which we might classify stars and how those are measured. These ideas are summarized in Table 18.2. We have also given an example of a relationship between two of these characteristics in the mass-luminosity relation. When the characteristics of large numbers of stars were measured at the beginning of the twentieth century, astronomers were able to begin a deeper search for patterns and relationships in these data.

Measuring the Characteristics of Stars

Characteristic	Technique	
Surface temperature	1. Determine the color (very rough). 2. Measure the spectrum and get the spectral type.	
Chemical composition	Determine which lines are present in the spectrum.	
Luminosity	Measure the apparent brightness and compensate for distance.	
Radial velocity	Measure the Doppler shift in the spectrum.	

Table 18.2

Characteristic	Technique
Rotation	Measure the width of spectral lines.
Mass	Measure the period and radial velocity curves of spectroscopic binary stars.
Diameter	1. Measure the way a star's light is blocked by the Moon. 2. Measure the light curves and Doppler shifts for eclipsing binary stars.

Table 18.2

To help understand what sorts of relationships might be found, let's look briefly at a range of data about human beings. If you want to understand humans by comparing and contrasting their characteristics—without assuming any previous knowledge of these strange creatures—you could try to determine which characteristics lead you in a fruitful direction. For example, you might plot the heights of a large sample of humans against their weights (which is a measure of their mass). Such a plot is shown in Figure 18.12 and it has some interesting features. In the way we have chosen to present our data, height increases upward, whereas weight increases to the left. Notice that humans are not randomly distributed in the graph. Most points fall along a sequence that goes from the upper left to the lower right.

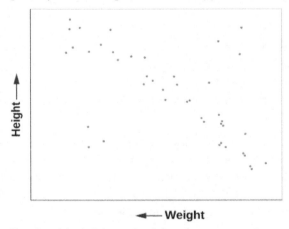

Figure 18.12 Height versus Weight. The plot of the heights and weights of a representative group of human beings. Most points lie along a "main sequence" representing most people, but there are a few exceptions.

We can conclude from this graph that human height and weight are related. Generally speaking, taller human beings weigh more, whereas shorter ones weigh less. This makes sense if you are familiar with the structure of human beings. Typically, if we have bigger bones, we have more flesh to fill out our larger frame. It's not mathematically exact—there is a wide range of variation—but it's not a bad overall rule. And, of course, there are some dramatic exceptions. You occasionally see a short human who is very overweight and would thus be more to the bottom left of our diagram than the average sequence of people. Or you might have a very tall, skinny fashion model with great height but relatively small weight, who would be found near the upper right.

A similar diagram has been found extremely useful for understanding the lives of stars. In 1913, American astronomer Henry Norris Russell plotted the luminosities of stars against their spectral classes (a way of denoting their surface temperatures). This investigation, and a similar independent study in 1911 by Danish astronomer Ejnar Hertzsprung, led to the extremely important discovery that the temperature and luminosity of stars are related (Figure 18.13).

(a)   (b)

Figure 18.13 Hertzsprung (1873–1967) and Russell (1877–1957). (a) Ejnar Hertzsprung and (b) Henry Norris Russell independently discovered the relationship between the luminosity and surface temperature of stars that is summarized in what is now called the H–R diagram.

VOYAGERS IN ASTRONOMY

Henry Norris Russell

When Henry Norris Russell graduated from Princeton University, his work had been so brilliant that the faculty decided to create a new level of honors degree beyond "summa cum laude" for him. His students later remembered him as a man whose thinking was three times faster than just about anybody else's. His memory was so phenomenal, he could correctly quote an enormous number of poems and limericks, the entire Bible, tables of mathematical functions, and almost anything he had learned about astronomy. He was nervous, active, competitive, critical, and very articulate; he tended to dominate every meeting he attended. In outward appearance, he was an old-fashioned product of the nineteenth century who wore high-top black shoes and high starched collars, and carried an umbrella every day of his life. His 264 papers were enormously influential in many areas of astronomy.

Born in 1877, the son of a Presbyterian minister, Russell showed early promise. When he was 12, his family sent him to live with an aunt in Princeton so he could attend a top preparatory school. He lived in the same house in that town until his death in 1957 (interrupted only by a brief stay in Europe for graduate work). He was fond of recounting that both his mother and his maternal grandmother had won prizes in mathematics, and that he probably inherited his talents in that field from their side of the family.

Before Russell, American astronomers devoted themselves mainly to surveying the stars and making impressive catalogs of their properties, especially their spectra (as described in Analyzing Starlight. Russell began to see that interpreting the spectra of stars required a much more sophisticated understanding of the physics of the atom, a subject that was being developed by European physicists in the 1910s and 1920s. Russell embarked on a lifelong quest to ascertain the physical conditions inside stars from the clues in their spectra; his work inspired, and was continued by, a generation of astronomers, many trained by Russell and his collaborators.

Russell also made important contributions in the study of binary stars and the measurement of star masses, the origin of the solar system, the atmospheres of planets, and the measurement of distances in astronomy, among other fields. He was an influential teacher and popularizer of astronomy, writing a column on astronomical topics for *Scientific American* magazine for more than 40 years. He and two colleagues wrote a textbook for college astronomy classes that helped train astronomers and astronomy enthusiasts over several decades. That book set the scene for the kind of textbook you are now reading, which not only lays out the facts of astronomy but also explains how they fit together. Russell gave lectures around the country, often emphasizing the importance of understanding modern physics in order to grasp what was happening in astronomy.

Harlow Shapley, director of the Harvard College Observatory, called Russell "the dean of American astronomers." Russell was certainly regarded as the leader of the field for many years and was consulted on many astronomical problems by colleagues from around the world. Today, one of the highest recognitions that an astronomer can receive is an award from the American Astronomical Society called the Russell Prize, set up in his memory.

Features of the H–R Diagram

Following Hertzsprung and Russell, let us plot the temperature (or spectral class) of a selected group of nearby stars against their luminosity and see what we find (Figure 18.14). Such a plot is frequently called the *Hertzsprung–Russell diagram*, abbreviated **H–R diagram**. It is one of the most important and widely used diagrams in astronomy, with applications that extend far beyond the purposes for which it was originally developed more than a century ago.

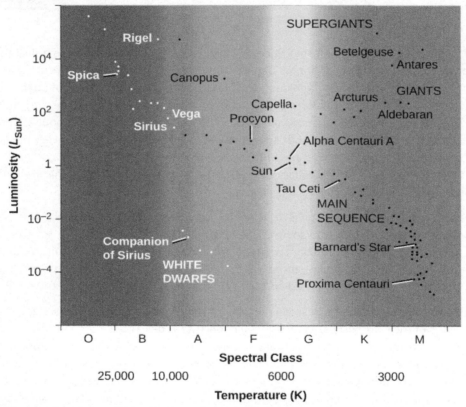

Figure 18.14 H–R Diagram for a Selected Sample of Stars. In such diagrams, luminosity is plotted along the vertical axis. Along the horizontal axis, we can plot either temperature or spectral type (also sometimes called spectral class). Several of the brightest stars are identified by name. Most stars fall on the main sequence.

It is customary to plot H–R diagrams in such a way that temperature increases toward the left and luminosity toward the top. Notice the similarity to our plot of height and weight for people (Figure 18.12). Stars, like people, are not distributed over the diagram at random, as they would be if they exhibited all combinations of luminosity and temperature. Instead, we see that the stars cluster into certain parts of the H–R diagram. The great majority are aligned along a narrow sequence running from the upper left (hot, highly luminous) to the lower right (cool, less luminous). This band of points is called the **main sequence**. It represents a relationship between *temperature* and *luminosity* that is followed by most stars. We can summarize this relationship by saying that hotter stars are more luminous than cooler ones.

A number of stars, however, lie above the main sequence on the H-R diagram, in the upper-right region, where stars have low temperature and high luminosity. How can a star be at once cool, meaning each square meter on the star does not put out all that much energy, and yet very luminous? The only way is for the star to be enormous—to have so many square meters on its surface that the *total* energy output is still large. These stars must be *giants* or *supergiants*, the stars of huge diameter we discussed earlier.

There are also some stars in the lower-left corner of the diagram, which have high temperature and low luminosity. If they have high surface temperatures, each square meter on that star puts out a lot of energy. How then can the overall star be dim? It must be that it has a very small total surface area; such stars are known as **white dwarfs** (white because, at these high temperatures, the colors of the electromagnetic radiation that they emit blend together to make them look bluish-white). We will say more about these puzzling objects in a moment. Figure 18.15 is a schematic H-R diagram for a large sample of stars, drawn to make the different types more apparent.

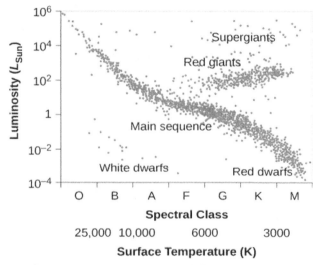

Figure 18.15 Schematic H-R Diagram for Many Stars. Ninety percent of all stars on such a diagram fall along a narrow band called the main sequence. A minority of stars are found in the upper right; they are both cool (and hence red) and bright, and must be giants. Some stars fall in the lower left of the diagram; they are both hot and dim, and must be white dwarfs.

Now, think back to our discussion of star surveys. It is difficult to plot an H–R diagram that is truly representative of all stars because most stars are so faint that we cannot see those outside our immediate neighborhood. The stars plotted in Figure 18.14 were selected because their distances are known. This sample omits many intrinsically faint stars that are nearby but have not had their distances measured, so it shows fewer faint main-sequence stars than a "fair" diagram would. To be truly representative of the stellar population, an H–R diagram should be plotted for all stars within a certain distance. Unfortunately, our knowledge is reasonably complete only for stars within 10 to 20 light-years of the Sun, among which there are no giants or supergiants. Still, from many surveys (and more can now be done with new, more powerful telescopes), we estimate that about 90% of the true stars overall (excluding brown dwarfs) in our part of space are main-sequence stars, about 10% are white dwarfs, and fewer than 1% are giants or supergiants.

These estimates can be used directly to understand the lives of stars. Permit us another quick analogy with people. Suppose we survey people just like astronomers survey stars, but we want to focus our attention on the location of young people, ages 6 to 18 years. Survey teams fan out and take data about where such youngsters are found at all times during a 24-hour day. Some are found in the local pizza parlor, others are asleep at home, some are at the movies, and many are in school. After surveying a very large number of young people, one of the things that the teams determine is that, averaged over the course of the 24 hours, one-third of all youngsters are found in school.

How can they interpret this result? Does it mean that two-thirds of students are truants and the remaining one-third spend all their time in school? No, we must bear in mind that the survey teams counted youngsters throughout the full 24-hour day. Some survey teams worked at night, when most youngsters were at home asleep, and others worked in the late afternoon, when most youngsters were on their way home from school (and more likely to be enjoying a pizza). If the survey was truly representative, we *can* conclude, however, that if an average of one-third of all youngsters are found in school, then humans ages 6 to 18 years must spend about one-third of *their time* in school.

We can do something similar for stars. We find that, on average, 90% of all stars are located on the main sequence of the H–R diagram. If we can identify some activity or life stage with the main sequence, then it follows that stars must spend 90% of their lives in that activity or life stage.

Understanding the Main Sequence

In The Sun: A Nuclear Powerhouse, we discussed the Sun as a representative star. We saw that what stars such as the Sun "do for a living" is to convert protons into helium deep in their interiors via the process of nuclear fusion, thus producing energy. The fusion of protons to helium is an excellent, long-lasting source of energy for a star because the bulk of every star consists of hydrogen atoms, whose nuclei are protons.

Our computer models of how stars evolve over time show us that a typical star will spend about 90% of its life fusing the abundant hydrogen in its core into helium. This then is a good explanation of why 90% of all stars are found on the main sequence in the H–R diagram. But if all the stars on the main sequence are doing the same thing (fusing hydrogen), why are they distributed along a sequence of points? That is, why do they differ in luminosity and surface temperature (which is what we are plotting on the H–R diagram)?

To help us understand how main-sequence stars differ, we can use one of the most important results from our studies of model stars. Astrophysicists have been able to show that the structure of stars that are in equilibrium and derive all their energy from nuclear fusion is completely and uniquely determined by just two quantities: the *total mass* and the *composition* of the star. This fact provides an interpretation of many features of the H–R diagram.

Imagine a cluster of stars forming from a cloud of interstellar "raw material" whose chemical composition is similar to the Sun's. (We'll describe this process in more detail in The Birth of Stars and Discovery of Planets outside the Solar System, but for now, the details will not concern us.) In such a cloud, all the clumps of gas and dust that become stars begin with the same chemical composition and differ from one another only in mass. Now suppose that we compute a model of each of these stars for the time at which it becomes stable and derives its energy from nuclear reactions, but before it has time to alter its composition appreciably as a result of these reactions.

The models calculated for these stars allow us to determine their luminosities, temperatures, and sizes. If we plot the results from the models—one point for each model star—on the H–R diagram, we get something that looks just like the main sequence we saw for real stars.

And here is what we find when we do this. The model stars with the largest masses are the hottest and most luminous, and they are located at the upper left of the diagram.

The least-massive model stars are the coolest and least luminous, and they are placed at the lower right of the plot. The other model stars all lie along a line running diagonally across the diagram. In other words, *the main sequence turns out to be a sequence of stellar masses.*

This makes sense if you think about it. The most massive stars have the most gravity and can thus compress their centers to the greatest degree. This means they are the hottest inside and the best at generating energy from nuclear reactions deep within. As a result, they shine with the greatest luminosity and have the hottest surface temperatures. The stars with lowest mass, in turn, are the coolest inside and least effective in generating energy. Thus, they are the least luminous and wind up being the coolest on the surface. Our Sun lies somewhere in the middle of these extremes (as you can see in Figure 18.14). The characteristics of representative main-sequence stars (excluding brown dwarfs, which are not true stars) are listed in Table 18.3.

Characteristics of Main-Sequence Stars

Spectral Type	Mass (Sun = 1)	Luminosity (Sun = 1)	Temperature	Radius (Sun = 1)
O5	40	7×10^5	40,000 K	18
B0	16	2.7×10^5	28,000 K	7
A0	3.3	55	10,000 K	2.5
F0	1.7	5	7500 K	1.4
G0	1.1	1.4	6000 K	1.1
K0	0.8	0.35	5000 K	0.8
M0	0.4	0.05	3500 K	0.6

Table 18.3

Note that we've seen this 90% figure come up before. This is exactly what we found earlier when we examined the mass-luminosity relation (Figure 18.9). We observed that 90% of all stars seem to follow the relationship; these are the 90% of all stars that lie on the main sequence in our H–R diagram. Our models and our observations agree.

What about the other stars on the H–R diagram—the giants and supergiants, and the white dwarfs? As we will see in the next few chapters, these are what main-sequence stars turn into as they age: They are the later stages in a star's life. As a star consumes its nuclear fuel, its source of energy changes, as do its chemical composition and interior structure. These changes cause the star to alter its luminosity and surface temperature so that it no longer lies on the main sequence on our diagram. Because stars spend much less time in these later stages of their lives, we see fewer stars in those regions of the H–R diagram.

Extremes of Stellar Luminosities, Diameters, and Densities

We can use the H–R diagram to explore the extremes in size, luminosity, and density found among the stars. Such extreme stars are not only interesting to fans of the *Guinness Book of World Records*; they can teach us a lot about how stars work. For example, we saw that the most massive main-sequence stars are the most luminous ones. We know of a few extreme stars that are a million times more luminous than the Sun, with masses that exceed 100 times the Sun's mass. These superluminous stars, which are at the upper left of the H–R diagram, are exceedingly hot, very blue stars of spectral type O. These are the stars that would be the most conspicuous at vast distances in space.

The cool supergiants in the upper corner of the H–R diagram are as much as 10,000 times as luminous as the Sun. In addition, these stars have diameters very much larger than that of the Sun. As discussed above, some supergiants are so large that if the solar system could be centered in one, the star's surface would lie beyond the orbit of Mars (see Figure 18.16). We will have to ask, in coming chapters, what process can make a star swell up to such an enormous size, and how long these "swollen" stars can last in their distended state.

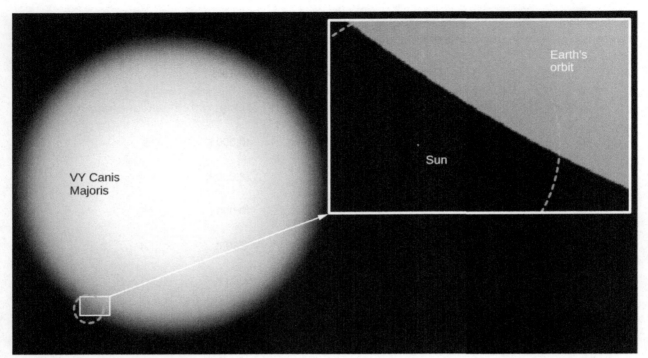

Figure 18.16 The Sun and a Supergiant. Here you see how small the Sun looks in comparison to one of the largest known stars: VY Canis Majoris, a supergiant.

In contrast, the very common red, cool, low-luminosity stars at the lower end of the main sequence are much smaller and more compact than the Sun. An example of such a red dwarf is Ross 614B, with a surface temperature of 2700 K and only 1/2000 of the Sun's luminosity. We call such a star a dwarf because its diameter is only 1/10 that of the Sun. A star with such a low luminosity also has a low mass (about 1/12 that of the Sun). This combination of mass and diameter means that it is so compressed that the star has an average density about 80 times that of the Sun. Its density must be higher, in fact, than that of any known solid found on the surface of Earth. (Despite this, the star is made of gas throughout because its center is so hot.)

The faint, red, main-sequence stars are not the stars of the most extreme densities, however. The white dwarfs, at the lower-left corner of the H–R diagram, have densities many times greater still.

The White Dwarfs

The first white dwarf star was detected in 1862. Called Sirius B, it forms a binary system with Sirius A, the brightest-appearing star in the sky. It eluded discovery and analysis for a long time because its faint light tends to be lost in the glare of nearby Sirius A (Figure 18.17). (Since Sirius is often called the Dog Star—being the brightest star in the constellation of Canis Major, the big dog—Sirius B is sometimes nicknamed the Pup.)

(a)   (b)

Figure 18.17 Two Views of Sirius and Its White Dwarf Companion. (a) The (visible light) image, taken with the Hubble Space Telescope, shows bright Sirius A, and, below it and off to its left, faint Sirius B. (b) This image of the Sirius star system was taken with the Chandra X-Ray Telescope. Now, the bright object is the white dwarf companion, Sirius B. Sirius A is the faint object above it; what we are seeing from Sirius is probably not actually X-ray radiation but rather ultraviolet light that has leaked into the detector. Note that the ultraviolet intensities of these two objects are completely reversed from the situation in visible light because Sirius B is hotter and emits more higher-frequency radiation. (credit a: modification of work by NASA, H.E. Bond and E. Nelan (Space Telescope Science Institute), M. Barstow and M. Burleigh (University of Leicester) and J.B. Holberg (University of Arizona); credit b: modification of work by NASA/SAO/CXC)

We have now found thousands of white dwarfs. Table 18.1 shows that about 7% of the true stars (spectral types O–M) in our local neighborhood are white dwarfs. A good example of a typical white dwarf is the nearby star 40 Eridani B. Its surface temperature is a relatively hot 12,000 K, but its luminosity is only 1/275 L_{Sun}. Calculations show that its radius is only 1.4% of the Sun's, or about the same as that of Earth, and its volume is 2.5×10^{-6} that of the Sun. Its mass, however, is 0.57 times the Sun's mass, just a little more than half. To fit such a substantial mass into so tiny a volume, the star's density must be about 210,000 times the density of the Sun, or more than 300,000 g/cm^3. A teaspoonful of this material would have a mass of some 1.6 tons! At such enormous densities, matter cannot exist in its usual state; we will examine the particular behavior of this type of matter in The Death of Stars. For now, we just note that white dwarfs are dying stars, reaching the end of their productive lives and ready for their stories to be over.

The British astrophysicist (and science popularizer) Arthur Eddington (1882–1944) described the first known white dwarf this way:

The message of the companion of Sirius, when decoded, ran: "I am composed of material three thousand times denser than anything you've ever come across. A ton of my material would be a little nugget you could put in a matchbox." What reply could one make to something like that? Well, the reply most of us made in 1914 was, "Shut up; don't talk nonsense."

Today, however, astronomers not only accept that stars as dense as white dwarfs exist but (as we will see) have found even denser and stranger objects in their quest to understand the evolution of different types of stars.

LINK TO LEARNING

Use this interactive H-R diagram (https://openstax.org/l/30interhr) to explore the sizes of stars based on their placement. Overlay the closest stars and the brightest stars, and consider where they lie on the figure. How many stars are among both the closest and brightest stars?

Key Terms

binary stars   two stars that revolve about each other

brown dwarf   an object intermediate in size between a planet and a star; the approximate mass range is from about 1/100 of the mass of the Sun up to the lower mass limit for self-sustaining nuclear reactions, which is about 1/12 the mass of the Sun

eclipsing binary   a binary star in which the plane of revolution of the two stars is nearly edge-on to our line of sight, so that the light of one star is periodically diminished by the other passing in front of it

H–R diagram   (Hertzsprung–Russell diagram) a plot of luminosity against surface temperature (or spectral type) for a group of stars

main sequence   a sequence of stars on the Hertzsprung–Russell diagram, containing the majority of stars, that runs diagonally from the upper left to the lower right

mass-luminosity relation   the observed relation between the masses and luminosities of many (90% of all) stars

selection effect   the selection of sample data in a nonrandom way, causing the sample data to be unrepresentative of the entire data set

spectroscopic binary   a binary star in which the components are not resolved but whose binary nature is indicated by periodic variations in radial velocity, indicating orbital motion

visual binary   a binary star in which the two components are telescopically resolved

white dwarf   a low-mass star that has exhausted most or all of its nuclear fuel and has collapsed to a very small size; such a star is near its final state of life

Summary

18.1 A Stellar Census

To understand the properties of stars, we must make wide-ranging surveys. We find the stars that appear brightest to our eyes are bright primarily because they are intrinsically very luminous, not because they are the closest to us. Most of the nearest stars are intrinsically so faint that they can be seen only with the aid of a telescope. Stars with low mass and low luminosity are much more common than stars with high mass and high luminosity. Most of the brown dwarfs in the local neighborhood have not yet been discovered.

18.2 Measuring Stellar Masses

The masses of stars can be determined by analysis of the orbit of binary stars—two stars that orbit a common center of mass. In visual binaries, the two stars can be seen separately in a telescope, whereas in a spectroscopic binary, only the spectrum reveals the presence of two stars. Stellar masses range from about 1/12 to more than 100 times the mass of the Sun (in rare cases, going to 250 times the Sun's mass). Objects with masses between 1/12 and 1/100 that of the Sun are called brown dwarfs. Objects in which no nuclear reactions can take place are planets. The most massive stars are, in most cases, also the most luminous, and this correlation is known as the mass-luminosity relation.

18.3 Diameters of Stars

The diameters of stars can be determined by measuring the time it takes an object (the Moon, a planet, or a companion star) to pass in front of it and block its light. Diameters of members of eclipsing binary systems (where the stars pass in front of each other) can be determined through analysis of their orbital motions.

18.4 The H–R Diagram

The Hertzsprung–Russell diagram, or H–R diagram, is a plot of stellar luminosity against surface temperature. Most stars lie on the main sequence, which extends diagonally across the H–R diagram from high temperature and high luminosity to low temperature and low luminosity. The position of a star along the main sequence is determined by its mass. High-mass stars emit more energy and are hotter than low-mass stars on the main sequence. Main-sequence stars derive their energy from the fusion of protons to helium. About 90% of the stars lie on the main sequence. Only about 10% of the stars are white dwarfs, and fewer than 1% are giants or supergiants.

 For Further Exploration

Articles

Croswell, K. "The Periodic Table of the Cosmos." *Scientific American* (July 2011):45–49. A brief introduction to the history and uses of the H–R diagram.

Davis, J. "Measuring the Stars." *Sky & Telescope* (October 1991): 361. The article explains direct measurements of stellar diameters.

DeVorkin, D. "Henry Norris Russell." *Scientific American* (May 1989): 126.

Kaler, J. "Journeys on the H–R Diagram." *Sky & Telescope* (May 1988): 483.

McAllister, H. "Twenty Years of Seeing Double." *Sky & Telescope* (November 1996): 28. An update on modern studies of binary stars.

Parker, B. "Those Amazing White Dwarfs." *Astronomy* (July 1984): 15. The article focuses on the history of their discovery.

Pasachoff, J. "The H–R Diagram's 100th Anniversary." *Sky & Telescope* (June 2014): 32.

Roth, J., and Sinnott, R. "Our Studies of Celestial Neighbors." *Sky & Telescope* (October 1996): 32. A discussion is provided on finding the nearest stars.

Websites

Eclipsing Binary Stars: http://www.midnightkite.com/index.aspx?URL=Binary (http://www.midnightkite.com/index.aspx?URL=Binary). Dan Bruton at Austin State University has created this collection of animations, articles, and links showing how astronomers use eclipsing binary light curves.

Henry Norris Russell: http://www.phys-astro.sonoma.edu/brucemedalists/russell/RussellBio.pdf (http://www.phys-astro.sonoma.edu/brucemedalists/russell/RussellBio.pdf). A Bruce Medal profile of Russell.

Hertzsprung–Russell Diagram: http://skyserver.sdss.org/dr1/en/proj/advanced/hr/ (http://skyserver.sdss.org/dr1/en/proj/advanced/hr/). This site from the Sloan Digital Sky Survey introduces the H–R diagram and gives you information for making your own. You can go step by step by using the menu at the left. Note that in the project instructions, the word "here" is a link and takes you to the data you need.

Stars of the Week: http://stars.astro.illinois.edu/sow/sowlist.html (http://stars.astro.illinois.edu/sow/sowlist.html). Astronomer James Kaler does "biographical summaries" of famous stars—not the Hollywood type, but ones in the real sky.

Videos

Constructing a Hertzsprung-Russell Diagram for a Globular Star Cluster: https://www.youtube.com/watch?v=HWQslu4S5eQ (https://www.youtube.com/watch?v=HWQslu4S5eQ).

WISE Mission Surveys Nearby Stars: http://www.jpl.nasa.gov/video/details.php?id=1089 (http://www.jpl.nasa.gov/video/details.php?id=1089). Short video about the WISE telescope survey of brown dwarfs and M dwarfs in our immediate neighborhood (1:21).

 Collaborative Group Activities

A. Two stars are seen close together in the sky, and your group is given the task of determining whether they are a visual binary or whether they just happen to be seen in nearly the same direction. You have access to a good observatory. Make a list of the types of measurements you would make to determine whether they orbit each other.

B. Your group is given information about five main sequence stars that are among the brightest-appearing stars in the sky and yet are pretty far away. Where would these stars be on the H–R diagram and why? Next, your group is given information about five main-sequence stars that are typical of the stars closest to us. Where would these stars be on the H–R diagram and why?

C. A very wealthy (but eccentric) alumnus of your college donates a lot of money for a fund that will help in the search for more brown dwarfs. Your group is the committee in charge of this fund. How would you spend the money? (Be as specific as you can, listing instruments and observing programs.)

D. Use the internet to search for information about the stars with the largest known diameter. What star is considered the record holder (this changes as new measurements are made)? Read about some of the largest stars on the web. Can your group list some reasons why it might be hard to know which star is the largest?

E. Use the internet to search for information about stars with the largest mass. What star is the current "mass champion" among stars? Try to research how the mass of one or more of the most massive stars was measured, and report to the group or the whole class.

📋 Exercises

Review Questions

1. How does the mass of the Sun compare with that of other stars in our local neighborhood?

2. Name and describe the three types of binary systems.

3. Describe two ways of determining the diameter of a star.

4. What are the largest- and smallest-known values of the mass, luminosity, surface temperature, and diameter of stars (roughly)?

5. You are able to take spectra of both stars in an eclipsing binary system. List all properties of the stars that can be measured from their spectra and light curves.

6. Sketch an H–R diagram. Label the axes. Show where cool supergiants, white dwarfs, the Sun, and main-sequence stars are found.

7. Describe what a typical star in the Galaxy would be like compared to the Sun.

8. How do we distinguish stars from brown dwarfs? How do we distinguish brown dwarfs from planets?

9. Describe how the mass, luminosity, surface temperature, and radius of main-sequence stars change in value going from the "bottom" to the "top" of the main sequence.

10. One method to measure the diameter of a star is to use an object like the Moon or a planet to block out its light and to measure the time it takes to cover up the object. Why is this method used more often with the Moon rather than the planets, even though there are more planets?

11. We discussed in the chapter that about half of stars come in pairs, or multiple star systems, yet the first eclipsing binary was not discovered until the eighteenth century. Why?

Thought Questions

12. Is the Sun an average star? Why or why not?

13. Suppose you want to determine the average educational level of people throughout the nation. Since it would be a great deal of work to survey every citizen, you decide to make your task easier by asking only the people on your campus. Will you get an accurate answer? Will your survey be distorted by a selection effect? Explain.

14. Why do most known visual binaries have relatively long periods and most spectroscopic binaries have relatively short periods?

15. Figure 18.11 shows the light curve of a hypothetical eclipsing binary star in which the light of one star is completely blocked by another. What would the light curve look like for a system in which the light of the smaller star is only partially blocked by the larger one? Assume the smaller star is the hotter one. Sketch the relative positions of the two stars that correspond to various portions of the light curve.

16. There are fewer eclipsing binaries than spectroscopic binaries. Explain why.

17. Within 50 light-years of the Sun, visual binaries outnumber eclipsing binaries. Why?

18. Which is easier to observe at large distances—a spectroscopic binary or a visual binary?

19. The eclipsing binary Algol drops from maximum to minimum brightness in about 4 hours, remains at minimum brightness for 20 minutes, and then takes another 4 hours to return to maximum brightness. Assume that we view this system exactly edge-on, so that one star crosses directly in front of the other. Is one star much larger than the other, or are they fairly similar in size? (Hint: Refer to the diagrams of eclipsing binary light curves.)

20. Review this spectral data for five stars.

Table A

Star	Spectrum
1	G, main sequence
2	K, giant
3	K, main sequence
4	O, main sequence
5	M, main sequence

Which is the hottest? Coolest? Most luminous? Least luminous? In each case, give your reasoning.

21. Which changes by the largest factor along the main sequence from spectral types O to M—mass or luminosity?

22. Suppose you want to search for brown dwarfs using a space telescope. Will you design your telescope to detect light in the ultraviolet or the infrared part of the spectrum? Why?

23. An astronomer discovers a type-M star with a large luminosity. How is this possible? What kind of star is it?

24. Approximately 9000 stars are bright enough to be seen without a telescope. Are any of these white dwarfs? Use the information given in this chapter to explain your reasoning.

25. Use the data in Appendix J to plot an H–R diagram for the brightest stars. Use the data from Table 18.3 to show where the main sequence lies. Do 90% of the brightest stars lie on or near the main sequence? Explain why or why not.

26. Use the diagram you have drawn for Exercise 18.25 to answer the following questions: Which star is more massive—Sirius or Alpha Centauri? Rigel and Regulus have nearly the same spectral type. Which is larger? Rigel and Betelgeuse have nearly the same luminosity. Which is larger? Which is redder?

27. Use the data in Appendix I to plot an H–R diagram for this sample of nearby stars. How does this plot differ from the one for the brightest stars in Exercise 18.25? Why?

28. If a visual binary system were to have two equal-mass stars, how would they be located relative to the center of the mass of the system? What would you observe as you watched these stars as they orbited the center of mass, assuming very circular orbits, and assuming the orbit was face on to your view?

29. Two stars are in a visual binary star system that we see face on. One star is very massive whereas the other is much less massive. Assuming circular orbits, describe their relative orbits in terms of orbit size, period, and orbital velocity.

30. Describe the spectra for a spectroscopic binary for a system comprised of an F-type and L-type star. Assume that the system is too far away to be able to easily observe the L-type star.

31. Figure 18.7 shows the velocity of two stars in a spectroscopic binary system. Which star is the most massive? Explain your reasoning.

32. You go out stargazing one night, and someone asks you how far away the brightest stars we see in the sky without a telescope are. What would be a good, general response? (Use Appendix J for more information.)

33. If you were to compare three stars with the same surface temperature, with one star being a giant, another a supergiant, and the third a main-sequence star, how would their radii compare to one another?

34. Are supergiant stars also extremely massive? Explain the reasoning behind your answer.

35. Consider the following data on four stars:

Table B		
Star	Luminosity (in L_{Sun})	Type
1	100	B, main sequence
2	1/100	B, white dwarf
3	1/100	M, main sequence
4	100	M, giant

Which star would have the largest radius? Which star would have the smallest radius? Which star is the most common in our area of the Galaxy? Which star is the least common?

Figuring for Yourself

36. If two stars are in a binary system with a combined mass of 5.5 solar masses and an orbital period of 12 years, what is the average distance between the two stars?

37. It is possible that stars as much as 200 times the Sun's mass or more exist. What is the luminosity of such a star based upon the mass-luminosity relation?

38. The lowest mass for a true star is 1/12 the mass of the Sun. What is the luminosity of such a star based upon the mass-luminosity relationship?

39. Spectral types are an indicator of temperature. For the first 10 stars in Appendix J, the list of the brightest stars in our skies, estimate their temperatures from their spectral types. Use information in the figures and/or tables in this chapter and describe how you made the estimates.

40. We can estimate the masses of most of the stars in Appendix J from the mass-luminosity relationship in Figure 18.9. However, remember this relationship works only for main sequence stars. Determine which of the first 10 stars in Appendix J are main sequence stars. Use one of the figures in this chapter. Make a table of stars' masses.

41. In Diameters of Stars, the relative diameters of the two stars in the Sirius system were determined. Let's use this value to explore other aspects of this system. This will be done through several steps, each in its own exercise. Assume the temperature of the Sun is 5800 K, and the temperature of Sirius A, the larger star of the binary, is
10,000 K. The luminosity of Sirius A can be found in Appendix J, and is given as about 23 times that of the Sun. Using the values provided, calculate the radius of Sirius A relative to that of the Sun.

42. Now calculate the radius of Sirius' white dwarf companion, Sirius B, to the Sun.

43. How does this radius of Sirius B compare with that of Earth?

44. From the previous calculations and the results from Diameters of Stars, it is possible to calculate the density of Sirius B relative to the Sun. It is worth noting that the radius of the companion is very similar to that of Earth, whereas the mass is very similar to the Sun's. How does the companion's density compare to that of the Sun? Recall that density = mass/volume, and the volume of a sphere = $(4/3)\pi R^3$. How does this density compare with that of water and other materials discussed in this text? Can you see why astronomers were so surprised and puzzled when they first determined the orbit of the companion to Sirius?

45. How much would you weigh if you were suddenly transported to the white dwarf Sirius B? You may use your own weight (or if don't want to own up to what it is, assume you weigh 70 kg or 150 lb). In this case, assume that the companion to Sirius has a mass equal to that of the Sun and a radius equal to that of Earth. Remember Newton's law of gravity:
$$F = GM_1M_2/R^2$$
and that your weight is proportional to the force that you feel. What kind of star should you travel to if you want to *lose* weight (and not gain it)?

46. The star Betelgeuse has a temperature of 3400 K and a luminosity of 13,200 L_{Sun}. Calculate the radius of Betelgeuse relative to the Sun.

47. Using the information provided in Table 18.1, what is the average stellar density in our part of the Galaxy? Use only the true stars (types O–M) and assume a spherical distribution with radius of 26 light-years.

48. Confirm that the angular diameter of the Sun of 1/2° corresponds to a linear diameter of 1.39 million km. Use the average distance of the Sun and Earth to derive the answer. (Hint: This can be solved using a trigonometric function.)

49. An eclipsing binary star system is observed with the following contact times for the main eclipse:

Table C		
Contact	**Time**	**Date**
First contact	12:00 p.m.	March 12
Second contact	4:00 p.m.	March 13
Third contact	9:00 a.m.	March 18
Fourth contact	1:00 p.m.	March 19

The orbital velocity of the smaller star relative to the larger is 62,000 km/h. Determine the diameters for each star in the system.

50. If a 100 solar mass star were to have a luminosity of 10^7 times the Sun's luminosity, how would such a star's density compare when it is on the main sequence as an O-type star, and when it is a cool supergiant (M-type)? Use values of temperature from Figure 18.14 or Figure 18.15 and the relationship between luminosity, radius, and temperature as given in Exercise 18.47.

51. If Betelgeuse had a mass that was 25 times that of the Sun, how would its average density compare to that of the Sun? Use the definition of $density = \frac{mass}{volume}$, where the volume is that of a sphere.

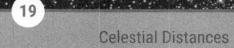

19

Celestial Distances

Figure 19.1 Globular Cluster M80. This beautiful image shows a giant cluster of stars called Messier 80, located about 28,000 light-years from Earth. Such crowded groups, which astronomers call globular clusters, contain hundreds of thousands of stars, including some of the RR Lyrae variables discussed in this chapter. Especially obvious in this picture are the bright red giants, which are stars similar to the Sun in mass that are nearing the ends of their lives. (credit: modification of work by The Hubble Heritage Team (AURA/ STScI/ NASA))

Chapter Outline

19.1 Fundamental Units of Distance

19.2 Surveying the Stars

19.3 Variable Stars: One Key to Cosmic Distances

19.4 The H–R Diagram and Cosmic Distances

 Thinking Ahead

How large is the universe? What is the most distant object we can see? These are among the most fundamental questions astronomers can ask. But just as babies must crawl before they can take their first halting steps, so too must we start with a more modest question: How far away are the stars? And even this question proves to be very hard to answer. After all, stars are mere points of light. Suppose you see a point of light in the darkness when you are driving on a country road late at night. How can you tell whether it is a nearby firefly, an oncoming motorcycle some distance away, or the porchlight of a house much farther down the road? It's not so easy, is it? Astronomers faced an even more difficult problem when they tried to estimate how far away the stars are.

In this chapter, we begin with the fundamental definitions of distances on Earth and then extend our reach outward to the stars. We will also examine the newest satellites that are surveying the night sky and discuss the special types of stars that can be used as trail markers to distant galaxies.

19.1 | Fundamental Units of Distance

Learning Objectives

By the end of this section, you will be able to:

> Understand the importance of defining a standard distance unit
> Explain how the meter was originally defined and how it has changed over time
> Discuss how radar is used to measure distances to the other members of the solar system

The first measures of distances were based on human dimensions—the inch as the distance between knuckles on the finger, or the yard as the span from the extended index finger to the nose of the British king. Later, the requirements of commerce led to some standardization of such units, but each nation tended to set up its own definitions. It was not until the middle of the eighteenth century that any real efforts were made to establish a uniform, international set of standards.

The Metric System

One of the enduring legacies of the era of the French emperor Napoleon is the establishment of the *metric system* of units, officially adopted in France in 1799 and now used in most countries around the world. The fundamental metric unit of length is the *meter*, originally defined as one ten-millionth of the distance along Earth's surface from the equator to the pole. French astronomers of the seventeenth and eighteenth centuries were pioneers in determining the dimensions of Earth, so it was logical to use their information as the foundation of the new system.

Practical problems exist with a definition expressed in terms of the size of Earth, since anyone wishing to determine the distance from one place to another can hardly be expected to go out and re-measure the planet. Therefore, an intermediate standard meter consisting of a bar of platinum-iridium metal was set up in Paris. In 1889, by international agreement, this bar was defined to be exactly one meter in length, and precise copies of the original meter bar were made to serve as standards for other nations.

Other units of length are derived from the meter. Thus, 1 kilometer (km) equals 1000 meters, 1 centimeter (cm) equals 1/100 meter, and so on. Even the old British and American units, such as the inch and the mile, are now defined in terms of the metric system.

Modern Redefinitions of the Meter

In 1960, the official definition of the meter was changed again. As a result of improved technology for generating spectral lines of precisely known wavelengths (see the chapter on Radiation and Spectra), the meter was redefined to equal 1,650,763.73 wavelengths of a particular atomic transition in the element krypton-86. The advantage of this redefinition is that anyone with a suitably equipped laboratory can reproduce a standard meter, without reference to any particular metal bar.

In 1983, the meter was defined once more, this time in terms of the velocity of light. Light in a vacuum can travel a distance of one meter in 1/299,792,458 second. Today, therefore, light travel time provides our basic unit of length. Put another way, a distance of *one light-second* (the amount of space light covers in one second) is defined to be 299,792,458 meters. That's almost 300 million meters that light covers in just one second; light really is *very* fast! We could just as well use the light-second as the fundamental unit of length, but for practical reasons (and to respect tradition), we have defined the meter as a small fraction of the light-second.

Distance within the Solar System

The work of Copernicus and Kepler established the *relative* distances of the planets—that is, how far from the Sun one planet is compared to another (see Observing the Sky: The Birth of Astronomy and Orbits and Gravity). But their work could not establish the *absolute* distances (in light-seconds or meters or other standard units of length). This is like knowing the height of all the students in your class only as compared to the height of your astronomy instructor, but not in inches or centimeters. Somebody's height has to be measured directly.

Similarly, to establish absolute distances, astronomers had to measure one distance in the solar system directly. Generally, the closer to us the object is, the easier such a measurement would be. Estimates of the distance to Venus were made as Venus crossed the face of the Sun in 1761 and 1769, and an international campaign was organized to estimate the distance to the asteroid Eros in the early 1930s, when its orbit brought it close to Earth. More recently, Venus crossed (or *transited*) the surface of the Sun in 2004 and 2012, and allowed us to make a modern distance estimate, although, as we will see below, by then it wasn't needed (Figure 19.2).

LINK TO LEARNING

If you would like more information on just how the motion of Venus across the Sun helped us pin down distances in the solar system, you can turn to a nice explanation (https://openstax.org/l/30VenusandSun) by a NASA astronomer.

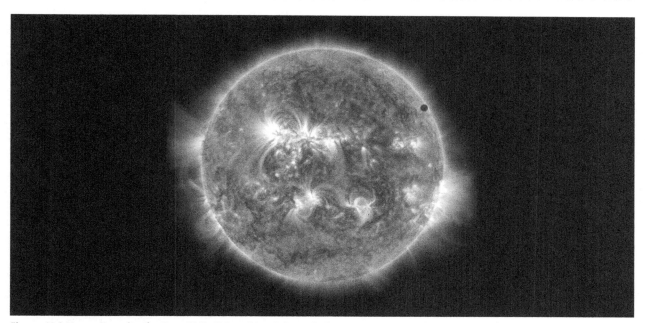

Figure 19.2 Venus Transits the Sun, 2012. This striking "picture" of Venus crossing the face of the Sun (it's the black dot at about 2 o'clock) is more than just an impressive image. Taken with the Solar Dynamics Observatory spacecraft and special filters, it shows a modern transit of Venus. Such events allowed astronomers in the 1800s to estimate the distance to Venus. They measured the time it took Venus to cross the face of the Sun from different latitudes on Earth. The differences in times can be used to estimate the distance to the planet. Today, radar is used for much more precise distance estimates. (credit: modification of work by NASA/SDO, AIA)

The key to our modern determination of solar system dimensions is radar, a type of radio wave that can bounce off solid objects (Figure 19.3). As discussed in several earlier chapters, by timing how long a radar beam (traveling at the speed of light) takes to reach another world and return, we can measure the distance involved very accurately. In 1961, radar signals were bounced off Venus for the first time, providing a direct measurement of the distance from Earth to Venus in terms of light-seconds (from the roundtrip travel time of the radar signal).

Subsequently, radar has been used to determine the distances to Mercury, Mars, the satellites of Jupiter, the rings of Saturn, and several asteroids. Note, by the way, that it is not possible to use radar to measure the distance to the Sun directly because the Sun does not reflect radar very efficiently. But we can measure the distance to many other solar system objects and use Kepler's laws to give us the distance to the Sun.

Figure 19.3 Radar Telescope. This dish-shaped antenna, part of the NASA Deep Space Network in California's Mojave Desert, is 70 meters wide. Nicknamed the "Mars antenna," this radar telescope can send and receive radar waves, and thus measure the distances to planets, satellites, and asteroids. (credit: NASA/JPL-Caltech)

From the various (related) solar system distances, astronomers selected the average distance from Earth to the Sun as our standard "measuring stick" within the solar system. When Earth and the Sun are closest, they are about 147.1 million kilometers apart; when Earth and the Sun are farthest, they are about 152.1 million kilometers apart. The average of these two distances is called the astronomical unit (AU). We then express all the other distances in the solar system in terms of the AU. Years of painstaking analyses of radar measurements have led to a determination of the length of the AU to a precision of about one part in a billion. The length of 1 AU can be expressed in light travel time as 499.004854 light-seconds, or about 8.3 light-minutes. If we use the definition of the meter given previously, this is equivalent to 1 AU = 149,597,870,700 meters.

These distances are, of course, given here to a much higher level of precision than is normally needed. In this text, we are usually content to express numbers to a couple of significant places and leave it at that. For our purposes, it will be sufficient to round off these numbers:

$$\text{speed of light: } c = 3 \times 10^8 \text{ m/s} = 3 \times 10^5 \text{ km/s}$$
$$\text{length of light-second: } ls = 3 \times 10^8 \text{ m} = 3 \times 10^5 \text{ km}$$
$$\text{astronomical unit: } AU = 1.50 \times 10^{11} \text{ m} = 1.50 \times 10^8 \text{ km} = 500 \text{ light-seconds}$$

We now know the absolute distance scale within our own solar system with fantastic accuracy. This is the first link in the chain of cosmic distances.

LINK TO LEARNING

The distances between the celestial bodies in our solar system are sometimes difficult to grasp or put into perspective. This interactive website (https://openstax.org/l/30DistanceScale) provides a "map" that shows the distances by using a scale at the bottom of the screen and allows you to scroll (using your arrow keys) through screens of "empty space" to get to the next planet—all while your current distance from the Sun is visible on the scale.

19.2 | Surveying the Stars

Learning Objectives

By the end of this section, you will be able to:

> Understand the concept of triangulating distances to distant objects, including stars
> Explain why space-based satellites deliver more precise distances than ground-based methods
> Discuss astronomers' efforts to determine the distance of the stars closest to the Sun

It is an enormous step to go from the planets to the stars. For example, our Voyager 1 probe, which was launched in 1977, has now traveled farther from Earth than any other spacecraft. As of 2021, Voyager 1 is 152 AU from the Sun.[1] The nearest star, however, is hundreds of thousands of AU from Earth. Even so, we can, in principle, survey distances to the stars using the same technique that a civil engineer employs to survey the distance to an inaccessible mountain or tree—the method of *triangulation*.

Triangulation in Space

A practical example of triangulation is your own depth perception. As you are pleased to discover every morning when you look in the mirror, your two eyes are located some distance apart. You therefore view the world from two different vantage points, and it is this dual perspective that allows you to get a general sense of how far away objects are.

To see what we mean, take a pen and hold it a few inches in front of your face. Look at it first with one eye (closing the other) and then switch eyes. Note how the pen seems to shift relative to objects across the room. Now hold the pen at arm's length: the shift is less. If you play with moving the pen for a while, you will notice that the farther away you hold it, the less it seems to shift. Your brain automatically performs such comparisons and gives you a pretty good sense of how far away things in your immediate neighborhood are.

If your arms were made of rubber, you could stretch the pen far enough away from your eyes that the shift would become imperceptible. This is because our depth perception fails for objects more than a few tens of meters away. In order to see the shift of an object a city block or more from you, your eyes would need to be spread apart a lot farther.

Let's see how surveyors take advantage of the same idea. Suppose you are trying to measure the distance to a tree across a deep river (Figure 19.4). You set up two observing stations some distance apart. That distance (line AB in Figure 19.4) is called the *baseline*. Now the direction to the tree (C in the figure) in relation to the baseline is observed from each station. Note that C appears in different directions from the two stations. This apparent change in direction of the remote object due to a change in vantage point of the observer is called **parallax**.

---

1   To have some basis for comparison, the dwarf planet Pluto orbits at an average distance of 40 AU from the Sun, and the dwarf planet Eris is currently roughly 96 AU from the Sun.

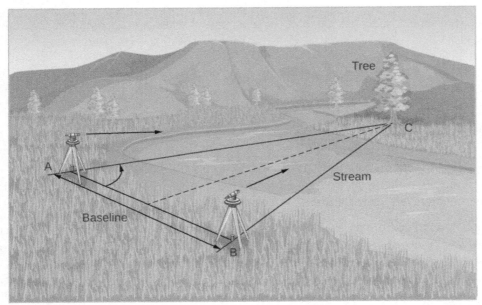

Figure 19.4 Triangulation. Triangulation allows us to measure distances to inaccessible objects. By getting the angle to a tree from two different vantage points, we can calculate the properties of the triangle they make and thus the distance to the tree.

The parallax is also the angle that lines AC and BC make—in mathematical terms, the angle subtended by the baseline. A knowledge of the angles at A and B and the length of the baseline, AB, allows the triangle ABC to be solved for any of its dimensions—say, the distance AC or BC. The solution could be reached by constructing a scale drawing or by using trigonometry to make a numerical calculation. If the tree were farther away, the whole triangle would be longer and skinnier, and the parallax angle would be smaller. Thus, we have the general rule that the smaller the parallax, the more distant the object we are measuring must be.

In practice, the kinds of baselines surveyors use for measuring distances on Earth are completely useless when we try to gauge distances in space. The farther away an astronomical object lies, the longer the baseline has to be to give us a reasonable chance of making a measurement. Unfortunately, nearly all astronomical objects are very far away. To measure their distances requires a very large baseline and highly precise angular measurements. The Moon is the only object near enough that its distance can be found fairly accurately with measurements made without a telescope. Ptolemy determined the distance to the Moon correctly to within a few percent. He used the turning Earth itself as a baseline, measuring the position of the Moon relative to the stars at two different times of night.

With the aid of telescopes, later astronomers were able to measure the distances to the nearer planets and asteroids using Earth's diameter as a baseline. This is how the AU was first established. To reach for the stars, however, requires a much longer baseline for triangulation and extremely sensitive measurements. Such a baseline is provided by Earth's annual trip around the Sun.

Distances to Stars

As Earth travels from one side of its orbit to the other, it graciously provides us with a baseline of 2 AU, or about 300 million kilometers. Although this is a much bigger baseline than the diameter of Earth, the stars are *so far away* that the resulting parallax shift is *still* not visible to the naked eye—not even for the closest stars.

In the chapter on Observing the Sky: The Birth of Astronomy, we discussed how this dilemma perplexed the ancient Greeks, some of whom had actually suggested that the Sun might be the center of the solar system, with Earth in motion around it. Aristotle and others argued, however, that Earth could not be revolving about the Sun. If it were, they said, we would surely observe the parallax of the nearer stars against the background of more distant objects as we viewed the sky from different parts of Earth's orbit (Figure 19.6). Tycho Brahe (1546–1601) advanced the same faulty argument nearly 2000 years later, when his careful measurements of stellar positions with the unaided eye revealed no such shift.

These early observers did not realize how truly distant the stars were and how small the change in their positions therefore was, even with the entire orbit of Earth as a baseline. The problem was that they did not have tools to measure parallax shifts too small to be seen with the human eye. By the eighteenth century, when there was no longer serious doubt about Earth's revolution, it became clear that the stars must be extremely distant. Astronomers equipped with telescopes began to devise instruments capable of measuring the tiny shifts of nearby stars relative to the background of more distant (and thus unshifting) celestial objects.

This was a significant technical challenge, since, even for the nearest stars, parallax angles are usually only a fraction of a second of arc. Recall that one second of arc (arcsec) is an angle of only 1/3600 of a degree. A coin the size of a US quarter would appear to have a diameter of 1 arcsecond if you were viewing it from a distance of about 5 kilometers (3 miles). Think about how small an angle that is. No wonder it took astronomers a long time before they could measure such tiny shifts.

The first successful detections of stellar parallax were in the year 1838, when Friedrich Bessel in Germany (Figure 19.5), Thomas Henderson, a Scottish astronomer working at the Cape of Good Hope, and Friedrich Struve in Russia independently measured the parallaxes of the stars 61 Cygni, Alpha Centauri, and Vega, respectively. Even the closest star, Alpha Centauri, showed a total displacement of only about 1.5 arcseconds during the course of a year.

(a)   (b)   (c)

Figure 19.5 Friedrich Wilhelm Bessel (1784–1846), Thomas J. Henderson (1798–1844), and Friedrich Struve (1793–1864). (a) Bessel made the first authenticated measurement of the distance to a star (61 Cygni) in 1838, a feat that had eluded many dedicated astronomers for almost a century. But two others, (b) Scottish astronomer Thomas J. Henderson and (c) Friedrich Struve, in Russia, were close on his heels.

Figure 19.6 shows how such measurements work. Seen from opposite sides of Earth's orbit, a nearby star shifts position when compared to a pattern of more distant stars. Astronomers actually define parallax to be *one-half* the angle that a star shifts when seen from opposite sides of Earth's orbit (the angle labeled *P* in Figure 19.6). The reason for this definition is just that they prefer to deal with a baseline of 1 AU instead of 2 AU.

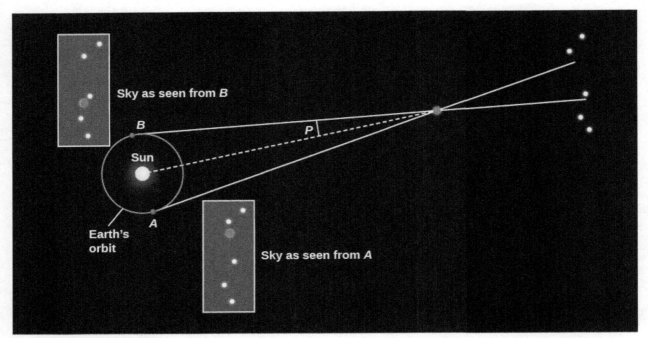

Figure 19.6 Parallax. As Earth revolves around the Sun, the direction in which we see a nearby star varies with respect to distant stars. We define the parallax of the nearby star to be one half of the total change in direction, and we usually measure it in arcseconds.

LINK TO LEARNING

Use the Astronomical Parallax model (https://openstax.org/l/30parallaxmod) to explore how the Earth's motion around the Sun causes nearby stars to appear to "wobble" back and forth compared to background stars. Click on the Orbit box, and use the view along the bottom to see the apparent motion of the nearby object.

Units of Stellar Distance

With a baseline of one AU, how far away would a star have to be to have a parallax of 1 arcsecond? The answer turns out to be 206,265 AU, or 3.26 light-years. This is equal to 3.1×10^{13} kilometers (in other words, 31 trillion kilometers). We give this unit a special name, the **parsec** (pc)—derived from "the distance at which we have a *par*allax of one *sec*ond." The distance (D) of a star in parsecs is just the reciprocal of its parallax (p) in arcseconds; that is,

$$D = \frac{1}{p}$$

Thus, a star with a parallax of 0.1 arcsecond would be found at a distance of 10 parsecs, and one with a parallax of 0.05 arcsecond would be 20 parsecs away.

Back in the days when most of our distances came from parallax measurements, a parsec was a useful unit of distance, but it is not as intuitive as the light-year. One advantage of the light-year as a unit is that it emphasizes the fact that, as we look out into space, we are also looking back into time. The light that we see from a star 100 light-years away left that star 100 years ago. What we study is not the star as it is now, but rather as it was in the past. The light that reaches our telescopes today from distant galaxies left them before Earth even existed.

In this text, we will use light-years as our unit of distance, but many astronomers still use parsecs when they write technical papers or talk with each other at meetings. To convert between the two distance units, just bear in mind: 1 parsec = 3.26 light-year, and 1 light-year = 0.31 parsec.

EXAMPLE 19.1

How Far Is a Light-Year?
A light-year is the distance light travels in 1 year. Given that light travels at a speed of 300,000 km/s, how many kilometers are there in a light-year?

Solution
We learned earlier that speed = distance/time. We can rearrange this equation so that distance = velocity × time. Now, we need to determine the number of seconds in a year.

There are approximately 365 days in 1 year. To determine the number of seconds, we must estimate the number of seconds in 1 day.

We can change units as follows (notice how the units of time cancel out):

$$1 \text{ day} \times 24 \text{ hr/day} \times 60 \text{ min/hr} \times 60 \text{ s/min} = 86{,}400 \text{ s/day}$$

Next, to get the number of seconds per year:

$$365 \text{ days/year} \times 86{,}400 \text{ s/day} = 31{,}536{,}000 \text{ s/year}$$

Now we can multiply the speed of light by the number of seconds per year to get the distance traveled by light in 1 year:

$$
\begin{aligned}
\text{distance} &= \text{velocity} \times \text{time} \\
&= 300{,}000 \text{ km/s} \times 31{,}536{,}000 \text{ s} \\
&= 9.46 \times 10^{12} \text{ km}
\end{aligned}
$$

That's almost 10,000,000,000,000 km that light covers in a year. To help you imagine how long this distance is, we'll mention that a string 1 light-year long could fit around the circumference of Earth 236 million times.

Check Your Learning
The number above is really large. What happens if we put it in terms that might be a little more understandable, like the diameter of Earth? Earth's diameter is about 12,700 km.

Answer:
$$
\begin{aligned}
1 \text{ light-year} &= 9.46 \times 10^{12} \text{ km} \\
&= 9.46 \times 10^{12} \text{ km} \times \frac{1 \text{ Earth diameter}}{12{,}700 \text{ km}} \\
&= 7.45 \times 10^{8} \text{ Earth diameters}
\end{aligned}
$$
That means that 1 light-year is about 745 million times the diameter of Earth.

ASTRONOMY BASICS

Naming Stars

You may be wondering why stars have such a confusing assortment of names. Just look at the first three stars to have their parallaxes measured: 61 Cygni, Alpha Centauri, and Vega. Each of these names comes from a different tradition of designating stars.

The brightest stars have names that derive from the ancients. Some are from the Greek, such as Sirius, which means "the scorched one"—a reference to its brilliance. A few are from Latin, but many of the best-known names are from Arabic because, as discussed in Observing the Sky: The Birth of Astronomy, much of Greek and Roman astronomy was "rediscovered" in Europe after the Dark Ages by means of Arabic translations. Vega, for example, means "swooping Eagle," and Betelgeuse (pronounced "Beetle-juice") means "right hand of the central one."

In 1603, German astronomer Johann Bayer (1572–1625) introduced a more systematic approach to naming stars. For each constellation, he assigned a Greek letter to the brightest stars, roughly in order of brightness. In the constellation of Orion, for example, Betelgeuse is the brightest star, so it got the first letter in the Greek alphabet—alpha—and is known as Alpha Orionis. ("Orionis" is the possessive form of Orion, so Alpha Orionis means "the first of Orion.") A star called Rigel, being the second brightest in that constellation, is called Beta Orionis (Figure 19.7). Since there are 24 letters in the Greek alphabet, this system allows the labeling of 24 stars in each constellation, but constellations have many more stars than that.

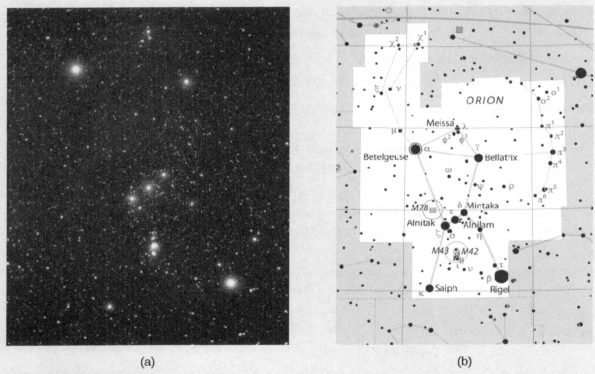

(a)   (b)

Figure 19.7 Objects in Orion. (a) This image shows the brightest objects in or near the star pattern of Orion, the hunter (of Greek mythology), in the constellation of Orion. (b) Note the Greek letters of Bayer's system in this diagram of the Orion constellation. The objects denoted M42, M43, and M78 are not stars but nebulae—clouds of gas and dust; these numbers come from a list of "fuzzy objects" made by Charles Messier in 1781. (credit a: modification of work by Matthew Spinelli; credit b: modification of work by ESO, IAU and *Sky & Telescope*)

In 1725, the English Astronomer Royal John Flamsteed introduced yet another system, in which the brighter stars eventually got a number in each constellation in order of their location in the sky or, more precisely, their right ascension. (The system of sky coordinates that includes right ascension was discussed in Earth, Moon, and Sky.) In this system, Betelgeuse is called 58 Orionis and 61 Cygni is the 61st star in the constellation of Cygnus, the swan.

It gets worse. As astronomers began to understand more and more about stars, they drew up a series of specialized star catalogs, and fans of those catalogs began calling stars by their catalog numbers. If you look at Appendix I—our list of the nearest stars (many of which are much too faint to get an ancient name,

Bayer letter, or Flamsteed number)—you will see references to some of these catalogs. An example is a set of stars labeled with a BD number, for "Bonner Durchmusterung." This was a mammoth catalog of over 324,000 stars in a series of zones in the sky, organized at the Bonn Observatory in the 1850s and 1860s. Keep in mind that this catalog was made before photography or computers came into use, so the position of each star had to be measured (at least twice) by eye, a daunting undertaking.

There is also a completely different system for keeping track of stars whose luminosity varies, and another for stars that brighten explosively at unpredictable times. Astronomers have gotten used to the many different star-naming systems, but students often find them bewildering and wish astronomers would settle on one. Don't hold your breath: in astronomy, as in many fields of human thought, tradition holds a powerful attraction. Still, with high-speed computer databases to aid human memory, names may become less and less necessary. Today's astronomers often refer to stars by their precise locations in the sky rather than by their names or various catalog numbers.

The Nearest Stars

No known star (other than the Sun) is within 1 light-year or even 1 parsec of Earth. The stellar neighbors nearest the Sun are three stars in the constellation of Centaurus. To the unaided eye, the brightest of these three stars is Alpha Centauri, which is only 30° from the south celestial pole and hence not visible from the mainland United States. Alpha Centauri itself is a binary star—two stars in mutual revolution—too close together to be distinguished without a telescope. These two stars are 4.4 light-years from us. Nearby is a third faint star, known as Proxima Centauri. Proxima, with a distance of 4.3 light-years, is slightly closer to us than the other two stars. If Proxima Centauri is part of a triple star system with the binary Alpha Centauri, as seems likely, then its orbital period may be longer than 500,000 years.

Proxima Centauri is an example of the most common type of star, and our most common type of stellar neighbor (as we saw in Stars: A Celestial Census.) Low-mass red M dwarfs make up about 70% of all stars and dominate the census of stars within 10 parsecs (33 light-years) of the Sun. For example, a recent survey of the solar neighborhood counted 357 stars and brown dwarfs within 10 parsecs, and 248 of these are red dwarfs. Yet, if you wanted to see an M dwarf with your naked eye, you would be out of luck. These stars only produce a fraction of the Sun's light, and nearly all of them require a telescope to be detected.

The nearest star visible without a telescope from most of the United States is the brightest appearing of all the stars, Sirius, which has a distance of a little more than 8 light-years. It too is a binary system, composed of a faint white dwarf orbiting a bluish-white, main-sequence star. It is an interesting coincidence of numbers that light reaches us from the Sun in about 8 minutes and from the next brightest star in the sky in about 8 years.

EXAMPLE 19.2

Calculating the Diameter of the Sun

For nearby stars, we can measure the apparent shift in their positions as Earth orbits the Sun. We wrote earlier that an object must be 206,265 AU distant to have a parallax of one second of arc. This must seem like a very strange number, but you can figure out why this is the right value. We will start by estimating the diameter of the Sun and then apply the same idea to a star with a parallax of 1 arcsecond. Make a sketch that has a round circle to represent the Sun, place Earth some distance away, and put an observer on it. Draw two lines from the point where the observer is standing, one to each side of the Sun. Sketch a circle centered at Earth with its circumference passing through the center of the Sun. Now think about proportions. The Sun spans about half a degree on the sky. A full circle has 360°. The circumference of the circle centered on Earth and passing through the Sun is given by:

$$\text{circumference} = 2\pi \times 93{,}000{,}000 \text{ miles}$$

Then, the following two ratios are equal:

$$\frac{0.5°}{360°} = \frac{\text{diameter of Sun}}{2\pi \times 93{,}000{,}000}$$

Calculate the diameter of the Sun. How does your answer compare to the actual diameter?

Solution

To solve for the diameter of the Sun, we can evaluate the expression above.

$$\text{diameter of the sun} = \frac{0.5°}{360°} \times 2\pi \times 93{,}000{,}000 \text{ miles}$$
$$= 811{,}577 \text{ miles}$$

This is very close to the true value of about 848,000 miles.

Check Your Learning

Now apply this idea to calculating the distance to a star that has a parallax of 1 arcsec. Draw a picture similar to the one we suggested above and calculate the distance in AU. (Hint: Remember that the parallax angle is defined by 1 AU, not 2 AU, and that 3600 arcseconds = 1 degree.)

Answer:

206,265 AU

Measuring Parallaxes in Space

The measurements of stellar parallax were revolutionized by the launch of the spacecraft Hipparcos in 1989, which measured distances for thousands of stars out to about 300 light-years with an accuracy of 10 to 20% (see Figure 19.8 and the feature on Parallax and Space Astronomy). However, even 300 light-years are less than 1% the size of our Galaxy's main disk.

In December 2013, the successor to Hipparcos, named *Gaia*, was launched by the European Space Agency. *Gaia* is measuring the position and distances to almost one billion stars with an accuracy of a few millionths of an arcsecond. *Gaia's* distance limit extends well beyond Hipparcos, studying stars out to 30,000 light-years (100 times farther than Hipparcos, covering nearly 1/3 of the galactic disk). *Gaia* is also able to measure proper motions[2] for thousands of stars in the halo of the Milky Way—something that can only be done for the brightest stars right now. At the end of *Gaia's* mission, we will not only have a three-dimensional map of a large fraction of our own Milky Way Galaxy, but we will also have a strong link in the chain of cosmic distances that we are discussing in this chapter. Yet, to extend this chain beyond *Gaia's* reach and explore distances to nearby galaxies, we need some completely new techniques.

2　Proper motion (as discussed in Analyzing Starlight, is the motion of a star across the sky (perpendicular to our line of sight.)

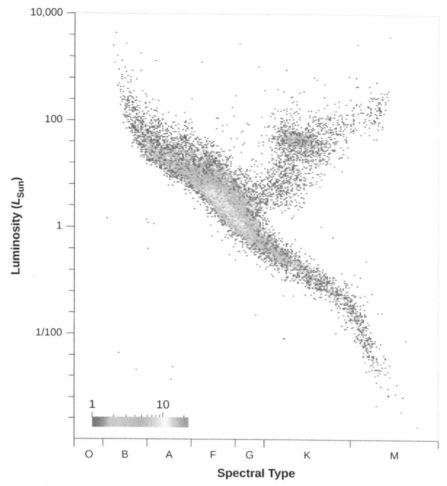

Figure 19.8 H–R Diagram of Stars Measured by *Gaia* and Hipparcos. This plot includes 16,631 stars for which the parallaxes have an accuracy of 10% or better. The colors indicate the numbers of stars at each point of the diagram, with red corresponding to the largest number and blue to the lowest. Luminosity is plotted along the vertical axis, with luminosity increasing upward. An infrared color is plotted as a proxy for temperature, with temperature decreasing to the right. Most of the data points are distributed along the diagonal running from the top left corner (high luminosity, high temperature) to the bottom right (low temperature, low luminosity). These are main sequence stars. The large clump of data points above the main sequence on the right side of the diagram is composed of red giant stars. (credit: modification of work by the European Space Agency)

MAKING CONNECTIONS

Parallax and Space Astronomy

One of the most difficult things about precisely measuring the tiny angles of parallax shifts from Earth is that you have to observe the stars through our planet's atmosphere. As we saw in Astronomical Instruments, the effect of the atmosphere is to spread out the points of starlight into fuzzy disks, making exact measurements of their positions more difficult. Astronomers had long dreamed of being able to measure parallaxes from space, and two orbiting observatories have now turned this dream into reality.

The name of the Hipparcos satellite, launched in 1989 by the European Space Agency, is both an abbreviation for High Precision Parallax Collecting Satellite and a tribute to Hipparchus, the pioneering Greek astronomer whose work we discussed in the Observing the Sky: The Birth of Astronomy. The satellite was designed to make the most accurate parallax measurements in history, from 36,000 kilometers above Earth. However, its onboard rocket motor failed to fire, which meant it did not get the needed boost to reach the desired altitude. Hipparcos ended up spending its 4-year life in an elliptical orbit that varied from

500 to 36,000 kilometers high. In this orbit, the satellite plunged into Earth's radiation belts every 5 hours or so, which finally took its toll on the solar panels that provided energy to power the instruments.

Nevertheless, the mission was successful, resulting in two catalogs. One gives positions of 120,000 stars to an accuracy of one-thousandth of an arcsecond—about the diameter of a golf ball in New York as viewed from Europe. The second catalog contains information for more than a million stars, whose positions have been measured to thirty-thousandths of an arcsecond. We now have accurate parallax measurements of stars out to distances of about 300 light-years. (With ground-based telescopes, accurate measurements were feasible out to only about 60 light-years.)

In order to build on the success of Hipparcos, in 2013, the European Space Agency launched a new satellite called *Gaia*. Because *Gaia* carries larger telescopes than Hipparcos, it can observe fainter stars and measure their positions 200 times more accurately. The main goal of the Gaia mission is to make an accurate three-dimensional map of that portion of the Galaxy within about 30,000 light-years by observing 1 billion stars 70 times each, measuring their positions and hence their parallaxes as well as their brightnesses.

For a long time, the measurement of parallaxes and accurate stellar positions was a backwater of astronomical research—mainly because the accuracy of measurements did not improve much for about 100 years. However, the ability to make measurements from space has revolutionized this field of astronomy and will continue to provide a critical link in our chain of cosmic distances.

LINK TO LEARNING

The European Space Agency (ESA) maintains a Gaia mission website (https://openstax.org/l/30GaiaMission) where you can learn more about the Gaia mission and to get the latest news on *Gaia* observations.

To learn more about Hipparcos, explore this European Space Agency webpage (https://openstax.org/l/30Hipparcos) with an ESA vodcast *Charting the Galaxy—from Hipparcos to Gaia.*

19.3 | Variable Stars: One Key to Cosmic Distances

Learning Objectives

By the end of this section, you will be able to:

> Describe how some stars vary their light output and why such stars are important
> Explain the importance of pulsating variable stars, such as cepheids and RR Lyrae-type stars, to our study of the universe

Let's briefly review the key reasons that measuring distances to the stars is such a struggle. As discussed in The Brightness of Stars, our problem is that stars come in a bewildering variety of intrinsic luminosities. (If stars were light bulbs, we'd say they come in a wide range of wattages.) Suppose, instead, that all stars had the same "wattage" or luminosity. In that case, the more distant ones would always look dimmer, and we could tell how far away a star is simply by how dim it appeared. In the real universe, however, when we look at a star in our sky (with eye or telescope) and measure its apparent brightness, we cannot know whether it looks dim because it's a low-wattage bulb or because it is far away, or perhaps some of each.

Astronomers need to discover something else about the star that allows us to "read off" its intrinsic luminosity—in effect, to know what the star's true wattage is. With this information, we can then attribute how dim it looks from Earth to its distance. Recall that the apparent brightness of an object decreases with the square of the distance to that object. If two objects have the same luminosity but one is three times farther than the other, the more distant one will look nine times fainter. Therefore, if we know the luminosity of a star and its apparent brightness, we can calculate how far away it is. Astronomers have long searched for techniques that would somehow allow us to determine the luminosity of a star—and it is to these techniques that we turn next.

Variable Stars

The breakthrough in measuring distances to remote parts of our Galaxy, and to other galaxies as well, came from the study of variable stars. Most stars are constant in their luminosity, at least to within a percent or two. Like the Sun, they generate a steady flow of energy from their interiors. However, some stars are seen to vary in brightness and, for this reason, are called *variable stars*. Many such stars vary on a regular cycle, like the flashing bulbs that decorate stores and homes during the winter holidays.

Let's define some tools to help us keep track of how a star varies. A graph that shows how the brightness of a variable star changes with time is called a **light curve** (Figure 19.9). The *maximum* is the point of the light curve where the star has its greatest brightness; the *minimum* is the point where it is faintest. If the light variations repeat themselves periodically, the interval between the two maxima is called the *period* of the star. (If this kind of graph looks familiar, it is because we introduced it in Diameters of Stars.)

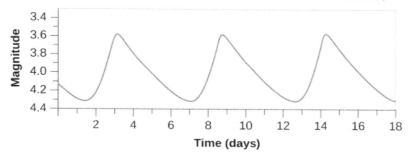

Figure 19.9 Cepheid Light Curve. This light curve shows how the brightness changes with time for a typical cepheid variable, with a period of about 6 days.

Pulsating Variables

There are two special types of variable stars for which—as we will see—measurements of the light curve give us accurate distances. These are called **cepheid** and **RR Lyrae** variables, both of which are **pulsating variable stars**. Such a star actually changes its diameter with time—periodically expanding and contracting, as your chest does when you breathe. We now understand that these stars are going through a brief unstable stage late in their lives.

The expansion and contraction of pulsating variables can be measured by using the Doppler effect. The lines in the spectrum shift toward the blue as the surface of the star moves toward us and then shift to the red as the surface shrinks back. As the star pulsates, it also changes its overall color, indicating that its temperature is also varying. And, most important for our purposes, the luminosity of the pulsating variable also changes in a regular way as it expands and contracts.

Cepheid Variables

Cepheids are large, yellow, pulsating stars named for the first-known star of the group, Delta Cephei. This, by the way, is another example of how confusing naming conventions get in astronomy; here, a whole class of stars is named after the constellation in which the first one happened to be found. (We textbook authors can only apologize to our readers for the whole mess!)

The variability of Delta Cephei was discovered in 1784 by the young English astronomer John Goodricke (see John Goodricke). The star rises rather rapidly to maximum light and then falls more slowly to minimum light, taking a total of 5.4 days for one cycle. The curve in Figure 19.9 represents a simplified version of the light curve of Delta Cephei.

Several hundred cepheid variables are known in our Galaxy. Most cepheids have periods in the range of 3 to 50 days and luminosities that are about 1000 to 10,000 times greater than that of the Sun. Their variations in luminosity range from a few percent to a factor of 10.

Polaris, the North Star, is a cepheid variable that, for a long time, varied by one tenth of a magnitude, or by about 10% in visual luminosity, in a period of just under 4 days. Recent measurements indicate that the amount by which the brightness of Polaris changes is decreasing and that, sometime in the future, this star will no longer be a pulsating variable. This is just one more piece of evidence that stars really do evolve and change in fundamental ways as they age, and that being a cepheid variable represents a stage in the life of the star.

The Period-Luminosity Relation

The importance of cepheid variables lies in the fact that their periods and average luminosities turn out to be directly related. The longer the period (the longer the star takes to vary), the greater the luminosity. This **period-luminosity relation** was a remarkable discovery, one for which astronomers still (pardon the expression) thank their lucky stars. The period of such a star is easy to measure: a good telescope and a good clock are all you need. Once you have the period, the relationship (which can be put into precise mathematical terms) will give you the luminosity of the star.

Let's be clear on what that means. The relation allows you to essentially "read off" how bright the star really is (how much energy it puts out). Astronomers can then compare this intrinsic brightness with the apparent brightness of the star. As we saw, the difference between the two allows them to calculate the distance.

The relation between period and luminosity was discovered in 1908 by Henrietta Leavitt (Figure 19.10), a staff member at the Harvard College Observatory (and one of a number of women working for low wages assisting Edward Pickering, the observatory's director; see Annie Cannon: Classifier of the Stars). Leavitt discovered hundreds of variable stars in the Large Magellanic Cloud and Small Magellanic Cloud, two great star systems that are actually neighboring galaxies (although they were not known to be galaxies then). A small fraction of these variables were cepheids (Figure 19.11).

Figure 19.10 Henrietta Swan Leavitt (1868–1921). Leavitt worked as an astronomer at the Harvard College Observatory. While studying photographs of the Magellanic Clouds, she found over 1700 variable stars, including 20 cepheids. Since all the cepheids in these systems were at roughly the same distance, she was able to compare their luminosities and periods of variation. She thus discovered a fundamental relationship between these characteristics that led to a new and much better way of estimating cosmic distances. (credit: modification of work by AIP)

These systems presented a wonderful opportunity to study the behavior of variable stars independent of their distance. For all practical purposes, the Magellanic Clouds are so far away that astronomers can assume that all the stars in them are at roughly the same distance from us. (In the same way, all the suburbs of Los Angeles are roughly the same distance from New York City. Of course, if you are *in* Los Angeles, you will notice annoying distances between the suburbs, but compared to how far away New York City is, the differences seem small.) If all the variable stars in the Magellanic Clouds are at roughly the same distance, then any difference in their apparent brightnesses must be caused by differences in their intrinsic luminosities.

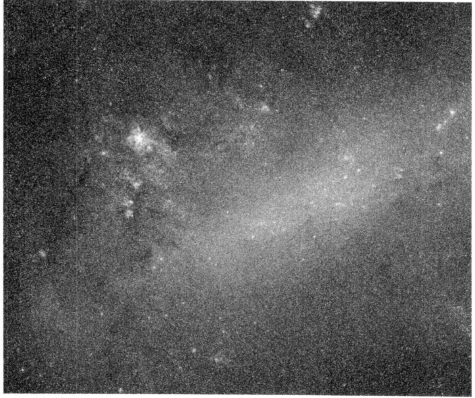

Figure 19.11 Large Magellanic Cloud. The Large Magellanic Cloud (so named because Magellan's crew were the first Europeans to record it) is a small, irregularly shaped galaxy near our own Milky Way. It was in this galaxy that Henrietta Leavitt discovered the cepheid period-luminosity relation. (credit: ESO)

Leavitt found that the brighter-appearing cepheids always have the longer periods of light variation. Thus, she reasoned, the period must be related to the luminosity of the stars. When Leavitt did this work, the distance to the Magellanic Clouds was not known, so she was only able to show that luminosity was related to period. She could not determine exactly what the relationship is.

To define the period-luminosity relation with actual numbers (to *calibrate* it), astronomers first had to measure the actual distances to a few nearby cepheids in another way. (This was accomplished by finding cepheids associated in clusters with other stars whose distances could be estimated from their spectra, as discussed in the next section of this chapter.) But once the relation was thus defined, it could give us the distance to any cepheid, wherever it might be located (Figure 19.12).

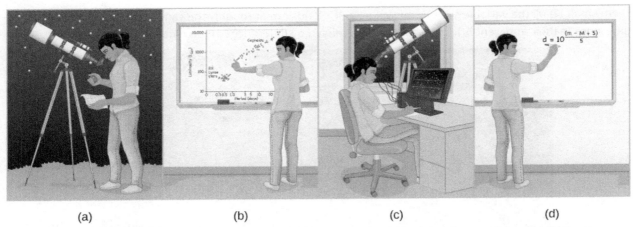

(a)   (b)   (c)   (d)

Figure 19.12 How to Use a Cepheid to Measure Distance. (a) Find a cepheid variable star and measure its period. (b) Use the period-luminosity relation to calculate the star's luminosity. (c) Measure the star's apparent brightness. (d) Compare the luminosity with the apparent brightness to calculate the distance.

Here at last was the technique astronomers had been searching for to break the confines of distance that parallax imposed on them. Cepheids can be observed and monitored, it turns out, in many parts of our own Galaxy and in other nearby galaxies as well. Astronomers, including Ejnar Hertzsprung and Harvard's Harlow Shapley, immediately saw the potential of the new technique; they and many others set to work exploring more distant reaches of space using cepheids as signposts. In the 1920s, Edwin Hubble made one of the most significant astronomical discoveries of all time using cepheids, when he observed them in nearby galaxies and discovered the expansion of the universe. As we will see, this work still continues, as the Hubble Space Telescope and other modern instruments try to identify and measure individual cepheids in galaxies farther and farther away. The most distant known variable stars are all cepheids, with some about 60 million light-years away.

VOYAGERS IN ASTRONOMY

John Goodricke

The brief life of John Goodricke (Figure 19.13) is a testament to the human spirit under adversity. Born deaf and unable to speak, Goodricke nevertheless made a number of pioneering discoveries in astronomy through patient and careful observations of the heavens.

Figure 19.13 John Goodricke (1764–1786). This portrait of Goodricke by artist J. Scouler hangs in the Royal Astronomical Society

in London. There is some controversy about whether this is actually what Goodricke looked like or whether the painting was much retouched to please his family. (credit: James Scouler)

Born in Holland, where his father was on a diplomatic mission, Goodricke was sent back to England at age eight to study at a special school for the deaf. He did sufficiently well to enter Warrington Academy, a secondary school that offered no special assistance for students with handicaps. His mathematics teacher there inspired an interest in astronomy, and in 1781, at age 17, Goodricke began observing the sky at his family home in York, England. Within a year, he had discovered the brightness variations of the star Algol (discussed in The Stars: A Celestial Census) and suggested that an unseen companion star was causing the changes, a theory that waited over 100 years for proof. His paper on the subject was read before the Royal Society (the main British group of scientists) in 1783 and won him a medal from that distinguished group.

In the meantime, Goodricke had discovered two other stars that varied regularly, Beta Lyrae and Delta Cephei, both of which continued to interest astronomers for years to come. Goodricke shared his interest in observing with his older cousin, Edward Pigott, who went on to discover other variable stars during his much longer life. But Goodricke's time was quickly drawing to a close; at age 21, only 2 weeks after he was elected to the Royal Society, he caught a cold while making astronomical observations and never recovered.

Today, the University of York has a building named Goodricke Hall and a plaque that honors his contributions to science. Yet if you go to the churchyard cemetery where he is buried, an overgrown tombstone has only the initials "J. G." to show where he lies. Astronomer Zdenek Kopal, who looked carefully into Goodricke's life, speculated on why the marker is so modest: perhaps the rather staid Goodricke relatives were ashamed of having a "deaf-mute" in the family and could not sufficiently appreciate how much a man who could not hear could nevertheless see.

Go to https://www.bslzone.co.uk/watch/deaf-history/deaf-history-john-goodricke (https://www.bslzone.co.uk/watch/deaf-history/deaf-history-john-goodricke) to see a short video on the life and work of John Goodricke, which is part of the "Deaf History" series and set up so both hearing and hearing-impaired viewers can enjoy it.

RR Lyrae Stars

A related group of stars, whose nature was understood somewhat later than that of the cepheids, are called RR Lyrae variables, named for the star RR Lyrae, the best-known member of the group. More common than the cepheids, but less luminous, thousands of these pulsating variables are known in our Galaxy. The periods of RR Lyrae stars are always less than 1 day, and their changes in brightness are typically less than about a factor of two.

Astronomers have observed that the RR Lyrae stars occurring in any particular cluster all have about the same apparent brightness. Since stars in a cluster are all at approximately the same distance, it follows that RR Lyrae variables must all have nearly the same intrinsic luminosity, which turns out to be about 50 L_{Sun}. In this sense, RR Lyrae stars are a little bit like standard light bulbs and can also be used to obtain distances, particularly within our Galaxy. Figure 19.14 displays the ranges of periods and luminosities for both the cepheids and the RR Lyrae stars.

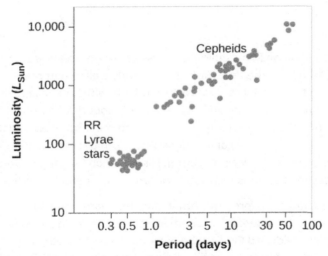

Figure 19.14 Period-Luminosity Relation for Cepheid Variables. In this class of variable stars, the time the star takes to go through a cycle of luminosity changes is related to the average luminosity of the star. Also shown are the period and luminosity for RR Lyrae stars.

19.4 | The H–R Diagram and Cosmic Distances

Learning Objectives

By the end of this section, you will be able to:

> Understand how spectral types are used to estimate stellar luminosities
> Examine how these techniques are used by astronomers today

Variable stars are not the only way that we can estimate the luminosity of stars. Another way involves the H–R diagram, which shows that the intrinsic brightness of a star can be estimated if we know its spectral type.

Distances from Spectral Types

As satisfying and productive as variable stars have been for distance measurement, these stars are rare and are not found near all the objects to which we wish to measure distances. Suppose, for example, we need the distance to a star that is not varying, or to a group of stars, none of which is a variable. In this case, it turns out the H–R diagram can come to our rescue.

If we can observe the spectrum of a star, we can estimate its distance from our understanding of the H–R diagram. As discussed in Analyzing Starlight, a detailed examination of a stellar spectrum allows astronomers to classify the star into one of the *spectral types* indicating surface temperature. (The types are O, B, A, F, G, K, M, L, T, and Y; each of these can be divided into numbered subgroups.) In general, however, the spectral type alone is not enough to allow us to estimate luminosity. Look again at Figure 18.15. A G2 star could be a main-sequence star with a luminosity of 1 L_{Sun}, or it could be a giant with a luminosity of 100 L_{Sun}, or even a supergiant with a still higher luminosity.

We can learn more from a star's spectrum, however, than just its temperature. Remember, for example, that we can detect pressure differences in stars from the details of the spectrum. This knowledge is very useful because giant stars are larger (and have lower pressures) than main-sequence stars, and supergiants are still larger than giants. If we look in detail at the spectrum of a star, we can determine whether it is a main-sequence star, a giant, or a supergiant.

Suppose, to start with the simplest example, that the spectrum, color, and other properties of a distant G2 star match those of the Sun exactly. It is then reasonable to conclude that this distant star is likely to be a main-sequence star just like the Sun and to have the same luminosity as the Sun. But if there are subtle differences between the solar spectrum and the spectrum of the distant star, then the distant star may be a giant or even a supergiant.

The most widely used system of star classification divides stars of a given spectral class into six categories called **luminosity classes**. These luminosity classes are denoted by Roman numbers as follows:

- Ia: Brightest supergiants
- Ib: Less luminous supergiants
- II: Bright giants
- III: Giants
- IV: Subgiants (intermediate between giants and main-sequence stars)
- V: Main-sequence stars

The full spectral specification of a star includes its luminosity class. For example, a main-sequence star with spectral class F3 is written as F3 V. The specification for an M2 giant is M2 III. Figure 19.15 illustrates the approximate position of stars of various luminosity classes on the H–R diagram. The dashed portions of the lines represent regions with very few or no stars.

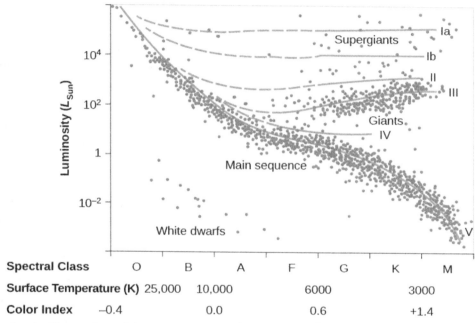

Figure 19.15 Luminosity Classes. Stars of the same temperature (or spectral class) can fall into different luminosity classes on the Hertzsprung-Russell diagram. By studying details of the spectrum for each star, astronomers can determine which luminosity class they fall in (whether they are main-sequence stars, giant stars, or supergiant stars).

With both its spectral and luminosity classes known, a star's position on the H–R diagram is uniquely determined. Since the diagram plots luminosity versus temperature, this means we can now read off the star's luminosity (once its spectrum has helped us place it on the diagram). As before, if we know how luminous the star really is and see how dim it looks, the difference allows us to calculate its distance. (For historical reasons, astronomers sometimes call this method of distance determination *spectroscopic parallax*, even though the method has nothing to do with parallax.)

The H–R diagram method allows astronomers to estimate distances to nearby stars, as well as some of the most distant stars in our Galaxy, but it is anchored by measurements of parallax. The distances measured using parallax are the gold standard for distances: they rely on no assumptions, only geometry. Once astronomers take a spectrum of a nearby star for which we also know the parallax, we know the luminosity that corresponds to that spectral type. Nearby stars thus serve as benchmarks for more distant stars because we can assume that two stars with identical spectra have the same intrinsic luminosity.

A Few Words about the Real World

Introductory textbooks such as ours work hard to present the material in a straightforward and simplified way.

In doing so, we sometimes do our students a disservice by making scientific techniques seem too clean and painless. In the real world, the techniques we have just described turn out to be messy and difficult, and often give astronomers headaches that last long into the day.

For example, the relationships we have described such as the period-luminosity relation for certain variable stars aren't exactly straight lines on a graph. The points representing many stars scatter widely when plotted, and thus, the distances derived from them also have a certain built-in scatter or uncertainty.

The distances we measure with the methods we have discussed are therefore only accurate to within a certain percentage of error—sometimes 10%, sometimes 25%, sometimes as much as 50% or more. A 25% error for a star estimated to be 10,000 light-years away means it could be anywhere from 7500 to 12,500 light-years away. This would be an unacceptable uncertainty if you were loading fuel into a spaceship for a trip to the star, but it is not a bad first figure to work with if you are an astronomer stuck on planet Earth.

Nor is the construction of H–R diagrams as easy as you might think at first. To make a good diagram, one needs to measure the characteristics and distances of many stars, which can be a time-consuming task. Since our own solar neighborhood is already well mapped, the stars astronomers most want to study to advance our knowledge are likely to be far away and faint. It may take hours of observing to obtain a single spectrum. Observers may have to spend many nights at the telescope (and many days back home working with their data) before they get their distance measurement. Fortunately, this is changing because surveys like Gaia will study billions of stars, producing public datasets that all astronomers can use.

Despite these difficulties, the tools we have been discussing allow us to measure a remarkable range of distances—parallaxes for the nearest stars, RR Lyrae variable stars; the H–R diagram for clusters of stars in our own and nearby galaxies; and cepheids out to distances of 60 million light-years. Table 19.1 describes the distance limits and overlap of each method.

Each technique described in this chapter builds on at least one other method, forming what many call the *cosmic distance ladder*. Parallaxes are the foundation of all stellar distance estimates, spectroscopic methods use nearby stars to calibrate their H–R diagrams, and RR Lyrae and cepheid distance estimates are grounded in H–R diagram distance estimates (and even in a parallax measurement to a nearby cepheid, Delta Cephei).

This chain of methods allows astronomers to push the limits when looking for even more distant stars. Recent work, for example, has used RR Lyrae stars to identify dim companion galaxies to our own Milky Way out at distances of 300,000 light-years. The H–R diagram method was recently used to identify the two most distant stars in the Galaxy: red giant stars way out in the halo of the Milky Way with distances of almost 1 million light-years.

We can combine the distances we find for stars with measurements of their composition, luminosity, and temperature—made with the techniques described in Analyzing Starlight and The Stars: A Celestial Census. Together, these make up the arsenal of information we need to trace the evolution of stars from birth to death, the subject to which we turn in the chapters that follow.

Distance Range of Celestial Measurement Methods

Method	Distance Range
Trigonometric parallax	4–30,000 light-years when the Gaia mission is complete
RR Lyrae stars	Out to 300,000 light-years
H–R diagram and spectroscopic distances	Out to 1,200,000 light-years

Table 19.1

Method	Distance Range
Cepheid stars	Out to 60,000,000 light-years

Table 19.1

🔑 Key Terms

cepheid   a star that belongs to a class of yellow supergiant pulsating stars; these stars vary periodically in brightness, and the relationship between their periods and luminosities is useful in deriving distances to them

light curve   a graph that displays the time variation of the light from a variable or eclipsing binary star or, more generally, from any other object whose radiation output changes with time

luminosity class   a classification of a star according to its luminosity within a given spectral class; our Sun, a G2V star, has luminosity class V, for example

parallax   an apparent displacement of a nearby star that results from the motion of Earth around the Sun

parsec   a unit of distance in astronomy, equal to 3.26 light-years; at a distance of 1 parsec, a star has a parallax of 1 arcsecond

period-luminosity relation   an empirical relation between the periods and luminosities of certain variable stars

pulsating variable star   a variable star that pulsates in size and luminosity

RR Lyrae   one of a class of giant pulsating stars with periods shorter than 1 day, useful for finding distances

📖 Summary

19.1 Fundamental Units of Distance

Early measurements of length were based on human dimensions, but today, we use worldwide standards that specify lengths in units such as the meter. Distances within the solar system are now determined by timing how long it takes radar signals to travel from Earth to the surface of a planet or other body and then return.

19.2 Surveying the Stars

For stars that are relatively nearby, we can "triangulate" the distances from a baseline created by Earth's annual motion around the Sun. Half the shift in a nearby star's position relative to very distant background stars, as viewed from opposite sides of Earth's orbit, is called the parallax of that star and is a measure of its distance. The units used to measure stellar distance are the light-year, the distance light travels in 1 year, and the parsec (pc), the distance of a star with a parallax of 1 arcsecond (1 parsec = 3.26 light-years). The closest star, a red dwarf, is over 1 parsec away. The first successful measurements of stellar parallaxes were reported in 1838. Parallax measurements are a fundamental link in the chain of cosmic distances. The Hipparcos satellite has allowed us to measure accurate parallaxes for stars out to about 300 light-years, and the Gaia mission will result in parallaxes out to 30,000 light-years.

19.3 Variable Stars: One Key to Cosmic Distances

Cepheids and RR Lyrae stars are two types of pulsating variable stars. Light curves of these stars show that their luminosities vary with a regularly repeating period. RR Lyrae stars can be used as standard bulbs, and cepheid variables obey a period-luminosity relation, so measuring their periods can tell us their luminosities. Then, we can calculate their distances by comparing their luminosities with their apparent brightnesses, and this can allow us to measure distances to these stars out to over 60 million light-years.

19.4 The H–R Diagram and Cosmic Distances

Stars with identical temperatures but different pressures (and diameters) have somewhat different spectra. Spectral classification can therefore be used to estimate the luminosity class of a star as well as its temperature. As a result, a spectrum can allow us to pinpoint where the star is located on an H–R diagram and establish its luminosity. This, with the star's apparent brightness, again yields its distance. The various distance methods can be used to check one against another and thus make a kind of distance ladder which allows us to find even larger distances.

For Further Exploration

Articles

Adams, A. "The Triumph of Hipparcos." *Astronomy* (December 1997): 60. Brief introduction.

Dambeck, T. "Gaia's Mission to the Milky Way." *Sky & Telescope* (March 2008): 36–39. An introduction to the mission to measure distances and positions of stars with unprecedented accuracy.

Hirshfeld, A. "The Absolute Magnitude of Stars." *Sky & Telescope* (September 1994): 35. Good review of how we measure luminosity, with charts.

Hirshfeld, A. "The Race to Measure the Cosmos." *Sky & Telescope* (November 2001): 38. On parallax.

Trefil, J. Puzzling Out Parallax." *Astronomy* (September 1998): 46. On the concept and history of parallax.

Turon, C. "Measuring the Universe." *Sky & Telescope* (July 1997): 28. On the Hipparcos mission and its results.

Zimmerman, R. "Polaris: The Code-Blue Star." *Astronomy* (March 1995): 45. On the famous cepheid variable and how it is changing.

Websites

ABCs of Distance: http://www.astro.ucla.edu/~wright/distance.htm (http://www.astro.ucla.edu/~wright/distance.htm). Astronomer Ned Wright (UCLA) gives a concise primer on many different methods of obtaining distances. This site is at a higher level than our textbook, but is an excellent review for those with some background in astronomy.

American Association of Variable Star Observers (AAVSO): https://www.aavso.org/ (https://www.aavso.org/). This organization of amateur astronomers helps to keep track of variable stars; its site has some background material, observing instructions, and links.

Friedrich Wilhelm Bessel: http://messier.seds.org/xtra/Bios/bessel.html (http://messier.seds.org/xtra/Bios/bessel.html). A brief site about the first person to detect stellar parallax, with references and links.

Gaia: http://sci.esa.int/gaia/ (http://sci.esa.int/gaia/). News from the Gaia mission, including images and a blog of the latest findings.

Hipparchos: http://sci.esa.int/hipparcos/ (http://sci.esa.int/hipparcos/). Background, results, catalogs of data, and educational resources from the Hipparchos mission to observe parallaxes from space. Some sections are technical, but others are accessible to students.

John Goodricke: The Deaf Astronomer: http://www.bbc.com/news/magazine-20725639 (http://www.bbc.com/news/magazine-20725639). A biographical article from the BBC.

Women in Astronomy: http://bit.ly/astronomywomen (http://bit.ly/astronomywomen). More about Henrietta Leavitt's and other women's contributions to astronomy and the obstacles they faced.

Videos

Gaia's Mission: Solving the Celestial Puzzle: https://www.youtube.com/watch?v=oGri4YNggoc (https://www.youtube.com/watch?v=oGri4YNggoc). Describes the Gaia mission and what scientists hope to learn, from Cambridge University (19:58).

Hipparcos: Route Map to the Stars: http://www.esa.int/spaceinvideos/Videos/1997/05/Hipparcos_Route_Maps_to_the_Stars_May_97 (http://www.esa.int/spaceinvideos/Videos/1997/05/Hipparcos_Route_Maps_to_the_Stars_May_97). This ESA video describes the mission to measure parallax and its results (14:32)

How Big Is the Universe: https://www.youtube.com/watch?v=K_xZuopg4Sk (https://www.youtube.com/

watch?v=K_xZuopg4Sk). Astronomer Pete Edwards from the British Institute of Physics discusses the size of the universe and gives a step-by-step introduction to the concepts of distances (6:22)

Henrietta Leavitt: Ahead of Her Time: https://www.youtube.com/watch?v=XQv03YqEPNM (https://www.youtube.com/watch?v=XQv03YqEPNM). A brief introduction to her life and work from HubbleCast (4:34).

Women in Astronomy: http://www.youtube.com/watch?v=5vMR7su4fi8 (http://www.youtube.com/watch?v=5vMR7su4fi8). Emily Rice (CUNY) gives a talk on the contributions of women to astronomy, with many historical and contemporary examples, and an analysis of modern trends (52:54).

Collaborative Group Activities

A. In this chapter, we explain the various measurements that have been used to establish the size of a standard meter. Your group should discuss why we have changed the definitions of our standard unit of measurement in science from time to time. What factors in our modern society contribute to the growth of technology? Does technology "drive" science, or does science "drive" technology? Or do you think the two are so intertwined that it's impossible to say which is the driver?

B. Cepheids are scattered throughout our own Milky Way Galaxy, but the period-luminosity relation was discovered from observations of the Magellanic Clouds, a satellite galaxy now known to be about 160,000 light-years away. What reasons can you give to explain why the relation was not discovered from observations of cepheids in our own Galaxy? Would your answer change if there were a small cluster in our own Galaxy that contained 20 cepheids? Why or why not?

C. You want to write a proposal to use the Hubble Space Telescope to look for the brightest cepheids in galaxy M100 and estimate their luminosities. What observations would you need to make? Make a list of all the reasons such observations are harder than it first might appear.

D. Why does your group think so many different ways of naming stars developed through history? (Think back to the days before everyone connected online.) Are there other fields where things are named confusingly and arbitrarily? How do stars differ from other phenomena that science and other professions tend to catalog?

E. Although cepheids and RR Lyrae variable stars tend to change their brightness pretty regularly (while they are in that stage of their lives), some variable stars are unpredictable or change their behavior even during the course of a single human lifetime. Amateur astronomers all over the world follow such variable stars patiently and persistently, sending their nightly observations to huge databases that are being kept on the behavior of many thousands of stars. None of the hobbyists who do this get paid for making such painstaking observations. Have your group discuss why they do it. Would you ever consider a hobby that involves so much work, long into the night, often on work nights? If observing variable stars doesn't pique your interest, is there something you think you could do as a volunteer after college that does excite you? Why?

F. In Figure 19.8, the highest concentration of stars occurs in the middle of the main sequence. Can your group give reasons why this might be so? Why are there fewer very hot stars and fewer very cool stars on this diagram?

G. In this chapter, we discuss two astronomers who were differently abled than their colleagues. John Goodricke could neither hear nor speak, and Henrietta Leavitt struggled with hearing impairment for all of her adult life. Yet they each made fundamental contributions to our understanding of the universe. Does your group know people who are handling a disability? What obstacles would people with different disabilities face in trying to do astronomy and what could be done to ease their way? For a set of resources in this area, see http://astronomerswithoutborders.org/gam2013/programs/1319-people-with-disabilities-astronomy-resources.html.

▣ Exercises

Review Questions

1. Explain how parallax measurements can be used to determine distances to stars. Why can we not make accurate measurements of parallax beyond a certain distance?

2. Suppose you have discovered a new cepheid variable star. What steps would you take to determine its distance?

3. Explain how you would use the spectrum of a star to estimate its distance.

4. Which method would you use to obtain the distance to each of the following?
A. An asteroid crossing Earth's orbit
B. A star astronomers believe to be no more than 50 light-years from the Sun
C. A tight group of stars in the Milky Way Galaxy that includes a significant number of variable stars
D. A star that is not variable but for which you can obtain a clearly defined spectrum

5. What are the luminosity class and spectral type of a star with an effective temperature of 5000 K and a luminosity of 100 L_{Sun}?

Thought Questions

6. The meter was redefined as a reference to Earth, then to krypton, and finally to the speed of light. Why do you think the reference point for a meter continued to change?

7. While a meter is the fundamental unit of length, most distances traveled by humans are measured in miles or kilometers. Why do you think this is?

8. Most distances in the Galaxy are measured in light-years instead of meters. Why do you think this is the case?

9. The AU is defined as the *average* distance between Earth and the Sun, not the distance between Earth and the Sun. Why does this need to be the case?

10. What would be the advantage of making parallax measurements from Pluto rather than from Earth? Would there be a disadvantage?

11. Parallaxes are measured in fractions of an arcsecond. One arcsecond equals 1/60 arcmin; an arcminute is, in turn, 1/60th of a degree (°). To get some idea of how big 1° is, go outside at night and find the Big Dipper. The two pointer stars at the ends of the bowl are 5.5° apart. The two stars across the top of the bowl are 10° apart. (Ten degrees is also about the width of your fist when held at arm's length and projected against the sky.) Mizar, the second star from the end of the Big Dipper's handle, appears double. The fainter star, Alcor, is about 12 arcmin from Mizar. For comparison, the diameter of the full moon is about 30 arcmin. The belt of Orion is about 3° long. Keeping all this in mind, why did it take until 1838 to make parallax measurements for even the nearest stars?

12. For centuries, astronomers wondered whether comets were true celestial objects, like the planets and stars, or a phenomenon that occurred in the atmosphere of Earth. Describe an experiment to determine which of these two possibilities is correct.

13. The Sun is much closer to Earth than are the nearest stars, yet it is not possible to measure accurately the diurnal parallax of the Sun relative to the stars by measuring its position relative to background objects in the sky directly. Explain why.

14. Parallaxes of stars are sometimes measured relative to the positions of galaxies or distant objects called quasars. Why is this a good technique?

15. Estimating the luminosity class of an M star is much more important than measuring it for an O star if you are determining the distance to that star. Why is that the case?

16. Figure 19.9 is the light curve for the prototype cepheid variable Delta Cephei. How does the luminosity of this star compare with that of the Sun?

17. Which of the following can you determine about a star without knowing its distance, and which can you not determine: radial velocity, temperature, apparent brightness, or luminosity? Explain.

18. A G2 star has a luminosity 100 times that of the Sun. What kind of star is it? How does its radius compare with that of the Sun?

19. A star has a temperature of 10,000 K and a luminosity of 10^{-2} L_{Sun}. What kind of star is it?

20. What is the advantage of measuring a parallax distance to a star as compared to our other distance measuring methods?

21. What is the disadvantage of the parallax method, especially for studying distant parts of the Galaxy?

22. Luhman 16 and WISE 0720 are brown dwarfs, also known as failed stars, and are some of the new closest neighbors to Earth, but were only discovered in the last decade. Why do you think they took so long to be discovered?

23. Most stars close to the Sun are red dwarfs. What does this tell us about the average star formation event in our Galaxy?

24. Why would it be easier to measure the characteristics of intrinsically less luminous cepheids than more luminous ones?

25. When Henrietta Leavitt discovered the period-luminosity relationship, she used cepheid stars that were all located in the Small Magellanic Cloud. Why did she need to use stars in another galaxy and not cepheids located in the Milky Way?

Figuring for Yourself

26. A radar astronomer who is new at the job claims she beamed radio waves to Jupiter and received an echo exactly 48 min later. Should you believe her? Why or why not?

27. The New Horizons probe flew past Pluto in July 2015. At the time, Pluto was about 32 AU from Earth. How long did it take for communication from the probe to reach Earth, given that the speed of light in km/hr is 1.08×10^9?

28. Estimate the maximum and minimum time it takes a radar signal to make the round trip between Earth and Venus, which has a semimajor axis of 0.72 AU.

29. The Apollo program (not the lunar missions with astronauts) being conducted at the Apache Point Observatory uses a 3.5-m telescope to direct lasers at retro-reflectors left on the Moon by the Apollo astronauts. If the Moon is 384,472 km away, approximately how long do the operators need to wait to see the laser light return to Earth?

30. In 1974, the Arecibo Radio telescope in Puerto Rico was used to transmit a signal to M13, a star cluster about 25,000 light-years away. How long will it take the message to reach M13, and how far has the message travelled so far (in light-years)?

31. Demonstrate that 1 pc equals 3.09×10^{13} km and that it also equals 3.26 light-years. Show your calculations.

32. The best parallaxes obtained with Hipparcos have an accuracy of 0.001 arcsec. If you want to measure the distance to a star with an accuracy of 10%, its parallax must be 10 times larger than the typical error. How far away can you obtain a distance that is accurate to 10% with Hipparcos data? The disk of our Galaxy is 100,000 light-years in diameter. What fraction of the diameter of the Galaxy's disk is the distance for which we can measure accurate parallaxes?

33. Astronomers are always making comparisons between measurements in astronomy and something that might be more familiar. For example, the Hipparcos web pages tell us that the measurement accuracy of 0.001 arcsec is equivalent to the angle made by a golf ball viewed from across the Atlantic Ocean, or to the angle made by the height of a person on the Moon as viewed from Earth, or to the length of growth of a human hair in 10 sec as seen from 10 meters away. Use the ideas in Example 19.2 to verify one of the first two comparisons.

34. *Gaia* will have greatly improved precision over the measurements of Hipparcos. The average uncertainty for most *Gaia* parallaxes will be about 50 microarcsec, or 0.00005 arcsec. How many times better than Hipparcos (see Exercise 19.32) is this precision?

35. Using the same techniques as used in Exercise 19.32, how far away can *Gaia* be used to measure distances with an uncertainty of 10%? What fraction of the Galactic disk does this correspond to?

36. The human eye is capable of an angular resolution of about one arcminute, and the average distance between eyes is approximately 2 in. If you blinked and saw something move about one arcmin across, how far away from you is it? (Hint: You can use the setup in Example 19.2 as a guide.)

37. How much better is the resolution of the *Gaia* spacecraft compared to the human eye (which can resolve about 1 arcmin)?

38. The most recently discovered system close to Earth is a pair of brown dwarfs known as Luhman 16. It has a distance of 6.5 light-years. How many parsecs is this?

39. What would the parallax of Luhman 16 (see Exercise 19.38) be as measured from Earth?

40. The New Horizons probe that passed by Pluto during July 2015 is one of the fastest spacecraft ever assembled. It was moving at about 14 km/s when it went by Pluto. If it maintained this speed, how long would it take New Horizons to reach the nearest star, Proxima Centauri, which is about 4.3 light-years away? (Note: It isn't headed in that direction, but you can pretend that it is.)

41. What physical properties are different for an M giant with a luminosity of 1000 L_{Sun} and an M dwarf with a luminosity of 0.5 L_{Sun}? What physical properties are the same?

20

Between the Stars: Gas and Dust in Space

Figure 20.1 **NGC 3603 and Its Parent Cloud.** This image, taken by the Hubble Space Telescope, shows the young star cluster NGC 3603 interacting with the cloud of gas from which it recently formed. The bright blue stars of the cluster have blown a bubble in the gas cloud. The remains of this cloud can be seen in the lower right part of the frame, glowing in response to the starlight illuminating it. In its darker parts, shielded from the harsh light of NGC 3603, new stars continue to form. Although the stars of NGC 3603 formed only recently, the most massive of them are already dying and ejecting their mass, producing the blue ring and streak features visible in the upper left part of the image. Thus, this image shows the full life cycle of stars, from formation out of interstellar gas, through life on the main sequence, to death and the return of stellar matter to interstellar space. (credit: modification of work by NASA, Wolfgang Brandner (JPL/IPAC), Eva K. Grebel (University of Washington), You-Hua Chu (University of Illinois Urbana-Champaign))

Chapter Outline

20.1 The Interstellar Medium
20.2 Interstellar Gas
20.3 Cosmic Dust
20.4 Cosmic Rays
20.5 The Life Cycle of Cosmic Material
20.6 Interstellar Matter around the Sun

 Thinking Ahead

Where do stars come from? We already know from earlier chapters that stars must die because ultimately they exhaust their nuclear fuel. We might hypothesize that new stars come into existence to replace the ones that die. In order to form new stars, however, we need the raw material to make them. It also turns out that stars eject mass throughout their lives (a kind of wind blows from their surface layers) and that material must go somewhere. What does this "raw material" of stars look like? How would you detect it, especially if it is not yet in the form of stars and cannot generate its own energy?

One of the most exciting discoveries of twentieth-century astronomy was that our Galaxy contains vast quantities of this "raw material"—atoms or molecules of gas and tiny solid dust particles found between the stars. Studying this diffuse matter between the stars helps us understand how new stars form and gives us important clues about our own origins billions of years ago.

20.1 | The Interstellar Medium

Learning Objectives

By the end of this section, you will be able to:

> Explain how much interstellar matter there is in the Milky Way, and what its typical density is
> Describe how the interstellar medium is divided into gaseous and solid components

Astronomers refer to all the material between stars as *interstellar* matter; the entire collection of interstellar matter is called the **interstellar medium (ISM)**. Some interstellar material is concentrated into giant clouds, each of which is known as a **nebula** (plural "nebulae," Latin for "clouds"). The best-known nebulae are the ones that we can see glowing or reflecting visible light; there are many pictures of these in this chapter.

Interstellar clouds do not last for the lifetime of the universe. Instead, they are like clouds on Earth, constantly shifting, merging with each other, growing, or dispersing. Some become dense and massive enough to collapse under their own gravity, forming new stars. When stars die, they, in turn, eject some of their material into interstellar space. This material can then form new clouds and begin the cycle over again.

About 99% of the material between the stars is in the form of a *gas*—that is, it consists of individual atoms or molecules. The most abundant elements in this gas are hydrogen and helium (which we saw are also the most abundant elements in the stars), but the gas also includes other elements. Some of the gas is in the form of molecules—combinations of atoms. The remaining 1% of the interstellar material is solid—frozen particles consisting of many atoms and molecules that are called *interstellar grains* or **interstellar dust** (Figure 20.2). A typical dust grain consists of a core of rocklike material (silicates) or graphite surrounded by a mantle of ices; water, methane, and ammonia are probably the most abundant ices.

Figure 20.2 Various Types of Interstellar Matter. The reddish nebulae in this spectacular photograph glow with light emitted by hydrogen atoms. The darkest areas are clouds of dust that block the light from stars behind them. The upper part of the picture is filled with the bluish glow of light reflected from hot stars embedded in the outskirts of a huge, cool cloud of dust and gas. The cool supergiant star Antares can be seen as a big, reddish patch in the lower-left part of the picture. The star is shedding some of its outer atmosphere and is surrounded by a cloud of its own making that reflects the red light of the star. The red nebula in the middle right partially surrounds the star Sigma Scorpii. (To the right of Antares, you can see M4, a much more distant cluster of extremely old

If all the interstellar gas within the Galaxy were spread out smoothly, there would be only about one atom of gas per cm^3 in interstellar space. (In contrast, the air in the room where you are reading this book has roughly 10^{19} atoms per cm^3.) The dust grains are even scarcer. A km^3 of space would contain only a few hundred to a few thousand tiny grains, each typically less than one ten-thousandth of a millimeter in diameter. These numbers are just averages, however, because the gas and·dust are distributed in a patchy and irregular way, much as water vapor in Earth's atmosphere is often concentrated into clouds.

In some interstellar clouds, the density of gas and dust may exceed the average by as much as a thousand times or more, but even this density is more nearly a vacuum than any we can make on Earth. To show what we mean, let's imagine a vertical tube of air reaching from the ground to the top of Earth's atmosphere with a cross-section of 1 square meter. Now let us extend the same-size tube from the top of the atmosphere all the way to the edge of the observable universe—over 10 billion light-years away. Long though it is, the second tube would still contain fewer atoms than the one in our planet's atmosphere.

While the *density* of interstellar matter is very low, the volume of space in which such matter is found is huge, and so its *total mass* is substantial. To see why, we must bear in mind that stars occupy only a tiny fraction of the volume of the Milky Way Galaxy. For example, it takes light only about four seconds to travel a distance equal to the diameter of the Sun, but more than four *years* to travel from the Sun to the nearest star. Even though the spaces among the stars are sparsely populated, there's a lot of space out there!

Astronomers estimate that the total mass of gas and dust in the Milky Way Galaxy is equal to about 15% of the mass contained in stars. This means that the mass of the interstellar matter in our Galaxy amounts to about 10 billion times the mass of the Sun. There is plenty of raw material in the Galaxy to make generations of new stars and planets (and perhaps even astronomy students).

EXAMPLE 20.1

Estimating Interstellar Mass

You can make a rough estimate of how much interstellar mass our Galaxy contains and also how many new stars could be made from this interstellar matter. All you need to know is how big the Galaxy is and the average density using this formula:

$$\text{total mass} = \text{volume} \times \text{density of atoms} \times \text{mass per atom}$$

You have to remember to use consistent units—such as meters and kilograms. We will assume that our Galaxy is shaped like a cylinder; the volume of a cylinder equals the area of its base times its height

$$V = \pi R^2 \, h$$

where R is the radius of the cylinder and h is its height.

Suppose that the average density of hydrogen gas in our Galaxy is one atom per cm^3. Each hydrogen atom has a mass of 1.7×10^{-27} kg. If the Galaxy is a cylinder with a diameter of 100,000 light-years and a height of 300 light-years, what is the mass of this gas? How many solar-mass stars (2.0×10^{30} kg) could be produced from this mass of gas if it were all turned into stars?

Solution

If the diameter of the Galaxy is 100,000 light-years, then the radius is 50,000 light-years. Recall that 1 light-year = 9.5×10^{12} km = 9.5×10^{17} cm, so the volume of the Galaxy is

$$V = \pi R^2 \, h = \pi (50{,}000 \times 9.5 \times 10^{17} \text{ cm})^2 \, (300 \times 9.5 \times 10^{17} \text{ cm}) = 2.0 \times 10^{66} \text{ cm}^3$$

The total mass is therefore

$$M = V \times \text{density of atoms} \times \text{mass per atom}$$

$$2.0 \times 10^{66} \text{ cm}^3 \times (1 \text{ atom/cm}^3) \times 1.7 \times 10^{-27} \text{ kg} = 3.5 \times 10^{39} \text{ kg}$$

This is sufficient to make

$$N = \frac{M}{(2.0 \times 10^{30} \text{ kg})} = 1.75 \times 10^9$$

stars equal in mass to the Sun. That's roughly 2 billion stars.

Check Your Learning

You can use the same method to estimate the mass of interstellar gas around the Sun. The distance from the Sun to the nearest other star, Proxima Centauri, is 4.2 light-years. We will see in Interstellar Matter around the Sun that the gas in the immediate vicinity of the Sun is less dense than average, about 0.1 atoms per cm^3. What is the total mass of interstellar hydrogen in a sphere centered on the Sun and extending out to Proxima Centauri? How does this compare to the mass of the Sun? It is helpful to remember that the volume of a sphere is related to its radius:

$$V = (4/3)\pi R^3$$

Answer:

The volume of a sphere stretching from the Sun to Proxima Centauri is:

$$V = (4/3)\,\pi R^3 = (4/3)\,\pi\left(4.2 \times 9.5 \times 10^{17} \text{ cm}\right)^3 = 2.7 \times 10^{56} \text{ cm}^3$$

Therefore, the mass of hydrogen in this sphere is:

$$M = V \times \left(0.1 \text{ atom/cm}^3\right) \times 1.7 \times 10^{-27} \text{ kg} = 4.5 \times 10^{28} \text{ kg}$$

This is only $(4.5 \times 10^{28} \text{ kg})/(2.0 \times 10^{30} \text{ kg}) = 2.2\%$ the mass of the Sun.

ASTRONOMY BASICS

Naming the Nebulae

As you look at the captions for some of the spectacular photographs in this chapter and The Birth of Stars and the Discovery of Planets outside the Solar System, you will notice the variety of names given to the nebulae. A few, which in small telescopes look like something recognizable, are sometimes named after the creatures or objects they resemble. Examples include the Crab, Tarantula, and Keyhole Nebulae. But most have only numbers that are entries in a catalog of astronomical objects.

Perhaps the best-known catalog of nebulae (as well as star clusters and galaxies) was compiled by the French astronomer Charles Messier (1730–1817). Messier's passion was discovering comets, and his devotion to this cause earned him the nickname "The Comet Ferret" from King Louis XV. When comets are first seen coming toward the Sun, they look like little fuzzy patches of light; in small telescopes, they are easy to confuse with nebulae or with groupings of many stars so far away that their light is all blended together. Time and again, Messier's heart leapt as he thought he had discovered one of his treasured comets, only to find that he had "merely" observed a nebula or cluster.

In frustration, Messier set out to catalog the position and appearance of over 100 objects that could be mistaken for comets. For him, this list was merely a tool in the far more important work of comet hunting. He would be very surprised if he returned today to discover that no one recalls his comets anymore, but his catalog of "fuzzy things that are not comets" is still widely used. When Figure 20.2 refers to M4, it denotes the fourth entry in Messier's list. Visit https://www.nasa.gov/content/goddard/hubble-s-messier-catalog

(https://www.nasa.gov/content/goddard/hubble-s-messier-catalog) for a gallery of M objects as photographed with the Hubble Space Telescope.

A far more extensive listing was compiled under the title of the *New General Catalog* (*NGC*) *of Nebulae and Star Clusters* in 1888 by John Dreyer, working at the observatory in Armagh, Ireland. He based his compilation on the work of William Herschel and his son John, plus many other observers who followed them. With the addition of two further listings (called the *Index Catalogs*), Dreyer's compilation eventually included 13,000 objects. Astronomers today still use his NGC numbers when referring to most nebulae and star groups.

20.2 Interstellar Gas

Learning Objectives

By the end of this section, you will be able to:

> Name the major types of interstellar gas
> Discuss how we can observe each type
> Describe the temperature and other major properties of each type

Interstellar gas, depending on where it is located, can be as cold as a few degrees above absolute zero or as hot as a million degrees or more. We will begin our voyage through the interstellar medium by exploring the different conditions under which we find gas.

Ionized Hydrogen (H II) Regions—Gas Near Hot Stars

Some of the most spectacular astronomical photographs show interstellar gas located near hot stars (Figure 20.3). The strongest line in the visible region of the hydrogen spectrum is the red line in the Balmer series[1] (as explained in the chapter on Radiation and Spectra); this emission line accounts for the characteristic red glow in images like Figure 20.3.

Figure 20.3 Orion Nebula. The red glow that pervades the great Orion Nebula is produced by the first line in the Balmer series of hydrogen. Hydrogen emission indicates that there are hot young stars nearby that ionize these clouds of gas. When electrons then recombine with protons and move back down into lower energy orbits, emission lines are produced. The blue color seen at the edges of some of the clouds is produced by small particles of dust that scatter the light from the hot stars. Dust can also be seen

---

1 Scientists also call this red Balmer line the H-alpha line, with alpha meaning it is the first spectral line in the Balmer series.

silhouetted against the glowing gas. (credit: NASA,ESA, M. Robberto (Space Telescope Science Institute/ESA) and the Hubble Space Telescope Orion Treasury Project Team)

Hot stars are able to heat nearby gas to temperatures close to 10,000 K. The ultraviolet radiation from the stars also ionizes the hydrogen (remember that during ionization, the electron is stripped completely away from the proton). Such a detached proton won't remain alone forever when attractive electrons are around; it will capture a free electron, becoming a neutral hydrogen once more. However, such a neutral atom can then absorb ultraviolet radiation again and start the cycle over. At a typical moment, most of the atoms near a hot star are in the ionized state.

Since hydrogen is the main constituent of interstellar gas, we often characterize a region of space according to whether its hydrogen is neutral or ionized. A cloud of ionized hydrogen is called an **H II region**. (Scientists who work with spectra use the Roman numeral I to indicate that an atom is neutral; successively higher Roman numerals are used for each higher stage of ionization. H II thus refers to hydrogen that has lost its one electron; Fe III is iron with two electrons missing.)

The electrons that are captured by the hydrogen nuclei cascade down through the various energy levels of the hydrogen atoms on their way to the lowest level, or ground state. During each transition downward, they give up energy in the form of light. The process of converting ultraviolet radiation into visible light is called *fluorescence*. Interstellar gas contains other elements besides hydrogen. Many of them are also ionized in the vicinity of hot stars; they then capture electrons and emit light, just as hydrogen does, allowing them to be observed by astronomers. But generally, the red hydrogen line is the strongest, and that is why H II regions look red.

A fluorescent light on Earth works using the same principles as a fluorescent H II region. When you turn on the current, electrons collide with atoms of mercury vapor in the tube. The mercury is excited to a high-energy state because of these collisions. When the electrons in the mercury atoms return to lower-energy levels, some of the energy they emit is in the form of ultraviolet photons. These, in turn, strike a phosphor-coated screen on the inner wall of the light tube. The atoms in the screen absorb the ultraviolet photons and emit visible light as they cascade downward among the energy levels. (The difference is that these atoms give off a wider range of light colors, which mix to give the characteristic white glow of fluorescent lights, whereas the hydrogen atoms in an H II region give off a more limited set of colors.)

Neutral Hydrogen Clouds

The very hot stars required to produce H II regions are rare, and only a small fraction of interstellar matter is close enough to such hot stars to be ionized by them. Most of the volume of the interstellar medium is filled with neutral (nonionized) hydrogen. How do we go about looking for it?

Unfortunately, neutral hydrogen atoms at temperatures typical of the gas in interstellar space neither emit nor absorb light in the visible part of the spectrum. Nor, for the most part, do the other trace elements that are mixed with the interstellar hydrogen. However, some of these other elements can *absorb* visible light even at typical interstellar temperatures. This means that when we observe a bright source such as a hot star or a galaxy, we can sometimes see additional lines in its spectrum produced when interstellar gas absorbs light at particular frequencies (see Figure 20.4). Some of the strongest interstellar absorption lines are produced by calcium and sodium, but many other elements can be detected as well in sufficiently sensitive observations (as discussed in Radiation and Spectra).

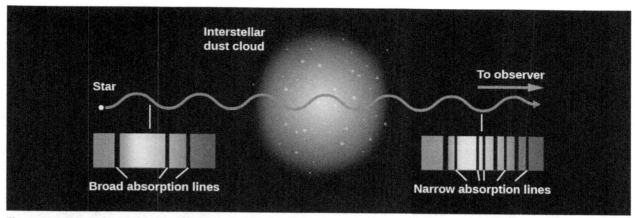

Figure 20.4 Absorption Lines though an Interstellar Dust Cloud. When there is a significant amount of cool interstellar matter (gas with some dust) between us and a star, we can see the absorption lines of the gas in the star's spectrum. We can distinguish the two kinds of lines because, whereas the star's lines are broad, the lines from the gas are narrower.

The first evidence for absorption by interstellar clouds came from the analysis of a spectroscopic binary star (see The Stars: A Celestial Census), published in 1904. While most of the lines in the spectrum of this binary shifted alternately from longer to shorter wavelengths and back again, as we would expect from the Doppler effect for stars in orbit around each other, a few lines in the spectrum remained fixed in wavelength. Since both stars are moving in a binary system, lines that showed no motion puzzled astronomers. The lines were also peculiar in that they were much, much narrower than the rest of the lines, indicating that the gas producing them was at a very low pressure. Subsequent work demonstrated that these lines were not formed in the star's atmosphere at all, but rather in a cold cloud of gas located between Earth and the binary star.

While these and similar observations proved there was interstellar gas, they could not yet detect hydrogen, the most common element, due to its lack of spectral features in the visible part of the spectrum. (The Balmer line of hydrogen is in the visible range, but only excited hydrogen atoms produce it. In the cold interstellar medium, the hydrogen atoms are all in the ground state and no electrons are in the higher-energy levels required to produce either emission or absorption lines in the Balmer series.) Direct detection of hydrogen had to await the development of telescopes capable of seeing very-low-energy changes in hydrogen atoms in other parts of the spectrum. The first such observations were made using radio telescopes, and radio emission and absorption by interstellar hydrogen remains one of our main tools for studying the vast amounts of cold hydrogen in the universe to this day.

In 1944, while he was still a student, the Dutch astronomer Hendrik van de Hulst predicted that hydrogen would produce a strong line at a wavelength of 21 centimeters. That's quite a long wavelength, implying that the wave has such a low frequency and low energy that it cannot come from electrons jumping between energy levels (as we discussed in Radiation and Spectra). Instead, energy is emitted when the electron does a flip, something like an acrobat in a circus flipping upright after standing on his head.

The flip works like this: a hydrogen atom consists of a proton and an electron bound together. Both the proton and the electron act is if they were spinning like tops, and spin axes of the two tops can either be pointed in the same direction (aligned) or in opposite directions (anti-aligned). If the proton and electron were spinning in opposite directions, the atom as a whole would have a very slightly lower energy than if the two spins were aligned (Figure 20.5). If an atom in the lower-energy state (spins opposed) acquired a small amount of energy, then the spins of the proton and electron could be aligned, leaving the atom in a slightly *excited state*. If the atom then lost that same amount of energy again, it would return to its ground state. The amount of energy involved corresponds to a wave with a wavelength of 21 centimeters; hence, it is known as the *21-centimeter line*.

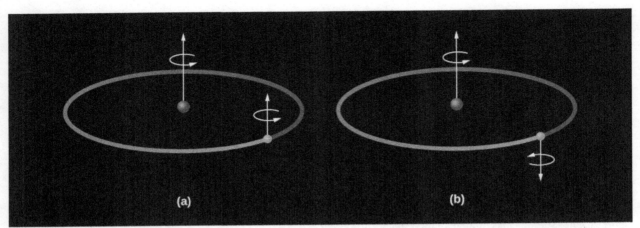

Figure 20.5 Formation of the 21-Centimeter Line. When the electron in a hydrogen atom is in the orbit closest to the nucleus, the proton and the electron may be spinning either (a) in the same direction or (b) in opposite directions. When the electron flips over, the atom gains or loses a tiny bit of energy by either absorbing or emitting electromagnetic energy with a wavelength of 21 centimeters.

Neutral hydrogen atoms can acquire small amounts of energy through collisions with other hydrogen atoms or with free electrons. Such collisions are extremely rare in the sparse gases of interstellar space. An individual atom may wait centuries before such an encounter aligns the spins of its proton and electron. Nevertheless, over many millions of years, a significant fraction of the hydrogen atoms are excited by a collision. (Out there in cold space, that's about as much excitement as an atom typically experiences.)

An excited atom can later lose its excess energy either by colliding with another particle or by giving off a radio wave with a wavelength of 21 centimeters. If there are no collisions, an excited hydrogen atom will wait an average of about 10 million years before emitting a photon and returning to its state of lowest energy. Even though the probability that any single atom will emit a photon is low, there are so many hydrogen atoms in a typical gas cloud that collectively they will produce an observable line at 21 centimeters.

Equipment sensitive enough to detect the 21-cm line of neutral hydrogen became available in 1951. Dutch astronomers had built an instrument to detect the 21-cm waves that they had predicted, but a fire destroyed it. As a result, two Harvard physicists, Harold Ewen and Edward Purcell, made the first detection (Figure 20.6), soon followed by confirmations from the Dutch and a group in Australia. Since the detection of the 21-cm line, many other radio lines produced by both atoms and molecules have been discovered (as we will discuss in a moment), and these have allowed astronomers to map out the neutral gas throughout our home Galaxy. Astronomers have also detected neutral interstellar gas, including hydrogen, at many other wavelengths from the infrared to the ultraviolet.

Figure 20.6 Harold Ewen (1922–2015) and Edward Purcell (1912–1997). We see Harold Ewen in 1952 working with the horn antenna (atop the physics laboratory at Harvard) that made the first detection of interstellar 21-cm radiation. The inset shows Edward Purcell, the winner of the 1952 Nobel Prize in physics, a few years later. (credit: modification of work by NRAO)

Modern radio observations show that most of the neutral hydrogen in our Galaxy is confined to an extremely flat layer, less than 300 light-years thick, that extends throughout the disk of the Milky Way Galaxy. This gas has densities ranging from about 0.1 to about 100 atoms per cm^3, and it exists at a wide range of temperatures, from as low as about 100 K (–173 °C) to as high as about 8000 K. These regions of warm and cold gas are interspersed with each other, and the density and temperature at any particular point in space are constantly changing.

Ultra-Hot Interstellar Gas

While the temperatures of 10,000 K found in H II regions might seem warm, they are not the hottest phase of the interstellar medium. Some of the interstellar gas is at a temperature of a *million* degrees, even though there is no visible source of heat nearby. The discovery of this ultra-hot interstellar gas was a big surprise. Before the launch of astronomical observatories into space, which could see radiation in the ultraviolet and X-ray parts of the spectrum, astronomers assumed that most of the region between stars was filled with hydrogen at temperatures no warmer than those found in H II regions. But telescopes launched above Earth's atmosphere obtained ultraviolet spectra that contained interstellar lines produced by oxygen atoms that have been ionized five times. To strip five electrons from their orbits around an oxygen nucleus requires a lot of energy. Subsequent observations with orbiting X-ray telescopes revealed that the Galaxy is filled with numerous bubbles of X-ray-emitting gas. To emit X-rays, and to contain oxygen atoms that have been ionized five times, gas must be heated to temperatures of a million degrees or more.

Theorists have now shown that the source of energy producing these remarkable temperatures is the explosion of massive stars at the ends of their lives (Figure 20.7). Such explosions, called *supernovae*, will be discussed in detail in the chapter on The Death of Stars. For now, we'll just say that some stars, nearing the ends of their lives, become unstable and literally explode. These explosions launch gas into interstellar space at velocities of tens of thousands of kilometers per second (up to about 30% the speed of light). When this ejected gas collides with interstellar gas, it produces shocks that heat the gas to millions or tens of millions of degrees.

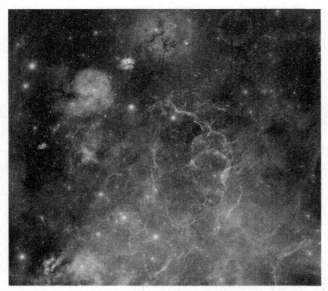

Figure 20.7 Vela Supernova Remnant. About 11,000 years ago, a dying star in the constellation of Vela exploded, becoming as bright as the full moon in Earth's skies. You can see the faint rounded filaments from that explosion in the center of this colorful image. The edges of the remnant are colliding with the interstellar medium, heating the gas they plow through to temperatures of millions of K. Telescopes in space also reveal a glowing sphere of X-ray radiation from the remnant. (credit: Digitized Sky Survey, ESA/ESO/NASA FITS Liberator, Davide De Martin)

Astronomers estimate that one supernova explodes roughly every 100 years somewhere in the Galaxy. On average, shocks launched by supernovae sweep through any given point in the Galaxy about once every few million years. These shocks keep some interstellar space filled with gas at temperatures of millions of degrees, and they continually disturb the colder gas, keeping it in constant, turbulent motion.

Molecular Clouds

A few simple molecules out in space, such as CN and CH, were discovered decades ago because they produce absorption lines in the visible-light spectra of stars behind them. When more sophisticated equipment for obtaining spectra in radio and infrared wavelengths became available, astronomers—to their surprise—found much more complex molecules in interstellar clouds as well.

Just as atoms leave their "fingerprints" in the spectrum of visible light, so the vibration and rotation of atoms within molecules can leave spectral fingerprints in radio and infrared waves. If we spread out the radiation at such longer wavelengths, we can detect emission or absorption lines in the spectra that are characteristic of specific molecules. Over the years, experiments in our laboratories have shown us the exact wavelengths associated with changes in the rotation and vibration of many common molecules, giving us a template of possible lines against which we can now compare our observations of interstellar matter.

The discovery of complex molecules in space came as a surprise because most of interstellar space is filled with ultraviolet light from stars, and this light is capable of *dissociating* molecules (breaking them apart into individual atoms). In retrospect, however, the presence of molecules is not surprising. As we will discuss further in the next section, and have already seen above, interstellar space also contains significant amounts of dust capable of blocking out starlight. When this dust accumulates in a single location, the result is a dark cloud where ultraviolet starlight is blocked and molecules can survive. The largest of these structures are created where gravity pulls interstellar gas together to form giant **molecular clouds**, structures as massive as a million times the mass of the Sun. Within these, most of the interstellar hydrogen has formed the molecule H_2 (molecular hydrogen). Other, more complex molecules are also present in much smaller quantities.

Giant molecular clouds have densities of hundreds to thousands of atoms per cm^3, much denser than interstellar space is on average. As a result, though they account for a very small fraction of the volume of interstellar space, they contain a significant fraction—20–30%—of the total mass of the Milky Way's gas.

Because of their high density, molecular clouds block ultraviolet starlight, the main agent for heating most interstellar gas. As a result, they tend to be extremely cold, with typical temperatures near 10 K (−263 °C). Giant molecular clouds are also the sites where new stars form, as we will discuss below.

It is in these dark regions of space, protected from starlight, that molecules can form. Chemical reactions occurring both in the gas and on the surface of dust grains lead to much more complex compounds, hundreds of which have been identified in interstellar space. Among the simplest of these are water (H_2O), carbon monoxide (CO), which is produced by fires on Earth, and ammonia (NH_3), whose smell you recognize in strong home cleaning products. Carbon monoxide is particularly abundant in interstellar space and is the primary tool that astronomers use to study giant molecular clouds. Unfortunately, the most abundant molecule, H_2, is particularly difficult to observe directly because in most giant molecular clouds, it is too cold to emit even at radio wavelengths. CO, which tends to be present wherever H_2 is found, is a much better emitter and is often used by astronomers to trace molecular hydrogen.

The more complex molecules astronomers have found are mostly combinations of hydrogen, oxygen, carbon, nitrogen, and sulfur atoms. Many of these molecules are *organic* (those that contain carbon and are associated with the carbon chemistry of life on Earth.) They include formaldehyde (used to preserve living tissues), alcohol (see the feature box on Cocktails in Space), and antifreeze.

In 1996, astronomers discovered acetic acid (the prime ingredient of vinegar) in a cloud lying in the direction of the constellation of Sagittarius. To balance the sour with the sweet, a simple sugar (glycolaldehyde) has also been found. The largest compounds yet discovered in interstellar space are *fullerenes*, molecules in which 60 or 70 carbon atoms are arranged in a cage-like configuration (see Figure 20.8). See Table 20.1 for a list of a few of the more interesting interstellar molecules that have been found so far.

Figure 20.8 Fullerene C60. This three-dimensional perspective shows the characteristic cage-like arrangement of the 60 carbon atoms in a molecule of fullerene C60. Fullerene C60 is also known as a "buckyball," or as its full name, buckminsterfullerene, because of its similarity to the multisided architectural domes designed by American inventor R. Buckminster Fuller.

Some Interesting Interstellar Molecules		
Name	**Chemical Formula**	**Use on Earth**
Ammonia	NH_3	Household cleansers
Formaldehyde	H_2CO	Embalming fluid
Acetylene	HC_2H	Fuel for a welding torch

Table 20.1

Some Interesting Interstellar Molecules		
Name	**Chemical Formula**	**Use on Earth**
Acetic acid	$C_2H_2O_4$	The essence of vinegar
Ethyl alcohol	CH_3CH_2OH	End-of-semester parties
Ethylene glycol	$HOCH_2CH_2OH$	Antifreeze ingredient
Benzene	C_6H_6	Carbon ring, ingredient in varnishes and dyes

Table 20.1

The cold interstellar clouds also contain cyanoacetylene (HC_3N) and acetaldehyde (CH_3CHO), generally regarded as starting points for *amino acid* formation. These are building blocks of proteins, which are among the fundamental chemicals from which living organisms on Earth are constructed. The presence of these organic molecules does not imply that life exists in space, but it does show that the chemical building blocks of life can form under a wide range of conditions in the universe. As we learn more about how complex molecules are produced in interstellar clouds, we gain an increased understanding of the kinds of processes that preceded the beginnings of life on Earth billions of years ago.

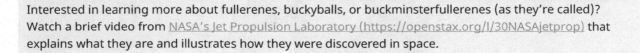

LINK TO LEARNING

Interested in learning more about fullerenes, buckyballs, or buckminsterfullerenes (as they're called)? Watch a brief video from NASA's Jet Propulsion Laboratory (https://openstax.org/l/30NASAjetprop) that explains what they are and illustrates how they were discovered in space.

MAKING CONNECTIONS

Cocktails in Space

Among the molecules astronomers have identified in interstellar clouds is alcohol, which comes in two varieties: methyl (or wood) alcohol and ethyl alcohol (the kind you find in cocktails). Ethyl alcohol is a pretty complex molecule, written by chemists as C_2H_5OH. It is quite plentiful in space (relatively speaking). In clouds where it has been identified, we detect up to one molecule for every m^3. The largest of the clouds (which can be several hundred light-years across) have enough ethyl alcohol to make 10^{28} fifths of liquor.

We need not fear, however, that future interstellar astronauts will become interstellar alcoholics. Even if a spaceship were equipped with a giant funnel 1 kilometer across and could scoop it through such a cloud at the speed of light, it would take about a thousand years to gather up enough alcohol for one standard martini.

Furthermore, the very same clouds also contain water (H_2O) molecules. Your scoop would gather them up as well, and there are a lot more of them because they are simpler and thus easier to form. For the fun of it, one astronomical paper actually calculated the proof of a typical cloud. *Proof* is the ratio of alcohol to water in a drink, where 0 proof means all water, 100 proof means half alcohol and half water, and 200 proof means all alcohol. The proof of the interstellar cloud was only 0.2, not enough to qualify as a stiff drink

20.3 | Cosmic Dust

Learning Objectives

By the end of this section, you will be able to:

> Describe how we can detect interstellar dust
> Understand the role and importance of infrared observations in studying dust
> Explain the terms extinction and interstellar reddening

Figure 20.9 shows a striking example of what is actually a common sight through large telescopes: a dark region on the sky that appears to be nearly empty of stars. For a long time, astronomers debated whether these dark regions were empty "tunnels" through which we looked beyond the stars of the Milky Way Galaxy into intergalactic space, or clouds of some dark material that blocked the light of the stars beyond. The astronomer William Herschel (discoverer of the planet Uranus) thought it was the former, once remarking after seeing one, "Here truly is a hole in heaven!" However, American astronomer E. E. Barnard is generally credited with showing from his extensive series of nebula photographs that the latter interpretation is the correct one (see the feature box on Edward Emerson Barnard).

Figure 20.9 Barnard 68. This object, first catalogued by E. E. Barnard, is a dark interstellar cloud. Its striking appearance is due to the fact that, since it is relatively close to Earth, there are no bright stars between us and it, and its dust obscures the light from the stars behind it. (It looks a little bit like a sideways heart; one astronomers sent a photo of this object to his sweetheart as a valentine.) (credit: modification of work by ESO)

Dusty clouds in space betray their presence in several ways: by blocking the light from distant stars, by emitting energy in the infrared part of the spectrum, by reflecting the light from nearby stars, and by making distant stars look redder than they really are.

VOYAGERS IN ASTRONOMY

Edward Emerson Barnard

Born in 1857 in Nashville, Tennessee, two months after his father died, Edward Barnard (Figure 20.10) grew up in such poor circumstances that he had to drop out of school at age nine to help support his ailing mother. He soon became an assistant to a local photographer, where he learned to love both photography and astronomy, destined to become the dual passions of his life. He worked as a photographer's aide for 17 years, studying astronomy on his own. In 1883, he obtained a job as an assistant at the Vanderbilt University Observatory, which enabled him at last to take some astronomy courses.

Married in 1881, Barnard built a house for his family that he could ill afford. But as it happened, a patent medicine manufacturer offered a $200 prize (a lot of money in those days) for the discovery of any new comet. With the determination that became characteristic of him, Barnard spent every clear night searching for comets. He discovered seven of them between 1881 and 1887, earning enough money to make the payments on his home; this "Comet House" later became a local attraction. (By the end of his life, Barnard had found 17 comets through diligent observation.)

In 1887, Barnard got a position at the newly founded Lick Observatory, where he soon locked horns with the director, Edward Holden, a blustering administrator who made Barnard's life miserable. (To be fair, Barnard soon tried to do the same for him.) Despite being denied the telescope time that he needed for his photographic work, in 1892, Barnard managed to discover the first new moon found around Jupiter since Galileo's day, a stunning observational feat that earned him world renown. Now in a position to demand more telescope time, he perfected his photographic techniques and soon began to publish the best images of the Milky Way taken up to that time. It was during the course of this work that he began to examine the dark regions among the crowded star lanes of the Galaxy and to realize that they must be vast clouds of obscuring material (rather than "holes" in the distribution of stars).

Astronomer-historian Donald Osterbrock has called Barnard an "observaholic:" his daily mood seemed to depend entirely on how clear the sky promised to be for his night of observing. He was a driven, neurotic man, concerned about his lack of formal training, fearful of being scorned, and afraid that he might somehow slip back into the poverty of his younger days. He had difficulty taking vacations and lived for his work: only serious illness could deter him from making astronomical observations.

In 1895, Barnard, having had enough of the political battles at Lick, accepted a job at the Yerkes Observatory near Chicago, where he remained until his death in 1923. He continued his photographic work, publishing compilations of his images that became classic photographic atlases, and investigating the varieties of nebulae revealed in his photographs. He also made measurements of the sizes and features of planets, participated in observations of solar eclipses, and carefully cataloged dark nebulae (see Figure 20.9). In 1916, he discovered the star with the largest proper motion, the second-closest star system to our own (see Analyzing Starlight). It is now called Barnard's Star in his honor.

Figure 20.10 Edward Emerson Barnard (1857–1923). Barnard's observations provided information that furthered many astronomical explorations. (credit: The Lick Observatory)

Detecting Dust

The dark cloud seen in Figure 20.9 blocks the light of the many stars that lie behind it; note how the regions in other parts of the photograph are crowded with stars. Barnard 68 is an example of a relatively dense cloud or *dark nebula* containing tiny, solid dust grains. Such opaque clouds are conspicuous on any photograph of the Milky Way, the galaxy in which the Sun is located (see the figures in The Milky Way Galaxy). The "dark rift," which runs lengthwise down a long part of the Milky Way in our sky and appears to split it in two, is produced by a collection of such obscuring clouds.

While dust clouds are too cold to radiate a measurable amount of energy in the visible part of the spectrum, they glow brightly in the infrared (Figure 20.11). The reason is that small dust grains absorb visible light and ultraviolet radiation very efficiently. The grains are heated by the absorbed radiation, typically to temperatures from 10 to about 500 K, and re-radiate this heat at infrared wavelengths.

(a)   (b)

Figure 20.11 Visible and Infrared Images of the Horsehead Nebula in Orion. This dark cloud is one of the best-known images in astronomy, probably because it really does resemble a horse's head. The horse-head shape is an extension of a large cloud of dust that fills the lower part of the picture. (a) Seen in visible light, the dust clouds are especially easy to see against the bright background. (b) This infrared radiation image from the region of the horse head was recorded by NASA's Wide-Field Infrared Survey Explorer. Note how the regions that appear dark in visible light appear bright in the infrared. The dust is heated by nearby stars and re-radiates this heat in the infrared. Only the top of the horse's head is visible in the infrared image. Bright dots seen in the nebula below and to the left and at the top of the horse head are young, newly formed stars. The insets show the horse head and the bright nebula in more detail. (credit a: modification of work by ESO and Digitized Sky Survey; credit b: modification of work by NASA/JPL-Caltech)

Thanks to their small sizes and low temperatures, interstellar grains radiate most of their energy at infrared to microwave frequencies, with wavelengths of tens to hundreds of microns. Earth's atmosphere is opaque to radiation at these wavelengths, so emission by interstellar dust is best measured from space. Observations from above Earth's atmosphere show that dust clouds are present throughout the plane of the Milky Way (Figure 20.12).

Figure 20.12 Infrared Emission from the Plane of the Milky Way. This infrared image taken by the Spitzer Space Telescope shows a field in the plane of the Milky Way Galaxy. (Our Galaxy is in the shape of a frisbee; the plane of the Milky Way is the flat disk of that frisbee. Since the Sun, Earth, and solar system are located in the plane of the Milky Way and at a large distance from its center, we view the Galaxy edge on, much as we might look at a glass plate from its edge.) This emission is produced by tiny dust grains, which emit at 3.6 microns (blue in this image), 8.0 microns (green), and 24 microns (red). The densest regions of dust are so cold and opaque that they appear as dark clouds even at these infrared wavelengths. The red bubbles visible throughout indicate regions where the dust has been warmed up by young stars. This heating increases the emission at 24 microns, leading to the redder color in this image. (credit: modification of work by NASA/JPL-Caltech/University of Wisconsin)

Some dense clouds of dust are close to luminous stars and scatter enough starlight to become visible. Such a cloud of dust, illuminated by starlight, is called a *reflection nebula*, since the light we see is starlight reflected off the grains of dust. One of the best-known examples is the nebulosity around each of the brightest stars in the Pleiades cluster (see Figure 20.1). The dust grains are small, and such small particles turn out to scatter light with blue wavelengths more efficiently than light at red wavelengths. A reflection nebula, therefore, usually appears bluer than its illuminating star (Figure 20.13).

Figure 20.13 Pleiades Star Cluster. The bluish light surrounding the stars in this image is an example of a reflection nebula. Like fog around a street lamp, a reflection nebula shines only because the dust within it scatters light from a nearby bright source. The Pleiades cluster is currently passing through an interstellar cloud that contains dust grains, which scatter the light from the hot blue stars in the cluster. The Pleiades cluster is about 400 light-years from the Sun. (credit: NASA, ESA and AURA/Caltech)

Gas and dust are generally intermixed in space, although the proportions are not exactly the same everywhere. The presence of dust is apparent in many photographs of *emission nebulae* in the constellation of Sagittarius, where we see an H II region surrounded by a blue reflection nebula. Which type of nebula appears brighter depends on the kinds of stars that cause the gas and dust to glow. Stars cooler than about 25,000 K have so little ultraviolet radiation of wavelengths shorter than 91.2 nanometers—which is the wavelength required to ionize hydrogen—that the reflection nebulae around such stars outshine the emission nebulae. Stars hotter than 25,000 K emit enough ultraviolet energy that the emission nebulae produced around them generally outshine the reflection nebulae.

Interstellar Reddening

The tiny interstellar dust grains absorb some of the starlight they intercept. But at least half of the starlight that interacts with a grain is merely scattered, that is, it is redirected rather than absorbed. Since neither the absorbed nor the scattered starlight reaches us directly, both absorption and scattering make stars look dimmer. The effects of both processes are called **interstellar extinction** ([Figure 20.14](#)).

Astronomers first came to understand interstellar extinction around the early 1930s, as the explanation of a puzzling observation. In the early part of the twentieth century, astronomers discovered that some stars look red even though their spectral lines indicate that they must be extremely hot (and thus should look blue). The solution to this seeming contradiction turned out to be that the light from these hot stars is not only dimmed but also reddened by interstellar dust, a phenomenon known as interstellar **reddening**.

Figure 20.14 Barnard 68 in Infrared. In this image, we see Barnard 68, the same object shown in [Figure 20.9](#). The difference is that, in the previous image, the blue, green, and red channels showed light in the visible (or very nearly visible) part of the spectrum. In this image, the red color shows radiation emitted in the infrared at a wavelength of 2.2 microns. Interstellar extinction is much smaller at infrared than at visible wavelengths, so the stars behind the cloud become visible in the infrared channel. (credit: ESO)

Dust does not interact with all the colors of light the same way. Much of the violet, blue, and green light from these stars has been scattered or absorbed by dust, so it does not reach Earth. Some of their orange and red light, with longer wavelengths, on the other hand, more easily penetrates the intervening dust and completes its long journey through space to enter Earth-based telescopes ([Figure 20.15](#)). Thus, the star looks redder from Earth than it would if you could see it from nearby. (Strictly speaking, *reddening* is not the most accurate term for this process, since no red color is added; instead, blues and related colors are subtracted, so it should more properly be called "deblueing.") In the most extreme cases, stars can be so reddened that they are entirely undetectable at visible wavelengths and can be seen only at infrared or longer wavelengths ([Figure 20.14](#)).

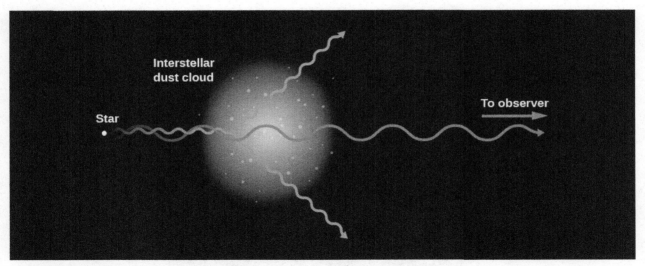

Figure 20.15 Scattering of Light by Dust. Interstellar dust scatters blue light more efficiently than red light, thereby making distant stars appear redder and giving clouds of dust near stars a bluish hue. Here, a red ray of light from a star comes straight through to the observer, whereas a blue ray is shown scattering. A similar scattering process makes Earth's sky look blue.

We have all seen an example of reddening on Earth. The Sun appears much redder at sunset than it does at noon. The lower the Sun is in the sky, the longer the path its light must travel through the atmosphere. Over this greater distance, there is a greater chance that sunlight will be scattered. Since red light is less likely to be scattered than blue light, the Sun appears more and more red as it approaches the horizon.

By the way, scattering of sunlight is also what causes our sky to look blue, even though the gases that make up Earth's atmosphere are transparent. As sunlight comes in, it scatters from the molecules of air. The small size of the molecules means that the blue colors scatter much more efficiently than the greens, yellows, and reds. Thus, the blue in sunlight is scattered out of the beam and all over the sky. The light from the Sun that comes to your eye, on the other hand, is missing some of its blue, so the Sun looks a bit yellower, even when it is high in the sky, than it would from space.

The fact that starlight is reddened by interstellar dust means that long-wavelength radiation is transmitted through the Galaxy more efficiently than short-wavelength radiation. Consequently, if we wish to see farther in a direction with considerable interstellar material, we should look at long wavelengths. This simple fact provides one of the motivations for the development of infrared astronomy. In the infrared region at 2 microns (2000 nanometers), for example, the obscuration is only one-sixth as great as in the visible region (500 nanometers), and we can therefore study stars that are more than twice as distant before their light is blocked by interstellar dust. This ability to see farther by observing in the infrared portion of the spectrum represents a major gain for astronomers trying to understand the structure of our Galaxy or probing its puzzling, but distant, center (see The Milky Way Galaxy).

Interstellar Grains

Before we get to the details about interstellar dust, we should perhaps get one concern out of the way. Why couldn't it be the interstellar *gas* that reddens distant stars and not the dust? We already know from everyday experience that atomic or molecular gas is almost transparent. Consider Earth's atmosphere. Despite its very high density compared with that of interstellar gas, it is so transparent as to be practically invisible. (Gas does have a few specific spectral lines, but they absorb only a tiny fraction of the light as it passes through.)The quantity of *gas* required to produce the observed absorption of light in interstellar space would have to be enormous. The gravitational attraction of so great a mass of gas would affect the motions of stars in ways that could easily be detected. Such motions are not observed, and thus, the interstellar absorption cannot be the result of gases.

Although gas does not absorb much light, we know from everyday experience that tiny solid or liquid particles can be very efficient absorbers. Water vapor in the air is quite invisible. When some of that vapor condenses into tiny water droplets, however, the resulting cloud is opaque. Dust storms, smoke, and smog offer familiar examples of the efficiency with which solid particles absorb light. On the basis of arguments like these, astronomers have concluded that widely scattered *solid* particles in interstellar space must be responsible for the observed dimming of starlight. What are these particles made of? And how did they form?

Observations like the pictures in this chapter show that a great deal of this dust exists; hence, it must be primarily composed of elements that are abundant in the universe (and in interstellar matter). After hydrogen and helium, the most abundant elements are oxygen, carbon, and nitrogen. These three elements, along with magnesium, silicon, iron—and perhaps hydrogen itself—turn out to be the most important components of interstellar dust.

Many of the dust particles can be characterized as sootlike (rich in carbon) or sandlike (containing silicon and oxygen). Grains of interstellar dust are found in meteorites and can be identified because the abundances of certain isotopes are different from what we see in other solar system material. Several different interstellar dust substances have been identified in this way in the laboratory, including graphite and diamonds. (Don't get excited; these diamonds are only a billionth of a meter in size and would hardly make an impressive engagement ring!)

The most widely accepted model pictures the grains with rocky cores that are either like soot (rich in carbon) or like sand (rich in silicates). In the dark clouds where molecules can form, these cores are covered by icy mantles (Figure 20.16). The most common ices in the grains are water (H_2O), methane (CH_4), and ammonia (NH_3)—all built out of atoms that are especially abundant in the realm of the stars. The ice mantles, in turn, are sites for some of the chemical reactions that produce complex organic molecules.

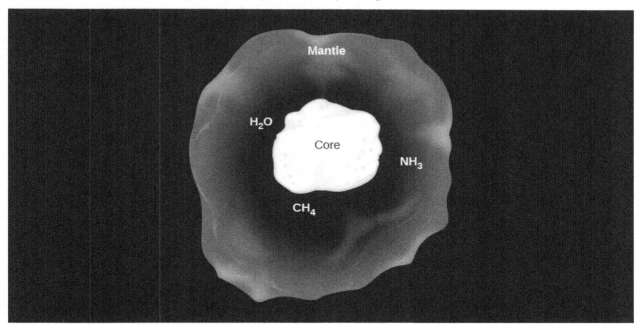

Figure 20.16 Model of an Interstellar Dust Grain. A typical interstellar grain is thought to consist of a core of rocky material (silicates) or graphite, surrounded by a mantle of ices. Typical grain sizes are 10^{-8} to 10^{-7} meters. (This is from 1/100 to 1/10 of a micron; by contrast, human hair is about 10–200 microns wide.)

Typical individual grains must be just slightly smaller than the wavelength of visible light. If the grains were a lot smaller, they would not block the light efficiently, as Figure 20.13 and other images in this chapter show that it does.

On the other hand, if the dust grains were much larger than the wavelength of light, then starlight would not be reddened. Things that are much larger than the wavelength of light would block both blue and red light with equal efficiency. In this way we can deduce that a characteristic interstellar dust grain contains 10^6 to 10^9

atoms and has a diameter of 10^{-8} to 10^{-7} meters (10 to 100 nanometers). This is actually more like the specks of solid matter in cigarette smoke than the larger grains of dust you might find under your desk when you are too busy studying astronomy to clean properly.

20.4 | Cosmic Rays

Learning Objectives

By the end of this section, you will be able to:

> Define cosmic rays and describe their composition
> Explain why it is hard to study the origin of cosmic rays, and the current leading hypotheses about where they might come from

In addition to gas and dust, a third class of particles, noteworthy for the high speeds with which they travel, is found in interstellar space. **Cosmic rays** were discovered in 1911 by an Austrian physicist, Victor Hess, who flew simple instruments aboard balloons and showed that high-speed particles arrive at Earth from space (Figure 20.17). The term "cosmic ray" is misleading, implying it might be like a ray of light, but we are stuck with the name. They are definitely particles and have nearly the same composition as ordinary interstellar gas. Their behavior, however, is radically different from the gas we have discussed so far.

Figure 20.17 Victor Hess (1883–1964). Cosmic-ray pioneer Victor Hess returns from a 1912 balloon flight that reached an altitude of 5.3 kilometers. It was on such balloon flights that Hess discovered cosmic rays.

The Nature of Cosmic Rays

Cosmic rays are mostly high-speed atomic nuclei and electrons. Speeds equal to 90% of the speed of light are typical. Almost 90% of the cosmic rays are hydrogen nuclei (protons) stripped of their accompanying electron. Helium and heavier nuclei constitute about 9% more. About 1% of cosmic rays have masses equal to the mass of the electron, and 10–20% of these carry positive charge rather than the negative charge that characterizes electrons. A positively charged particle with the mass of an electron is called a *positron* and is a form of *antimatter* (we discussed antimatter in The Sun: A Nuclear Powerhouse).

The abundances of various atomic nuclei in cosmic rays mirror the abundances in stars and interstellar gas, with one important exception. The light elements lithium, beryllium, and boron are far more abundant in cosmic rays than in the Sun and stars. These light elements are formed when high-speed, cosmic-ray nuclei of carbon, nitrogen, and oxygen collide with protons in interstellar space and break apart. (By the way, if you, like most readers, have not memorized all the elements and want to see how any of those we mention fit into the sequence of elements, you will find them all listed in Appendix K in order of the number of protons they contain.)

Cosmic rays reach Earth in substantial numbers, and we can determine their properties either by capturing them directly or by observing the reactions that occur when they collide with atoms in our atmosphere. The total energy deposited by cosmic rays in Earth's atmosphere is only about one-billionth the energy received from the Sun, but it is comparable to the energy received in the form of starlight. Some of the cosmic rays come to Earth from the surface of the Sun, but most come from outside the solar system.

Where Do They Come From?

There is a serious problem in identifying the source of cosmic rays. Since light travels in straight lines, we can tell where it comes from simply by looking. Cosmic rays are charged particles, and their direction of motion can be changed by magnetic fields. The paths of cosmic rays are curved both by magnetic fields in interstellar space and by Earth's own field. Calculations show that low-energy cosmic rays may spiral many times around Earth before entering the atmosphere where we can detect them. If an airplane circles an airport many times before landing, it is difficult for an observer to determine the direction from which it originated. So, too, after a cosmic ray circles Earth several times, it is impossible to know where its journey began.

There are a few clues, however, about where cosmic rays might be generated. We know, for example, that magnetic fields in interstellar space are strong enough to keep all but the most energetic cosmic rays from escaping the Galaxy. It therefore seems likely that they are produced somewhere inside the Galaxy. The only likely exceptions are those with the very highest energy. Such cosmic rays move so rapidly that they are not significantly influenced by interstellar magnetic fields, and thus, they could escape our Galaxy. By analogy, they could escape other galaxies as well, so some of the highest-energy cosmic rays that we detect may have been created in some distant galaxy. Still, most cosmic rays must have their source inside the Milky Way Galaxy.

We can also estimate how far typical cosmic rays travel before striking Earth. The light elements lithium, beryllium, and boron hold the key. Since these elements are formed when carbon, nitrogen, and oxygen strike interstellar protons, we can calculate how long, on average, cosmic rays must travel through space in order to experience enough collisions to account for the amount of lithium and the other light elements that they contain. It turns out that the required distance is about 30 times around the Galaxy. At speeds near the speed of light, it takes perhaps 3–10 million years for the average cosmic ray to travel this distance. This is only a small fraction of the age of the Galaxy or the universe, so cosmic rays must have been created fairly recently on a cosmic timescale.

The best candidates for a source of cosmic rays are the supernova explosions, which mark the violent deaths of some stars (and which we will discuss in The Death of Stars). The material ejected by the explosion produces a shock wave, which travels through the interstellar medium. Charged particles can become trapped, bouncing back and forth across the front of the shock wave many times. With each pass through the shock, the magnetic fields inside it accelerate the particles more and more. Eventually, they are traveling at close to the speed of light and can escape from the shock to become cosmic rays. Some collapsed stars (including star remnants left over from supernova explosions) may, under the right circumstances, also serve as accelerators of particles. In any case, we again find that the raw material of the Galaxy is enriched by the life cycle of stars. In the next section, we will look at this enrichment process in more detail.

LINK TO LEARNING

You can watch a brief video about the Calorimetric Electron Telescope (CALET) (https://openstax.org/l/30CALETvid) mission, a cosmic ray detector at the International Space Station. The link takes you to NASA Johnson's "Space Station Live: Cosmic Ray Detector for ISS."

20.5 The Life Cycle of Cosmic Material

Learning Objectives

By the end of this section, you will be able to:

> Explain how interstellar matter flows into and out of our Galaxy and transforms from one phase to another, and understand how star formation and evolution affects the properties of the interstellar medium

> Explain how the heavy elements and dust grains found in interstellar space got there and describe how dust grains help produce molecules that eventually find their way into planetary systems

Flows of Interstellar Gas

The most important thing to understand about the interstellar medium is that it is not static. Interstellar gas orbits through the Galaxy, and as it does so, it can become more or less dense, hotter and colder, and change its state of ionization. A particular parcel of gas may be neutral hydrogen at some point, then find itself near a young, hot star and become part of an H II region. The star may then explode as a supernova, heating the nearby gas up to temperatures of millions of degrees. Over millions of years, the gas may cool back down and become neutral again, before it collects into a dense region that gravity gathers into a giant molecular cloud (Figure 20.18)

1 kpc (3200 LY)

Figure 20.18 Large-Scale Distribution of Interstellar Matter. This image is from a computer simulation of the Milky Way Galaxy's interstellar medium as a whole. The majority of gas, visible in greenish colors, is neutral hydrogen. In the densest regions in the spiral arms, shown in yellow, the gas is collected into giant molecular clouds. Low-density holes in the spiral arms, shown in blue, are the result of supernova explosions. (credit: modification of work by Mark Krumholz)

At any given time in the Milky Way, the majority of the interstellar gas by mass and volume is in the form of atomic hydrogen. The much-denser molecular clouds occupy a tiny fraction of the volume of interstellar space but add roughly 30% to the total mass of gas between the stars. Conversely, the hot gas produced by supernova explosions contributes a negligible mass but occupies a significant fraction of the volume of interstellar space. H II regions, though they are visually spectacular, constitute only a very small fraction of either the mass or volume of interstellar material.

However, the interstellar medium is not a closed system. Gas from intergalactic space constantly falls onto the Milky Way due to its gravity, adding new gas to the interstellar medium. Conversely, in giant molecular clouds where gas collects together due to gravity, the gas can collapse to form new stars, as discussed in The Birth of Stars and the Discovery of Planets outside the Solar System. This process locks interstellar matter into stars. As the stars age, evolve, and eventually die, massive stars lose a large fraction of their mass, and low-mass stars lose very little. On average, roughly one-third of the matter incorporated into stars goes back into interstellar space. Supernova explosions have so much energy that they can drive interstellar mass out of the Galaxy and back into intergalactic space. Thus, the total amount of mass of the interstellar medium is set by a competition between the gain of mass from intergalactic space, the conversion of interstellar mass into stars, and the loss of interstellar mass back into intergalactic space due to supernovae. This entire process is known as the **baryon cycle**—baryon is from the Greek word for "heavy," and the cycle has this name because it is the repeating process that the heavier components of the universe—the atoms—undergo.

The Cycle of Dust and Heavy Elements

While much of the mass of the interstellar medium is material accreted during the last few billion years from intergalactic space, this is not true of the elements heavier than hydrogen and helium, or of the dust. Instead, these components of the interstellar medium were made inside stars in the Milky Way, which returned them to the interstellar medium at the end of their lives. We will talk more about this process in later chapters, but for now just bear in mind what we learned in The Sun: A Nuclear Powerhouse. What stars "do for a living" is fuse heavier elements from lighter ones, producing energy in the process. As stars mature, they begin to lose some of the newly made elements to the reservoir of interstellar matter.

The same is true of dust grains. Dust forms when grains can condense in regions where gas is dense and cool. One place where the right conditions are found is the winds from luminous cool stars (the red giants and supergiants we discussed in The Stars: A Celestial Census). Grains can also condense in the matter thrown off by a supernova explosion as the ejected gases begin to cool.

The dust grains produced by stars may grow even further when they spend time in the dense parts of the interstellar medium, inside molecular clouds. In these environments, grains can stick together or gather additional atoms from the gas around them, growing larger. They also facilitate the production of other compounds, including some of the more complex molecules we discussed earlier.

The surfaces of the dust grains (see Cosmic Dust)—which would seem very large if you were an atom—provide "nooks and crannies" where these atoms can stick long enough to find partners and form molecules. (Think of the dust grains as "interstellar social clubs" where lonely atoms can meet and form meaningful relationships.) Eventually, the dust grains become coated with ices. The presence of the dust shields the molecules inside the clouds from ultraviolet radiation and cosmic rays that would break them up.

When stars finally begin to form within the cloud, they heat the grains and evaporate the ices. The gravitational attraction of the newly forming stars also increases the density of the surrounding cloud material. Many more chemical reactions take place on the surfaces of grains in the gas surrounding the newly forming stars, and these areas are where organic molecules are formed. These molecules can be incorporated into newly formed planetary systems, and the early Earth may have been seeded in just such a way.

Indeed, scientists speculate that some of the water on Earth may have come from interstellar grains. Recent observations from space have shown that water is abundant in dense interstellar clouds. Since stars are formed from this material, water must be present when solar systems, including our own, come into existence. The water in our oceans and lakes may have come initially from water locked into the rocky material that accreted to form Earth. Alternatively, the water may have been brought to Earth when asteroids and comets (formed from the same cloud that made the planets) later impacted it. Scientists estimate that one comet impact every thousand years during Earth's first billion years would have been enough to account for the water we see today. Of course, both sources may have contributed to the water we now enjoy drinking and swimming in.

Any interstellar grains that are incorporated into newly forming stars (instead of the colder planets and smaller bodies around them) will be destroyed by their high temperatures. But eventually, each new generation of stars will evolve to become red giants, with stellar winds of their own. Some of these stars will also become supernovae and explode. Thus, the process of recycling cosmic material can start all over again.

20.6 | Interstellar Matter around the Sun

Learning Objectives

By the end of this section, you will be able to:

> Describe how interstellar matter is arranged around our solar system
> Explain why scientists think that the Sun is located in a hot bubble

We want to conclude our discussion of interstellar matter by asking how this material is organized in our immediate neighborhood. As we discussed above, orbiting X-ray observatories have shown that the Galaxy is full of bubbles of hot, X-ray-emitting gas. They also revealed a diffuse background of X-rays that appears to fill the entire sky from our perspective (Figure 20.19). While some of this emission comes from the interaction of the solar wind with the interstellar medium, a majority of it comes from beyond the solar system. The natural explanation for why there is X-ray-emitting gas all around us is that the Sun is itself inside one of the bubbles. We therefore call our "neighborhood" the Local Hot Bubble, or **Local Bubble** for short. The Local Bubble is much less dense—an average of approximately 0.01 atoms per cm^3—than the average interstellar density of about 1 atom per cm^3. This local gas has a temperature of about a million degrees, just like the gas in the other superbubbles that spread throughout our Galaxy, but because there is so little hot material, this high temperature does not affect the stars or planets in the area in any way.

What caused the Local Bubble to form? Scientists are not entirely sure, but the leading candidate is winds from stars and supernova explosions. In a nearby region in the direction of the constellations Scorpius and Centaurus, a lot of star formation took place about 15 million years ago. The most massive of these stars evolved very quickly until they produced strong winds, and some ended their lives by exploding. These processes filled the region around the Sun with hot gas, driving away cooler, denser gas. The rim of this expanding superbubble reached the Sun about 7.6 million years ago and now lies more than 200 light-years past the Sun in the general direction of the constellations of Orion, Perseus, and Auriga.

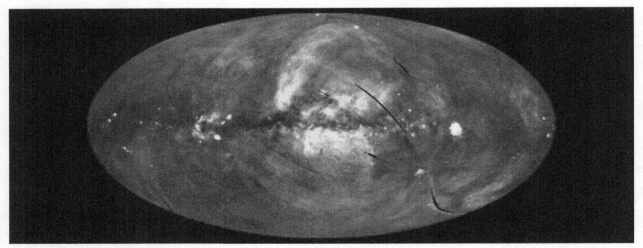

Figure 20.19 Sky in X-Rays. This image, made by the ROSAT satellite, shows the whole sky in X-rays as seen from Earth. Different colors indicate different X-ray energies: red is 0.25 kiloelectron volts, green is 0.75 kiloelectron volts, and blue is 1.5 kiloelectron volts. The image is oriented so the plane of the Galaxy runs across the middle of the image. The ubiquitous red color, which does not disappear completely even in the galactic plane, is evidence for a source of X-rays all around the Sun. (credit: modification of work by NASA)

A few clouds of interstellar matter do exist within the Local Bubble. The Sun itself seems to have entered a cloud about 10,000 years ago. This cloud is warm (with a temperature of about 7000 K) and has a density of 0.3 hydrogen atom per cm^3—higher than most of the Local Bubble but still so tenuous that it is also referred to as **Local Fluff** (Figure 20.20). (Aren't these astronomical names fun sometimes?)

While this is a pretty thin cloud, we estimate that it contributes 50 to 100 times more particles than the solar wind to the diffuse material between the planets in our solar system. These interstellar particles have been detected and their numbers counted by the spacecraft traveling between the planets. Perhaps someday, scientists will devise a way to collect them without destroying them and to return them to Earth, so that we can touch—or at least study in our laboratories—these messengers from distant stars.

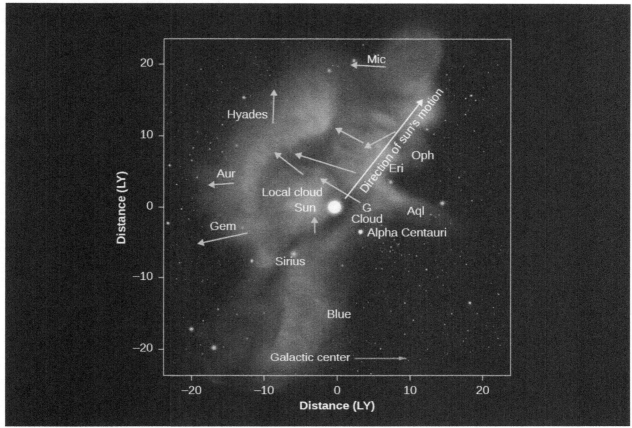

Figure 20.20 Local Fluff. The Sun and planets are currently moving through the Local Interstellar Cloud, which is also called the Local Fluff. Fluff is an appropriate description because the density of this cloud is only about 0.3 atom per cm^3. In comparison, Earth's atmosphere at the edge of space has around 1.2×10^{13} molecules per cm^3. This image shows the patches of interstellar matter (mostly hydrogen gas) within about 20 light-years of the Sun. The temperature of the Local Interstellar Cloud is about 7000 K. The arrows point toward the directions that different parts of the cloud are moving. The names associated with each arrow indicate the constellations located on the sky toward which the parts of the cloud are headed. The solar system is thought to have entered the Local Interstellar Cloud, which is a small cloud located within a much larger superbubble that is expanding outward from the Scorpius-Centaurus region of the sky, at some point between 44,000 and 150,000 years ago and is expected to remain within it for another 10,000 to 20,000 years. (credit: modification of work by NASA/Goddard/Adler/University Chicago/Wesleyan)

🔑 Key Terms

baryon cycle   the cycling of mass in and out of the interstellar medium, including accretion of gas from intergalactic space, loss of gas back into intergalactic space, and conversion of interstellar gas into stars

cosmic rays   atomic nuclei (mostly protons) and electrons that are observed to strike Earth's atmosphere with exceedingly high energies.

H II region   the region of ionized hydrogen in interstellar space

interstellar dust   tiny solid grains in interstellar space thought to consist of a core of rocklike material (silicates) or graphite surrounded by a mantle of ices; water, methane, and ammonia are probably the most abundant ices

interstellar extinction   the attenuation or absorption of light by dust in the interstellar medium

interstellar medium (ISM)   (or interstellar matter) the gas and dust between the stars in a galaxy

Local Bubble   (or Local Hot Bubble) a region of low-density, million degree gas in which the Sun and solar system are currently located

Local Fluff   a slightly denser cloud inside the Local Bubble, inside which the Sun also lies

molecular cloud   a large, dense, cold interstellar cloud; because of its size and density, this type of cloud can keep ultraviolet radiation from reaching its interior, where molecules are able to form

nebula   a cloud of interstellar gas or dust; the term is most often used for clouds that are seen to glow with visible light or infrared

reddening (interstellar)   the reddening of starlight passing through interstellar dust because dust scatters blue light more effectively than red

📖 Summary

20.1 The Interstellar Medium

About 15% of the visible matter in the Galaxy is in the form of gas and dust, serving as the raw material for new stars. About 99% of this interstellar matter is in the form of gas—individual atoms or molecules. The most abundant elements in the interstellar gas are hydrogen and helium. About 1% of the interstellar matter is in the form of solid interstellar dust grains.

20.2 Interstellar Gas

Interstellar gas may be hot or cold. Gas found near hot stars emits light by fluorescence, that is, light is emitted when an electron is captured by an ion and cascades down to lower-energy levels. Glowing clouds (nebulae) of ionized hydrogen are called H II regions and have temperatures of about 10,000 K. Most hydrogen in interstellar space is not ionized and can best be studied by radio measurements of the 21-centimeter line. Some of the gas in interstellar space is at a temperature of a million degrees, even though it is far away in hot stars; this ultra-hot gas is probably heated when rapidly moving gas ejected in supernova explosions sweeps through space. In some places, gravity gathers interstellar gas into giant clouds, within which the gas is protected from starlight and can form molecules; more than 200 different molecules have been found in space, including the basic building blocks of proteins, which are fundamental to life as we know it here on Earth.

20.3 Cosmic Dust

Interstellar dust can be detected: (1) when it blocks the light of stars behind it, (2) when it scatters the light from nearby stars, and (3) because it makes distant stars look both redder and fainter. These effects are called interstellar extinction and reddening. Dust can also be detected in the infrared because it emits heat radiation. Dust is found throughout the plane of the Milky Way. The dust particles are about the same size as the wavelength of light and consist of rocky cores that are either sootlike (carbon-rich) or sandlike (silicates) with mantles made of ices such as water, ammonia, and methane.

20.4 Cosmic Rays

Cosmic rays are particles that travel through interstellar space at a typical speed of 90% of the speed of light. The most abundant elements in cosmic rays are the nuclei of hydrogen and helium, but electrons and positrons are also found. It is likely that many cosmic rays are produced in supernova shocks.

20.5 The Life Cycle of Cosmic Material

Interstellar matter is constantly flowing through the Galaxy and changing from one phase to another. At the same time, gas is constantly being added to the Galaxy by accretion from extragalactic space, while mass is removed from the interstellar medium by being locked in stars. Some of the mass in stars is, in turn, returned to the interstellar medium when those stars evolve and die. In particular, the heavy elements in interstellar space were all produced inside stars, while the dust grains are made in the outer regions of stars that have swelled to be giants. These elements and grains, in turn, can then be incorporated into new stars and planetary systems that form out of the interstellar medium.

20.6 Interstellar Matter around the Sun

The Sun is located at the edge of a low-density cloud called the Local Fluff. The Sun and this cloud are located within the Local Bubble, a region extending to at least 300 light-years from the Sun, within which the density of interstellar material is extremely low. Astronomers think this bubble was blown by some nearby stars that experienced a strong wind and some supernova explosions.

 For Further Exploration

Articles

Goodman, A. "Recycling the Universe." *Sky & Telescope* November (2000): 44. Review of how stellar evolution, the interstellar medium, and supernovae all work together to recycle cosmic material.

Greenberg, J. "The Secrets of Stardust." *Scientific American* December (2000): 70. The makeup and evolutionary role of solid particles between the stars.

Knapp, G. "The Stuff between the Stars." *Sky & Telescope* May (1995): 20. An introduction to the interstellar medium.

Nadis, S. "Searching for the Molecules of Life in Space." *Sky & Telescope* January (2002): 32. Recent observations of water in the interstellar medium by satellite telescopes.

Olinto, A. "Solving the Mystery of Cosmic Rays." *Astronomy* April (2014): 30. What accelerates them to such high energies.

Reynolds, R. "The Gas between the Stars." *Scientific American* January (2002): 34. On the interstellar medium.

Websites and Apps

Barnard, E. E., Biographical Memoir: http://www.nasonline.org/publications/biographical-memoirs/memoir-pdfs/barnard-edward.pdf (http://www.nasonline.org/publications/biographical-memoirs/memoir-pdfs/barnard-edward.pdf).

Cosmicopia: https://cosmicopia.gsfc.nasa.gov/ (https://cosmicopia.gsfc.nasa.gov/). NASA's learning site explains about the history and modern understanding of cosmic rays.

DECO: https://wipac.wisc.edu/deco (https://wipac.wisc.edu/deco). A smart-phone app for turning your phone into a cosmic-ray detector.

Hubble Space Telescope Images of Nebulae: https://esahubble.org/images/archive/category/nebulae (https://esahubble.org/images/archive/category/nebulae). Click on any of the beautiful images in this collection, and you are taken to a page with more information; while looking at these images, you may also want to browse through the slide sequence on the meaning of colors in the Hubble pictures (https://hubblesite.org/contents/articles/the-meaning-of-light-and-color (https://hubblesite.org/contents/articles/the-meaning-of-light-and-color)).

Interstellar Medium Online Tutorial: http://www-ssg.sr.unh.edu/ism/ (http://www-ssg.sr.unh.edu/ism/). Nontechnical introduction to the interstellar medium (ISM) and how we study it; by the University of New Hampshire astronomy department.

Messier Catalog of Nebulae, Clusters, and Galaxies: http://astropixels.com/messier/messiercat.html (http://astropixels.com/messier/messiercat.html). Astronomer Fred Espenak provides the full catalog, with information and images. (The Wikipedia list does something similar: https://en.wikipedia.org/wiki/List_of_Messier_objects (https://en.wikipedia.org/wiki/List_of_Messier_objects).

Nebulae: What Are They?: http://www.universetoday.com/61103/what-is-a-nebula/ (http://www.universetoday.com/61103/what-is-a-nebula/). Concise introduction by Matt Williams.

Videos

Barnard 68: The Hole in the Sky: https://www.youtube.com/watch?v=Faz5rgdQSo4 (https://www.youtube.com/watch?v=Faz5rgdQSo4). About this dark cloud and dark clouds in interstellar space in general (02:08).

Horsehead Nebula in New Light: http://www.esa.int/spaceinvideos/Videos/2013/04/The_Horsehead_Nebula_in_new_light (http://www.esa.int/spaceinvideos/Videos/2013/04/The_Horsehead_Nebula_in_new_light). Tour of the dark nebula in different wavelengths; no audio narration, just music, but explanatory material appears on the screen (03:03).

Hubblecast 65: A Whole New View of the Horsehead Nebula: http://www.spacetelescope.org/videos/heic1307a/ (http://www.spacetelescope.org/videos/heic1307a/). Report on nebulae in general and about the Horsehead specifically, with ESO astronomer Joe Liske (06:03).

Interstellar Reddening: https://www.youtube.com/watch?v=H2M80RAQB6k (https://www.youtube.com/watch?v=H2M80RAQB6k). Video demonstrating how reddening works, with Scott Miller of Penn State; a bit nerdy but useful (03:45).

Hubble Field Guide to the Nebulae: https://www.youtube.com/watch?v=PYRDiR7peLw (https://www.youtube.com/watch?v=PYRDiR7peLw). Video briefly explaining the difference between types of nebulae (04:24).

👥 Collaborative Group Activities

A. The Sun is located in a region where the density of interstellar matter is low. Suppose that instead it were located in a dense cloud 20 light-years in diameter that dimmed the visible light from stars lying outside it by a factor of 100. Have your group discuss how this would have affected the development of civilization on Earth. For example, would it have presented a problem for early navigators?

B. Your group members should look through the pictures in this chapter. How big are the nebulae you see in the images? Are there any clues either in the images or in the captions? Are the clouds they are part of significantly bigger than the nebulae we can see? Why? Suggest some ways that we can determine the sizes of nebulae.

C. How do the members of your group think astronomers are able to estimate the distances of such nebulae in our own Galaxy? (Hint: Look at the images. Can you see anything between us and the nebula in some cases. Review Celestial Distances, if you need to remind yourself about methods of measuring distances.)

D. The text suggests that a tube of air extending from the surface of Earth to the top of the atmosphere contains more atoms than a tube of the same diameter extending from the top of the atmosphere to the edge of the observable universe. Scientists often do what they call "back of the envelope calculations," in which they make very rough approximations just to see whether statements or ideas are true. Try doing such a "quick and dirty" estimate for this statement with your group. What are the steps in comparing the numbers of atoms contained in the two different tubes? What information do you need to make the approximations? Can you find it in this text? And is the statement true?

E. If your astronomy course has involved learning about the solar system before you got to this chapter, have your group discuss where else besides interstellar clouds astronomers have been discovering organic molecules (the chemical building blocks of life). How might the discoveries of such molecules in our own solar system be related to the molecules in the clouds discussed in this chapter?

F. Two stars both have a reddish appearance in telescopes. One star is actually red; the other's light has been reddened by interstellar dust on its way to us. Have your group make a list of the observations you could perform to determine which star is which.

G. You have been asked to give a talk to your little brother's middle school class on astronomy, and you decide to talk about how nature recycles gas and dust. Have your group discuss what images from this book you would use in your talk. In what order? What is the one big idea you would like the students to remember when the class is over?

H. This chapter and the next (on The Birth of Stars) include some of the most beautiful images of nebulae that glow with the light produced when starlight interacts with gas and dust. Have your group select one to four of your favorite such nebulae and prepare a report on them to share with the rest of the class. (Include such things as their location, distance, size, way they are glowing, and what is happening within them.)

Exercises

Review Questions

1. Identify several dark nebulae in photographs in this chapter. Give the figure numbers of the photographs, and specify where the dark nebulae are to be found on them.

2. Why do nebulae near hot stars look red? Why do dust clouds near stars usually look blue?

3. Describe the characteristics of the various kinds of interstellar gas (HII regions, neutral hydrogen clouds, ultra-hot gas clouds, and molecular clouds).

4. Prepare a table listing the different ways in which dust and gas can be detected in interstellar space.

5. Describe how the 21-cm line of hydrogen is formed. Why is this line such an important tool for understanding the interstellar medium?

6. Describe the properties of the dust grains found in the space between stars.

7. Why is it difficult to determine where cosmic rays come from?

8. What causes reddening of starlight? Explain how the reddish color of the Sun's disk at sunset is caused by the same process.

9. Why do molecules, including H_2 and more complex organic molecules, only form inside dark clouds? Why don't they fill all interstellar space?

10. Why can't we use visible light telescopes to study molecular clouds where stars and planets form? Why do infrared or radio telescopes work better?

11. The mass of the interstellar medium is determined by a balance between sources (which add mass) and sinks (which remove it). Make a table listing the major sources and sinks, and briefly explain each one.

12. Where does interstellar dust come from? How does it form?

Thought Questions

13. Figure 20.2 shows a reddish glow around the star Antares, and yet the caption says that is a dust cloud. What observations would you make to determine whether the red glow is actually produced by dust or whether it is produced by an H II region?

14. If the red glow around Antares is indeed produced by reflection of the light from Antares by dust, what does its red appearance tell you about the likely temperature of Antares? Look up the spectral type of Antares in Appendix J. Was your estimate of the temperature about right? In most of the images in this chapter, a red glow is associated with ionized hydrogen. Would you expect to find an H II region around Antares? Explain your answer.

15. Even though neutral hydrogen is the most abundant element in interstellar matter, it was detected first with a radio telescope, not a visible light telescope. Explain why. (The explanation given in Analyzing Starlight for the fact that hydrogen lines are not strong in stars of all temperatures may be helpful.)

16. The terms H II and H_2 are both pronounced "H two." What is the difference in meaning of those two terms? Can there be such a thing as H III?

17. Suppose someone told you that she had discovered H II around the star Aldebaran. Would you believe her? Why or why not?

18. Describe the spectrum of each of the following:
A. starlight reflected by dust,
B. a star behind invisible interstellar gas, and
C. an emission nebula.

19. According to the text, a star must be hotter than about 25,000 K to produce an H II region. Both the hottest white dwarfs and main-sequence O stars have temperatures hotter than 25,000 K. Which type of star can ionize more hydrogen? Why?

20. From the comments in the text about which kinds of stars produce emission nebulae and which kinds are associated with reflection nebulae, what can you say about the temperatures of the stars that produce NGC 1999 (Figure 20.13)?

21. One way to calculate the size and shape of the Galaxy is to estimate the distances to faint stars just from their observed apparent brightnesses and to note the distance at which stars are no longer observable. The first astronomers to try this experiment did not know that starlight is dimmed by interstellar dust. Their estimates of the size of the Galaxy were much too small. Explain why.

22. New stars form in regions where the density of gas and dust is relatively high. Suppose you wanted to search for some recently formed stars. Would you more likely be successful if you observed at visible wavelengths or at infrared wavelengths? Why?

23. Thinking about the topics in this chapter, here is an Earth analogy. In big cities, you can see much farther on days without smog. Why?

24. Stars form in the Milky Way at a rate of about 1 solar mass per year. At this rate, how long would it take for all the interstellar gas in the Milky Way to be turned into stars if there were no fresh gas coming in from outside? How does this compare to the estimated age of the universe, 14 billion years? What do you conclude from this?

25. The 21-cm line can be used not just to find out where hydrogen is located in the sky, but also to determine how fast it is moving toward or away from us. Describe how this might work.

26. Astronomers recently detected light emitted by a supernova that was originally observed in 1572, just reaching Earth now. This light was reflected off a dust cloud; astronomers call such a reflected light a "light echo" (just like reflected sound is called an echo). How would you expect the spectrum of the light echo to compare to that of the original supernova?

27. We can detect 21-cm emission from other galaxies as well as from our own Galaxy. However, 21-cm emission from our own Galaxy fills most of the sky, so we usually see both at once. How can we distinguish the extragalactic 21-cm emission from that arising in our own Galaxy? (Hint: Other galaxies are generally moving relative to the Milky Way.)

28. We have said repeatedly that blue light undergoes more extinction than red light, which is true for visible and shorter wavelengths. Is the same true for X-rays? Look at Figure 20.19. The most dust is in the galactic plane in the middle of the image, and the red color in the image corresponds to the reddest (lowest-energy) light. Based on what you see in the galactic plane, are X-rays experiencing more extinction at redder or bluer colors? You might consider comparing Figure 20.19 to Figure 20.14.

29. Suppose that, instead of being inside the Local Bubble, the Sun were deep inside a giant molecular cloud. What would the night sky look like as seen from Earth at various wavelengths?

30. Suppose that, instead of being inside the Local Bubble, the Sun were inside an H II region. What would the night sky look like at various wavelengths?

Figuring for Yourself

31. A molecular cloud is about 1000 times denser than the average of the interstellar medium. Let's compare this difference in densities to something more familiar. Air has a density of about 1 kg/m^3, so something 1000 times denser than air would have a density of about 1000 kg/m^3. How does this compare to the typical density of water? Of granite? (You can find figures for these densities on the internet.) Is the density difference between a molecular cloud and the interstellar medium larger or smaller than the density difference between air and water or granite?

32. Would you expect to be able to detect an H II region in X-ray emission? Why or why not? (Hint: You might apply Wien's law)

33. Suppose that you gathered a ball of interstellar gas that was equal to the size of Earth (a radius of about 6000 km). If this gas has a density of 1 hydrogen atom per cm^3, typical of the interstellar medium, how would its mass compare to the mass of a bowling ball (5 or 6 kg)? How about if it had the typical density of the Local Bubble, about 0.01 atoms per cm^3? The volume of a sphere is $V = (4/3)\pi R^3$.

34. At the average density of the interstellar medium, 1 atom per cm^3, how big a volume of material must be used to make a star with the mass of the Sun? What is the radius of a sphere this size? Express your answer in light-years.

35. Consider a grain of sand that contains 1 mg of oxygen (a typical amount for a medium-sized sand grain, since sand is mostly SiO_2). How many oxygen atoms does the grain contain? What is the radius of the sphere you would have to spread them out over if you wanted them to have the same density as the interstellar medium, about 1 atom per cm^3? You can look up the mass of an oxygen atom.

36. H II regions can exist only if there is a nearby star hot enough to ionize hydrogen. Hydrogen is ionized only by radiation with wavelengths shorter than 91.2 nm. What is the temperature of a star that emits its maximum energy at 91.2 nm? (Use Wien's law from Radiation and Spectra.) Based on this result, what are the spectral types of those stars likely to provide enough energy to produce H II regions?

37. In the text, we said that the five-times ionized oxygen (OVI) seen in hot gas must have been produced by supernova shocks that heated the gas to millions of degrees, and not by starlight, the way H II is produced. Producing OVI by light requires wavelengths shorter than 10.9 nm. The hottest observed stars have surface temperatures of about 50,000 K. Could they produce OVI?

38. Dust was originally discovered because the stars in certain clusters seemed to be fainter than expected. Suppose a star is behind a cloud of dust that dims its brightness by a factor of 100. Suppose you do not realize the dust is there. How much in error will your distance estimate be? Can you think of any measurement you might make to detect the dust?

39. How would the density inside a cold cloud ($T = 10$ K) compare with the density of the ultra-hot interstellar gas ($T = 10^6$ K) if they were in pressure equilibrium? (It takes a large cloud to be able to shield its interior from heating so that it can be at such a low temperature.) (Hint: In pressure equilibrium, the two regions must have nT equal, where n is the number of particles per unit volume and T is the temperature.) Which region do you think is more suitable for the creation of new stars? Why?

40. The text says that the Local Fluff, which surrounds the Sun, has a temperature of 7500 K and a density 0.1 atom per cm^3. The Local Fluff is embedded in hot gas with a temperature of 10^6 K and a density of about 0.01 atom per cm^3. Are they in equilibrium? (Hint: In pressure equilibrium, the two regions must have nT equal, where n is the number of particles per unit volume and T is the temperature.) What is likely to happen to the Local Fluff?

The Birth of Stars and the Discovery of Planets outside the Solar System

Figure 21.1 Where Stars Are Born. We see a close-up of part of the Carina Nebula taken with the Hubble Space Telescope. This image reveals jets powered by newly forming stars embedded in a great cloud of gas and dust. Parts of the clouds are glowing from the energy of very young stars recently formed within them. (credit: modification of work by NASA, ESA, and M. Livio and the Hubble 20th Anniversary Team (STScI))

Chapter Outline

21.1 Star Formation
21.2 The H–R Diagram and the Study of Stellar Evolution
21.3 Evidence That Planets Form around Other Stars
21.4 Planets beyond the Solar System: Search and Discovery
21.5 Exoplanets Everywhere: What We Are Learning
21.6 New Perspectives on Planet Formation

Thinking Ahead

There are countless suns and countless earths all rotating round their suns in exactly the same way as the planets of our system. We see only the suns because they are the largest bodies and are luminous, but their planets remain invisible to us because they are smaller and non-luminous. . . . The unnumbered worlds in the universe are all similar in form and rank and subject to the same forces and the same laws.
>—Giordano Bruno in *On the Infinite Universe and Worlds* (1584)

Bruno was tried for heresy by the Roman Inquisition and burned at the stake in 1600.

We've discussed stars as nuclear furnaces that convert light elements into heavier ones. A star's nuclear evolution begins when hydrogen is fused into helium, but that can only occur when the core temperature exceeds 10 to 12 million K. Since stars form from cold interstellar material, we must understand how they collapse and eventually reach this "ignition temperature" to explain the birth of stars. Star formation is a continuous process, from the birth of our Galaxy right up to today. We estimate that every year in our Galaxy, on average, three solar masses of interstellar matter are converted into stars. This may sound like a small amount of mass for an object as large as a galaxy, but only three new stars (out of billions in the *Galaxy*) are formed each year.

Do planets orbit other stars or is ours the only planetary system? In the past few decades, new technology has enabled us to answer that question by revealing nearly 4300 exoplanets in over 3200 planetary systems. Even before planets were detected, astronomers had predicted that planetary systems were likely to be byproducts of the star-formation process. In this chapter, we look at how interstellar matter is transformed into stars and planets.

21.1 Star Formation

Learning Objectives

By the end of this section, you will be able to:

> Identify the sometimes-violent processes by which parts of a molecular cloud collapse to produce stars
> Recognize some of the structures seen in images of molecular clouds like the one in Orion
> Explain how the environment of a molecular cloud enables the formation of stars
> Describe how advancing waves of star formation cause a molecular cloud to evolve

As we begin our exploration of how stars are formed, let's review some basics about stars discussed in earlier chapters:

- Stable (main-sequence) stars such as our Sun maintain equilibrium by producing energy through nuclear fusion in their cores. The ability to generate energy by fusion defines a star.
- Each second in the Sun, approximately 600 million tons of hydrogen undergo fusion into helium, with about 4 million tons turning into energy in the process. This rate of hydrogen use means that eventually the Sun (and all other stars) will run out of central fuel.
- Stars come with many different masses, ranging from 1/12 solar masses (M_{Sun}) to roughly 100–200 M_{Sun}. There are far more low-mass than high-mass stars.
- The most massive main-sequence stars (spectral type O) are also the most luminous and have the highest surface temperature. The lowest-mass stars on the main sequence (spectral type M or L) are the least luminous and the coolest.
- A galaxy of stars such as the Milky Way contains enormous amounts of gas and dust—enough to make billions of stars like the Sun.

If we want to find stars still in the process of formation, we must look in places that have plenty of the raw material from which stars are assembled. Since stars are made of gas, we focus our attention (and our telescopes) on the dense and cold clouds of gas and dust that dot the Milky Way (see Figure 21.1 and Figure 21.2).

(a)   (b)

Figure 21.2 Pillars of Dust and Dense Globules in M16. (a) This Hubble Space Telescope image of the central regions of M16 (also known as the Eagle Nebula) shows huge columns of cool gas, (including molecular hydrogen, H2) and dust. These columns are of higher density than the surrounding regions and have resisted evaporation by the ultraviolet radiation from a cluster of hot stars just beyond the upper-right corner of this image. The tallest pillar is about 1 light-year long, and the M16 region is about 7000 light-years away from us. (b) This close-up view of one of the pillars shows some very dense globules, many of which harbor embryonic stars. Astronomers coined the term *evaporating gas globules* (EGGs) for these structures, in part so that they could say we found EGGs inside the Eagle Nebula. It is possible that because these EGGs are exposed to the relentless action of the radiation from nearby hot stars, some may not yet have collected enough material to form a star. (credit a : modification of work by NASA, ESA, and the Hubble Heritage Team (STScI/AURA); credit b: modification of work by NASA, ESA, STScI, J. Hester and P. Scowen (Arizona State University))

Molecular Clouds: Stellar Nurseries

As we saw in Between the Stars: Gas and Dust in Space, the most massive reservoirs of interstellar matter—and some of the most massive objects in the Milky Way Galaxy—are the **giant molecular clouds**. These clouds have cold interiors with characteristic temperatures of only 10–20 K; most of their gas atoms are bound into molecules. These clouds turn out to be the birthplaces of most stars in our Galaxy.

The masses of molecular clouds range from a thousand times the mass of the Sun to about 3 million solar masses. Molecular clouds have a complex filamentary structure, similar to cirrus clouds in Earth's atmosphere, but much less dense. The molecular cloud filaments can be up to 1000 light-years long. Within the clouds are cold, dense regions with typical masses of 50 to 500 times the mass of the Sun; we give these regions the highly technical name *clumps*. Within these clumps, there are even denser, smaller regions called cores. The cores are the embryos of stars. The conditions in these cores—low temperature and high density—are just what is required to make stars. Remember that the essence of the life story of any star is the ongoing competition between two forces: *gravity* and *pressure*. The force of gravity, pulling inward, tries to make a star collapse. Internal pressure produced by the motions of the gas atoms, pushing outward, tries to force the star to expand. When a star is first forming, low temperature (and hence, low pressure) and high density (hence, greater gravitational attraction) both work to give gravity the advantage. In order to form a star—that is, a dense, hot ball of matter capable of starting nuclear reactions deep within—we need a typical core of interstellar atoms and molecules to shrink in radius and increase in density by a factor of nearly 10^{20}. It is the force of gravity that produces this drastic collapse.

The Orion Molecular Cloud

Let's discuss what happens in regions of star formation by considering a nearby site where stars are forming right now. One of the best-studied stellar nurseries is in the constellation of Orion, The Hunter, about 1500 light-years away (Figure 21.3). The pattern of the hunter is easy to recognize by the conspicuous "belt" of three stars that mark his waist. The Orion molecular cloud is much larger than the star pattern and is truly an impressive structure. In its long dimension, it stretches over a distance of about 100 light-years. The total quantity of molecular gas is about 200,000 times the mass of the Sun. Most of the cloud does not glow with visible light but betrays its presence by the radiation that the dusty gas gives off at infrared and radio wavelengths.

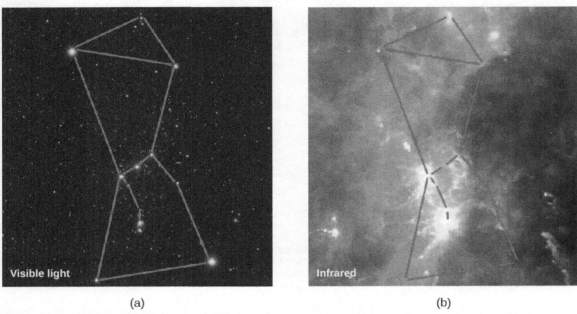

(a)   (b)

Figure 21.3 Orion in Visible and Infrared. (a) The Orion star group was named after the legendary hunter in Greek mythology. Three stars close together in a link mark Orion's belt. The ancients imagined a sword hanging from the belt; the object at the end of the blue line in this sword is the Orion Nebula. (b) This wide-angle, infrared view of the same area was taken with the Infrared Astronomical Satellite. Heated dust clouds dominate in this false-color image, and many of the stars that stood out on part (a) are now invisible. An exception is the cool, red-giant star Betelgeuse, which can be seen as a yellowish point at the left vertex of the blue triangle (at Orion's left armpit). The large, yellow ring to the right of Betelgeuse is the remnant of an exploded star. The infrared image lets us see how large and full of cooler material the Orion molecular cloud really is. On the visible-light image at left, you see only two colorful regions of interstellar matter—the two, bright yellow splotches at the left end of and below Orion's belt. The lower one is the Orion Nebula and the higher one is the region of the Horsehead Nebula. (credit: modification of work by NASA, visible light: Akira Fujii; infrared: Infrared Astronomical Satellite)

The stars in Orion's belt are typically about 5 million years old, whereas the stars near the middle of the "sword" hanging from Orion's belt are only 300,000 to 1 million years old. The region about halfway down the sword where star formation is still taking place is called the Orion Nebula. About 2200 young stars are found in this region, which is only slightly larger than a dozen light-years in diameter. The Orion Nebula also contains a tight cluster of stars called the Trapezium (Figure 21.5). The brightest Trapezium stars can be seen easily with a small telescope.

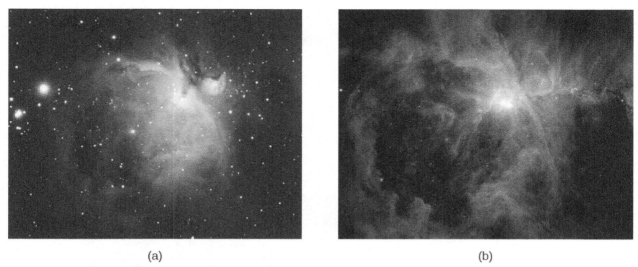

(a)   (b)

Figure 21.4 Orion Nebula. (a) The Orion Nebula is shown in visible light. (b) With near-infrared radiation, we can see more detail within the dusty nebula since infrared can penetrate dust more easily than can visible light. (credit a: modification of work by Filip Lolić; credit b: modification of work by NASA/JPL-Caltech/T. Megeath (University of Toledo, Ohio))

Compare this with our own solar neighborhood, where the typical spacing between stars is about 3 light-years. Only a small number of stars in the Orion cluster can be seen with visible light, but infrared images—which penetrate the dust better—detect the more than 2000 stars that are part of the group (Figure 21.5).

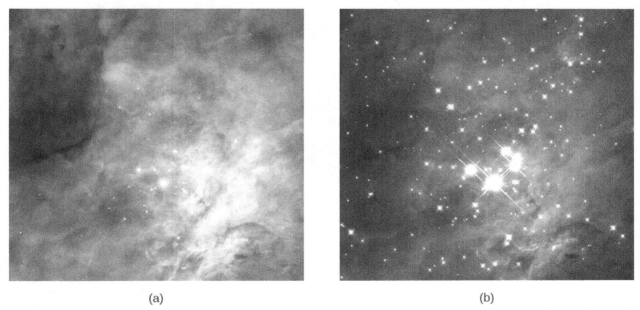

(a)   (b)

Figure 21.5 Central Region of the Orion Nebula. The Orion Nebula harbors some of the youngest stars in the solar neighborhood. At the heart of the nebula is the Trapezium cluster, which includes four very bright stars that provide much of the energy that causes the nebula to glow so brightly. In these images, we see a section of the nebula in (a) visible light and (b) infrared. The four bright stars in the center of the visible-light image are the Trapezium stars. Notice that most of the stars seen in the infrared are completely hidden by dust in the visible-light image. (credit a: modification of work by NASA, C.R. O'Dell and S.K. Wong (Rice University); credit b: modification of work by NASA; K.L. Luhman (Harvard-Smithsonian Center for Astrophysics); and G. Schneider, E. Young, G. Rieke, A. Cotera, H. Chen, M. Rieke, R. Thompson (Steward Observatory, University of Arizona))

Studies of Orion and other star-forming regions show that star formation is not a very efficient process. In the region of the Orion Nebula, about 1% of the material in the cloud has been turned into stars. That is why we still see a substantial amount of gas and dust near the Trapezium stars. The leftover material is eventually heated, either by the radiation and winds from the hot stars that form or by explosions of the most massive stars. (We will see in later chapters that the most massive stars go through their lives very quickly and end by exploding.)

LINK TO LEARNING

Take a journey through the Orion Nebula (https://openstax.org/l/30OriNebula) to view a nice narrated video tour of this region.

Whether gently or explosively, the material in the neighborhood of the new stars is blown away into interstellar space. Older groups or clusters of stars can now be easily observed in visible light because they are no longer shrouded in dust and gas (Figure 21.6).

Figure 21.6 Westerlund 2. This young cluster of stars known as Westerlund 2 formed within the Carina star-forming region about 2 million years ago. Stellar winds and pressure produced by the radiation from the hot stars within the cluster are blowing and sculpting the surrounding gas and dust. The nebula still contains many globules of dust. Stars are continuing to form within the denser globules and pillars of the nebula. This Hubble Space Telescope image includes near-infrared exposures of the star cluster and visible-light observations of the surrounding nebula. Colors in the nebula are dominated by the red glow of hydrogen gas, and blue-green emissions from glowing oxygen. (credit: NASA, ESA, the Hubble Heritage Team (STScI/AURA), A. Nota (ESA/STScI), and the Westerlund 2 Science Team)

Although we do not know what initially caused stars to begin forming in Orion, there is good evidence that the first generation of stars triggered the formation of additional stars, which in turn led to the formation of still more stars (Figure 21.7).

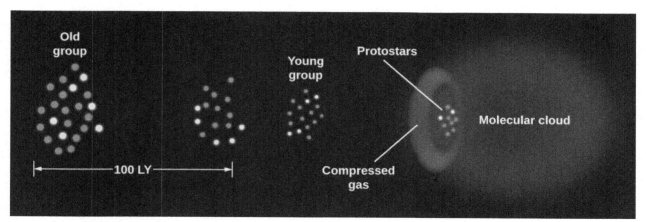

Figure 21.7 Propagating Star Formation. Star formation can move progressively through a molecular cloud. The oldest group of stars lies to the left of the diagram and has expanded because of the motions of individual stars. Eventually, the stars in the group will disperse and no longer be recognizable as a cluster. The youngest group of stars lies to the right, next to the molecular cloud. This group of stars is only 1 to 2 million years old. The pressure of the hot, ionized gas surrounding these stars compresses the material in the nearby edge of the molecular cloud and initiates the gravitational collapse that will lead to the formation of more stars.

The basic idea of triggered star formation is this: when a massive star is formed, it emits a large amount of ultraviolet radiation and ejects high-speed gas in the form of a stellar wind. This injection of energy heats the gas around the stars and causes it to expand. When massive stars exhaust their supply of fuel, they explode, and the energy of the explosion also heats the gas. The hot gases pile into the surrounding cold molecular cloud, compressing the material in it and increasing its density. If this increase in density is large enough, gravity will overcome pressure, and stars will begin to form in the compressed gas. Such a chain reaction—where the brightest and hottest stars of one area become the cause of star formation "next door"—seems to have occurred not only in Orion but also in many other molecular clouds.

There are many molecular clouds that form only (or mainly) low-mass stars. Because low-mass stars do not have strong winds and do not die by exploding, triggered star formation cannot occur in these clouds. There are also stars that form in relative isolation in small cores. Therefore, not all star formation is originally triggered by the death of massive stars. However, there are likely to be other possible triggers, such as spiral density waves and other processes we do not yet understand.

The Birth of a Star

Although regions such as Orion give us clues about how star formation begins, the subsequent stages are still shrouded in mystery (and a lot of dust). There is an enormous difference between the density of a molecular cloud core and the density of the youngest stars that can be detected. Direct observations of this collapse to higher density are nearly impossible for two reasons. First, the dust-shrouded interiors of molecular clouds where stellar births take place cannot be observed with visible light. Second, the timescale for the initial collapse—thousands of years—is very short, astronomically speaking. Since each star spends such a tiny fraction of its life in this stage, relatively few stars are going through the collapse process at any given time. Nevertheless, through a combination of theoretical calculations and the limited observations available, astronomers have pieced together a picture of what the earliest stages of stellar evolution are likely to be.

The first step in the process of creating stars is the formation of dense cores within a clump of gas and dust (Figure 21.8(a)). It is generally thought that all the material for the star comes from the core, the larger structure surrounding the forming star. Eventually, the gravitational force of the infalling gas becomes strong enough to overwhelm the pressure exerted by the cold material that forms the dense cores. The material then undergoes a rapid collapse, and the density of the core increases greatly as a result. During the time a dense core is contracting to become a true star, but before the fusion of protons to produce helium begins, we call the object a **protostar**.

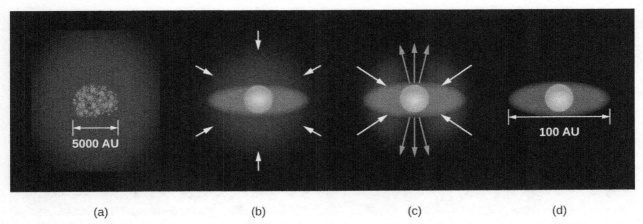

(a)   (b)   (c)   (d)

Figure 21.8 Formation of a Star. (a) Dense cores form within a molecular cloud. (b) A protostar with a surrounding disk of material forms at the center of a dense core, accumulating additional material from the molecular cloud through gravitational attraction. (c) A stellar wind breaks out but is confined by the disk to flow out along the two poles of the star. (d) Eventually, this wind sweeps away the cloud material and halts the accumulation of additional material, and a newly formed star, surrounded by a disk, becomes observable. These sketches are not drawn to the same scale. The diameter of a typical envelope that is supplying gas to the newly forming star is about 5000 AU. The typical diameter of the disk is about 100 AU or slightly larger than the diameter of the orbit of Pluto.

The natural turbulence inside a clump tends to give any portion of it some initial spinning motion (even if it is very slow). As a result, each collapsing core is expected to spin. According to the law of conservation of angular momentum (discussed in the chapter on Orbits and Gravity), a rotating body spins more rapidly as it decreases in size. In other words, if the object can turn its material around a smaller circle, it can move that material more quickly—like a figure skater spinning more rapidly as she brings her arms in tight to her body. This is exactly what happens when a core contracts to form a protostar: as it shrinks, its rate of spin increases.

But all directions on a spinning sphere are not created equal. As the protostar rotates, it is much easier for material to fall right onto the poles (which spin most slowly) than onto the equator (where material moves around most rapidly). Therefore, gas and dust falling in toward the protostar's equator are "held back" by the rotation and form a whirling extended disk around the equator (part b in Figure 21.8). You may have observed this same "equator effect" on the amusement park ride in which you stand with your back to a cylinder that is spun faster and faster. As you spin really fast, you are pushed against the wall so strongly that you cannot possibly fall toward the center of the cylinder. Gas can, however, fall onto the protostar easily from directions away from the star's equator.

The protostar and disk at this stage are embedded in an envelope of dust and gas from which material is still falling onto the protostar. This dusty envelope blocks visible light, but infrared radiation can get through. As a result, in this phase of its evolution, the protostar itself is emitting infrared radiation and so is observable only in the infrared region of the spectrum. Once almost all of the available material has been accreted and the central protostar has reached nearly its final mass, it is given a special name: it is called a *T Tauri star*, named after one of the best studied and brightest members of this class of stars, which was discovered in the constellation of Taurus. (Astronomers have a tendency to name types of stars after the first example they discover or come to understand. It's not an elegant system, but it works.) Only stars with masses less than or similar to the mass of the Sun become T Tauri stars. Massive stars do not go through this stage, although they do appear to follow the formation scenario illustrated in Figure 21.8.

Winds and Jets

Recent observations suggest that T Tauri stars may actually be stars in a middle stage between protostars and hydrogen-fusing stars such as the Sun. High-resolution infrared images have revealed jets of material as well as *stellar winds* coming from some T Tauri stars, proof of interaction with their environment. A **stellar wind** consists mainly of protons (hydrogen nuclei) and electrons streaming away from the star at speeds of a few hundred kilometers per second (several hundred thousand miles per hour). When the wind first starts up, the disk of material around the star's equator blocks the wind in its direction. Where the wind particles *can* escape most effectively is in the direction of the star's poles.

Astronomers have actually seen evidence of these beams of particles shooting out in opposite directions from the polar regions of newly formed stars. In many cases, these beams point back to the location of a protostar that is still so completely shrouded in dust that we cannot yet see it (Figure 21.9).

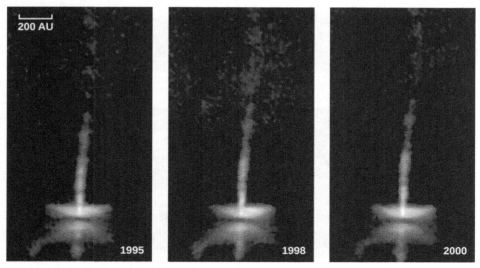

Figure 21.9 Gas Jets Flowing away from a Protostar. Here we see the neighborhood of a protostar, known to us as HH 34 because it is a Herbig-Haro object. The star is about 450 light-years away and only about 1 million years old. Light from the star itself is blocked by a disk, which is larger than 60 billion kilometers in diameter and is seen almost edge-on. Jets are seen emerging perpendicular to the disk. The material in these jets is flowing outward at speeds up to 580,000 kilometers per hour. The series of three images shows changes during a period of 5 years. Every few months, a compact clump of gas is ejected, and its motion outward can be followed. The changes in the brightness of the disk may be due to motions of clouds within the disk that alternately block some of the light and then let it through. This image corresponds to the stage in the life of a protostar shown in part (c) of Figure 21.8. (credit: modification of work by Hubble Space Telescope, NASA, ESA)

On occasion, the jets of high-speed particles streaming away from the protostar collide with a somewhat-denser lump of gas nearby, excite its atoms, and cause them to emit light. These glowing regions, each of which is known as a **Herbig-Haro (HH) object** after the two astronomers who first identified them, allow us to trace the progress of the jet to a distance of a light-year or more from the star that produced it. Figure 21.10 shows two spectacular images of HH objects.

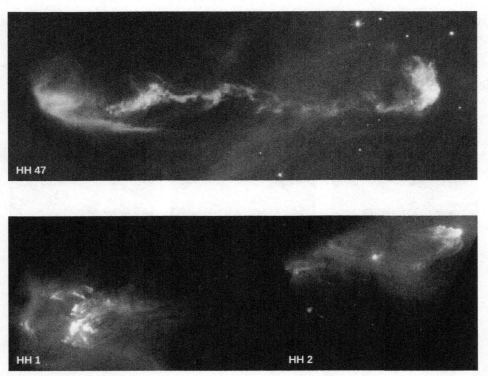

Figure 21.10 Outflows from Protostars. These images were taken with the Hubble Space Telescope and show jets flowing outward from newly formed stars. In the HH47 image, a protostar 1500 light-years away (invisible inside a dust disk at the left edge of the image) produces a very complicated jet. The star may actually be wobbling, perhaps because it has a companion. Light from the star illuminates the white region at the left because light can emerge perpendicular to the disk (just as the jet does). At right, the jet is plowing into existing clumps of interstellar gas, producing a shock wave that resembles an arrowhead. The HH1/2 image shows a double-beam jet emanating from a protostar (hidden in a dust disk in the center) in the constellation of Orion. Tip to tip, these jets are more than 1 light-year long. The bright regions (first identified by Herbig and Haro) are places where the jet is a slamming into a clump of interstellar gas and causing it to glow. (credit "HH 47": modification of work by NASA, ESA, and P. Hartigan (Rice University); credit "HH 1 and HH 2: modification of work by J. Hester, WFPC2 Team, NASA)

The wind from a forming star will ultimately sweep away the material that remains in the obscuring envelope of dust and gas, leaving behind the naked disk and protostar, which can then be seen with visible light. We should note that at this point, the protostar itself is still contracting slowly and has not yet reached the main-sequence stage on the H–R diagram (a concept introduced in the chapter The Stars: A Celestial Census). The disk can be detected directly when observed at infrared wavelengths or when it is seen silhouetted against a bright background (Figure 21.11).

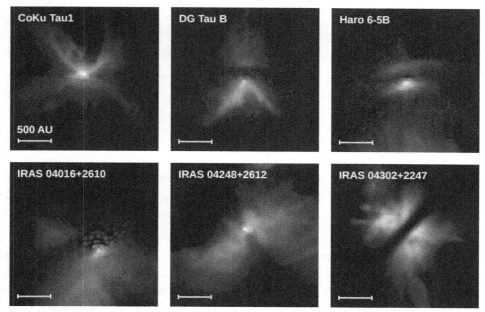

Figure 21.11 Disks around Protostars. These Hubble Space Telescope infrared images show disks around young stars in the constellation of Taurus, in a region about 450 light-years away. In some cases, we can see the central star (or stars—some are binaries). In other cases, the dark, horizontal bands indicate regions where the dust disk is so thick that even infrared radiation from the star embedded within it cannot make its way through. The brightly glowing regions are starlight reflected from the upper and lower surfaces of the disk, which are less dense than the central, dark regions. (Credit: modification of work by D. Padgett (IPAC/Caltech), W. Brandner (IPAC), K. Stapelfeldt (JPL) and NASA)

This description of a protostar surrounded by a rotating disk of gas and dust sounds very much like what happened in our solar system when the Sun and planets formed. Indeed, one of the most important discoveries from the study of star formation in the last decade of the twentieth century was that disks are an inevitable byproduct of the process of creating stars. The next questions that astronomers set out to answer was: will the disks around protostars also form planets? And if so, how often? We will return to these questions later in this chapter.

To keep things simple, we have described the formation of single stars. Many stars, however, are members of binary or triple systems, where several stars are born together. In this case, the stars form in nearly the same way. Widely separated binaries may each have their own disk; close binaries may share a single disk.

21.2 | The H–R Diagram and the Study of Stellar Evolution

Learning Objectives

By the end of this section, you will be able to:

> Determine the age of a protostar using an H–R diagram and the protostar's luminosity and temperature

> Explain the interplay between gravity and pressure, and how the contracting protostar changes its position in the H–R diagram as a result

One of the best ways to summarize all of these details about how a star or protostar changes with time is to use a Hertzsprung-Russell (H–R) diagram. Recall from The Stars: A Celestial Census that, when looking at an H–R diagram, the temperature (the horizontal axis) is plotted increasing toward the left. As a star goes through the stages of its life, its luminosity and temperature change. Thus, its position on the H–R diagram, in which luminosity is plotted against temperature, also changes. As a star ages, we must replot it in different places on the diagram. Therefore, astronomers often speak of a star *moving* on the H–R diagram, or of its evolution tracing out a path on the diagram. In this context, "tracing out a path" has nothing to do with the star's motion through space; this is just a shorthand way of saying that its temperature and luminosity change as it evolves.

LINK TO LEARNING

Watch an animation (https://openstax.org/l/30aniomelen) of the stars in the Omega Centauri cluster as they rearrange according to luminosity and temperature, forming a Hertzsprung-Russell (H–R) diagram.

To estimate just how much the luminosity and temperature of a star change as it ages, we must resort to calculations. Theorists compute a series of models for a star, with each successive model representing a later point in time. Stars may change for a variety of reasons. Protostars, for example, change in size because they are contracting, and their temperature and luminosity change as they do so. After nuclear fusion begins in the star's core (see Stars from Adolescence to Old Age), main-sequence stars change because they are using up their nuclear fuel.

Given a model that represents a star at one stage of its evolution, we can calculate what it will be like at a slightly later time. At each step, the model predicts the luminosity and size of the star, and from these values, we can figure out its surface temperature. A series of points on an H–R diagram, calculated in this way, allows us to follow the life changes of a star and hence is called its *evolutionary track*.

Evolutionary Tracks

Let's now use these ideas to follow the evolution of protostars that are on their way to becoming main-sequence stars. The evolutionary tracks of newly forming stars with a range of stellar masses are shown in Figure 21.12. These young stellar objects are not yet producing energy by nuclear reactions, but they derive energy from gravitational contraction—through the sort of process proposed for the Sun by Helmhotz and Kelvin in this last century (see the chapter on The Sun: A Nuclear Powerhouse).

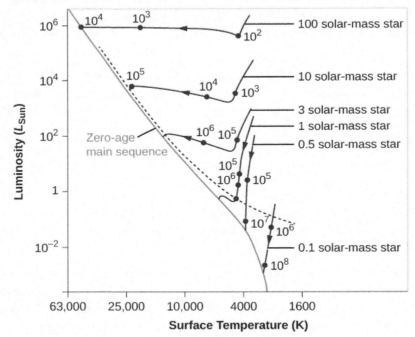

Figure 21.12 Evolutionary Tracks for Contracting Protostars. Tracks are plotted on the H–R diagram to show how stars of different masses change during the early parts of their lives. The number next to each dark point on a track is the rough number of years it takes an embryo star to reach that stage (the numbers are the result of computer models and are therefore not well known). Note that the surface temperature (K) on the horizontal axis increases toward the left. You can see that the more mass a star has, the shorter time it takes to go through each stage. Stars above the dashed line are typically still surrounded by infalling material and are hidden by it.

Initially, a protostar remains fairly cool with a very large radius and a very low density. It is transparent to infrared radiation, and the heat generated by gravitational contraction can be radiated away freely into space. Because heat builds up slowly inside the protostar, the gas pressure remains low, and the outer layers fall almost unhindered toward the center. Thus, the protostar undergoes very rapid collapse, a stage that corresponds to the roughly vertical lines at the right of Figure 21.12. As the star shrinks, its surface area gets smaller, and so its total luminosity decreases. The rapid contraction stops only when the protostar becomes dense and opaque enough to trap the heat released by gravitational contraction.

When the star begins to retain its heat, the contraction becomes much slower, and changes inside the contracting star keep the luminosity of stars like our Sun roughly constant. The surface temperatures start to build up, and the star "moves" to the left in the H–R diagram. Stars first become visible only after the stellar wind described earlier clears away the surrounding dust and gas. This can happen during the rapid-contraction phase for low-mass stars, but high-mass stars remain shrouded in dust until they end their early phase of gravitational contraction (see the dashed line in Figure 21.12).

To help you keep track of the various stages that stars go through in their lives, it can be useful to compare the development of a star to that of a human being. (Clearly, you will not find an exact correspondence, but thinking through the stages in human terms may help you remember some of the ideas we are trying to emphasize.) Protostars might be compared to human embryos—as yet unable to sustain themselves but drawing resources from their environment as they grow. Just as the birth of a child is the moment it is called upon to produce its own energy (through eating and breathing), so astronomers say that a star is born when it is able to sustain itself through nuclear reactions (by making its own energy.)

When the star's central temperature becomes high enough (about 12 million K) to fuse hydrogen into helium, we say that the star has reached the main sequence (a concept introduced in The Stars: A Celestial Census). It is now a full-fledged star, more or less in equilibrium, and its rate of change slows dramatically. Only the gradual depletion of hydrogen as it is transformed into helium in the core slowly changes the star's properties.

The mass of a star determines exactly where it falls on the main sequence. As Figure 21.12 shows, massive stars on the main sequence have high temperatures and high luminosities. Low-mass stars have low temperatures and low luminosities.

Objects of extremely low mass never achieve high-enough central temperatures to ignite nuclear reactions. The lower end of the main sequence stops where stars have a mass just barely great enough to sustain nuclear reactions at a sufficient rate to stop gravitational contraction. This critical mass is calculated to be about 0.075 times the mass of the Sun. As we discussed in the chapter on Analyzing Starlight, objects below this critical mass are called either brown dwarfs or planets. At the other extreme, the upper end of the main sequence terminates at the point where the energy radiated by the newly forming massive star becomes so great that it halts the accretion of additional matter. The upper limit of stellar mass is between 100 and 200 solar masses.

Evolutionary Timescales

How long it takes a star to form depends on its mass. The numbers that label the points on each track in Figure 21.12 are the times, in years, required for the embryo stars to reach the stages we have been discussing. Stars of masses much higher than the Sun's reach the main sequence in a few thousand to a million years. The Sun required millions of years before it was born. Tens of millions of years are required for stars of lower mass to evolve to the lower main sequence. (We will see that this turns out to be a general principle: massive stars go through *all* stages of evolution faster than low-mass stars do.)

We will take up the subsequent stages in the life of a star in Stars from Adolescence to Old Age, examining what happens after stars arrive in the main sequence and begin a "prolonged adolescence" and "adulthood" of fusing hydrogen to form helium. But now we want to examine the connection between the formation of stars and planets.

21.3 Evidence That Planets Form around Other Stars

Learning Objectives

By the end of this section, you will be able to:

> Trace the evolution of dust surrounding a protostar, leading to the development of rocky planets and gas giants

> Estimate the timescale for growth of planets using observations of the disks surrounding young stars

> Evaluate evidence for planets around forming stars based on the structures seen in images of the circumstellar dust disks

Having developed on a planet and finding it essential to our existence, we have a special interest in how planets fit into the story of star formation. Yet planets outside the solar system are extremely difficult to detect. Recall that we see planets in our own system only because they reflect sunlight and are close by. When we look to the other stars, we find that the amount of light a planet reflects is a depressingly tiny fraction of the light its star gives off. Furthermore, from a distance, planets are lost in the glare of their much-brighter parent stars.

Disks around Protostars: Planetary Systems in Formation

It is a lot easier to detect the spread-out raw material from which planets might be assembled than to detect planets after they are fully formed. From our study of the solar system, we understand that planets form by the gathering together of gas and dust particles in orbit around a newly created star. Each dust particle is heated by the young protostar and radiates in the infrared region of the spectrum. Before any planets form, we can detect such radiation from all of the spread-out individual dust particles that are destined to become parts of planets. We can also detect the silhouette of the disk if it blocks bright light coming from a source behind it (Figure 21.13).

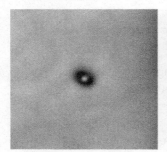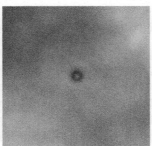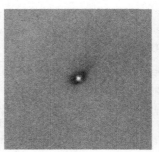

Figure 21.13 Disks around Protostars. These Hubble Space Telescope images show four disks around young stars in the Orion Nebula. The dark, dusty disks are seen silhouetted against the bright backdrop of the glowing gas in the nebula. The size of each image is about 30 times the diameter of our planetary system; this means the disks we see here range in size from two to eight times the orbit of Pluto. The red glow at the center of each disk is a young star, no more than a million years old. These images correspond to the stage in the life of a protostar shown in part (d) of Figure 21.8. (credit: modification of work by Mark McCaughrean (Max-Planck-Institute for Astronomy), C. Robert O'Dell (Rice University), and NASA)

Once the dust particles gather together and form a few planets (and maybe some moons), the overwhelming majority of the dust is hidden in the interiors of the planets where we cannot see it. All we can now detect is the radiation from the outside surfaces, which cover a drastically smaller area than the huge, dusty disk from which they formed. The amount of infrared radiation is therefore greatest before the dust particles combine into planets. For this reason, our search for planets begins with a search for infrared radiation from the material required to make them.

A disk of gas and dust appears to be an essential part of star formation. Observations show that nearly all very young protostars have disks and that the disks range in size from 10 to 1000 AU. (For comparison, the average diameter of the orbit of Pluto, which can be considered the rough size of our own planetary system, is 80 AU, whereas the outer diameter of the Kuiper belt of smaller icy bodies is about 100 AU.) The mass contained in these disks is typically 1–10% of the mass of our own Sun, which is more than the mass of all the planets in our solar system put together. Such observations already demonstrate that a large fraction of stars begin their lives with enough material in the right place to form a planetary system.

The Timing of Planet Formation and Growth

We can use observations of how the disks change with time to estimate how long it takes for planets to form. If we measure the temperature and luminosity of a protostar, then, as we saw, we can place it in an H–R diagram like the one shown in Figure 21.12. By comparing the real star with our models of how protostars should evolve with time, we can estimate its age. We can then look at how the disks we observe change with the ages of the stars that they surround.

What such observations show is that if a protostar is less than about 1 to 3 million years old, its disk extends all the way from very close to the surface of the star out to tens or hundreds of AU away. In older protostars, we find disks with outer parts that still contain large amounts of dust, but the inner regions have lost most of their dust. In these objects, the disk looks like a donut, with the protostar centered in its hole. The inner, dense parts of most disks have disappeared by the time the stars are 10 million years old (Figure 21.14).

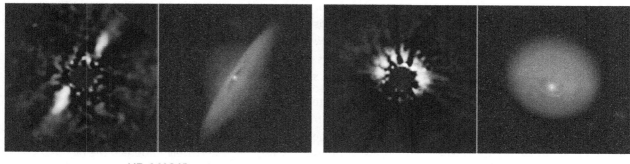

HD 141943   HD 191089

Figure 21.14 Protoplanetary Disks around Two Stars. The left view of each star shows infrared observations by the Hubble Space Telescope of their protoplanetary disks. The central star is much brighter than the surrounding disk, so the instrument includes a coronagraph, which has a small shield that blocks the light of the central star but allows the surrounding disk to be imaged. The right image of each star shows models of the disks based on the observations. The star HD 141943 has an age of about 17 million years, while HD 191089 is about 12 million years old. (credit: modification of work by NASA, ESA, R. Soummer and M. Perrin (STScI), L. Pueyo (STScI/Johns Hopkins University), C. Chen and D. Golimowski (STScI), J.B. Hagan (STScI/Purdue University), T. Mittal (University of California, Berkeley/Johns Hopkins University), E. Choquet, M. Moerchen, and M. N'Diaye (STScI), A. Rajan (Arizona State University), S. Wolff (STScI/Purdue University), J. Debes and D. Hines (STScI), and G. Schneider (Steward Observatory/University of Arizona))

Calculations show that the formation of one or more planets could produce such a donut-like distribution of dust. Suppose a planet forms a few AU away from the protostar, presumably due to the gathering together of matter from the disk. As the planet grows in mass, the process clears out a dust-free region in its immediate neighborhood. Calculations also show that any small dust particles and gas that were initially located in the region between the protostar and the planet, and that are not swept up by the planet, will then fall onto the star very quickly in about 50,000 years.

Matter lying outside the planet's orbit, in contrast, is prevented from moving into the hole by the gravitational forces exerted by the planet. (We saw something similar in Saturn's rings, where the action of the shepherd moons keeps the material near the edge of the rings from spreading out.) If the formation of a planet is indeed what produces and sustains holes in the disks that surround very young stars, then planets must form in 3 to 30 million years. This is a short period compared with the lifetimes of most stars and shows that the formation of planets may be a quick byproduct of the birth of stars.

Calculations show that *accretion* can drive the rapid growth of planets—small, dust-grain-size particles orbiting in the disk collide and stick together, with the larger collections growing more rapidly as they attract and capture smaller ones. Once these clumps grow to about 10 centimeters in size or so, they enter a perilous stage in their development. At that size, unless they can grow to larger than about 100 meters in diameter, they are subject to drag forces produced by friction with the gas in the disk—and their orbits can rapidly decay, plunging them into the host star. Therefore, these bodies must rapidly grow to nearly 1 kilometer in size in diameter to avoid a fiery fate. At this stage, they are considered planetesimals (the small chunks of solid matter—ice and dust particles—that you learned about in Other Worlds: An Introduction to the Solar System). Once they survive to those sizes, the largest survivors will continue to grow by accreting smaller planetesimals; ultimately, this process results in a few large planets.

If the growing planets reach a mass bigger than about 10 times the mass of Earth, their gravity is strong enough to capture and hold on to hydrogen gas that remains in the disk. At that point, they will grow in mass and radius rapidly, reaching giant planet dimensions. However, to do so requires that the rapidly evolving central star hasn't yet driven away the gas in the disk with its increasingly vigorous wind (see the earlier section on Star Formation). From observations, we see that the disk can be blown away within 10 million years, so growth of a giant planet must also be a very fast process, astronomically speaking.

Debris Disks and Shepherd Planets

The dust around newly formed stars is gradually either incorporated into the growing planets in the newly forming planetary system or ejected through gravitational interactions with the planets into space. The dust will disappear after about 30 million years unless the disk is continually supplied with new material. Local comets and asteroids are the most likely sources of new dust. As the planet-size bodies grow, they stir up the orbits of smaller objects in the area. These small bodies collide at high speeds, shatter, and produce tiny particles of silicate dust and ices that can keep the disk supplied with the debris from these collisions.

Over several hundred million years, the comets and asteroids will gradually be reduced in number, the frequency of collisions will go down, and the supply of fresh dust will diminish. Remember that the heavy bombardment in the early solar system ended when the Sun was only about 500 million years old. Observations show that the dusty "debris disks" around stars also become largely undetectable by the time the stars reach an age of 400 to 500 million years. It is likely, however, that some small amount of cometary material will remain in orbit, much like our Kuiper belt, a flattened disk of comets outside the orbit of Neptune.

In a young planetary system, even if we cannot see the planets directly, the planets can concentrate the dust particles into clumps and arcs that are much larger than the planets themselves and more easily imaged. This is similar to how the tiny moons of Saturn shepherd the particles in the rings and produce large arcs and structures in Saturn's rings.

Debris disks—many with just such clumps and arcs—have now been found around many stars, such as HL Tau, located about 450 light-years from Earth in the constellation Taurus (Figure 21.15). In some stars, the brightness of the rings varies with position; around other stars, there are bright arcs and gaps in the rings. The brightness indicates the relative concentration of dust, since what we are seeing is infrared (heat radiation) from the dust particles in the rings. More dust means more radiation.

Figure 21.15 Dust Ring around a Young Star. This image was made by ALMA (the Atacama Large Millimeter/Submillimeter Array) at a wavelength of 1.3 millimeters and shows the young star HL Tau and its protoplanetary disk. It reveals multiple rings and gaps that indicate the presence of emerging planets, which are sweeping their orbits clear of dust and gas. (credit: modification of work by ALMA (ESO/NAOJ/NRAO))

LINK TO LEARNING

Watch a short video clip (https://openstax.org/l/30vidNRAOdir) of the director of NRAO (National Radio Astronomy Observatory) describing the high-resolution observations of the young star HL Tau. While you're there, watch an artist's animation of a protoplanetary disk to see newly formed planets traveling around a host (parent) star.

21.4 | Planets beyond the Solar System: Search and Discovery

Learning Objectives

By the end of this section, you will be able to:

> Describe the orbital motion of planets in our solar system using Kepler's laws
> Compare the indirect and direct observational techniques for exoplanet detection

For centuries, astronomers have dreamed of finding planets around other stars, including other planets like Earth. Direct observations of such distant planets are very difficult, however. You might compare a planet orbiting a star to a mosquito flying around one of those giant spotlights at a shopping center opening. From close up, you might spot the mosquito. But imagine viewing the scene from some distance away—say, from an airplane. You could see the spotlight just fine, but what are your chances of catching the mosquito in that light? Instead of making direct images, astronomers have relied on indirect observations and have now succeeded in detecting a multitude of planets around other stars.

In 1995, after decades of effort, we found the first such **exoplanet** (a planet outside our solar system) orbiting a main-sequence star, and today we know that most stars form with planets. This is an example of how persistence and new methods of observation advance the knowledge of humanity. By studying exoplanets, astronomers hope to better understand our solar system in context of the rest of the universe. For instance, how does the arrangement of our solar system compare to planetary systems in the rest of the universe?

What do exoplanets tell us about the process of planet formation? And how does knowing the frequency of exoplanets influence our estimates of whether there is life elsewhere?

Searching for Orbital Motion

Most exoplanet detections are made using techniques where we observe the *effect* that the planet exerts on the host star. For example, the gravitational tug of an unseen planet will cause a small wobble in the host star. Or, if its orbit is properly aligned, a planet will periodically cross in front of the star, causing the brightness of the star to dim.

To understand how a planet can move its host star, consider a single Jupiter-like planet. Both the planet and the star actually revolve about their *common center of mass*. Remember that gravity is a mutual attraction. The star and the planet each exert a force on the other, and we can find a stable point, the center of mass, between them about which both objects move. The smaller the mass of a body in such a system, the larger its orbit. A massive star barely swings around the center of mass, while a low-mass planet makes a much larger "tour."

Suppose the planet is like Jupiter and has a mass about one-thousandth that of its star; in this case, the size of the star's orbit is one-thousandth the size of the planet's. To get a sense of how difficult observing such motion might be, let's see how hard Jupiter would be to detect in this way from the distance of a nearby star. Consider an alien astronomer trying to observe our own system from Alpha Centauri, the closest star system to our own (about 4.3 light-years away). There are two ways this astronomer could try to detect the orbital motion of the Sun. One way would be to look for changes in the Sun's position on the sky. The second would be to use the Doppler effect to look for changes in its velocity. Let's discuss each of these in turn.

The diameter of Jupiter's apparent orbit viewed from Alpha Centauri is 10 seconds of arc, and that of the Sun's orbit is 0.010 seconds of arc. (Remember, 1 second of arc is 1/3600 degree.) If they could measure the apparent position of the Sun (which is bright and easy to detect) to sufficient precision, they would describe an orbit of diameter 0.010 seconds of arc with a period equal to that of Jupiter, which is 12 years.

In other words, if they watched the Sun for 12 years, they would see it wiggle back and forth in the sky by this minuscule fraction of a degree. From the observed motion and the period of the "wiggle," they could deduce the mass of Jupiter and its distance using Kepler's laws. (To refresh your memory about these laws, see the chapter on Orbits and Gravity.)

Measuring positions in the sky this accurately is extremely difficult, and so far, astronomers have not made any confirmed detections of planets using this technique. However, we have been successful in using spectrometers to measure the changing velocity of stars with planets around them.

As the star and planet orbit each other, part of their motion will be in our line of sight (toward us or away from us). Such motion can be measured using the *Doppler effect* and the star's spectrum. As the star moves back and forth in orbit around the system's center of mass in response to the gravitational tug of an orbiting planet, the lines in its spectrum will shift back and forth.

Let's again consider the example of the Sun. Its *radial velocity* (motion toward or away from us) changes by about 13 meters per second with a period of 12 years because of the gravitational pull of Jupiter. This corresponds to about 30 miles per hour, roughly the speed at which many of us drive around town. Detecting motion at this level in a star's spectrum presents an enormous technical challenge, but several groups of astronomers around the world, using specialized spectrographs designed for this purpose, have succeeded. Note that the change in speed does not depend on the distance of the star from the observer. Using the Doppler effect to detect planets will work at any distance, as long as the star is bright enough to provide a good spectrum and a large telescope is available to make the observations (Figure 21.16).

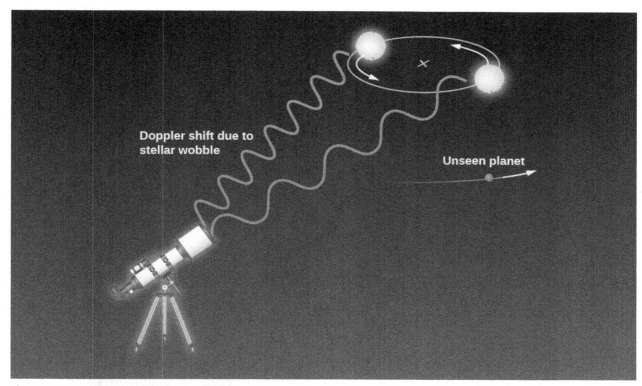

Figure 21.16 Doppler Method of Detecting Planets. The motion of a star around a common center of mass with an orbiting planet can be detected by measuring the changing speed of the star. When the star is moving away from us, the lines in its spectrum show a tiny redshift; when it is moving toward us, they show a tiny blueshift. The change in color (wavelength) has been exaggerated here for illustrative purposes. In reality, the Doppler shifts we measure are extremely small and require sophisticated equipment to be detected.

The first successful use of the Doppler effect to find a planet around another star was in 1995. Michel Mayor and Didier Queloz of the Geneva Observatory (Figure 21.17) used this technique to find a planet orbiting a star resembling our Sun called 51 Pegasi, about 40 light-years away. (The star can be found in the sky near the great square of Pegasus, the flying horse of Greek mythology, one of the easiest-to-find star patterns.) To everyone's surprise, the planet takes a mere 4.2 days to orbit around the star. (Remember that Mercury, the innermost planet in our solar system, takes 88 days to go once around the Sun, so 4.2 days seems fantastically short.)

Figure 21.17 Planet Discoverers. In 1995, Didier Queloz and Michel Mayor of the Geneva Observatory were the first to discover a planet around a regular star (51 Pegasi), for which they received the 2019 Nobel Prize in physics. They are seen here at an observatory in Chile where they are continuing their planet hunting. (credit: Weinstein/Ciel et Espace Photos)

Mayor and Queloz's findings mean the planet must be very close to 51 Pegasi, circling it about 7 million kilometers away (Figure 21.18). At that distance, the energy of the star should heat the planet's surface to a temperature of a few thousand degrees Celsius (a bit hot for future tourism). From its motion, astronomers calculate that it has at least half the mass of Jupiter[1], making it clearly a jovian and not a terrestrial-type planet.

Figure 21.18 Hot Jupiter. Artist Greg Bacon painted this impression of a hot, Jupiter-type planet orbiting close to a sunlike star. The artist shows bands on the planet like Jupiter, but we only estimate the mass of most hot, Jupiter-type planets from the Doppler method and don't know what conditions on the planet are like. (credit: ESO)

Since that initial planet discovery, the rate of progress has been breathtaking. Almost a thousand giant planets have been discovered using the Doppler technique. Many of these giant planets are orbiting close to their stars—astronomers have called these *hot Jupiters*.

The existence of giant planets so close to their stars was a surprise, and these discoveries have forced us to rethink our ideas about how planetary systems form. But for now, bear in mind that the Doppler-shift method—which relies on the pull of a planet making its star "wiggle" back and forth around the center of mass—is most effective at finding planets that are both close to their stars and massive. These planets cause the biggest "wiggles" in the motion of their stars and the biggest Doppler shifts in the spectrum. Plus, they will be found sooner, since astronomers like to monitor the star for at least one full orbit (and perhaps more) and hot Jupiters take the shortest time to complete their orbit.

So if such planets exist, we would expect to be finding this type first. Scientists call this a *selection effect*—where our technique of discovery selects certain kinds of objects as "easy finds." As an example of a selection effect in everyday life, imagine you decide you are ready for a new romantic relationship in your life. To begin with, you only attend social events on campus, all of which require a student ID to get in. Your selection of possible partners will then be limited to students at your college. That may not give you as diverse a group to choose from as you want. In the same way, when we first used the Doppler technique, it selected massive planets close to their stars as the most likely discoveries. As we spend longer times watching target stars and as our ability to measure smaller Doppler shifts improves, this technique can reveal more distant and less massive planets too.

---

1 The Doppler method only allows us to find the minimum mass of a planet. To determine the exact mass using the Doppler shift and Kepler's laws, we must also have the angle at which the planet's orbit is oriented to our view—something we don't have any independent way of knowing in most cases. Still, if the minimum mass is half of Jupiter's, the actual mass can only be larger than that, and we are sure that we are dealing with a jovian planet.

Transiting Planets

The second method for indirect detection of exoplanets is based not on the motion of the star but on its brightness. When the orbital plane of the planet is tilted or inclined so that it is viewed edge-on, we will see the planet cross in front of the star once per orbit, causing the star to dim slightly; this event is known as **transit**. Figure 21.19 shows a sketch of the transit at three time steps: (1) out of transit, (2) the start of transit, and (3) full transit, along with a sketch of the light curve, which shows the drop in the brightness of the host star. The amount of light blocked—the depth of the transit—depends on the area of the planet (its size) compared to the star. If we can determine the size of the star, the transit method tells us the size of the planet.

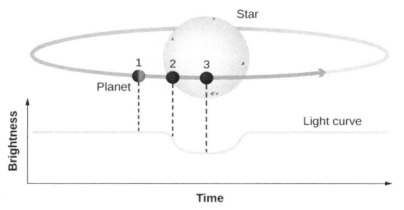

Figure 21.19 Planet Transits. As the planet transits, it blocks out some of the light from the star, causing a temporary dimming in the brightness of the star. The top figure shows three moments during the transit event and the bottom panel shows the corresponding light curve: (1) out of transit, (2) transit ingress, and (3) the full drop in brightness.

The interval between successive transits is the length of the year for that planet, which can be used (again using Kepler's laws) to find its distance from the star. Larger planets like Jupiter block out more starlight than small earthlike planets, making transits by giant planets easier to detect, even from ground-based observatories. But by going into space, above the distorting effects of Earth's atmosphere, the transit technique has been extended to exoplanets as small as Mars.

EXAMPLE 21.1

Transit Depth
In a transit, the planet's circular disk blocks the light of the star's circular disk. The area of a circle is πR^2. The amount of light the planet blocks, called the transit depth, is then given by

$$\frac{\pi R^2_{\text{planet}}}{\pi R^2_{\text{star}}} = \frac{R^2_{\text{planet}}}{R^2_{\text{star}}} = \left(\frac{R_{\text{planet}}}{R_{\text{star}}}\right)^2$$

Now calculate the transit depth for a star the size of the Sun with a gas giant planet the size of Jupiter.

Solution

The radius of Jupiter is 71,400 km, while the radius of the Sun is 695,700 km. Substituting into the equation,

we get $\left(\dfrac{R_{\text{planet}}}{R_{\text{star}}} \right)^2 = \left(\dfrac{71{,}400 \text{ km}}{695{,}700 \text{ km}} \right)^2 = 0.01$ or 1%, which can easily be detected with the instruments on board the Kepler spacecraft.

Check Your Learning

What is the transit depth for a star half the size of the Sun with a much smaller planet, like the size of Earth?

Answer:

The radius of Earth is 6371 km. Therefore,

$$\left(\frac{R_{\text{planet}}}{R_{\text{star}}} \right)^2 = \left(\frac{6371 \text{ km}}{695{,}700/2 \text{ km}} \right)^2 = \left(\frac{6371 \text{ km}}{347{,}850 \text{ km}} \right)^2 = 0.0003, \text{ or significantly less than 1\%.}$$

The Doppler method allows us to estimate the mass of a planet. If the same object can be studied by both the Doppler and transit techniques, we can measure both the mass and the size of the exoplanet. This is a powerful combination that can be used to derive the average density (mass/volume) of the planet. In 1999, using measurements from ground-based telescopes, the first transiting planet was detected orbiting the star HD 209458. The planet transits its parent star for about 3 hours every 3.5 days as we view it from Earth. Doppler measurements showed that the planet around HD 209458 has about 70% the mass of Jupiter, but its radius is about 35% larger than Jupiter's. This was the first case where we could determine what an exoplanet was made of—with that mass and radius, HD 209458 must be a gas and liquid world like Jupiter or Saturn.

It is even possible to learn something about the planet's atmosphere. When the planet passes in front of HD 209458, the atoms in the planet's atmosphere absorb starlight. Observations of this absorption were first made at the wavelengths of yellow sodium lines and showed that the atmosphere of the planet contains sodium; now, other elements can be measured as well.

LINK TO LEARNING

Try a transit simulator (https://openstax.org/l/30transimul) that demonstrates how a planet passing in front of its parent star can lead to the planet's detection. Follow the instructions to run the animation on your computer.

Transiting planets reveal such a wealth of information that the French Space Agency (CNES) and the European Space Agency (ESA) launched the CoRoT space telescope in 2007 to detect transiting exoplanets. CoRoT discovered 32 transiting exoplanets, including the first transiting planet with a size and density similar to Earth. In 2012, the spacecraft suffered an onboard computer failure, ending the mission. Meanwhile, NASA built a much more powerful transit observatory called Kepler.

In 2009, NASA launched the Kepler space telescope, dedicated to the discovery of transiting exoplanets. This spacecraft stared continuously at more than 150,000 stars in a small patch of sky near the constellation of Cygnus—just above the plane of our Milky Way Galaxy (Figure 21.20). Kepler's cameras and ability to measure small changes in brightness very precisely enabled the discovery of thousands of exoplanets, including many multi-planet systems. The spacecraft required three reaction wheels—a type of wheel used to help control slight rotation of the spacecraft—to stabilize the pointing of the telescope and monitor the brightness of the same group of stars over and over again. Kepler was launched with four reaction wheels (one a spare), but by

May 2013, two wheels had failed and the telescope could no longer be accurately pointed toward the target area. Kepler had been designed to operate for 4 years, and ironically, the pointing failure occurred exactly 4 years and 1 day after it began observing.

However, this failure did not end the mission. The Kepler telescope continued to observe for two more years, looking for short-period transits in different parts of the sky. We discuss the Kepler results in the next section Exoplanets Everywhere: What We Are Learning. A new NASA mission called TESS (Transiting Exoplanet Survey Satellite) is now carrying out a transit survey of the nearer (and therefore brighter) stars all over the sky. TESS completed its first planned survey of the sky in August 2020 and reported 66 exoplanet discoveries and almost 2100 candidates that need further observations to confirm.

Figure 21.20 Kepler's Field of View. The boxes show the region where the Kepler spacecraft cameras took images of over 150,000 stars regularly, to find transiting planets. (credit "field of view": modification of work by NASA/Kepler mission; credit "spacecraft": modification of work by NASA/Kepler mission/Wendy Stenzel)

What do we mean, exactly, by "discovery" of transiting exoplanets? A single transit shows up as a very slight drop in the brightness of the star, lasting several hours. However, astronomers must be on guard against other factors that might produce a false transit, especially when working at the limit of precision of the telescope. We must wait for a second transit of similar depth. But when another transit is observed, we don't initially know whether it might be due to another planet in a different orbit. The "discovery" occurs only when a third transit is found with similar depth and the same spacing in time as the first pair.

Computers normally conduct the analysis, which involves searching for tiny, periodic dips in the light from each star, extending over 4 years of observation. But the Kepler mission also has a program in which non-astronomers—citizen scientists—can examine the data. These dedicated volunteers have found several transits that were missed by the computer analyses, showing that the human eye and brain sometimes recognize unusual events that a computer was not programmed to look for.

Measuring three or four evenly spaced transits is normally enough to "discover" an exoplanet. But in a new field like exoplanet research, we would like to find further independent verification. The strongest confirmation happens when ground-based telescopes are also able to detect a Doppler shift with the same period as the transits. However, this is generally not possible for Earth-size planets. One of the most convincing ways to verify that a dip in brightness is due to a planet is to find more planets orbiting the same star—a *planetary system*. Multi-planet systems also provide alternative ways to estimate the masses of the planets, as we will discuss in the next section.

The selection effects (or biases) in the Kepler data are similar to those in Doppler observations. Large planets are easier to find than small ones, and short-period planets are easier than long-period planets. If we require three transits to establish the presence of a planet, we are of course limited to discovering planets with orbital periods less than one-third of the observing interval. Thus, it was only in its fourth and final year of operation that Kepler was able to find planets with orbits like Earth's that require 1 year to go around their star.

Direct Detection

The best possible evidence for an earthlike planet elsewhere would be an image. After all, "seeing is believing" is a very human prejudice. But imaging a distant planet is a formidable challenge indeed. Suppose, for example, you were a great distance away and wished to detect reflected light from Earth. Earth intercepts and reflects less than one billionth of the Sun's radiation, so its apparent brightness in visible light is less than one billionth that of the Sun. Compounding the challenge of detecting such a faint speck of light, the planet is swamped by the blaze of radiation from its parent star.

Even today, the best telescope mirrors' optics have slight imperfections that prevent the star's light from coming into focus in a completely sharp point.

Direct imaging works best for young gas giant planets that emit infrared light and reside at large separations from their host stars. Young giant planets emit more infrared light because they have more internal energy, stored from the process of planet formation. Even then, clever techniques must be employed to subtract out the light from the host star. In 2008, three such young planets were discovered orbiting HR 8799, a star in the constellation of Pegasus (Figure 21.21). Two years later, a fourth planet was detected closer to the star. Additional planets may reside even closer to HR 8799, but if they exist, they are currently lost in the glare of the star.

Since then, a number of planets around other stars have been found using direct imaging. However, one challenge is to tell whether the objects we are seeing are indeed planets or if they are brown dwarfs (failed stars) in orbit around a star.

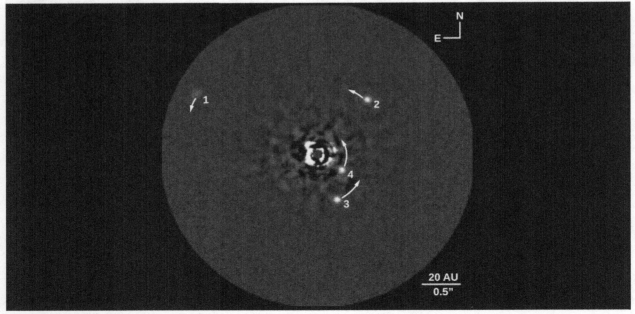

Figure 21.21 Exoplanets around HR 8799. This image shows Keck telescope observations of four directly imaged planets orbiting

HR 8799. A size scale for the system gives the distance in AU (remember that one astronomical unit is the distance between Earth and the Sun.) (credit: modification of work by Ben Zuckerman)

Direct imaging is an important technique for characterizing an exoplanet. The brightness of the planet can be measured at different wavelengths. These observations provide an estimate for the temperature of the planet's atmosphere; in the case of HR 8799 planet 1, the color suggests the presence of thick clouds. Spectra can also be obtained from the faint light to analyze the atmospheric constituents. A spectrum of HR 8799 planet 1 indicates a hydrogen-rich atmosphere, while the closer planet 4 shows evidence for methane in the atmosphere.

Another way to overcome the blurring effect of Earth's atmosphere is to observe from space. Infrared may be the optimal wavelength range in which to observe because planets get brighter in the infrared while stars like our Sun get fainter, thereby making it easier to detect a planet against the glare of its star. Special optical techniques can be used to suppress the light from the central star and make it easier to see the planet itself. However, even if we go into space, it will be difficult to obtain images of Earth-size planets.

21.5 | Exoplanets Everywhere: What We Are Learning

Learning Objectives

By the end of this section, you will be able to:

> Explain what we have learned from our discovery of exoplanets
> Identify which kind of exoplanets appear to be the most common in the Galaxy
> Discuss the kinds of planetary systems we are finding around other stars

Before the discovery of exoplanets, most astronomers expected that other planetary systems would be much like our own—planets following roughly circular orbits, with the most massive planets several AU from their parent star. Such systems do exist in large numbers, but many exoplanets and planetary systems are very different from those in our solar system. Another surprise is the existence of whole classes of exoplanets that we simply don't have in our solar system: planets with masses between the mass of Earth and Neptune, and planets that are several times more massive than Jupiter.

Kepler Results

The Kepler telescope has been responsible for the discovery of most exoplanets, especially at smaller sizes, as illustrated in Figure 21.22, where the Kepler discoveries are plotted in yellow. You can see the wide range of sizes, including planets substantially larger than Jupiter and smaller than Earth. The absence of Kepler-discovered exoplanets with orbital periods longer than a few hundred days is a consequence of the 4-year lifetime of the mission. (Remember that three evenly spaced transits must be observed to register a discovery.) At the smaller sizes, the absence of planets much smaller than one earth radius is due to the difficulty of detecting transits by very small planets. In effect, the "discovery space" for Kepler was limited to planets with orbital periods less than 400 days and sizes larger than Mars.

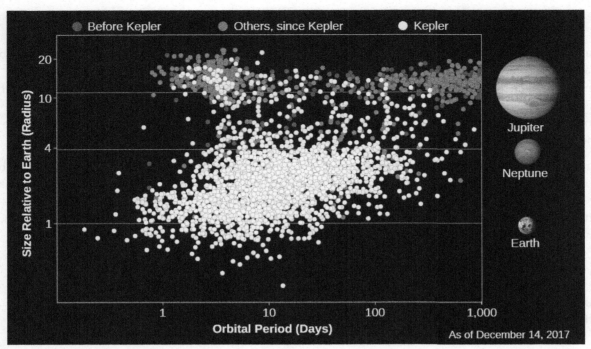

Figure 21.22 Exoplanet Discoveries through 2017. The vertical axis shows the radius of each planet compared to Earth. Horizontal lines show the size of Earth, Neptune, and Jupiter. The horizontal axis shows the time each planet takes to make one orbit (and is given in Earth days). Recall that Mercury takes 88 days and Earth takes a little more than 365 days to orbit the Sun. The yellow dots show planets discovered by the Kepler mission using transits, and the blue dots are the discoveries by other projects and techniques. When this graph was made, astronomers had discovered just over 3,500 exoplanets; as of the start of 2022, that number has risen to roughly 5,000. (credit: modification of work by NASA/Ames Research Center/Jessie Dotson and Wendy Stenzel)

One of the primary objectives of the Kepler mission was to find out how many stars hosted planets and especially to estimate the frequency of earthlike planets. Although Kepler looked at only a very tiny fraction of the stars in the Galaxy, the sample size was large enough to draw some interesting conclusions. While the observations apply only to the stars observed by Kepler, those stars are reasonably representative, and so astronomers can extrapolate to the entire Galaxy.

Figure 21.23 shows that the Kepler discoveries include many rocky, Earth-size planets, far more than Jupiter-size gas planets. This immediately tells us that the initial Doppler discovery of many hot Jupiters was a biased sample, in effect, finding the odd planetary systems because they were the easiest to detect. However, there is one huge difference between this observed size distribution and that of planets in our solar system. The most common planets have radii between 1.4 and 2.8 that of Earth, sizes for which we have no examples in the solar system. These have been nicknamed **super-Earths**, while the other large group with sizes between 2.8 and 4 that of Earth are often called **mini-Neptunes**.

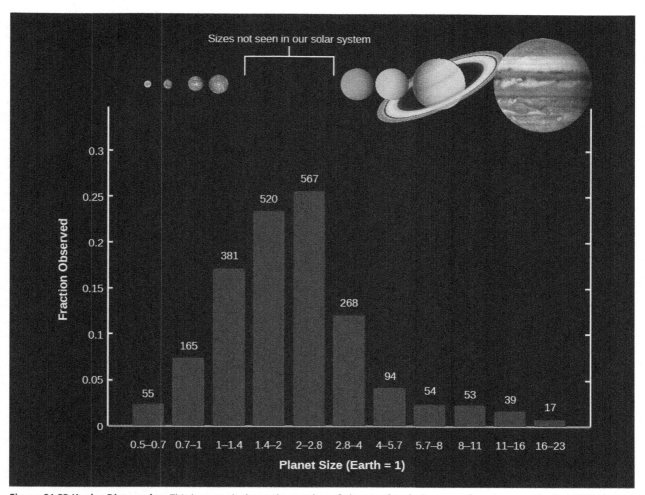

Figure 21.23 Kepler Discoveries. This bar graph shows the number of planets of each size range found among the first 2213 Kepler planet discoveries. Sizes range from half the size of Earth to 20 times that of Earth. On the vertical axis, you can see the fraction that each size range makes up of the total. Note that planets that are between 1.4 and 4 times the size of Earth make up the largest fractions, yet this size range is not represented among the planets in our solar system. (credit: modification of work by NASA/Kepler mission)

What a remarkable discovery it is that the most common types of planets in the Galaxy are completely absent from our solar system and were unknown until Kepler's survey. However, recall that really small planets were difficult for the Kepler instruments to find. So, to estimate the frequency of Earth-size exoplanets, we need to correct for this sampling bias. The result is the corrected size distribution shown in Figure 21.24. Notice that in this graph, we have also taken the step of showing not the number of Kepler detections but the average number of planets per star for solar-type stars (spectral types F, G, and K).

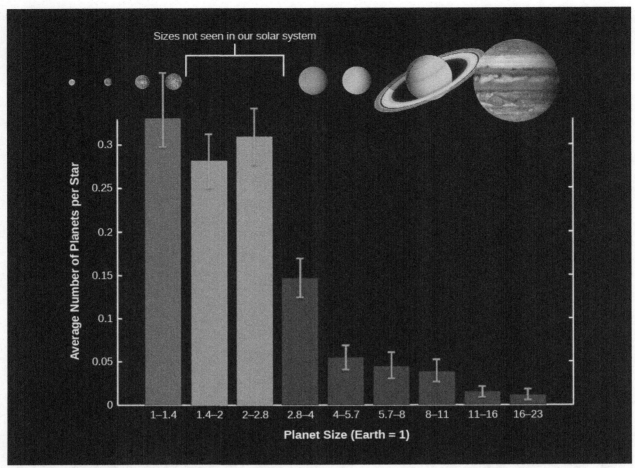

Figure 21.24 Size Distribution of Planets for Stars Similar to the Sun. We show the average number of planets per star in each planet size range. (The average is less than one because some stars will have zero planets of that size range.) This distribution, corrected for biases in the Kepler data, shows that Earth-size planets may actually be the most common type of exoplanets. (credit: modification of work by NASA/Kepler mission)

We see that the most common planet sizes of are those with radii from 1 to 3 times that of Earth—what we have called "Earths" and "super-Earths." Each group occurs in about one-third to one-quarter of stars. In other words, if we group these sizes together, we can conclude there is nearly one such planet per star! And remember, this census includes primarily planets with orbital periods less than 2 years. We do not yet know how many undiscovered planets might exist at larger distances from their star.

To estimate the number of Earth-size planets in our Galaxy, we need to remember that there are approximately 100 billion stars of spectral types F, G, and K. Therefore, we estimate that there are about 30 billion Earth-size planets in our Galaxy. If we include the super-Earths too, then there could be one hundred billion in the whole Galaxy. This idea—that planets of roughly Earth's size are so numerous—is surely one of the most important discoveries of modern astronomy.

Planets with Known Densities

For several hundred exoplanets, we have been able to measure both the size of the planet from transit data and its mass from Doppler data, yielding an estimate of its density. Comparing the average density of exoplanets to the density of planets in our solar system helps us understand whether they are rocky or gaseous in nature. This has been particularly important for understanding the structure of the new categories of super-Earths and mini-Neptunes with masses between 3–10 times the mass of Earth. A key observation so far is that planets that are more than 10 times the mass of Earth have substantial gaseous envelopes (like Uranus and Neptune) whereas lower-mass planets are predominately rocky in nature (like the terrestrial planets).

Figure 21.25 compares all the exoplanets that have both mass and radius measurements. The dependence of the radius on planet mass is also shown for a few illustrative cases—hypothetical planets made of pure iron, rock, water, or hydrogen.

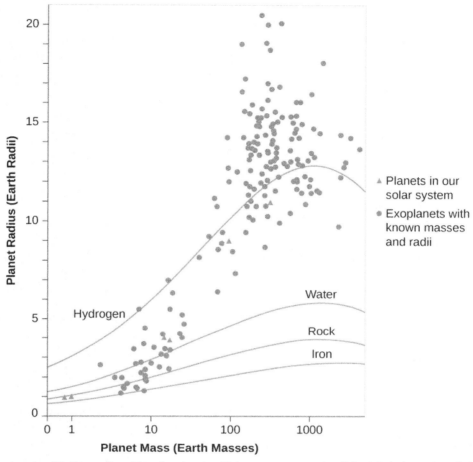

Figure 21.25 Exoplanets with Known Densities. Exoplanets with known masses and radii (red circles) are plotted along with solid lines that show the theoretical size of pure iron, rock, water, and hydrogen planets with increasing mass. Masses are given in multiples of Earth's mass. (For comparison, Jupiter contains enough mass to make 320 Earths.) The green triangles indicate planets in our solar system.

At lower masses, notice that as the mass of these hypothetical planets increases, the radius also increases. That makes sense—if you were building a model of a planet out of clay, your toy planet would increase in size as you added more clay. However, for the highest mass planets ($M > 1000 \, M_{Earth}$) in Figure 21.25, notice that the radius stops increasing and the planets with greater mass are actually smaller. This occurs because increasing the mass also increases the gravity of the planet, so that compressible materials (even rock is compressible) will become more tightly packed, shrinking the size of the more massive planet.

In reality, planets are not pure compositions like the hypothetical water or iron planet. Earth is composed of a solid iron core, an outer liquid-iron core, a rocky mantle and crust, and a relatively thin atmospheric layer. Exoplanets are similarly likely to be differentiated into compositional layers. The theoretical lines in Figure 21.25 are simply guides that suggest a range of possible compositions.

Astronomers who work on the complex modeling of the interiors of rocky planets make the simplifying assumption that the planet consists of two or three layers. This is not perfect, but it is a reasonable approximation and another good example of how science works. Often, the first step in understanding something new is to narrow down the range of possibilities. This sets the stage for refining and deepening our knowledge. In Figure 21.25, the two green triangles with roughly 1 M_{Earth} and 1 R_{Earth} represent Venus and Earth. Notice that these planets fall between the models for a pure iron and a pure rock planet, consistent with what we would expect for the known mixed-chemical composition of Venus and Earth.

In the case of gaseous planets, the situation is more complex. Hydrogen is the lightest element in the periodic table, yet many of the detected exoplanets in Figure 21.25 with masses greater than 100 M_{Earth} have radii that suggest they are lower in density than a pure hydrogen planet. Hydrogen is the lightest element, so what is happening here? Why do some gas giant planets have inflated radii that are larger than the fictitious pure hydrogen planet? Many of these planets reside in short-period orbits close to the host star where they intercept a significant amount of radiated energy. If this energy is trapped deep in the planet atmosphere, it can cause the planet to expand.

Planets that orbit close to their host stars in slightly eccentric orbits have another source of energy: the star will raise tides in these planets that tend to circularize the orbits. This process also results in tidal dissipation of energy that can inflate the atmosphere. It would be interesting to measure the size of gas giant planets in wider orbits where the planets should be cooler—the expectation is that unless they are very young, these cooler gas giant exoplanets (sometimes called "cold Jupiters") should not be inflated. But we don't yet have data on these more distant exoplanets.

Exoplanetary Systems

As we search for exoplanets, we don't expect to find only one planet per star. Our solar system has eight major planets, half a dozen dwarf planets, and millions of smaller objects orbiting the Sun. The evidence we have of planetary systems in formation also suggest that they are likely to produce multi-planet systems.

The first planetary system was found around the star Upsilon Andromedae in 1999 using the Doppler method, and many others have been found since then. If such exoplanetary system are common, let's consider which systems we expect to find in the Kepler transit data.

A planet will transit its star only if Earth lies in the plane of the planet's orbit. If the planets in other systems do not have orbits in the same plane, we are unlikely to see multiple transiting objects. Also, as we have noted before, Kepler was sensitive only to planets with orbital periods less than about 4 years. What we expect from Kepler data, then, is evidence of coplanar planetary systems confined to what would be the realm of the terrestrial planets in our solar system.

By 2022, astronomers gathered data on over 800 such exoplanet systems. Many have only two known planets, but a few have as many as five, and one has eight (the same number of planets as our own solar system). For the most part, these are very compact systems with most of their planets closer to their star than Mercury is to the Sun. The figure below shows one of the largest exoplanet systems: that of the star called Kepler-62 (Figure 21.26). Our solar system is shown to the same scale, for comparison (note that the Kepler-62 planets are drawn with artistic license; we have no detailed images of any exoplanets).

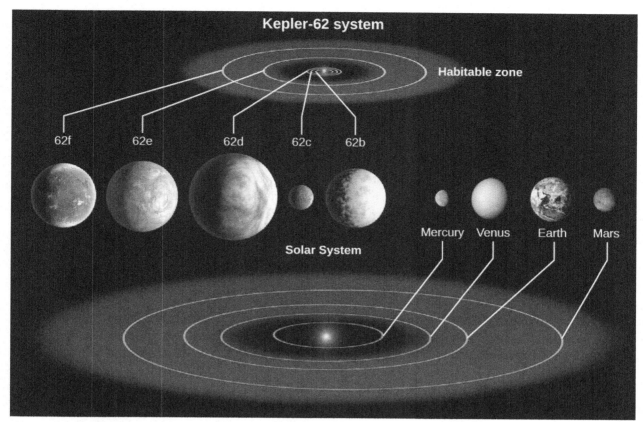

Figure 21.26 Exoplanet System Kepler-62, with the Solar System Shown to the Same Scale. The green areas are the "habitable zones," the range of distance from the star where surface temperatures are likely to be consistent with liquid water. (credit: modification of work by NASA/Ames/JPL-Caltech)

All but one of the planets in the K-62 system are larger than Earth. These are super-Earths, and one of them (62d) is in the size range of a mini-Neptune, where it is likely to be largely gaseous. The smallest planet in this system is about the size of Mars. The three inner planets orbit very close to their star, and only the outer two have orbits larger than Mercury in our system. The green areas represent each star's "habitable zone," which is the distance from the star where we calculate that surface temperatures would be consistent with liquid water. The Kepler-62 habitable zone is much smaller than that of the Sun because the star is intrinsically fainter.

With closely spaced systems like this, the planets can interact gravitationally with each other. The result is that the observed transits occur a few minutes earlier or later than would be predicted from simple orbits. These gravitational interactions have allowed the Kepler scientists to calculate masses for the planets, providing another way to learn about exoplanets.

Kepler has discovered some interesting and unusual planetary systems. For example, most astronomers expected planets to be limited to single stars. But we have found planets orbiting close double stars, so that the planet would see two suns in its sky, like those of the fictional planet Tatooine in the *Star Wars* films. At the opposite extreme, planets can orbit one star of a wide, double-star system without major interference from the second star.

21.6 New Perspectives on Planet Formation

Learning Objectives

By the end of this section, you will be able to:

> Explain how exoplanet discoveries have revised our understanding of planet formation
> Discuss how planetary systems quite different from our solar system might have come about

Traditionally, astronomers have assumed that the planets in our solar system formed at about their current distances from the Sun and have remained there ever since. The first step in the formation of a giant planet is to build up a solid core, which happens when planetesimals collide and stick. Eventually, this core becomes massive enough to begin sweeping up gaseous material in the disk, thereby building the gas giants Jupiter and Saturn.

How to Make a Hot Jupiter

The traditional model for the formation of planets works only if the giant planets are formed far from the central star (about 5–10 AU), where the disk is cold enough to have a fairly high density of solid matter. It cannot explain the hot Jupiters, which are located very close to their stars where any rocky raw material would be completely vaporized. It also cannot explain the elliptical orbits we observe for some exoplanets because the orbit of a protoplanet, whatever its initial shape, will quickly become circular through interactions with the surrounding disk of material and will remain that way as the planet grows by sweeping up additional matter.

So we have two options: either we find a new model for forming planets close to the searing heat of the parent star, or we find a way to change the orbits of planets so that cold Jupiters can travel inward *after* they form. Most research now supports the latter explanation.

Calculations show that if a planet forms while a substantial amount of gas remains in the disk, then some of the planet's orbital angular momentum can be transferred to the disk. As it loses momentum (through a process that reminds us of the effects of friction), the planet will spiral inward. This process can transport giant planets, initially formed in cold regions of the disk, closer to the central star—thereby producing hot Jupiters. Gravitational interactions between planets in the chaotic early solar system can also cause planets to slingshot inward from large distances. But for this to work, the other planet has to carry away the angular momentum and move to a more distant orbit.

In some cases, we can use the combination of transit plus Doppler measurements to determine whether the planets orbit in the same plane and in the same direction as the star. For the first few cases, things seemed to work just as we anticipated: like the solar system, the gas giant planets orbited in their star's equatorial plane and in the same direction as the spinning star.

Then, some startling discoveries were made of gas giant planets that orbited at right angles or even in the opposite sense as the spin of the star. How could this happen? Again, there must have been interactions between planets. It's possible that before the system settled down, two planets came close together, so that one was kicked into an unusual orbit. Or perhaps a passing star perturbed the system after the planets were newly formed.

Forming Planetary Systems

When the Milky Way Galaxy was young, the stars that formed did not contain many heavy elements like iron. Several generations of star formation and star death were required to enrich the interstellar medium for subsequent generations of stars. Since planets seem to form "inside out," starting with the accretion of the materials that can make the rocky cores with which planets start, astronomers wondered when in the history of the Galaxy, planet formation would turn on.

The star Kepler-444 has shed some light on this question. This is a tightly packed system of five planets—the smallest comparable in size to Mercury and the largest similar in size to Venus. All five planets were detected with the Kepler spacecraft as they transited their parent star. All five planets orbit their host star in less than the time it takes Mercury to complete one orbit about the Sun. Remarkably, the host star Kepler-444 is more than 11 billion years old and formed when the Milky Way was only 2 billion years old. So the heavier elements needed to make rocky planets must have already been available then. This ancient planetary system sets the clock on the beginning of rocky planet formation to be relatively soon after the formation of our Galaxy.

Kepler data demonstrate that while rocky planets inside Mercury's orbit are missing from our solar system, they are common around other stars, like Kepler-444. When the first systems packed with close-in rocky planets were discovered, we wondered why they were so different from our solar system. When many such systems were discovered, we began to wonder if it was our solar system that was different. This led to speculation that additional rocky planets might once have existed close to the Sun in our solar system.

There is some evidence from the motions in the outer solar system that Jupiter may have migrated inward long ago. If correct, then gravitational perturbations from Jupiter could have dislodged the orbits of close-in rocky planets, causing them to fall into the Sun. Consistent with this picture, astronomers now think that Uranus and Neptune probably did not form at their present distances from the Sun but rather closer to where Jupiter and Saturn are now. The reason for this idea is that density in the disk of matter surrounding the Sun at the time the planets formed was so low outside the orbit of Saturn that it would take several billion years to build up Uranus and Neptune. Yet we saw earlier in the chapter that the disks around protostars survive only a few million years.

Therefore, scientists have developed computer models demonstrating that Uranus and Neptune could have formed near the current locations of Jupiter and Saturn, and then been kicked out to larger distances through gravitational interactions with their neighbors. All these wonderful new observations illustrate how dangerous it can be to draw conclusions about a phenomenon in science (in this case, how planetary systems form and arrange themselves) when you are only working with a single example.

Exoplanets have given rise to a new picture of planetary system formation—one that is much more chaotic than we originally thought. If we think of the planets as being like skaters in a rink, our original model (with only our own solar system as a guide) assumed that the planets behaved like polite skaters, all obeying the rules of the rink and all moving in nearly the same direction, following roughly circular paths. The new picture corresponds more to a roller derby, where the skaters crash into one another, change directions, and sometimes are thrown entirely out of the rink.

Habitable Exoplanets

While thousands of exoplanets have been discovered in the past two decades, every observational technique has fallen short of finding more than a few candidates that resemble Earth (Figure 21.27). Astronomers are not sure exactly what properties would define another Earth. Do we need to find a planet that is *exactly* the same size and mass as Earth? That may be difficult and may not be important from the perspective of habitability. After all, we have no reason to think that life could not have arisen on Earth if our planet had been a little bit smaller or larger. And, remember that how habitable a planet is depends on both its distance from its star and the nature of its atmosphere. The greenhouse effect can make some planets warmer (as it did for Venus and is doing more and more for Earth).

Figure 21.27 Many Earthlike Planets. This painting, commissioned by NASA, conveys the idea that there may be many planets resembling Earth out there as our methods for finding them improve. (credit: NASA/JPL-Caltech/R. Hurt (SSC-Caltech))

We can ask other questions to which we don't yet know the answers. Does this "twin" of Earth need to orbit a solar-type star, or can we consider as candidates the numerous exoplanets orbiting K- and M-class stars? (In the summer of 2016, astronomers reported the discovery of a planet with at least 1.3 times the mass of Earth around the nearest star, Proxima Centauri, which is spectral type M and located 4.2 light years from us.) We have a special interest in finding planets that could support life like ours, in which case, we need to find exoplanets within their star's habitable zone, where surface temperatures are consistent with liquid water on the surface. This is probably the most important characteristic defining an Earth-analog exoplanet.

The search for potentially habitable worlds is one of the prime drivers for exoplanet research in the next decade. Astronomers are beginning to develop realistic plans for new instruments that can even look for signs of life on distant worlds (examining their atmospheres for gases associated with life, for example). If we require telescopes in space to find such worlds, we need to recognize that years are required to plan, build, and launch such space observatories. The discovery of exoplanets and the knowledge that most stars have planetary systems are transforming our thinking about life beyond Earth. We are closer than ever to knowing whether habitable (and inhabited) planets are common. This work lends a new spirit of optimism to the search for life elsewhere, a subject to which we will return in Life in the Universe.

🔑 Key Terms

exoplanet   a planet orbiting a star other than our Sun

giant molecular clouds   large, cold interstellar clouds with diameters of dozens of light-years and typical masses of 10^5 solar masses; found in the spiral arms of galaxies, these clouds are where stars form

Herbig-Haro (HH) object   luminous knots of gas in an area of star formation that are set to glow by jets of material from a protostar

mini-Neptune   a planet that is intermediate between the largest terrestrial planet in our solar system (Earth) and the smallest jovian planet (Neptune); generally, mini-Neptunes have sizes between 2.8 and 4 times Earth's size

protostar   a very young star still in the process of formation, before nuclear fusion begins

stellar wind   the outflow of gas, sometimes at speeds as high as hundreds of kilometers per second, from a star

super-Earth   a planet larger than Earth, generally between 1.4 and 2.8 times the size of our planet

transit   when one astronomical object moves in front of another

📖 Summary

21.1 Star Formation

Most stars form in giant molecular clouds with masses as large as 3×10^6 solar masses. The most well-studied molecular cloud is Orion, where star formation is currently taking place. Molecular clouds typically contain regions of higher density called clumps, which in turn contain several even-denser cores of gas and dust, each of which may become a star. A star can form inside a core if its density is high enough that gravity can overwhelm the internal pressure and cause the gas and dust to collapse. The accumulation of material halts when a protostar develops a strong stellar wind, leading to jets of material being observed coming from the star. These jets of material can collide with the material around the star and produce regions that emit light that are known as Herbig-Haro objects.

21.2 The H–R Diagram and the Study of Stellar Evolution

The evolution of a star can be described in terms of changes in its temperature and luminosity, which can best be followed by plotting them on an H–R diagram. Protostars generate energy (and internal heat) through gravitational contraction that typically continues for millions of years, until the star reaches the main sequence.

21.3 Evidence That Planets Form around Other Stars

Observational evidence shows that most protostars are surrounded by disks with large-enough diameters and enough mass (as much as 10% that of the Sun) to form planets. After a few million years, the inner part of the disk is cleared of dust, and the disk is then shaped like a donut with the protostar centered in the hole—something that can be explained by the formation of planets in that inner zone. Around a few older stars, we see disks formed from the debris produced when small bodies (comets and asteroids) collide with each other. The distribution of material in the rings of debris disks is probably determined by shepherd planets, just as Saturn's shepherd moons affect the orbits of the material in its rings. Protoplanets that grow to be 10 times the mass of Earth or bigger while there is still considerable gas in their disk can then capture more of that gas and become giant planets like Jupiter in the solar system.

21.4 Planets beyond the Solar System: Search and Discovery

Several observational techniques have successfully detected planets orbiting other stars. These techniques fall into two general categories—direct and indirect detection. The Doppler and transit techniques are our most powerful indirect tools for finding exoplanets. Some planets are also being found by direct imaging.

21.5 Exoplanets Everywhere: What We Are Learning

Although the Kepler mission is finding thousands of new exoplanets, these are limited to orbital periods of less than 400 days and sizes larger than Mars. Still, we can use the Kepler discoveries to extrapolate the distribution of planets in our Galaxy. The data so far imply that planets like Earth are the most common type of planet, and that there may be 100 billion Earth-size planets around Sun-like stars in the Galaxy. More than 800 planetary systems have been discovered around other stars. In many of them, planets are arranged differently than in our solar system.

21.6 New Perspectives on Planet Formation

The ensemble of exoplanets is incredibly diverse and has led to a revision in our understanding of planet formation that includes the possibility of vigorous, chaotic interactions, with planet migration and scattering. It is possible that the solar system is unusual (and not representative) in how its planets are arranged. Many systems seem to have rocky planets farther inward than we do, for example, and some even have "hot Jupiters" very close to their star. Ambitious space experiments should make it possible to image earthlike planets outside the solar system and even to obtain information about their habitability as we search for life elsewhere.

 For Further Exploration

Articles

Star Formation

Blaes, O. "A Universe of Disks." *Scientific American* (October 2004): 48. On accretion disks and jets around young stars and black holes.

Croswell, K. "The Dust Belt Next Door [Tau Ceti]." *Scientific American* (January 2015): 24. Short intro to recent observations of planets and a wide dust belt.

Frank, A. "Starmaker: The New Story of Stellar Birth." *Astronomy* (July 1996): 52.

Jayawardhana, R. "Spying on Stellar Nurseries." *Astronomy* (November 1998): 62. On protoplanetary disks.

O'Dell, C. R. "Exploring the Orion Nebula." *Sky & Telescope* (December 1994): 20. Good review with Hubble results.

Ray, T. "Fountains of Youth: Early Days in the Life of a Star." *Scientific American* (August 2000): 42. On outflows from young stars.

Young, E. "Cloudy with a Chance of Stars." *Scientific American* (February 2010): 34. On how clouds of interstellar matter turn into star systems.

Young, Monica "Making Massive Stars." *Sky & Telescope* (October 2015): 24. Models and observations on how the most massive stars form.

Exoplanets

Billings, L. "In Search of Alien Jupiters." *Scientific American* (August 2015): 40–47. The race to image jovian planets with current instruments and why a direct image of a terrestrial planet is still in the future.

Heller, R. "Better Than Earth." *Scientific American* (January 2015): 32–39. What kinds of planets may be habitable; super-Earths and jovian planet moons should also be considered.

Laughlin, G. "How Worlds Get Out of Whack." *Sky & Telescope* (May 2013): 26. On how planets can migrate from the places they form in a star system.

Marcy, G. "The New Search for Distant Planets." *Astronomy* (October 2006): 30. Fine brief overview. (The same issue has a dramatic fold-out visual atlas of extrasolar planets, from that era.)

Redd, N. "Why Haven't We Found Another Earth?" *Astronomy* (February 2016): 25. Looking for terrestrial planets in the habitable zone with evidence of life.

Seager, S. "Exoplanets Everywhere." *Sky & Telescope* (August 2013): 18. An excellent discussion of some of the frequently asked questions about the nature and arrangement of planets out there.

Seager, S. "The Hunt for Super-Earths." *Sky & Telescope* (October 2010): 30. The search for planets that are up to 10 times the mass of Earth and what they can teach us.

Villard, R. "Hunting for Earthlike Planets." *Astronomy* (April 2011): 28. How we expect to find and characterize super-Earth (planets somewhat bigger than ours) using new instruments and techniques that could show us what their atmospheres are made of.

Websites

Exoplanet Exploration: http://planetquest.jpl.nasa.gov/ (http://planetquest.jpl.nasa.gov/). PlanetQuest (from the Navigator Program at the Jet Propulsion Lab) is probably the best site for students and beginners, with introductory materials and nice illustrations; it focuses mostly on NASA work and missions.

Exoplanets: http://www.planetary.org/exoplanets/ (http://www.planetary.org/exoplanets/). Planetary Society's exoplanets pages with a dynamic catalog of planets found and good explanations.

Extrasolar Planets Encyclopedia: http://exoplanet.eu/ (http://exoplanet.eu/). Maintained by Jean Schneider of the Paris Observatory, has the largest catalog of planet discoveries and useful background material (some of it more technical).

Formation of Stars: https://www.spacetelescope.org/science/formation_of_stars/ (https://www.spacetelescope.org/science/formation_of_stars/). Star Formation page from the Hubble Space Telescope, with links to images and information.

Kepler Mission: https://www.nasa.gov/mission_pages/kepler/overview/index.html (https://www.nasa.gov/mission_pages/kepler/overview/index.html). The public website for the remarkable telescope in space that is searching planets using the transit technique and is our best hope for finding earthlike planets.

NASA Exoplanet Exploration Dashboard: https://exoplanets.nasa.gov/discovery/discoveries-dashboard/ (https://exoplanets.nasa.gov/discovery/discoveries-dashboard/). A handy place to find up-to-date information of exoplanet discoveries. This page is a summary of all discoveries by method and type. One click takes you to the full exoplanet catalog.

Proxima Centauri Planet Discovery: http://www.eso.org/public/news/eso1629/ (http://www.eso.org/public/news/eso1629/).

Apps

Exoplanet: http://exoplanetapp.com/ (http://exoplanetapp.com/). Allows you to browse through a regularly updated visual catalog of exoplanets that have been found so far.

Videos

A Star Is Born: https://www.youtube.com/watch?v=mkktE_fs4NA (https://www.youtube.com/watch?v=mkktE_fs4NA). Discovery Channel video with astronomer Michelle Thaller (2:25).

Are We Alone: An Evening Dialogue with the Kepler Mission Leaders: http://www.youtube.com/watch?v=O7ItAXfl0Lw (http://www.youtube.com/watch?v=O7ItAXfl0Lw). A non-technical panel discussion on Kepler results and ideas about planet formation with Bill Borucki, Natalie Batalha, and Gibor Basri (moderated by Andrew Fraknoi) at the University of California, Berkeley (2:07:01).

Finding the Next Earth: The Latest Results from Kepler: https://www.youtube.com/watch?v=ZbijeR_AALo (https://www.youtube.com/watch?v=ZbijeR_AALo). Natalie Batalha (San Jose State University & NASA Ames) public talk in the Silicon Valley Astronomy Lecture Series (1:28:38).

From Hot Jupiters to Habitable Worlds: https://vimeo.com/37696087 (https://vimeo.com/37696087) (Part 1) and https://vimeo.com/37700700 (https://vimeo.com/37700700) (Part 2). Debra Fischer (Yale University) public talk in Hawaii sponsored by the Keck Observatory (15:20 Part 1, 21:32 Part 2).

Search for Habitable Exoplanets: http://www.youtube.com/watch?v=RLWb_T9yaDU (http://www.youtube.com/watch?v=RLWb_T9yaDU). Sara Seeger (MIT) public talk at the SETI Institute, with Kepler results (1:10:35).

Collaborative Group Activities

A. Your group is a subcommittee of scientists examining whether any of the "hot Jupiters" (giant planets closer to their stars than Mercury is to the Sun) could have life on or near them. Can you come up with places on, in, or near such planets where life could develop or where some forms of life might survive?

B. A wealthy couple (who are alumni of your college or university and love babies) leaves the astronomy program several million dollars in their will, to spend in the best way possible to search for "infant stars in our section of the Galaxy." Your group has been assigned the task of advising the dean on how best to spend the money. What kind of instruments and search programs would you recommend, and why?

C. Some people consider the discovery of any planets (even hot Jupiters) around other stars one of the most important events in the history of astronomical research. Some astronomers have been surprised that the public is not more excited about the planet discoveries. One reason that has been suggested for this lack of public surprise and excitement is that science fiction stories have long prepared us for there being planets around other stars. (The Starship Enterprise on the 1960s *Star Trek* TV series found some in just about every weekly episode.) What does your group think? Did you know about the discovery of planets around other stars before taking this course? Do you consider it exciting? Were you surprised to hear about it? Are science fiction movies and books good or bad tools for astronomy education in general, do you think?

D. What if future space instruments reveal an earthlike exoplanet with significant amounts of oxygen and methane in its atmosphere? Suppose the planet and its star are 50 light-years away. What does your group suggest astronomers do next? How much effort and money would you recommend be put into finding out more about this planet and why?

E. Discuss with your group the following question: which is easier to find orbiting a star with instruments we have today: a jovian planet or a proto-planetary disk? Make a list of arguments for each side of this question.

F. (This activity should be done when your group has access to the internet.) Go to the page which indexes all the publicly released Hubble Space Telescope images by subject: http://hubblesite.org/newscenter/archive/browse/image/. Under "Star," go to "Protoplanetary Disk" and find a system—not mentioned in this chapter—that your group likes, and prepare a short report to the class about why you find it interesting. Then, under "Nebula," go to "Emission" and find a region of star formation not mentioned in this chapter, and prepare a short report to the class about what you find interesting about it.

G. There is a "citizen science" website called Planet Hunters (http://www.planethunters.org/) where you can participate in identifying exoplanets from the data that Kepler provided. Your group should access the site, work together to use it, and classify two light curves. Report back to the class on what you have done.

H. Yuri Milner, a Russian-American billionaire, recently pledged $100 million to develop the technology to send many miniaturized probes to a star in the Alpha Centauri triple star system (which includes Proxima Centauri, the nearest star to us, now known to have at least one planet.) Each tiny probe will be propelled by powerful lasers at 20% the speed of light, in the hope that one or more might arrive safely and be able to send back information about what it's like there. Your group should search online for more information about this project (called "Breakthrough: Starshot") and discuss your reactions to this project. Give specific reasons for your arguments.

Exercises

Review Questions

1. Give several reasons the Orion molecular cloud is such a useful "laboratory" for studying the stages of star formation.

2. Why is star formation more likely to occur in cold molecular clouds than in regions where the temperature of the interstellar medium is several hundred thousand degrees?

3. Why have we learned a lot about star formation since the invention of detectors sensitive to infrared radiation?

4. Describe what happens when a star forms. Begin with a dense core of material in a molecular cloud and trace the evolution up to the time the newly formed star reaches the main sequence.

5. Describe how the T Tauri star stage in the life of a low-mass star can lead to the formation of a Herbig-Haro (H-H) object.

6. Look at the four stages shown in Figure 21.8. In which stage(s) can we see the star in visible light? In infrared radiation?

7. The evolutionary track for a star of 1 solar mass remains nearly vertical in the H–R diagram for a while (see Figure 21.12). How is its luminosity changing during this time? Its temperature? Its radius?

8. Two protostars, one 10 times the mass of the Sun and one half the mass of the Sun are born at the same time in a molecular cloud. Which one will be first to reach the main sequence stage, where it is stable and getting energy from fusion?

9. Compare the scale (size) of a typical dusty disk around a forming star with the scale of our solar system.

10. Why is it so hard to see planets around other stars and so easy to see them around our own?

11. Why did it take astronomers until 1995 to discover the first exoplanet orbiting another star like the Sun?

12. Which types of planets are most easily detected by Doppler measurements? By transits?

13. List three ways in which the exoplanets we have detected have been found to be different from planets in our solar system.

14. List any similarities between discovered exoplanets and planets in our solar system.

15. What revisions to the theory of planet formation have astronomers had to make as a result of the discovery of exoplanets?

16. Why are young Jupiters easier to see with direct imaging than old Jupiters?

Thought Questions

17. A friend of yours who did not do well in her astronomy class tells you that she believes all stars are old and none could possibly be born today. What arguments would you use to persuade her that stars are being born somewhere in the Galaxy during your lifetime?

18. Observations suggest that it takes more than 3 million years for the dust to begin clearing out of the inner regions of the disks surrounding protostars. Suppose this is the minimum time required to form a planet. Would you expect to find a planet around a 10-M_{Sun} star? (Refer to Figure 21.12.)

19. Suppose you wanted to observe a planet around another star with direct imaging. Would you try to observe in visible light or in the infrared? Why? Would the planet be easier to see if it were at 1 AU or 5 AU from its star?

20. Why were giant planets close to their stars the first ones to be discovered? Why has the same technique not been used yet to discover giant planets at the distance of Saturn?

21. Exoplanets in eccentric orbits experience large temperature swings during their orbits. Suppose you had to plan for a mission to such a planet. Based on Kepler's second law, does the planet spend more time closer or farther from the star? Explain.

Figuring for Yourself

22. When astronomers found the first giant planets with orbits of only a few days, they did not know whether those planets were gaseous and liquid like Jupiter or rocky like Mercury. The observations of HD 209458 settled this question because observations of the transit of the star by this planet made it possible to determine the radius of the planet. Use the data given in the text to estimate the density of this planet, and then use that information to explain why it must be a gas giant.

23. An exoplanetary system has two known planets. Planet X orbits in 290 days and Planet Y orbits in 145 days. Which planet is closest to its host star? If the star has the same mass as the Sun, what is the semi-major axis of the orbits for Planets X and Y?

24. Kepler's third law says that the orbital period (in years) is proportional to the square root of the cube of the mean distance (in AU) from the Sun ($P \propto a^{1.5}$). For mean distances from 0.1 to 32 AU, calculate and plot a curve showing the expected Keplerian period. For each planet in our solar system, look up the mean distance from the Sun in AU and the orbital period in years and overplot these data on the theoretical Keplerian curve.

25. Calculate the transit depth for an M dwarf star that is 0.3 times the radius of the Sun with a gas giant planet the size of Jupiter.

26. If a transit depth of 0.00001 can be detected with the Kepler spacecraft, what is the smallest planet that could be detected around a 0.3 R_{sun} M dwarf star?

27. What fraction of gas giant planets seems to have inflated radii?

22 Stars from Adolescence to Old Age

Figure 22.1 Ant Nebula. During the later phases of stellar evolution, stars expel some of their mass, which returns to the interstellar medium to form new stars. This Hubble Space Telescope image shows a star losing mass. Known as Menzel 3, or the Ant Nebula, this beautiful region of expelled gas is about 3000 light-years away from the Sun. We see a central star that has ejected mass preferentially in two opposite directions. The object is about 1.6 light-years long. The image is color coded—red corresponds to an emission line of sulfur, green to nitrogen, blue to hydrogen, and blue/violet to oxygen. (credit: modification of work by NASA, ESA and The Hubble Heritage Team (STScI/AURA))

Chapter Outline

22.1 Evolution from the Main Sequence to Red Giants
22.2 Star Clusters
22.3 Checking Out the Theory
22.4 Further Evolution of Stars
22.5 The Evolution of More Massive Stars

Thinking Ahead

The Sun and other stars cannot last forever. Eventually they will exhaust their nuclear fuel and cease to shine. But how do they change during their long lifetimes? And what do these changes mean for the future of Earth?

We now turn from the birth of stars to the rest of their life stories. This is not an easy task since stars live much longer than astronomers. Thus, we cannot hope to see the life story of any single star unfold before our eyes or telescopes. To learn about their lives, we must survey as many of the stellar inhabitants of the Galaxy as possible. With thoroughness and a little luck, we can catch at least a few of them in each stage of their lives. As you've learned, stars have many different characteristics, with the differences sometimes resulting from their different masses, temperatures, and luminosities, and at other times derived from changes that occur as they age. Through a combination of observation and theory, we can use these differences to piece together the life story of a star.

22.1 | Evolution from the Main Sequence to Red Giants

Learning Objectives

By the end of this section, you will be able to:

> Explain the zero-age main sequence
> Describe what happens to main-sequence stars of various masses as they exhaust their hydrogen supply

One of the best ways to get a "snapshot" of a group of stars is by plotting their properties on an H–R diagram. We have already used the H–R diagram to follow the evolution of protostars up to the time they reach the main sequence. Now we'll see what happens next.

Once a star has reached the main-sequence stage of its life, it derives its energy almost entirely from the conversion of hydrogen to helium via the process of nuclear fusion in its core (see The Sun: A Nuclear Powerhouse). Since hydrogen is the most abundant element in stars, this process can maintain the star's equilibrium for a long time. Thus, all stars remain on the main sequence for most of their lives. Some astronomers like to call the main-sequence phase the star's "prolonged adolescence" or "adulthood" (continuing our analogy to the stages in a human life).

The left-hand edge of the main-sequence band in the H–R diagram is called the **zero-age main sequence** (see Figure 18.15). We use the term *zero-age* to mark the time when a star stops contracting, settles onto the main sequence, and begins to fuse hydrogen in its core. The zero-age main sequence is a continuous line in the H–R diagram that shows where stars of different masses but similar chemical composition can be found when they begin to fuse hydrogen.

Since only 0.7% of the hydrogen used in fusion reactions is converted into energy, fusion does not change the *total* mass of the star appreciably during this long period. It does, however, change the chemical composition in its central regions where nuclear reactions occur: hydrogen is gradually depleted, and helium accumulates. This change of composition changes the luminosity, temperature, size, and interior structure of the star. When a star's luminosity and temperature begin to change, the point that represents the star on the H–R diagram moves away from the zero-age main sequence.

Calculations show that the temperature and density in the inner region slowly increase as helium accumulates in the center of a star. As the temperature gets hotter, each proton acquires more energy of motion on average; this means it is more likely to interact with other protons, and as a result, the rate of fusion also increases. For the proton-proton cycle described in The Sun: A Nuclear Powerhouse, the rate of fusion goes up roughly as the temperature to the fourth power.

If the rate of fusion goes up, the rate at which energy is being generated also increases, and the luminosity of the star gradually rises. Initially, however, these changes are small, and stars remain within the main-sequence band on the H–R diagram for most of their lifetimes.

EXAMPLE 22.1

Star Temperature and Rate of Fusion

If a star's temperature were to double, by what factor would its rate of fusion increase?

Solution

Since the rate of fusion (like temperature) goes up to the fourth power, it would increase by a factor of 2^4, or 16 times.

Check Your Learning

If the rate of fusion of a star increased 256 times, by what factor would the temperature increase?

Answer:

The temperature would increase by a factor of $256^{0.25}$ (that is, the 4th root of 256), or 4 times.

Lifetimes on the Main Sequence

How many years a star remains in the main-sequence band depends on its mass. You might think that a more massive star, having more fuel, would last longer, but it's not that simple. The lifetime of a star in a particular stage of evolution depends on how much nuclear fuel it has and on *how quickly* it uses up that fuel. (In the same way, how long people can keep spending money depends not only on how much money they have but also on how quickly they spend it. This is why many lottery winners who go on spending sprees quickly wind up poor again.) In the case of stars, more massive ones use up their fuel much more quickly than stars of low mass.

The reason massive stars are such spendthrifts is that, as we saw above, the rate of fusion depends *very* strongly on the star's core temperature. And what determines how hot a star's central regions get? It is the *mass* of the star—the weight of the overlying layers determines how high the pressure in the core must be: higher mass requires higher pressure to balance it. Higher pressure, in turn, is produced by higher temperature. The higher the temperature in the central regions, the faster the star races through its storehouse of central hydrogen. Although massive stars have more fuel, they burn it so prodigiously that their lifetimes are much shorter than those of their low-mass counterparts. You can also understand now why the most massive main-sequence stars are also the most luminous. Like new rock stars with their first platinum album, they spend their resources at an astounding rate.

The main-sequence lifetimes of stars of different masses are listed in Table 22.1. This table shows that the most massive stars spend only a few million years on the main sequence. A star of 1 solar mass remains there for roughly 10 billion years, while a star of about 0.4 solar mass has a main-sequence lifetime of some 200 billion years, which is longer than the current age of the universe. (Bear in mind, however, that every star spends *most* of its total lifetime on the main sequence. Stars devote an average of 90% of their lives to peacefully fusing hydrogen into helium.)

Lifetimes of Main-Sequence Stars

Spectral Type	Surface Temperature (K)	Mass (Mass of Sun = 1)	Lifetime on Main Sequence (years)
O5	54,000	40	1 million
B0	29,200	16	10 million
A0	9600	3.3	500 million
F0	7350	1.7	2.7 billion
G0	6050	1.1	9 billion
K0	5240	0.8	14 billion
M0	3750	0.4	200 billion

Table 22.1

These results are not merely of academic interest. Human beings developed on a planet around a G-type star. This means that the Sun's stable main-sequence lifetime is so long that it afforded life on Earth plenty of time to evolve. When searching for intelligent life like our own on planets around other stars, it would be a pretty big waste of time to search around O- or B-type stars. These stars remain stable for such a short time that the development of creatures complicated enough to take astronomy courses is very unlikely.

From Main-Sequence Star to Red Giant

Eventually, all the hydrogen in a star's core, where it is hot enough for fusion reactions, is used up. The core then contains only helium, "contaminated" by whatever small percentage of heavier elements the star had to begin with. The helium in the core can be thought of as the accumulated "ash" from the nuclear "burning" of hydrogen during the main-sequence stage.

Energy can no longer be generated by hydrogen fusion in the stellar core because the hydrogen is all gone and, as we will see, the fusion of helium requires much higher temperatures. Since the central temperature is not yet high enough to fuse helium, there is no nuclear energy source to supply heat to the central region of the star. The long period of stability now ends, gravity again takes over, and the core begins to contract. Once more, the star's energy is partially supplied by gravitational energy, in the way described by Kelvin and Helmholtz (see Sources of Sunshine: Thermal and Gravitational Energy). As the star's core shrinks, the energy of the inward-falling material is converted to heat.

The heat generated in this way, like all heat, flows outward to where it is a bit cooler. In the process, the heat raises the temperature of a layer of hydrogen that spent the whole long main-sequence time just outside the core. Like an understudy waiting in the wings of a hit Broadway show for a chance at fame and glory, this hydrogen was almost (but not quite) hot enough to undergo fusion and take part in the main action that sustains the star. Now, the additional heat produced by the shrinking core puts this hydrogen "over the limit," and a shell of hydrogen nuclei just outside the core becomes hot enough for hydrogen fusion to begin.

New energy produced by fusion of this hydrogen now pours outward from this shell and begins to heat up layers of the star farther out, causing them to expand. Meanwhile, the helium core continues to contract, producing more heat right around it. This leads to more fusion in the shell of fresh hydrogen outside the core (Figure 22.2). The additional fusion produces still more energy, which also flows out into the upper layer of the star.

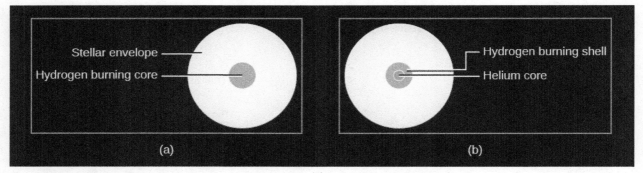

Figure 22.2 Star Layers during and after the Main Sequence. (a) During the main sequence, a star has a core where fusion takes place and a much larger envelope that is too cold for fusion. (b) When the hydrogen in the core is exhausted (made of helium, not hydrogen), the core is compressed by gravity and heats up. The additional heat starts hydrogen fusion in a layer just outside the core. Note that these parts of the Sun are not drawn to scale.

Most stars actually generate more energy each second when they are fusing hydrogen in the shell surrounding the helium core than they did when hydrogen fusion was confined to the central part of the star; thus, they increase in luminosity. With all the new energy pouring outward, the outer layers of the star begin to expand, and the star eventually grows and grows until it reaches enormous proportions (Figure 22.3).

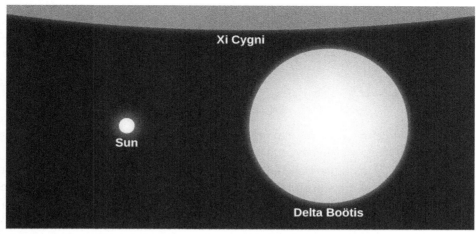

Figure 22.3 Relative Sizes of Stars. This image compares the size of the Sun to that of Delta Boötis, a giant star, and Xi Cygni, a supergiant. Note that Xi Cygni is so large in comparison to the other two stars that only a small portion of it is visible at the top of the frame.

When you take the lid off a pot of boiling water, the steam can expand and it cools down. In the same way, the expansion of a star's outer layers causes the temperature at the surface to decrease. As it cools, the star's overall color becomes redder. (We saw in Radiation and Spectra that a red color corresponds to cooler temperature.)

So the star becomes simultaneously more luminous and cooler. On the H–R diagram, the star therefore leaves the main-sequence band and moves upward (brighter) and to the right (cooler surface temperature). Over time, massive stars become red supergiants, and lower-mass stars like the Sun become red giants. (We first discussed such giant stars in The Stars: A Celestial Census; here we see how such "swollen" stars originate.) You might also say that these stars have "split personalities": their cores are contracting while their outer layers are expanding. (Note that red giant stars do not actually look deep red; their colors are more like orange or orange-red.)

Just how different are these red giants and supergiants from a main-sequence star? Table 22.2 compares the Sun with the red supergiant Betelgeuse, which is visible above Orion's belt as the bright red star that marks the hunter's armpit. Relative to the Sun, this supergiant has a much larger radius, a much lower average density, a cooler surface, and a much hotter core.

Comparing a Supergiant with the Sun

Property	Sun	Betelgeuse
Mass (2×10^{33} g)	1	16
Radius (km)	700,000	500,000,000
Surface temperature (K)	5,800	3,600
Core temperature (K)	15,000,000	160,000,000
Luminosity (4×10^{26} W)	1	46,000
Average density (g/cm^3)	1.4	1.3×10^{-7}

Table 22.2

Property	Sun	Betelgeuse
Age (millions of years)	4,500	10

Table 22.2

Red giants can become so large that if we were to replace the Sun with one of them, its outer atmosphere would extend to the orbit of Mars or even beyond (Figure 22.4). This is the next stage in the life of a star as it moves (to continue our analogy to human lives) from its long period of "youth" and "adulthood" to "old age." By considering the relative ages of the Sun and Betelgeuse, we can also see that the idea that "bigger stars die faster" is indeed true here. Betelgeuse is a mere 10 million years old, which is relatively young compared with our Sun's 4.5 billion years, but it is already nearing its death throes as a red supergiant.

Figure 22.4 Betelgeuse. Betelgeuse is in the constellation Orion, the hunter; in the right image, it is marked with a yellow "X" near the top left. In the left image, we see it in ultraviolet with the Hubble Space Telescope, in the first direct image ever made of the surface of another star. As shown by the scale at the bottom, Betelgeuse has an extended atmosphere so large that, if it were at the center of our solar system, it would stretch past the orbit of Jupiter. (credit: Modification of work by Andrea Dupree (Harvard-Smithsonian CfA), Ronald Gilliland (STScI), NASA and ESA)

Models for Evolution to the Giant Stage

As we discussed earlier, astronomers can construct computer models of stars with different masses and compositions to see how stars change throughout their lives. Figure 22.5, which is based on theoretical calculations by University of Illinois astronomer Icko Iben, shows an H–R diagram with several tracks of evolution from the main sequence to the giant stage. Tracks are shown for stars with different masses (from 0.5 to 15 times the mass of our Sun) and with chemical compositions similar to that of the Sun. The red line is the initial or zero-age main sequence. The numbers along the tracks indicate the time, in years, required for each star to reach those points in their evolution after leaving the main sequence. Once again, you can see that the more massive a star is, the more quickly it goes through each stage in its life.

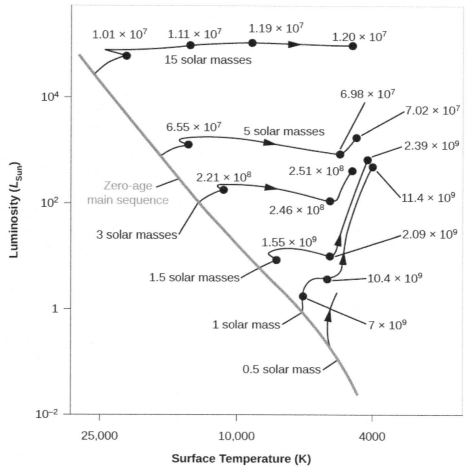

Figure 22.5 Evolutionary Tracks of Stars of Different Masses. The solid black lines show the predicted evolution from the main sequence through the red giant or supergiant stage on the H–R diagram. Each track is labeled with the mass of the star it is describing. The numbers show how many years each star takes to become a giant. The red line is the zero-age main sequence. While theorists debate the exact number of years shown here, our main point should be clear. The more massive the star, the shorter time it takes for each stage in its life.

Note that the most massive star in this diagram has a mass similar to that of Betelgeuse, and so its evolutionary track shows approximately the history of Betelgeuse. The track for a 1-solar-mass star shows that the Sun is still in the main-sequence phase of evolution, since it is only about 4.5 billion years old. It will be billions of years before the Sun begins its own "climb" away from the main sequence—the expansion of its outer layers that will make it a red giant.

LINK TO LEARNING

Use the Star in a Box simulation (https://openstax.org/l/30starinbox) to explore the evolution of stars of different masses. Select the star's mass, and then hit play to see how its luminosity, temperature, and size change during its lifetime.

22.2 | Star Clusters

Learning Objectives

By the end of this section, you will be able to:

> Explain how star clusters help us understand the stages of stellar evolution
> List the different types of star clusters and describe how they differ in number of stars, structure, and age
> Explain why the chemical composition of globular clusters is different from that of open clusters

The preceding description of stellar evolution is based on calculations. However, no star completes its main-sequence lifetime or its evolution to a red giant quickly enough for us to observe these structural changes as they happen. Fortunately, nature has provided us with an indirect way to test our calculations.

Instead of observing the evolution of a single star, we can look at a group or *cluster* of stars. We look for a group of stars that is very close together in space, held together by gravity, often moving around a common center. Then it is reasonable to assume that the individual stars in the group all formed at nearly the same time, from the same cloud, and with the same composition. We expect that these stars will differ only in mass. And their masses determine how quickly they go through each stage of their lives.

Since stars with higher masses evolve more quickly, we can find clusters in which massive stars have already completed their main-sequence phase of evolution and become red giants, while stars of lower mass in the same cluster are still on the main sequence, or even—if the cluster is very young—undergoing pre-main-sequence gravitational contraction. We can see many stages of stellar evolution among the members of a single cluster, and we can see whether our models can explain why the H–R diagrams of clusters of different ages look the way they do.

The three basic types of clusters astronomers have discovered are globular clusters, open clusters, and stellar associations. Their properties are summarized in Table 22.3. As we will see in the next section of this chapter, globular clusters contain only very old stars, whereas open clusters and associations contain young stars.

Characteristics of Star Clusters

Characteristic	Globular Clusters	Open Clusters	Associations
Number in the Galaxy	150	Thousands	Thousands
Location in the Galaxy	Halo and central bulge	Disk (and spiral arms)	Spiral arms
Diameter (in light-years)	50–450	<30	100–500
Mass M_{Sun}	10^4–10^6	10^2–10^3	10^2–10^3
Number of stars	10^4–10^6	50–1000	10^2–10^4

Table 22.3

Characteristic	Globular Clusters	Open Clusters		Associations
Color of brightest stars	Red	Red or blue		Blue
Luminosity of cluster (L_{Sun})	10^4–10^6	10^2–10^6		10^4–10^7
Typical ages	Billions of years	A few hundred million years to, in the case of unusually large clusters, more than a billion years		Up to about 10^7 years

Table 22.3

Globular Clusters

Globular clusters were given this name because they are nearly symmetrical round systems of, typically, hundreds of thousands of stars. The most massive globular cluster in our own Galaxy is Omega Centauri, which is about 16,000 light-years away and contains several million stars (Figure 22.6). Note that the brightest stars in this cluster, which are red giants that have already completed the main-sequence phase of their evolution, are red-orange in color. These stars have typical surface temperatures around 4000 K. As we will see, globular clusters are among the oldest parts of our Milky Way Galaxy.

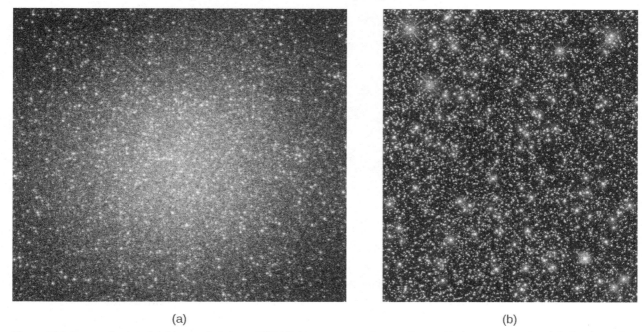

(a)   (b)

Figure 22.6 Omega Centauri. (a) Located at about 16,000 light-years away, Omega Centauri is the most massive globular cluster in our Galaxy. It contains several million stars. (b) This image, taken with the Hubble Space Telescope, zooms in near the center of Omega Centauri. The image is about 6.3 light-years wide. The most numerous stars in the image, which are yellow-white in color, are main-sequence stars similar to our Sun. The brightest stars are red giants that have begun to exhaust their hydrogen fuel and have expanded to about 100 times the diameter of our Sun. The blue stars have started helium fusion. (credit a: modification of work by NASA, ESA and the Hubble Heritage Team (STScI/AURA); credit b: modification of work by NASA, ESA, and the Hubble SM4 ERO Team)

What would it be like to live inside a globular cluster? In the dense central regions, the stars would be roughly a million times closer together than in our own neighborhood. If Earth orbited one of the inner stars in a globular cluster, the nearest stars would be light-months, not light-years, away. They would still appear as points of light, but would be brighter than any of the stars we see in our own sky. The Milky Way would probably be difficult to see through the bright haze of starlight produced by the cluster.

About 150 globular clusters are known in our Galaxy. Most of them are in a spherical halo (or cloud) surrounding the flat disk formed by the majority of our Galaxy's stars. All the globular clusters are very far from the Sun, and some are found at distances of 60,000 light-years or more from the main disk of the Milky Way. The diameters of globular star clusters range from 50 light-years to more than 450 light-years.

Open Clusters

Open clusters are found in the disk of the Galaxy. They have a range of ages, some as old as, or even older than, our Sun. The youngest open clusters are still associated with the interstellar matter from which they formed. Open clusters are smaller than globular clusters, usually having diameters of less than 30 light-years, and they typically contain only several dozen to several hundreds of stars (Figure 22.7). The stars in open clusters usually appear well separated from one another, even in the central regions, which explains why they are called "open." Our Galaxy contains thousands of open clusters, but we can see only a small fraction of them. Interstellar dust, which is also concentrated in the disk, dims the light of more distant clusters so much that they are undetectable.

Figure 22.7 Jewel Box (NGC 4755). This open cluster of young, bright stars is about 6400 light-years away from the Sun. Note the contrast in color between the bright yellow supergiant and the hot blue main-sequence stars. The name comes from John Herschel's nineteenth-century description of it as "a casket of variously colored precious stones." (credit: ESO/Y. Beletsky)

Although the individual stars in an open cluster can survive for billions of years, they typically remain together as a cluster for only a few million years, or at most, a few hundred million years. There are several reasons for this. In small open clusters, the average speed of the member stars within the cluster may be higher than the cluster's escape velocity,[1] and the stars will gradually "evaporate" from the cluster. Close encounters of member stars may also increase the velocity of one of the members beyond the escape velocity. Every few hundred million years or so, the cluster may have a close encounter with a giant molecular cloud, and the gravitational force exerted by the cloud may tear the cluster apart.

Several open clusters are visible to the unaided eye. Most famous among them is the Pleiades (Figure 20.13), which appears as a tiny group of six stars (some people can see even more than six, and the Pleiades is sometimes called the Seven Sisters). This cluster is arranged like a small dipping spoon and is seen in the constellation of Taurus, the bull. A good pair of binoculars shows dozens of stars in the cluster, and a telescope

---

1 Escape velocity is the speed needed to overcome the gravity of some object or group of objects. The rockets we send up from Earth, for example, must travel faster than the escape velocity of our planet to be able to get to other worlds.

reveals hundreds. (A car company, Subaru, takes its name from the Japanese term for this cluster; you can see the star group on the Subaru logo.)

The Hyades is another famous open cluster in the constellation of Taurus the bull. To the naked eye, it appears as a V-shaped group of faint stars marking the face of the bull. Telescopes show that Hyades actually contains more than 200 stars.

Stellar Associations

An **association** is a group of extremely young stars, typically containing 5 to 50 hot, bright O and B stars scattered over a region of space some 100–500 light-years in diameter. As an example, most of the stars in the constellation Orion form one of the nearest stellar associations. Associations also contain hundreds to thousands of low-mass stars, but these are much fainter and less conspicuous. The presence of really hot, luminous stars indicates that star formation in the association has occurred in the last million years or so. Since O stars go through their entire lives in only about a million years, they would not still be around unless star formation has occurred recently. It is therefore not surprising that associations are found in regions rich in the gas and dust required to form new stars. It's like a brand new building still surrounded by some of the construction materials used to build it and with the landscape still showing signs of construction. On the other hand, because associations, like ordinary open clusters, lie in regions occupied by dusty interstellar matter, many are hidden from our view.

22.3 | Checking Out the Theory

Learning Objectives

By the end of this section, you will be able to:
  > Explain how the H–R diagram of a star cluster can be related to the cluster's age and the stages of evolution of its stellar members
  > Describe how the main-sequence turnoff of a cluster reveals its age

In the previous section, we indicated that open clusters are younger than globular clusters, and associations are typically even younger. In this section, we will show how we determine the ages of these star clusters. The key observation is that the stars in these different types of clusters are found in different places in the H–R diagram, and we can use their locations in the diagram in combination with theoretical calculations to estimate how long they have lived.

H–R Diagrams of Young Clusters

What does theory predict for the H–R diagram of a cluster whose stars have recently condensed from an interstellar cloud? Remember that at every stage of evolution, massive stars evolve more quickly than their lower-mass counterparts. After a few million years ("recently" for astronomers), the most massive stars should have completed their contraction phase and be on the main sequence, while the less massive ones should be off to the right, still on their way to the main sequence. These ideas are illustrated in Figure 22.8, which shows the H–R diagram calculated by R. Kippenhahn and his associates at Munich University for a hypothetical cluster with an age of 3 million years.

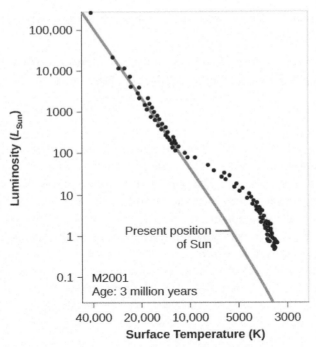

Figure 22.8 Young Cluster H–R Diagram. We see an H–R diagram for a hypothetical young cluster with an age of 3 million years. Note that the high-mass (high-luminosity) stars have already arrived at the main-sequence stage of their lives, while the lower-mass (lower-luminosity) stars are still contracting toward the zero-age main sequence (the red line) and are not yet hot enough to derive all of their energy from the fusion of hydrogen.

There are real star clusters that fit this description. The first to be studied (in about 1950) was NGC 2264, which is still associated with the region of gas and dust from which it was born (Figure 22.9).

Figure 22.9 Young Cluster NGC 2264. Located about 2600 light-years from us, this region of newly formed stars, known as the Christmas Tree Cluster, is a complex mixture of hydrogen gas (which is ionized by hot embedded stars and shown in red), dark obscuring dust lanes, and brilliant young stars. The image shows a scene about 30 light-years across. (credit: ESO)

The NGC 2264 cluster's H–R diagram is shown in Figure 22.10. The cluster in the middle of the Orion Nebula (shown in Figure 21.4 and Figure 21.5) is in a similar stage of evolution.

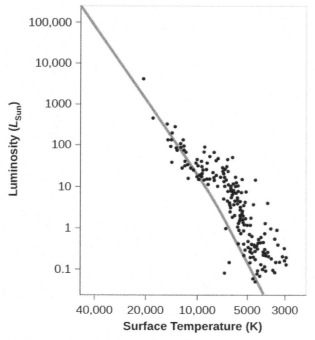

Figure 22.10 NGC 2264 H–R Diagram. Compare this H–R diagram to that in Figure 22.8; although the points scatter a bit more here, the theoretical and observational diagrams are remarkably, and satisfyingly, similar.

As clusters get older, their H–R diagrams begin to change. After a short time (less than a million years after they reach the main sequence), the most massive stars use up the hydrogen in their cores and evolve off the main sequence to become red giants and supergiants. As more time passes, stars of lower mass begin to leave the main sequence and make their way to the upper right of the H–R diagram.

LINK TO LEARNING

To see the evolution of a star cluster in a dwarf galaxy, you can watch this brief animation (https://openstax.org/l/30StarCluster) of how its H–R diagram changes.

Figure 22.11 is a photograph of NGC 3293, a cluster that is about 10 million years old. The dense clouds of gas and dust are gone. One massive star has evolved to become a red giant and stands out as an especially bright orange member of the cluster.

Figure 22.11 NGC 3293. All the stars in an open star cluster like NGC 3293 form at about the same time. The most massive stars, however, exhaust their nuclear fuel more rapidly and hence evolve more quickly than stars of low mass. As stars evolve, they become redder. The bright orange star in NGC 3293 is the member of the cluster that has evolved most rapidly. (credit: ESO/G. Beccari)

Figure 22.12 shows the H–R diagram of the open cluster M41, which is roughly 100 million years old; by this time, a significant number of stars have moved off to the right and become red giants. Note the gap that appears in this H–R diagram between the stars near the main sequence and the red giants. A gap does not necessarily imply that stars avoid a region of certain temperatures and luminosities. In this case, it simply represents a domain of temperature and luminosity through which stars evolve very quickly. We see a gap for M41 because at this particular moment, we have not caught a star in the process of scurrying across this part of the diagram.

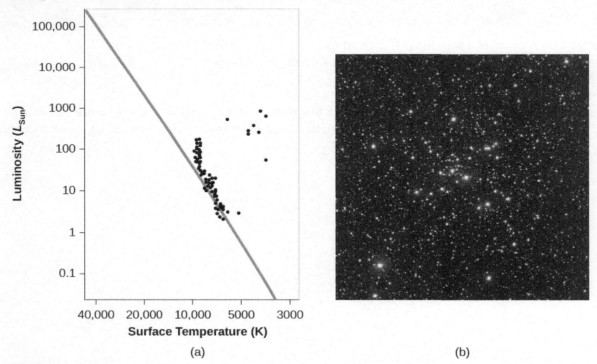

(a)   (b)

Figure 22.12 Cluster M41. (a) Cluster M41 is older than NGC 2264 (see Figure 22.10) and contains several red giants. Some of its more massive stars are no longer close to the zero-age main sequence (red line). (b) This ground-based photograph shows the open cluster M41. Note that it contains several orange-color stars. These are stars that have exhausted hydrogen in their centers, and have swelled up to become red giants. (credit b: modification of work by NOAO/AURA/NSF)

H–R Diagrams of Older Clusters

After 4 billion years have passed, many more stars, including stars that are only a few times more massive than the Sun, have left the main sequence (Figure 22.13). This means that no stars are left near the top of the main sequence; only the low-mass stars near the bottom remain. The older the cluster, the lower the point on the main sequence (and the lower the mass of the stars) where stars begin to move toward the red giant region. The location in the H–R diagram where the stars have begun to leave the main sequence is called the **main-sequence turnoff**.

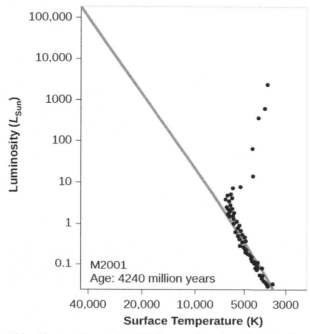

Figure 22.13 H–R Diagram for an Older Cluster. We see the H–R diagram for a hypothetical older cluster at an age of 4.24 billion years. Note that most of the stars on the upper part of the main sequence have turned off toward the red-giant region. And the most massive stars in the cluster have already died and are no longer on the diagram.

The oldest clusters of all are the globular clusters. Figure 22.14 shows the H–R diagram of globular cluster 47 Tucanae. Notice that the luminosity and temperature scales are different from those of the other H–R diagrams in this chapter. In Figure 22.13, for example, the luminosity scale on the left side of the diagram goes from 0.1 to 100,000 times the Sun's luminosity. But in Figure 22.14, the luminosity scale has been significantly reduced in extent. So many stars in this old cluster have had time to turn off the main sequence that only the very bottom of the main sequence remains.

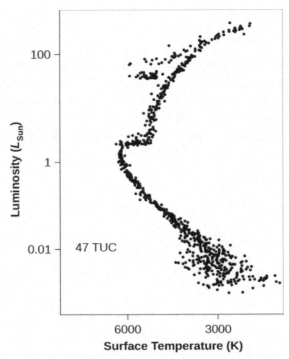

Figure 22.14 Cluster 47 Tucanae. This H–R diagram is for the globular cluster 47. Note that the scale of luminosity differs from that of the other H–R diagrams in this chapter. We are only focusing on the lower portion of the main sequence, the only part where stars still remain in this old cluster.

LINK TO LEARNING

Check out this brief NASA video with a 3-D visualization (https://openstax.org/l/30HRDiagram) of how an H–R diagram is created for the globular cluster Omega Centauri.

Just how old are the different clusters we have been discussing? To get their actual ages (in years), we must compare the appearances of our *calculated* H–R diagrams of different ages to *observed* H–R diagrams of real clusters. In practice, astronomers use the position at the top of the main sequence (that is, the luminosity at which stars begin to move off the main sequence to become red giants) as a measure of the age of a cluster (the main-sequence turnoff we discussed previously). For example, we can compare the luminosities of the brightest stars that are still on the main sequence in Figure 22.10 and Figure 22.13.

Using this method, some associations and open clusters turn out to be as young as 1 million years old, while others are several hundred million years old. Once all of the interstellar matter surrounding a cluster has been used to form stars or has dispersed and moved away from the cluster, star formation ceases, and stars of progressively lower mass move off the main sequence, as shown in Figure 22.10, Figure 22.12, and Figure 22.13.

To our surprise, even the youngest of the globular clusters in our Galaxy are found to be older than the oldest open cluster. All of the globular clusters have main sequences that turn off at a luminosity less than that of the Sun. Star formation in these crowded systems ceased billions of years ago, and no new stars are coming on to the main sequence to replace the ones that have turned off (see Figure 22.15).

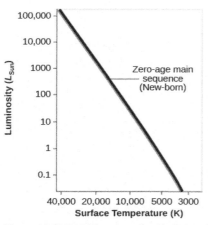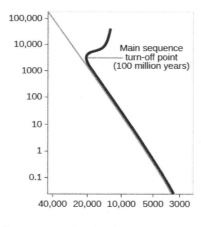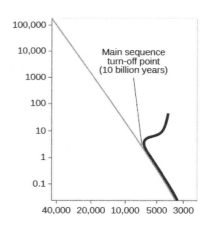

Figure 22.15 H–R Diagrams for Clusters of Different Ages. This sketch shows how the turn-off point from the main sequence gets lower as we make H–R diagrams for clusters that are older and older.

Indeed, the globular clusters are the oldest structures in our Galaxy (and in other galaxies as well). The youngest have ages of about 11 billion years and some appear to be even older. Since these are the oldest objects we know of, this estimate is one of the best limits we have on the age of the universe itself—it must be at least 11 billion years old. We will return to the fascinating question of determining the age of the entire universe in the chapter on The Big Bang.

22.4 | Further Evolution of Stars

Learning Objectives

By the end of this section, you will be able to:

> Explain what happens in a star's core when all of the hydrogen has been used up
> Define "planetary nebulae" and discuss their origin
> Discuss the creation of new chemical elements during the late stages of stellar evolution

The "life story" we have related so far applies to almost all stars: each starts as a contracting protostar, then lives most of its life as a stable main-sequence star, and eventually moves off the main sequence toward the red-giant region.

As we have seen, the pace at which each star goes through these stages depends on its mass, with more massive stars evolving more quickly. But after this point, the life stories of stars of different masses diverge, with a wider range of possible behavior according to their masses, their compositions, and the presence of any nearby companion stars.

Because we have written this book for students taking their first astronomy course, we will recount a simplified version of what happens to stars as they move toward the final stages in their lives. We will (perhaps to your heartfelt relief) not delve into all the possible ways aging stars can behave and the strange things that happen when a star is orbited by a second star in a binary system. Instead, we will focus only on the key stages in the evolution of single stars and show how the evolution of high-mass stars differs from that of low-mass stars (such as our Sun).

Helium Fusion

Let's begin by considering stars with composition like that of the Sun and whose *initial* masses are comparatively low—no more than about twice the mass of our Sun. (Such mass may not seem too low, but stars with masses less than this all behave in a fairly similar fashion. We will see what happens to more massive stars in the next section.) Because there are many more low-mass stars than high-mass stars in the Milky Way, the vast majority of stars—including our Sun—follow the scenario we are about to relate. By the way, we carefully used the term *initial masses* of stars because, as we will see, stars can lose quite a bit of mass in the process of aging and dying.

Remember that red giants start out with a helium core where no energy generation is taking place, surrounded by a shell where hydrogen is undergoing fusion. The core, having no source of energy to oppose the inward pull of gravity, is shrinking and growing hotter. As time goes on, the temperature in the core can rise to much hotter values than it had in its main-sequence days. Once it reaches a temperature of 100 million K (but not before such point), three helium atoms can begin to fuse to form a single carbon nucleus. This process is called the **triple-alpha process**, so named because physicists call the nucleus of the helium atom an alpha particle.

When the triple-alpha process begins in low-mass (about 0.8 to 2.0 solar masses) stars, calculations show that the entire core is ignited in a quick burst of fusion called a **helium flash**. (More massive stars also ignite helium but more gradually and not with a flash.) As soon as the temperature at the center of the star becomes high enough to start the triple-alpha process, the extra energy released is transmitted quickly through the entire helium core, producing very rapid heating. The heating speeds up the nuclear reactions, which provide more heating, and which accelerates the nuclear reactions even more. We have runaway generation of energy, which reignites the entire helium core in a flash.

You might wonder why the next major step in nuclear fusion in stars involves three helium nuclei and not just two. Although it is a lot easier to get two helium nuclei to collide, the product of this collision is not stable and falls apart very quickly. It takes three helium nuclei coming together simultaneously to make a stable nuclear structure. Given that each helium nucleus has two positive protons and that such protons repel one another, you can begin to see the problem. It takes a temperature of 100 million K to slam three helium nuclei (six protons) together and make them stick. But when that happens, the star produces a carbon nucleus.

ASTRONOMY BASICS

Stars in Your Little Finger

Stop reading for a moment and look at your little finger. It's full of carbon atoms because carbon is a fundamental chemical building block for life on Earth. Each of those carbon atoms was once inside a red giant star and was fused from helium nuclei in the triple-alpha process. All the carbon on Earth—in you, in the charcoal you use for barbecuing, and in the diamonds you might exchange with a loved one—was "cooked up" by previous generations of stars. How the carbon atoms (and other elements) made their way from inside some of those stars to become part of Earth is something we will discuss in the next chapter. For now, we want to emphasize that our description of stellar evolution is, in a very real sense, the story of our own cosmic "roots"—the history of how our own atoms originated among the stars. We are made of "star-stuff."

Becoming a Giant Again

After the helium flash, the star, having survived the "energy crisis" that followed the end of the main-sequence stage and the exhaustion of the hydrogen fuel at its center, finds its balance again. As the star readjusts to the release of energy from the triple-alpha process in its core, its internal structure changes once more: its surface temperature increases and its overall luminosity decreases. The point that represents the star on the H–R diagram thus moves to a new position to the left of and somewhat below its place as a red giant (Figure 22.16). The star then continues to fuse the helium in its core for a while, returning to the kind of equilibrium between pressure and gravity that characterized the main-sequence stage. During this time, a newly formed carbon nucleus at the center of the star can sometimes be joined by another helium nucleus to produce a nucleus of oxygen—another building block of life.

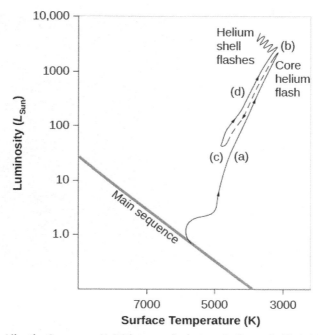

Figure 22.16 Evolution of a Star Like the Sun on an H–R Diagram. Each stage in the star's life is labeled. (a) The star evolves from the main sequence to be a red giant, decreasing in surface temperature and increasing in luminosity. (b) A helium flash occurs, leading to a readjustment of the star's internal structure and to (c) a brief period of stability during which helium is fused to carbon and oxygen in the core (in the process the star becomes hotter and less luminous than it was as a red giant). (d) After the central helium is exhausted, the star becomes a giant again and moves to higher luminosity and lower temperature. By this time, however, the star has exhausted its inner resources and will soon begin to die. Where the evolutionary track becomes a dashed line, the changes are so rapid that they are difficult to model.

However, at a temperature of 100 million K, the inner core is converting its helium fuel to carbon (and a bit of oxygen) at a rapid rate. Thus, the new period of stability cannot last very long: it is far shorter than the main-sequence stage. Soon, all the helium hot enough for fusion will be used up, just like the hot hydrogen that was used up earlier in the star's evolution. Once again, the inner core will not be able to generate energy via fusion. Once more, gravity will take over, and the core will start to shrink again. We can think of stellar evolution as a story of a constant struggle against gravitational collapse. A star can avoid collapsing as long as it can tap energy sources, but once any particular fuel is used up, it starts to collapse again.

The star's situation is analogous to the end of the main-sequence stage (when the central hydrogen got used up), but the star now has a somewhat more complicated structure. Again, the star's core begins to collapse under its own weight. Heat released by the shrinking of the carbon and oxygen core flows into a shell of helium just above the core. This helium, which had not been hot enough for fusion into carbon earlier, is heated just enough for fusion to begin and to generate a new flow of energy.

Farther out in the star, there is also a shell where fresh hydrogen has been heated enough to fuse helium. The star now has a multi-layered structure like an onion: a carbon-oxygen core, surrounded by a shell of helium fusion, a layer of helium, a shell of hydrogen fusion, and finally, the extended outer layers of the star (see Figure 22.17). As energy flows outward from the two fusion shells, once again the outer regions of the star begin to expand. Its brief period of stability is over; the star moves back to the red-giant domain on the H–R diagram for a short time (see Figure 22.16). But this is a brief and final burst of glory.

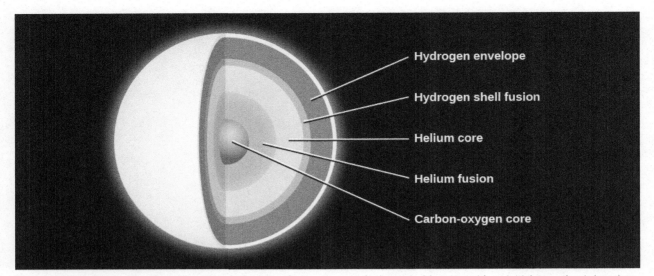

Figure 22.17 Layers inside a Low-Mass Star before Death. Here we see the layers inside a star with an initial mass that is less than twice the mass of the Sun. These include, from the center outward, the carbon-oxygen core, a layer of helium hot enough to fuse, a layer of cooler helium, a layer of hydrogen hot enough to fuse, and then cooler hydrogen beyond.

Recall that the last time the star was in this predicament, helium fusion came to its rescue. The temperature at the star's center eventually became hot enough for the *product* of the previous step of fusion (helium) to become the *fuel* for the next step (helium fusing into carbon). But the step after the fusion of helium nuclei requires a temperature so hot that the kinds of lower-mass stars (less than 2 solar masses) we are discussing simply cannot compress their cores to reach it. No further types of fusion are possible for such a star.

In a star with a mass similar to that of the Sun, the formation of a carbon-oxygen core thus marks the end of the generation of nuclear energy at the center of the star. The star must now confront the fact that its death is near. We will discuss how stars like this end their lives in The Death of Stars, but in the meantime, Table 22.4 summarizes the stages discussed so far in the life of a star with the same mass as that of the Sun. One thing that gives us confidence in our calculations of stellar evolution is that when we make H–R diagrams of older clusters, we actually see stars in each of the stages that we have been discussing.

The Evolution of a Star with the Sun's Mass

Stage	Time in This Stage (years)	Surface Temperature (K)	Luminosity (L_{Sun})	Diameter (Sun = 1)
Main sequence	11 billion	6000	1	1
Becomes red giant	1.3 billion	3100 at minimum	2300 at maximum	165
Helium fusion	100 million	4800	50	10
Giant again	20 million	3100	5200	180

Table 22.4

Mass Loss from Red-Giant Stars and the Formation of Planetary Nebulae

When stars swell up to become red giants, they have very large radii and therefore a low escape velocity.[2] Radiation pressure, stellar pulsations, and violent events like the helium flash can all drive atoms in the outer atmosphere away from the star, and cause it to lose a substantial fraction of its mass into space. Astronomers estimate that by the time a star like the Sun reaches the point of the helium flash, for example, it will have lost as much as 25% of its mass. And it can lose still more mass when it ascends the red-giant branch for the second time. As a result, aging stars are surrounded by one or more expanding shells of gas, each containing as much as 10–20% of the Sun's mass (or 0.1–0.2 M_{Sun}).

When nuclear energy generation in the carbon-oxygen core ceases, the star's core begins to shrink again and to heat up as it gets more and more compressed. (Remember that this compression will not be halted by another type of fusion in these low-mass stars.) The whole star follows along, shrinking and also becoming very hot—reaching surface temperatures as high as 100,000 K. Such hot stars are very strong sources of stellar winds and ultraviolet radiation, which sweep outward into the shells of material ejected when the star was a red giant. The winds and the ultraviolet radiation heat the shells, ionize them, and set them aglow (just as ultraviolet radiation from hot, young stars produces H II regions; see Between the Stars: Gas and Dust in Space).

The result is the creation of some of the most beautiful objects in the cosmos (see the gallery in Figure 22.18 and Figure 22.1). These objects were given an extremely misleading name when first found in the eighteenth century: **planetary nebulae**. The name is derived from the fact that a few planetary nebulae, when viewed through a small telescope, have a round shape bearing a superficial resemblance to planets. Actually, they have nothing to do with planets, but once names are put into regular use in astronomy, it is extremely difficult to change them. There are tens of thousands of planetary nebulae in our own Galaxy, although many are hidden from view because their light is absorbed by interstellar dust.

---

2 Recall that the force of gravity depends not only on the mass doing the pulling, but also on our distance from the center of gravity. As a red giant star gets a lot bigger, a point on the surface of the star is now farther from the center, and thus has less gravity. That's why the speed needed to escape the star goes down.

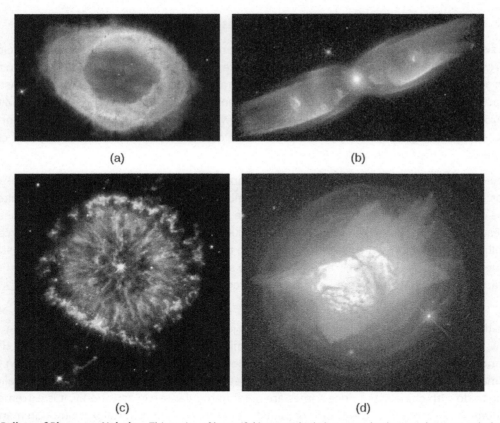

Figure 22.18 Gallery of Planetary Nebulae. This series of beautiful images depicting some intriguing planetary nebulae highlights the capabilities of the Hubble Space Telescope. (a) Perhaps the best known planetary nebula is the Ring Nebula (M57), located about 2000 light-years away in the constellation of Lyra. The ring is about 1 light-year in diameter, and the central star has a temperature of about 120,000 °C. Careful study of this image has shown scientists that, instead of looking at a spherical shell around this dying star, we may be looking down the barrel of a tube or cone. The blue region shows emission from very hot helium, which is located very close to the star; the red region isolates emission from ionized nitrogen, which is radiated by the coolest gas farthest from the star; and the green region represents oxygen emission, which is produced at intermediate temperatures and is at an intermediate distance from the star. (b) This planetary nebula, M2-9, is an example of a butterfly nebula. The central star (which is part of a binary system) has ejected mass preferentially in two opposite directions. In other images, a disk, perpendicular to the two long streams of gas, can be seen around the two stars in the middle. The stellar outburst that resulted in the expulsion of matter occurred about 1200 years ago. Neutral oxygen is shown in red, once-ionized nitrogen in green, and twice-ionized oxygen in blue. The planetary nebula is about 2100 light-years away in the constellation of Ophiuchus. (c) In this image of the planetary nebula NGC 6751, the blue regions mark the hottest gas, which forms a ring around the central star. The orange and red regions show the locations of cooler gas. The origin of these cool streamers is not known, but their shapes indicate that they are affected by radiation and stellar winds from the hot star at the center. The temperature of the star is about 140,000 °C. The diameter of the nebula is about 600 times larger than the diameter of our solar system. The nebula is about 6500 light-years away in the constellation of Aquila. (d) This image of the planetary nebula NGC 7027 shows several stages of mass loss. The faint blue concentric shells surrounding the central region identify the mass that was shed slowly from the surface of the star when it became a red giant. Somewhat later, the remaining outer layers were ejected but not in a spherically symmetric way. The dense clouds formed by this late ejection produce the bright inner regions. The hot central star can be seen faintly near the center of the nebulosity. NGC 7027 is about 3000 light-years away in the direction of the constellation of Cygnus. (credit a: modification of work by NASA, ESA, and the Hubble Heritage (STScI/AURA)-ESA/ Hubble Collaboration; credit b: modification of work by Bruce Balick (University of Washington), Vincent Icke (Leiden University, The Netherlands), Garrelt Mellema (Stockholm University), and NASA; credit c: modification of work by NASA, The Hubble Heritage Team (STScI/AURA); credit d: modification of work by H. Bond (STScI) and NASA)

As Figure 22.18 shows, sometimes a planetary nebula appears to be a simple ring. Others have faint shells surrounding the bright ring, which is evidence that there were multiple episodes of mass loss when the star was a red giant (see image (d) in Figure 22.18). In a few cases, we see two lobes of matter flowing in opposite directions. Many astronomers think that a considerable number of planetary nebulae basically consist of the same structure, but that the shape we see depends on the viewing angle (Figure 22.19). According to this idea, the dying star is surrounded by a very dense, doughnut-shaped disk of gas. (Theorists do not yet have a definite explanation for why the dying star should produce this ring, but many believe that binary stars, which are common, are involved.)

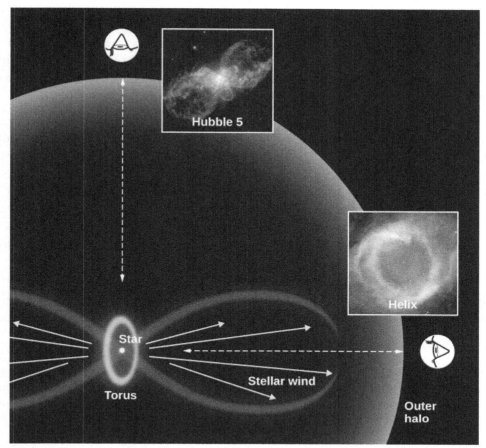

Figure 22.19 Model to Explain the Different Shapes of Planetary Nebulae. The range of different shapes that we see among planetary nebulae may, in many cases, arise from the same geometric shape, but seen from a variety of viewing directions. The basic shape is a hot central star surrounded by a thick torus (or doughnut-shaped disk) of gas. The star's wind cannot flow out into space very easily in the direction of the torus, but can escape more freely in the two directions perpendicular to it. If we view the nebula along the direction of the flow (Helix Nebula), it will appear nearly circular (like looking directly down into an empty ice-cream cone). If we look along the equator of the torus, we see both outflows and a very elongated shape (Hubble 5). Current research on planetary nebulae focuses on the reasons for having a torus around the star in the first place. Many astronomers suggest that the basic cause may be that many of the central stars are actually close binary stars, rather than single stars. (credit "Hubble 5": modification of work by Bruce Balick (University of Washington), Vincent Icke (Leiden University, The Netherlands), Garrelt Mellema (Stockholm University), and NASA/ESA; credit "Helix": modification of work by NASA, ESA, C.R. O'Dell (Vanderbilt University), and M. Meixner, P. McCullough)

As the star continues to lose mass, any less dense gas that leaves the star cannot penetrate the torus, but the gas *can* flow outward in directions perpendicular to the disk. If we look perpendicular to the direction of outflow, we see the disk and both of the outward flows. If we look "down the barrel" and into the flows, we see a ring. At intermediate angles, we may see wonderfully complex structures. Compare the viewpoints in Figure 22.19 with the images in Figure 22.18.

Planetary nebula shells usually expand at speeds of 20–30 km/s, and a typical planetary nebula has a diameter of about 1 light-year. If we assume that the gas shell has expanded at a constant speed, we can calculate that the shells of all the planetary nebulae visible to us were ejected within the past 50,000 years at most. After this amount of time, the shells have expanded so much that they are too thin and tenuous to be seen. That's a pretty short time that each planetary nebula can be observed (when compared to the whole lifetime of the star). Given the number of such nebulae we nevertheless see, we must conclude that a large fraction of all stars evolve through the planetary nebula phase. Since we saw that low-mass stars are much more common than high-mass stars, this confirms our view of planetary nebulae as sort of "last gasp" of low-mass star evolution.

Cosmic Recycling

The loss of mass by dying stars is a key step in the gigantic cosmic recycling scheme we discussed in Between the Stars: Gas and Dust in Space. Remember that stars form from vast clouds of gas and dust. As they end their lives, stars return part of their gas to the galactic reservoirs of raw material. Eventually, some of the expelled material from aging stars will participate in the formation of new star systems.

However, the atoms returned to the Galaxy by an aging star are not necessarily the same ones it received initially. The star, after all, has fused hydrogen and helium to form new elements over the course of its life. And during the red-giant stage, material from the star's central regions is dredged up and mixed with its outer layers, which can cause further nuclear reactions and the creation of still more new elements. As a result, the winds that blow outward from such stars include atoms that were "newly minted" inside the stars' cores. (As we will see, this mechanism is even more effective for high-mass stars, but it does work for stars with masses like that of the Sun.) In this way, the raw material of the Galaxy is not only resupplied but also receives infusions of new elements. You might say this cosmic recycling plan allows the universe to get more "interesting" all the time.

MAKING CONNECTIONS

The Red Giant Sun and the Fate of Earth

How will the evolution of the Sun affect conditions on Earth in the future? Although the Sun has appeared reasonably steady in size and luminosity over recorded human history, that brief span means nothing compared with the timescales we have been discussing. Let's examine the long-term prospects for our planet.

The Sun took its place on the zero-age main sequence approximately 4.5 billion years ago. At that time, it emitted only about 70% of the energy that it radiates today. One might expect that Earth would have been a lot colder than it is now, with the oceans frozen solid. But if this were the case, it would be hard to explain why simple life forms existed when Earth was less than a billion years old. Scientists now think that the explanation may be that much more carbon dioxide was present in Earth's atmosphere when it was young, and that a much stronger greenhouse effect kept Earth warm. (In the greenhouse effect, gases like carbon dioxide or water vapor allow the Sun's light to come in but do not allow the infrared radiation from the ground to escape back into space, so the temperature near Earth's surface increases.)

Carbon dioxide in Earth's atmosphere has steadily declined as the Sun has increased in luminosity. As the brighter Sun increases the temperature of Earth, rocks weather faster and react with carbon dioxide, removing it from the atmosphere. The warmer Sun and the weaker greenhouse effect have kept Earth at a nearly constant temperature for most of its life. This remarkable coincidence, which has resulted in fairly stable climatic conditions, has been the key in the development of complex life-forms on our planet.

As a result of changes caused by the buildup of helium in its core, the Sun will continue to increase in luminosity as it grows older, and more and more radiation will reach Earth. For a while, the amount of carbon dioxide will continue to decrease. (Note that this effect counteracts increases in carbon dioxide from human activities, but on a much-too-slow timescale to undo the changes in climate that are likely to occur in the next 100 years.)

Eventually, the heating of Earth will melt the polar caps and increase the evaporation of the oceans. Water vapor is also an efficient greenhouse gas and will more than compensate for the decrease in carbon dioxide. Sooner or later (atmospheric models are not yet good enough to say exactly when, but estimates range from 500 million to 2 billion years), the increased water vapor will cause a runaway greenhouse effect.

About 1 billion years from now, Earth will lose its water vapor. In the upper atmosphere, sunlight will break down water vapor into hydrogen, and the fast-moving hydrogen atoms will escape into outer space. Like Humpty Dumpty, the water molecules cannot be put back together again. Earth will start to resemble the Venus of today, and temperatures will become much too high for life as we know it.

All of this will happen before the Sun even becomes a red giant. Then the bad news really starts. The Sun, as it expands, will swallow Mercury and Venus, and friction with our star's outer atmosphere will make these planets spiral inward until they are completely vaporized. It is not completely clear whether Earth will escape a similar fate. As described in this chapter, the Sun will lose some of its mass as it becomes a red giant. The gravitational pull of the Sun decreases when it loses mass. The result would be that the diameter of Earth's orbit would increase (remember Kepler's third law). However, recent calculations also show that forces due to the tides raised on the Sun by Earth will act in the opposite direction, causing Earth's orbit to shrink. Thus, many astrophysicists conclude that Earth will be vaporized along with Mercury and Venus. Whether or not this dire prediction is true, there is little doubt that all life on Earth will surely be incinerated. But don't lose any sleep over this—we are talking about events that will occur billions of years from now.

What then are the prospects for preserving Earth life as we know it? The first strategy you might think of would be to move humanity to a more distant and cooler planet. However, calculations indicate that there are long periods of time (several hundred million years) when no planet is habitable. For example, Earth becomes far too warm for life long before Mars warms up enough.

A better alternative may be to move the entire Earth progressively farther from the Sun. The idea is to use gravity in the same way NASA has used it to send spacecraft to distant planets. When a spacecraft flies near a planet, the planet's motion can be used to speed up the spacecraft, slow it down, or redirect it. Calculations show that if we were to redirect an asteroid so that it follows just the right orbit between Earth and Jupiter, it could transfer orbital energy from Jupiter to Earth and move Earth slowly outward, pulling us away from the expanding Sun on each flyby. Since we have hundreds of millions of years to change Earth's orbit, the effect of each flyby need not be large. (Of course, the people directing the asteroid had better get the orbit exactly right and not cause the asteroid to hit Earth.)

It may seem crazy to think about projects to move an entire planet to a different orbit. But remember that we are talking about the distant future. If, by some miracle, human beings are able to get along for all that time and don't blow ourselves to bits, our technology is likely to be far more sophisticated than it is today. It may also be that if humans survive for hundreds of millions of years, we may spread to planets or habitats around other stars. Indeed, Earth, by then, might be a museum world to which youngsters from other planets return to learn about the origin of our species. It is also possible that evolution will by then have changed us in ways that allow us to survive in very different environments. Wouldn't it be exciting to see how the story of the human race turns out after all those billions of years?

22.5 | The Evolution of More Massive Stars

Learning Objectives

By the end of this section, you will be able to:

> Explain how and why massive stars evolve much more rapidly than lower-mass stars like our Sun
> Discuss the origin of the elements heavier than carbon within stars

If what we have described so far were the whole story of the evolution of stars and elements, we would have a big problem on our hands. We will see in later chapters that in our best models of the first few minutes of the universe, everything starts with the two simplest elements—hydrogen and helium (plus a tiny bit of lithium). All the predictions of the models imply that no heavier elements were produced at the beginning of the universe. Yet when we look around us on Earth, we see lots of other elements besides hydrogen and helium.

These elements must have been made (fused) somewhere in the universe, *and the only place hot enough to make them is inside stars*. One of the fundamental discoveries of twentieth-century astronomy is that the stars are the source of most of the chemical richness that characterizes our world and our lives.

We have already seen that carbon, oxygen, and some other elements are made inside lower-mass stars during the red-giant stage. But where do the heavier elements we know and love (such as the silicon and iron inside Earth, and the gold and silver in our jewelry) come from? The kinds of stars we have been discussing so far never get hot enough at their centers to make these elements. It turns out that these heavier elements are formed later in the lives *more massive* stars, as well as during explosions that can mark the end of the lives of some stars.

Making New Elements in Massive Stars

Massive stars evolve in much the same way that the Sun does (but always more quickly)—up to the formation of a carbon-oxygen core. One difference is that for stars with more than about twice the mass of the Sun, helium begins fusion more gradually, rather than with a sudden flash. Also, when more massive stars become red giants, they become so bright and large that we call them *supergiants*. Such stars can expand until their outer regions become as large as the orbit of Jupiter, which is precisely what the Hubble Space Telescope has shown for the star Betelgeuse (see Figure 22.4). They also lose mass very effectively, producing dramatic winds and outbursts as they age. Figure 22.20 shows a wonderful image of the very massive star Eta Carinae, with a great deal of ejected material clearly visible.

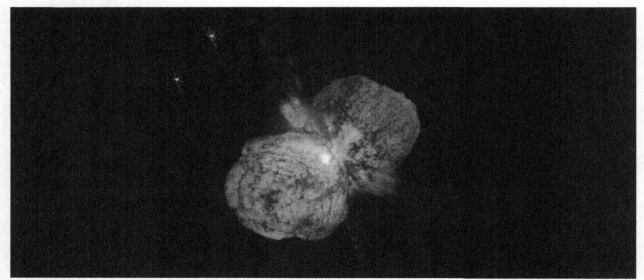

Figure 22.20 Eta Carinae. With a mass at least 100 times that of the Sun, the hot supergiant Eta Carinae is one of the most massive stars known. This Hubble Space Telescope image records the two giant lobes and equatorial disk of material it has ejected in the course of its evolution. The pink outer region is material ejected in an outburst seen in 1843, the largest of such mass loss event that any star is known to have survived. Moving away from the star at a speed of about 1000 km/s, the material is rich in nitrogen and other elements formed in the interior of the star. The inner blue-white region is the material ejected at lower speeds and is thus still closer to the star. It appears blue-white because it contains dust and reflects the light of Eta Carinae, whose luminosity is 4 million times that of our Sun. (credit: modification of work by Jon Morse (University of Colorado) & NASA)

But the crucial way that massive stars diverge from the story we have outlined is that they can start additional kinds of fusion in their centers and in the shells surrounding their central regions. The outer layers of a star with a mass greater than about 8 solar masses have a weight that is enough to compress the carbon-oxygen core until it becomes hot enough to ignite fusion of carbon nuclei. Carbon can fuse into still more oxygen, and at still higher temperatures, oxygen and then neon, magnesium, and finally silicon can build even heavier elements. Iron is, however, the endpoint of this process. The fusion of iron atoms produces products that are *more* massive than the nuclei that are being fused and therefore the process *requires* energy, as opposed to releasing energy, which all fusion reactions up to this point have done. This required energy comes at the expense of the star itself, which is now on the brink of death (Figure 22.21). What happens next will be described in the chapter on The Death of Stars.

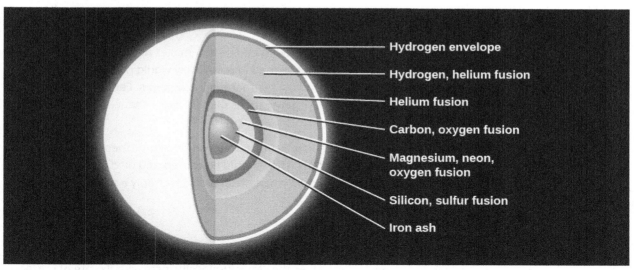

Figure 22.21 Interior Structure of a Massive Star Just before It Exhausts Its Nuclear Fuel. High-mass stars can fuse elements heavier than carbon. As a massive star nears the end of its evolution, its interior resembles an onion. Hydrogen fusion is taking place in an outer shell, and progressively heavier elements are undergoing fusion in the higher-temperature layers closer to the center. All of these fusion reactions generate energy and enable the star to continue shining. Iron is different. The fusion of iron requires energy, and when iron is finally created in the core, the star has only minutes to live.

Physicists have now found nuclear pathways whereby virtually all chemical elements of atomic weights up to that of iron can be built up by this **nucleosynthesis** (the making of new atomic nuclei) in the centers of the more massive red giant stars. This still leaves the question of where elements *heavier* than iron come from. We will see in the next chapter that when massive stars finally exhaust their nuclear fuel, they most often die in a spectacular explosion—a supernova. Heavier elements can be synthesized in the stunning violence of such explosions.

Not only can we explain in this way where the elements that make up our world and others come from, but our theories of nucleosynthesis inside stars are even able to predict the *relative abundances* with which the elements occur in nature. The way stars build up elements during various nuclear reactions really can explain why some elements (oxygen, carbon, and iron) are common and others are quite rare (gold, silver, and uranium).

Elements in Globular Clusters and Open Clusters Are Not the Same

The fact that the elements are made in stars over time explains an important difference between globular and open clusters. Hydrogen and helium, which are the most abundant elements in stars in the solar neighborhood, are also the most abundant constituents of stars in both kinds of clusters. However, the abundances of the elements *heavier* than helium are very different.

In the Sun and most of its neighboring stars, the combined abundance (by mass) of the elements heavier than hydrogen and helium is 1–4% of the star's mass. Spectra show that most open-cluster stars also have 1–4% of their matter in the form of heavy elements. Globular clusters, however, are a different story. The heavy-element abundance of stars in typical globular clusters is found to be only 1/10 to 1/100 that of the Sun. A few very old stars not in clusters have been discovered with even lower abundances of heavy elements.

The differences in chemical composition are a direct consequence of the formation of a cluster of stars. The very first generation of stars initially contained only hydrogen and helium. We have seen that these stars, in order to generate energy, created heavier elements in their interiors. In the last stages of their lives, they ejected matter, now enriched in heavy elements, into the reservoirs of raw material between the stars. Such matter was then incorporated into a new generation of stars.

This means that the relative abundance of the heavy elements must be less and less as we look further into the past. We saw that the globular clusters are much older than the open clusters. Since globular-cluster stars formed much earlier (that is, they are an earlier generation of stars) than those in open clusters, they have only a relatively small abundance of elements heavier than hydrogen and helium.

As time passes, the proportion of heavier elements in the "raw material" that makes new stars and planets increases. This means that the first generation of stars that formed in our Galaxy would not have been accompanied by a planet like Earth, full of silicon, iron, and many other heavy elements. Earth (and the astronomy students who live on it) was possible only after generations of stars had a chance to make and recycle their heavier elements.

Now the search is on for true *first*-generation stars, made only of hydrogen and helium. Theories predict that such stars should be very massive, live fast, and die quickly. They should have lived and died long ago. The place to look for them is in very distant galaxies that formed when the universe was only a few hundred million years old, but whose light is only arriving at Earth now.

Approaching Death

Compared with the main-sequence lifetimes of stars, the events that characterize the last stages of stellar evolution pass very quickly (especially for massive stars). As the star's luminosity increases, its rate of nuclear fuel consumption goes up rapidly—just at that point in its life when its fuel supply is beginning to run down.

After the prime fuel—hydrogen—is exhausted in a star's core, we saw that other sources of nuclear energy are available to the star in the fusion of, first, helium, and then of other more complex elements. But the energy yield of these reactions is much less than that of the fusion of hydrogen to helium. And to trigger these reactions, the central temperature must be higher than that required for the fusion of hydrogen to helium, leading to even more rapid consumption of fuel. Clearly this is a losing game, and very quickly the star reaches its end. As it does so, however, some remarkable things can happen, as we will see in The Death of Stars.

🔑 Key Terms

association   a loose group of young stars whose spectral types, motions, and positions in the sky indicate a common origin

globular cluster   one of about 150 large, spherical star clusters (each with hundreds of thousands of stars) that form a spherical halo around the center of our Galaxy

helium flash   a nearly explosive ignition of helium in the triple-alpha process in the dense core of a red giant star

main-sequence turnoff   location in the H–R diagram where stars begin to leave the main sequence

nucleosynthesis   the building up of heavy elements from lighter ones by nuclear fusion

open cluster   a comparatively loose cluster of stars, containing from a few dozen to a few thousand members, located in the spiral arms or disk of our Galaxy; sometimes referred to as a galactic cluster

planetary nebula   a shell of gas ejected by and expanding away from an extremely hot low-mass star that is nearing the end of its life (the nebulae glow because of the ultra-violet energy of the central star)

triple-alpha process   a nuclear reaction by which three helium nuclei are built up (fused) into one carbon nucleus

zero-age main sequence   a line denoting the main sequence on the H–R diagram for a system of stars that have completed their contraction from interstellar matter and are now deriving all their energy from nuclear reactions, but whose chemical composition has not yet been altered substantially by nuclear reactions

📄 Summary

22.1 Evolution from the Main Sequence to Red Giants

When stars first begin to fuse hydrogen to helium, they lie on the zero-age main sequence. The amount of time a star spends in the main-sequence stage depends on its mass. More massive stars complete each stage of evolution more quickly than lower-mass stars. The fusion of hydrogen to form helium changes the interior composition of a star, which in turn results in changes in its temperature, luminosity, and radius. Eventually, as stars age, they evolve away from the main sequence to become red giants or supergiants. The core of a red giant is contracting, but the outer layers are expanding as a result of hydrogen fusion in a shell outside the core. The star gets larger, redder, and more luminous as it expands and cools.

22.2 Star Clusters

Star clusters provide one of the best tests of our calculations of what happens as stars age. The stars in a given cluster were formed at about the same time and have the same composition, so they differ mainly in mass, and thus, in their life stage. There are three types of star clusters: globular, open, and associations. Globular clusters have diameters of 50–450 light-years, contain hundreds of thousands of stars, and are distributed in a halo around the Galaxy. Open clusters typically contain hundreds of stars, are located in the plane of the Galaxy, and have diameters less than 30 light-years. Associations are found in regions of gas and dust and contain extremely young stars.

22.3 Checking Out the Theory

The H–R diagram of stars in a cluster changes systematically as the cluster grows older. The most massive stars evolve most rapidly. In the youngest clusters and associations, highly luminous blue stars are on the main sequence; the stars with the lowest masses lie to the right of the main sequence and are still contracting toward it. With passing time, stars of progressively lower masses evolve away from (or turn off) the main sequence. In globular clusters, which are all at least 11 billion years old, there are no luminous blue stars at all. Astronomers can use the turnoff point from the main sequence to determine the age of a cluster.

22.4 Further Evolution of Stars

After stars become red giants, their cores eventually become hot enough to produce energy by fusing helium to form carbon (and sometimes a bit of oxygen.) The fusion of three helium nuclei produces carbon through the triple-alpha process. The rapid onset of helium fusion in the core of a low-mass star is called the helium flash. After this, the star becomes stable and reduces its luminosity and size briefly. In stars with masses about twice the mass of the Sun or less, fusion stops after the helium in the core has been exhausted. Fusion of hydrogen and helium in shells around the contracting core makes the star a bright red giant again, but only temporarily. When the star is a red giant, it can shed its outer layers and thereby expose hot inner layers. Planetary nebulae (which have nothing to do with planets) are shells of gas ejected by such stars, set glowing by the ultraviolet radiation of the dying central star.

22.5 The Evolution of More Massive Stars

In stars with masses higher than about 8 solar masses, nuclear reactions involving carbon, oxygen, and still heavier elements can build up nuclei as heavy as iron. The creation of new chemical elements is called nucleosynthesis. The late stages of evolution occur very quickly. Ultimately, all stars must use up all of their available energy supplies. In the process of dying, most stars eject some matter, enriched in heavy elements, into interstellar space where it can be used to form new stars. Each succeeding generation of stars therefore contains a larger proportion of elements heavier than hydrogen and helium. This progressive enrichment explains why the stars in open clusters (which formed more recently) contain more heavy elements than do those in ancient globular clusters, and it tells us where most of the atoms on Earth and in our bodies come from.

 For Further Exploration

Articles

Balick, B. & Frank, A. "The Extraordinary Deaths of Ordinary Stars." *Scientific American* (July 2004): 50. About planetary nebulae, the last gasps of low-mass stars, and the future of our own Sun.

Djorgovsky, G. "The Dynamic Lives of Globular Clusters." *Sky & Telescope* (October 1998): 38. Cluster evolution and blue straggler stars.

Frank, A. "Angry Giants of the Universe." *Astronomy* (October 1997): 32. On luminous blue variables like Eta Carinae.

Garlick, M. "The Fate of the Earth." *Sky & Telescope* (October 2002): 30. What will happen when our Sun becomes a red giant.

Harris, W. & Webb, J. "Life Inside a Globular Cluster." *Astronomy* (July 2014): 18. What would night sky be like there?

Iben, I. & Tutokov, A. "The Lives of the Stars: From Birth to Death and Beyond." *Sky & Telescope* (December 1997): 36.

Kaler, J. "The Largest Stars in the Galaxy." *Astronomy* (October 1990): 30. On red supergiants.

Kalirai, J. "New Light on Our Sun's Fate." *Astronomy* (February 2014): 44. What will happen to stars like our Sun between the main sequence and the white dwarf stages.

Kwok, S. "What Is the Real Shape of the Ring Nebula?" *Sky & Telescope* (July 2000): 33. On seeing planetary nebulae from different angles.

Kwok, S. "Stellar Metamorphosis." *Sky & Telescope* (October 1998): 30. How planetary nebulae form.

Stahler, S. "The Inner Life of Star Clusters." *Scientific American* (March 2013): 44–49. How all stars are born in clusters, but different clusters evolve differently.

Subinsky, R. "All About 47 Tucanae." *Astronomy* (September 2014): 66. What we know about this globular cluster and how to see it.

Websites

Encylopedia Britannica Article on Star Clusters: http://www.britannica.com/topic/star-cluster (http://www.britannica.com/topic/star-cluster). Written by astronomer Helen Sawyer Hogg-Priestley.

Hubble Image Gallery: Planetary Nebulae: Go to https://www.spacetelescope.org/images/ (https://www.spacetelescope.org/images/) and in the search box enter "planetary nebulae". See also a similar gallery from NOIRLab: https://noirlab.edu/public/images/ (https://noirlab.edu/public/images/). (Again, enter "planetary nebulae" in the search box.)

"Hubble Image Gallery: Star Clusters: Go to https://www.spacetelescope.org/images/ (https://www.spacetelescope.org/images/) and in the search box enter "star clusters". (See also a similar European Southern Observatory Gallery at: https://www.eso.org/public/images/archive/category/ starclusters/ (https://www.eso.org/public/images/archive/category/starclusters/)).

Measuring the Age of a Star Cluster: https://www.e-education.psu.edu/astro801/content/l7_p6.html (https://www.e-education.psu.edu/astro801/content/l7_p6.html). From Penn State.

Videos

Life Cycle of Stars: https://www.youtube.com/watch?v=PM9CQDIQI0A (https://www.youtube.com/ watch?v=PM9CQDIQI0A). Short summary of stellar evolution from the Institute of Physics in Great Britain, with astronomer Tim O'Brien (4:58).

Missions Take an Unparalleled Look into Superstar Eta Carinae: https://www.youtube.com/ watch?v=0rJQi6oaZf0 (https://www.youtube.com/watch?v=0rJQi6oaZf0). NASA Goddard video about observations in 2014 and what we know about the pair of stars in this complicated system (4:00).

Star Clusters: Open and Globular Clusters: https://www.youtube.com/watch?v=rGPRLxrYbYA (https://www.youtube.com/watch?v=rGPRLxrYbYA). Three Short Hubblecast Videos from 2007–2008 on discoveries involving star clusters (12:24).

Tour of Planetary Nebula NGC 5189: https://www.youtube.com/watch?v=1D2cwiZld0o (https://www.youtube.com/watch?v=1D2cwiZld0o). Brief Hubblecast episode with Joe Liske, explaining planetary nebulae in general and one example in particular (5:22).

Collaborative Group Activities

A. Have your group take a look at the list of the brightest stars in the sky in Appendix J. What fraction of them are past the main-sequence phase of evolution? The text says that stars spend 90% of their lifetimes in the main-sequence phase of evolution. This suggests that if we have a fair (or representative) sample of stars, 90% of them should be main-sequence stars. Your group should brainstorm why 90% of the brightest stars are not in the main-sequence phase of evolution.

B. Reading an H–R diagram can be tricky. Suppose your group is given the H–R diagram of a star cluster. Stars above and to the right of the main sequence could be either red giants that had evolved away from the main sequence or very young stars that are still evolving toward the main sequence. Discuss how you would decide which they are.

C. In the chapter on Life in the Universe, we discuss some of the efforts now underway to search for radio signals from possible intelligent civilizations around other stars. Our present resources for carrying out such searches are very limited and there are many stars in our Galaxy. Your group is a committee set up by the International Astronomical Union to come up with a list of the best possible stars with which such a search should begin. Make a list of criteria for choosing the stars on the list, and explain the reasons behind each entry (keeping in mind some of the ideas about the life story of stars and timescales that we discuss in the present chapter.)

D. Have your group make a list of the reasons why a star that formed at the very beginning of the universe (soon after the Big Bang) could not have a planet with astronomy students reading astronomy textbooks (even if the star has the same mass as that of our Sun).

E. Since we are pretty sure that when the Sun becomes a giant star, all life on Earth will be wiped out, does your group think that we should start making preparations of any kind? Let's suppose that a political leader who fell asleep during large parts of his astronomy class suddenly hears about this problem from a large donor and appoints your group as a task force to make suggestions on how to prepare for the end of Earth. Make a list of arguments for why such a task force is not really necessary.

F. Use star charts to identify at least one open cluster visible at this time of the year. (Such charts can be found in *Sky & Telescope* and *Astronomy* magazines each month and their websites; see Appendix B.) The Pleiades and Hyades are good autumn subjects, and Praesepe is good for springtime viewing. Go out and look at these clusters with binoculars and describe what you see.

G. Many astronomers think that planetary nebulae are among the most attractive and interesting objects we can see in the Galaxy. In this chapter, we could only show you a few examples of the pictures of these objects taken with the Hubble or large telescopes on the ground. Have members of your group search further for planetary nebula images online, and make a "top ten" list of your favorite ones (do not include more than three that were featured in this chapter.) Make a report (with images) for the whole class and explain why you found your top five especially interesting. (You may want to check Figure 22.19 in the process.)

Exercises

Review Questions

1. Compare the following stages in the lives of a human being and a star: prenatal, birth, adolescence/adulthood, middle age, old age, and death. What does a star with the mass of our Sun do in each of these stages?

2. What is the first event that happens to a star with roughly the mass of our Sun that exhausts the hydrogen in its core and stops the generation of energy by the nuclear fusion of hydrogen to helium? Describe the sequence of events that the star undergoes.

3. Astronomers find that 90% of the stars observed in the sky are on the main sequence of an H–R diagram; why does this make sense? Why are there far fewer stars in the giant and supergiant region?

4. Describe the evolution of a star with a mass similar to that of the Sun, from the protostar stage to the time it first becomes a red giant. Give the description in words and then sketch the evolution on an H–R diagram.

5. Describe the evolution of a star with a mass similar to that of the Sun, from just after it first becomes a red giant to the time it exhausts the last type of fuel its core is capable of fusing.

6. A star is often described as "moving" on an H–R diagram; why is this description used and what is actually happening with the star?

7. On which edge of the main sequence band on an H–R diagram would the zero-age main sequence be?

8. How do stars typically "move" through the main sequence band on an H–R diagram? Why?

9. Certain stars, like Betelgeuse, have a lower surface temperature than the Sun and yet are more luminous. How do these stars produce so much more energy than the Sun?

10. Gravity always tries to collapse the mass of a star toward its center. What mechanism can oppose this gravitational collapse for a star? During what stages of a star's life would there be a "balance" between them?

11. Why are star clusters so useful for astronomers who want to study the evolution of stars?

12. Would the Sun more likely have been a member of a globular cluster or open cluster in the past?

13. Suppose you were handed two H–R diagrams for two different clusters: diagram A has a majority of its stars plotted on the upper left part of the main sequence with the rest of the stars off the main sequence; and diagram B has a majority of its stars plotted on the lower right part of the main sequence with the rest of the stars off the main sequence. Which diagram would be for the older cluster? Why?

14. Referring to the H–R diagrams in Exercise 22.13, which diagram would more likely be the H–R diagram for an association?

15. The nuclear process for fusing helium into carbon is often called the "triple-alpha process." Why is it called as such, and why must it occur at a much higher temperature than the nuclear process for fusing hydrogen into helium?

16. Pictures of various planetary nebulae show a variety of shapes, but astronomers believe a majority of planetary nebulae have the same basic shape. How can this paradox be explained?

17. Describe the two "recycling" mechanisms that are associated with stars (one during each star's life and the other connecting generations of stars).

18. In which of these star groups would you mostly likely find the least heavy-element abundance for the stars within them: open clusters, globular clusters, or associations?

19. Explain how an H–R diagram of the stars in a cluster can be used to determine the age of the cluster.

20. Where did the carbon atoms in the trunk of a tree on your college campus come from originally? Where did the neon in the fabled "neon lights of Broadway" come from originally?

21. What is a planetary nebula? Will we have one around the Sun?

Thought Questions

22. Is the Sun on the zero-age main sequence? Explain your answer.

23. How are planetary nebulae comparable to a fluorescent light bulb in your classroom?

24. Which of the planets in our solar system have orbits that are smaller than the photospheric radius of Betelgeuse listed in Table 22.2?

25. Would you expect to find an earthlike planet (with a solid surface) around a very low-mass star that formed right at the beginning of a globular cluster's life? Explain.

26. In the H–R diagrams for some young clusters, stars of both very low and very high luminosity are off to the right of the main sequence, whereas those of intermediate luminosity are on the main sequence. Can you offer an explanation for that? Sketch an H–R diagram for such a cluster.

27. If the Sun were a member of the cluster NGC 2264, would it be on the main sequence yet? Why or why not?

28. If all the stars in a cluster have nearly the same age, why are clusters useful in studying evolutionary effects (different stages in the lives of stars)?

29. Suppose a star cluster were at such a large distance that it appeared as an unresolved spot of light through the telescope. What would you expect the overall color of the spot to be if it were the image of the cluster immediately after it was formed? How would the color differ after 10^{10} years? Why?

30. Suppose an astronomer known for joking around told you she had found a type-O main-sequence star in our Milky Way Galaxy that contained no elements heavier than helium. Would you believe her? Why?

31. Stars that have masses approximately 0.8 times the mass of the Sun take about 18 billion years to turn into red giants. How does this compare to the current age of the universe? Would you expect to find a globular cluster with a main-sequence turnoff for stars of 0.8 solar mass or less? Why or why not?

32. Automobiles are often used as an analogy to help people better understand how more massive stars have much shorter main-sequence lifetimes compared to less massive stars. Can you explain such an analogy using automobiles?

Figuring for Yourself

33. The text says a star does not change its mass very much during the course of its main-sequence lifetime. While it is on the main sequence, a star converts about 10% of the hydrogen initially present into helium (remember it's only the core of the star that is hot enough for fusion). Look in earlier chapters to find out what percentage of the hydrogen mass involved in fusion is lost because it is converted to energy. By how much does the mass of the whole star change as a result of fusion? Were we correct to say that the mass of a star does not change significantly while it is on the main sequence?

34. The text explains that massive stars have shorter lifetimes than low-mass stars. Even though massive stars have more fuel to burn, they use it up faster than low-mass stars. You can check and see whether this statement is true. The lifetime of a star is directly proportional to the amount of mass (fuel) it contains and inversely proportional to the rate at which it uses up that fuel (i.e., to its luminosity). Since the lifetime of the Sun is about 10^{10} y, we have the following relationship:
$$T = 10^{10} \frac{M}{L} \text{ y}$$
where T is the lifetime of a main-sequence star, M is its mass measured in terms of the mass of the Sun, and L is its luminosity measured in terms of the Sun's luminosity.
A. Explain in words why this equation works.
B. Use the data in Table 18.3 to calculate the ages of the main-sequence stars listed.
C. Do low-mass stars have longer main-sequence lifetimes?
D. Do you get the same answers as those in Table 22.1?

35. You can use the equation in Exercise 22.34 to estimate the approximate ages of the clusters in Figure 22.10, Figure 22.12, and Figure 22.13. Use the information in the figures to determine the luminosity of the most massive star still on the main sequence. Now use the data in Table 18.3 to estimate the mass of this star. Then calculate the age of the cluster. This method is similar to the procedure used by astronomers to obtain the ages of clusters, except that they use actual data and model calculations rather than simply making estimates from a drawing. How do your ages compare with the ages in the text?

36. You can estimate the age of the planetary nebula in image (c) in Figure 22.18. The diameter of the nebula is 600 times the diameter of our own solar system, or about 0.8 light-year. The gas is expanding away from the star at a rate of about 25 mi/s. Considering that distance = velocity × time, calculate how long ago the gas left the star if its speed has been constant the whole time. Make sure you use consistent units for time, speed, and distance.

37. If star A has a core temperature T, and star B has a core temperature $3T$, how does the rate of fusion of star A compare to the rate of fusion of star B?

23

The Death of Stars

Figure 23.1 Stellar Life Cycle. This remarkable picture of NGC 3603, a nebula in the Milky Way Galaxy, was taken with the Hubble Space Telescope. This image illustrates the life cycle of stars. In the bottom half of the image, we see clouds of dust and gas, where it is likely that star formation will take place in the near future. Near the center, there is a cluster of massive, hot young stars that are only a few million years old. Above and to the right of the cluster, there is an isolated star surrounded by a ring of gas. Perpendicular to the ring and on either side of it, there are two bluish blobs of gas. The ring and the blobs were ejected by the star, which is nearing the end of its life. (credit: modification of work by NASA, Wolfgang Brandner (JPL/IPAC), Eva K. Grebel (University of Washington), You-Hua Chu (University of Illinois Urbana-Champaign))

Chapter Outline

23.1 The Death of Low-Mass Stars
23.2 Evolution of Massive Stars: An Explosive Finish
23.3 Supernova Observations
23.4 Pulsars and the Discovery of Neutron Stars
23.5 The Evolution of Binary Star Systems
23.6 The Mystery of the Gamma-Ray Bursts

Thinking Ahead

Do stars die with a bang or a whimper? In the preceding two chapters, we followed the life story of stars, from the process of birth to the brink of death. Now we are ready to explore the ways that stars end their lives. Sooner or later, each star exhausts its store of nuclear energy. Without a source of internal pressure to balance the weight of the overlying layers, every star eventually gives way to the inexorable pull of gravity and collapses under its own weight.

Following the rough distinction made in the last chapter, we will discuss the end-of-life evolution of stars of lower and higher mass separately. What determines the outcome—bang or whimper—is the mass of the star *when it is ready to die*, not the mass it was born with. As we noted in the last chapter, stars can lose a significant amount of mass in their middle and old age.

23.1 | The Death of Low-Mass Stars

Learning Objectives

By the end of this section, you will be able to:

> Describe the physical characteristics of degenerate matter and explain how the mass and radius of degenerate stars are related

>> Plot the future evolution of a white dwarf and show how its observable features will change over time

>> Distinguish which stars will become white dwarfs

Let's begin with those stars whose final mass just before death is less than about 1.4 times the mass of the Sun (M_{Sun}). (We will explain why this mass is the crucial dividing line in a moment.) Note that most stars in the universe fall into this category. The number of stars decreases as mass increases; really massive stars are rare (see The Stars: A Celestial Census). This is similar to the music business where only a few musicians ever become superstars. Furthermore, many stars with an initial mass much greater than 1.4 M_{Sun} will be reduced to that level by the time they die. For example, we now know that stars that start out with masses of at least 8.0 M_{Sun} (and possibly as much as 10 M_{Sun}) manage to lose enough mass during their lives to fit into this category (an accomplishment anyone who has ever attempted to lose weight would surely envy).

A Star in Crisis

In the last chapter, we left the life story of a star with a mass like the Sun's just after it had climbed up to the red-giant region of the H–R diagram for a second time and had shed some of its outer layers to form a planetary nebula. Recall that during this time, the *core* of the star was undergoing an "energy crisis." Earlier in its life, during a brief stable period, helium in the core had gotten hot enough to fuse into carbon (and oxygen). But after this helium was exhausted, the star's core had once more found itself without a source of pressure to balance gravity and so had begun to contract.

This collapse is the final event in the life of the core. Because the star's mass is relatively low, it cannot push its core temperature high enough to begin another round of fusion (in the same way larger-mass stars can). The core continues to shrink until it reaches a density equal to nearly a million times the density of water! That is 200,000 times greater than the average density of Earth. At this extreme density, a new and different way for matter to behave kicks in and helps the star achieve a final state of equilibrium. In the process, what remains of the star becomes one of the strange *white dwarfs* that we met in The Stars: A Celestial Census.

Degenerate Stars

Because white dwarfs are far denser than any substance on Earth, the matter inside them behaves in a very unusual way—unlike anything we know from everyday experience. At this high density, gravity is incredibly strong and tries to shrink the star still further, but all the *electrons* resist being pushed closer together and set up a powerful pressure inside the core. This pressure is the result of the fundamental rules that govern the behavior of electrons (the quantum physics you were introduced to in The Sun: A Nuclear Powerhouse). According to these rules (known to physicists as the Pauli exclusion principle), which have been verified in studies of atoms in the laboratory, no two electrons can be in the same place at the same time doing the same thing. We specify the *place* of an electron by its position in space, and we specify what it is doing by its motion and the way it is spinning.

The temperature in the interior of a star is always so high that the atoms are stripped of virtually all their electrons. For most of a star's life, the density of matter is also relatively low, and the electrons in the star are moving rapidly. This means that no two of them will be in the same place moving in exactly the same way at the same time. But this all changes when a star exhausts its store of nuclear energy and begins its final collapse.

As the star's core contracts, electrons are squeezed closer and closer together. Eventually, a star like the Sun becomes so dense that further contraction would in fact require two or more electrons to violate the rule against occupying the same place and moving in the same way. Such a dense gas is said to be degenerate (a term coined by physicists and not related to the electron's moral character). The electrons in a **degenerate gas** resist further crowding with tremendous pressure. (It's as if the electrons said, "You can press inward all you want, but there is simply no room for any other electrons to squeeze in here without violating the rules of our existence.")

The degenerate electrons do not require an input of heat to maintain the pressure they exert, and so a star with this kind of structure, if nothing disturbs it, can last essentially forever. (Note that the repulsive force between degenerate electrons is different from, and much stronger than, the normal electrical repulsion between charges that have the same sign.)

The electrons in a degenerate gas do move about, as do particles in any gas, but not with a lot of freedom. A particular electron cannot change position or momentum until another electron in an adjacent stage gets out of the way. The situation is much like that in the parking lot after a big football game. Vehicles are closely packed, and a given car cannot move until the one in front of it moves, leaving an empty space to be filled.

Of course, the dying star also has atomic nuclei in it, not just electrons, but it turns out that the nuclei must be squeezed to much higher densities before their quantum nature becomes apparent. As a result, in white dwarfs, the nuclei do not exhibit degeneracy pressure. Hence, in the white dwarf stage of stellar evolution, it is the degeneracy pressure of the electrons, and not of the nuclei, that halts the collapse of the core.

White Dwarfs

White dwarfs, then, are stable, compact objects with electron-degenerate cores that cannot contract any further. Calculations showing that white dwarfs are the likely end state of low-mass stars were first carried out by the Indian-American astrophysicist Subrahmanyan Chandrasekhar. He was able to show how much a star will shrink before the degenerate electrons halt its further contraction and hence what its final diameter will be (Figure 23.2).

When Chandrasekhar made his calculation about white dwarfs, he found something very surprising: the radius of a white dwarf shrinks as the mass in the star increases (the larger the mass, the more tightly packed the electrons can become, resulting in a smaller radius). According to the best theoretical models, a white dwarf with a mass of about 1.4 M_{Sun} or larger would have a radius of zero. What the calculations are telling us is that even the force of degenerate electrons cannot stop the collapse of a star with more mass than this. The maximum mass that a star can end its life with and still become a white dwarf—1.4 M_{Sun}—is called the **Chandrasekhar limit**. Stars with end-of-life masses that exceed this limit have a different kind of end in store—one that we will explore in the next section.

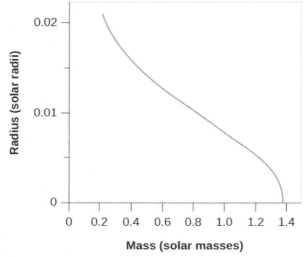

Figure 23.2 Relating Masses and Radii of White Dwarfs. Models of white-dwarf structure predict that as the mass of the star

increases (toward the right), its radius gets smaller and smaller.

VOYAGERS IN ASTRONOMY

Subrahmanyan Chandrasekhar

Born in 1910 in Lahore, India, Subrahmanyan Chandrasekhar (known as Chandra to his friends and colleagues) grew up in a home that encouraged scholarship and an interest in science (Figure 23.3). His uncle, C. V. Raman, was a physicist who won the 1930 Nobel Prize. A precocious student, Chandra tried to read as much as he could about the latest ideas in physics and astronomy, although obtaining technical books was not easy in India at the time. He finished college at age 19 and won a scholarship to study in England. It was during the long boat voyage to get to graduate school that he first began doing calculations about the structure of white dwarf stars.

Chandra developed his ideas during and after his studies as a graduate student, showing—as we have discussed—that white dwarfs with masses greater than 1.4 times the mass of the Sun cannot exist and that the theory predicts the existence of other kinds of stellar corpses. He wrote later that he felt very shy and lonely during this period, isolated from students, afraid to assert himself, and sometimes waiting for hours to speak with some of the famous professors he had read about in India. His calculations soon brought him into conflict with certain distinguished astronomers, including Sir Arthur Eddington, who publicly ridiculed Chandra's ideas. At a number of meetings of astronomers, such leaders in the field as Henry Norris Russell refused to give Chandra the opportunity to defend his ideas, while allowing his more senior critics lots of time to criticize them.

Yet Chandra persevered, writing books and articles elucidating his theories, which turned out not only to be correct, but to lay the foundation for much of our modern understanding of the death of stars. In 1983, he received the Nobel Prize in physics for this early work.

In 1937, Chandra came to the United States and joined the faculty at the University of Chicago, where he remained for the rest of his life. There he devoted himself to research and teaching, making major contributions to many fields of astronomy, from our understanding of the motions of stars through the Galaxy to the behavior of the bizarre objects called black holes (see Black Holes and Curved Spacetime). In 1999, NASA named its sophisticated orbiting X-ray telescope (designed in part to explore such stellar corpses) the Chandra X-ray Observatory.

Figure 23.3 S. Chandrasekhar (1910–1995). Chandra's research provided the basis for much of what we now know about stellar corpses. (credit: modification of work by American Institute of Physics)

Chandra spent a great deal of time with his graduate students, supervising the research of more than 50 PhDs during his life. He took his teaching responsibilities very seriously: during the 1940s, while based at the Yerkes Observatory, he willingly drove the more than 100-mile trip to the university each week to teach a class of only a few students.

Chandra also had a deep devotion to music, art, and philosophy, writing articles and books about the relationship between the humanities and science. He once wrote that "one can learn science the way one enjoys music or art. . . . Heisenberg had a marvelous phrase 'shuddering before the beautiful'. . . that is the kind of feeling I have."

LINK TO LEARNING

Using the Hubble Space Telescope, astronomers were able to detect images (https://openstax.org/l/30hubimgwhidwa) of faint white dwarf stars and other "stellar corpses" in the M4 star cluster, located about 7200 light-years away.

The Ultimate Fate of White Dwarfs

If the birth of a main-sequence star is defined by the onset of fusion reactions, then we must consider the end of all fusion reactions to be the time of a star's death. As the core is stabilized by degeneracy pressure, a last shudder of fusion passes through the outside of the star, consuming the little hydrogen still remaining. Now the star is a true white dwarf: nuclear fusion in its interior has ceased. Figure 23.4 shows the path of a star like the Sun on the H–R diagram during its final stages.

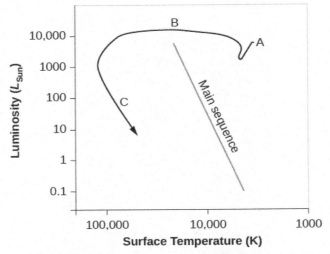

Figure 23.4 Evolutionary Track for a Star Like the Sun. This diagram shows the changes in luminosity and surface temperature for a star with a mass like the Sun's as it nears the end of its life. After the star becomes a giant again (point A on the diagram), it will lose more and more mass as its core begins to collapse. The mass loss will expose the hot inner core, which will appear at the center of a planetary nebula. In this stage, the star moves across the diagram to the left as it becomes hotter and hotter during its collapse (point B). At first, the luminosity remains nearly constant, but as the star begins to cool off, it becomes less and less bright (point C). It is now a white dwarf and will continue to cool slowly for billions of years until all of its remaining store of energy is radiated away. (This assumes the Sun will lose between 46–50% of its mass during the giant stages, based upon various theoretical models).

Since a stable white dwarf can no longer contract or produce energy through fusion, its only energy source is the heat represented by the motions of the atomic nuclei in its interior. The light it emits comes from this internal stored heat, which is substantial. Gradually, however, the white dwarf radiates away all its heat into space. After many billions of years, the nuclei will be moving much more slowly, and the white dwarf will no longer shine (Figure 23.5). It will then be a black dwarf—a cold stellar corpse with the mass of a star and the size of a planet. It will be composed mostly of carbon, oxygen, and neon, the products of the most advanced fusion reactions of which the star was capable.

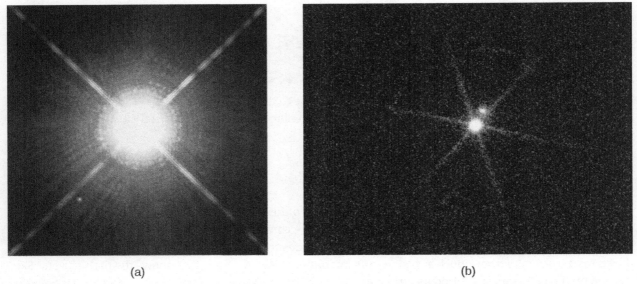

(a)   (b)

Figure 23.5 Visible Light and X-Ray Images of the Sirius Star System. (a) This image taken by the Hubble Space Telescope shows Sirius A (the large bright star), and its companion star, the white dwarf known as Sirius B (the tiny, faint star at the lower left). Sirius A and B are 8.6 light-years from Earth and are our fifth-closest star system. Note that the image has intentionally been overexposed to allow us to see Sirius B. (b) The same system is shown in X-ray taken with the Chandra Space Telescope. Note that Sirius A is fainter in X-rays than the hot white dwarf that is Sirius B. (credit a: modification of work by NASA, ESA, H. Bond, M. Barstow(University of Leicester); credit b: modification of work by NASA/SAO/CXC)

We have one final surprise as we leave our low-mass star in the stellar graveyard. Calculations show that as a degenerate star cools, the atoms inside it in essence "solidify" into a giant, highly compact lattice (organized rows of atoms, just like in a crystal). When carbon is compressed and crystallized in this way, it becomes a giant *diamond-like* star. A white dwarf star might make the most impressive engagement present you could ever see, although any attempt to mine the diamond-like material inside would crush an ardent lover instantly!

Evidence That Stars Can Shed a Lot of Mass as They Evolve

Whether or not a star will become a white dwarf depends on how much mass is lost in the red-giant and earlier phases of evolution. All stars that have masses below the Chandrasekhar limit when they run out of fuel will become white dwarfs, no matter what mass they were born with. But which stars shed enough mass to reach this limit?

One strategy for answering this question is to look in young, open clusters (which were discussed in Star Clusters). The basic idea is to search for young clusters that contain one or more white dwarf stars. Remember that more massive stars go through all stages of their evolution more rapidly than less massive ones. Suppose we find a cluster that has a white dwarf member and also contains stars on the main sequence that have 6 times the mass of the Sun. This means that only those stars with masses greater than 6 M_{Sun} have had time to exhaust their supply of nuclear energy and complete their evolution to the white dwarf stage. The star that turned into the white dwarf must therefore have had a main-sequence mass of more than 6 M_{Sun}, since stars with lower masses have not yet had time to use up their stores of nuclear energy. The star that became the white dwarf must, therefore, have gotten rid of at least 4.6 M_{Sun} so that its mass at the time nuclear energy generation ceased could be less than 1.4 M_{Sun}.

Astronomers continue to search for suitable clusters to make this test, and the evidence so far suggests that stars with masses up to about 8 M_{Sun} can shed enough mass to end their lives as white dwarfs. Stars like the Sun will probably lose about 45% of their initial mass and become white dwarfs with masses less than 1.4 M_{Sun}.

23.2 | Evolution of Massive Stars: An Explosive Finish

Learning Objectives

By the end of this section, you will be able to:

> Describe the interior of a massive star before a supernova
> Explain the steps of a core collapse and explosion
> List the hazards associated with nearby supernovae

Thanks to mass loss, then, stars with starting masses of at least 8 M_{Sun} (and perhaps even more) probably end their lives as white dwarfs. But we know stars can have masses as large as 150 (or more) M_{Sun}. They have a different kind of death in store for them. As we will see, these stars die with a bang.

Nuclear Fusion of Heavy Elements

After the helium in its core is exhausted (see The Evolution of More Massive Stars), the evolution of a massive star takes a significantly different course from that of lower-mass stars. In a massive star, the weight of the outer layers is sufficient to force the carbon core to contract until it becomes hot enough to fuse carbon into oxygen, neon, and magnesium. This cycle of contraction, heating, and the ignition of another nuclear fuel repeats several more times. After each of the possible nuclear fuels is exhausted, the core contracts again until it reaches a new temperature high enough to fuse still-heavier nuclei. The products of carbon fusion can be further converted into silicon, sulfur, calcium, and argon. And these elements, when heated to a still-higher temperature, can combine to produce iron. Massive stars go through these stages very, very quickly. In really massive stars, some fusion stages toward the very end can take only months or even days! This is a far cry from the millions of years they spend in the main-sequence stage.

At this stage of its evolution, a massive star resembles an onion with an iron core. As we get farther from the center, we find shells of decreasing temperature in which nuclear reactions involve nuclei of progressively lower mass—silicon and sulfur, oxygen, neon, carbon, helium, and finally, hydrogen (Figure 23.6).

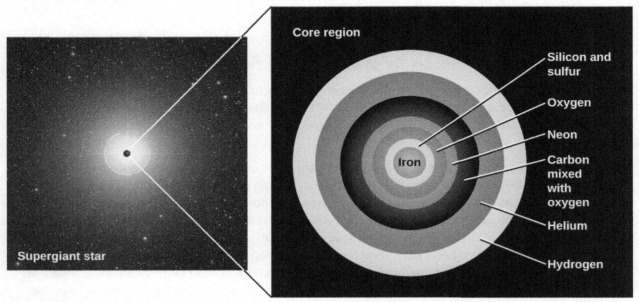

Figure 23.6 Structure of an Old Massive Star. Just before its final gravitational collapse, the core of a massive star resembles an onion. The iron core is surrounded by layers of silicon and sulfur, oxygen, neon, carbon mixed with some oxygen, helium, and finally hydrogen. Outside the core, the composition is mainly hydrogen and helium. (Note that this diagram is not precisely to scale but is just meant to convey the general idea of what such a star would be like.) (credit: modification of work by ESO, Digitized Sky Survey)

But there is a limit to how long this process of building up elements by fusion can go on. The fusion of silicon into iron turns out to be the last step in the sequence of nonexplosive element production. Up to this point, each fusion reaction has *produced* energy because the nucleus of each fusion product has been a bit more stable than the nuclei that formed it. As discussed in The Sun: A Nuclear Powerhouse, light nuclei give up some of their binding energy in the process of fusing into more tightly bound, heavier nuclei. It is this released energy that maintains the outward pressure in the core so that the star does not collapse. But of all the nuclei known, iron is the most tightly bound and thus the most stable.

You might think of the situation like this: all smaller nuclei want to "grow up" to be like iron, and they are willing to pay (*produce* energy) to move toward that goal. But iron is a mature nucleus with good self-esteem, perfectly content being iron; it requires payment (must *absorb* energy) to change its stable nuclear structure. This is the exact opposite of what has happened in each nuclear reaction so far: instead of *providing* energy to balance the inward pull of gravity, any nuclear reactions involving iron would *remove* some energy from the core of the star.

Unable to generate energy, the star now faces catastrophe.

Collapse into a Ball of Neutrons

When nuclear reactions stop, the core of a massive star is supported by degenerate electrons, just as a white dwarf is. For stars that begin their evolution with masses of at least 10 M_{Sun}, this core is likely made mainly of iron. (For stars with initial masses in the range 8 to 10 M_{Sun}, the core is likely made of oxygen, neon, and magnesium, because the star never gets hot enough to form elements as heavy as iron. The exact composition of the cores of stars in this mass range is very difficult to determine because of the complex physical characteristics in the cores, particularly at the very high densities and temperatures involved.) We will focus on the more massive iron cores in our discussion.

While no energy is being generated within the white dwarf core of the star, fusion still occurs in the shells that surround the core. As the shells finish their fusion reactions and stop producing energy, the ashes of the last reaction fall onto the white dwarf core, increasing its mass. As Figure 23.2 shows, a higher mass means a smaller core. The core can contract because even a degenerate gas is still mostly empty space. Electrons and atomic nuclei are, after all, extremely small. The electrons and nuclei in a stellar core may be crowded compared to the air in your room, but there is still lots of space between them.

The electrons at first resist being crowded closer together, and so the core shrinks only a small amount. Ultimately, however, the iron core reaches a mass so large that even degenerate electrons can no longer support it. When the density reaches 4×10^{11} g/cm^3 (400 billion times the density of water), some electrons are actually squeezed into the atomic nuclei, where they combine with protons to form neutrons and neutrinos. This transformation is not something that is familiar from everyday life, but becomes very important as such a massive star core collapses.

Some of the electrons are now gone, so the core can no longer resist the crushing mass of the star's overlying layers. The core begins to shrink rapidly. More and more electrons are now pushed into the atomic nuclei, which ultimately become so saturated with neutrons that they cannot hold onto them.

At this point, the neutrons are squeezed out of the nuclei and can exert a new force. As is true for electrons, it turns out that the neutrons strongly resist being in the same place and moving in the same way. The force that can be exerted by such *degenerate neutrons* is much greater than that produced by degenerate electrons, so unless the core is too massive, they can ultimately stop the collapse.

This means the collapsing core can reach a stable state as a crushed ball made mainly of neutrons, which astronomers call a **neutron star**. We don't have an exact number (a "Chandrasekhar limit") for the maximum mass of a neutron star, but calculations tell us that the upper mass limit of a body made of neutrons might only be about 3 M_{Sun}. So if the mass of the core were greater than this, then even neutron degeneracy would not be able to stop the core from collapsing further. The dying star must end up as something even more extremely compressed, which until recently was believed to be only one possible type of object—the state of ultimate compaction known as a black hole (which is the subject of our next chapter). This is because no force was believed to exist that could stop a collapse beyond the neutron star stage.

Collapse and Explosion

When the collapse of a high-mass star's core is stopped by degenerate neutrons, the core is saved from further destruction, but it turns out that the rest of the star is literally blown apart. Here's how it happens.

The collapse that takes place when electrons are absorbed into the nuclei is very rapid. In less than a second, a core with a mass of about 1 M_{Sun}, which originally was approximately the size of Earth, collapses to a diameter of less than 20 kilometers. The speed with which material falls inward reaches one-fourth the speed of light. The collapse halts only when the density of the core exceeds the density of an atomic nucleus (which is the densest form of matter we know). A typical neutron star is so compressed that to duplicate its density, we would have to squeeze all the people in the world into a single sugar cube! This would give us one sugar cube's worth (one cubic centimeter's worth) of a neutron star.

The neutron degenerate core strongly resists further compression, abruptly halting the collapse. The shock of the sudden jolt initiates a shock wave that starts to propagate outward. However, this shock alone is not enough to create a star explosion. The energy produced by the outflowing matter is quickly absorbed by atomic nuclei in the dense, overlying layers of gas, where it breaks up the nuclei into individual neutrons and protons.

Our understanding of nuclear processes indicates (as we mentioned above) that each time an electron and a proton in the star's core merge to make a neutron, the merger releases a *neutrino*. These ghostly subatomic particles, introduced in The Sun: A Nuclear Powerhouse, carry away some of the nuclear energy. It is their presence that launches the final disastrous explosion of the star. The total energy contained in the neutrinos is huge. In the initial second of the star's explosion, the power carried by the neutrinos (10^{46} watts) is greater than the power put out by all the stars in over a billion galaxies.

While neutrinos ordinarily do not interact very much with ordinary matter (we earlier accused them of being downright antisocial), matter near the center of a collapsing star is so dense that the neutrinos do interact with it to some degree. They deposit some of this energy in the layers of the star just outside the core. This huge, sudden input of energy reverses the infall of these layers and drives them explosively outward. Most of the mass of the star (apart from that which went into the neutron star in the core) is then ejected outward into space. As we saw earlier, such an explosion requires a star of at least 8 M_{Sun}, and the neutron star can have a mass of at most 3 $M_{Sun.}$ Consequently, at least five times the mass of our Sun is ejected into space in each such explosive event!

The resulting explosion is called a supernova (Figure 23.7). When these explosions happen close by, they can be among the most spectacular celestial events, as we will discuss in the next section. (Actually, there are at least two different types of supernova explosions: the kind we have been describing, which is the collapse of a massive star, is called, for historical reasons, a **type II supernova**. We will describe how the types differ later in this chapter).

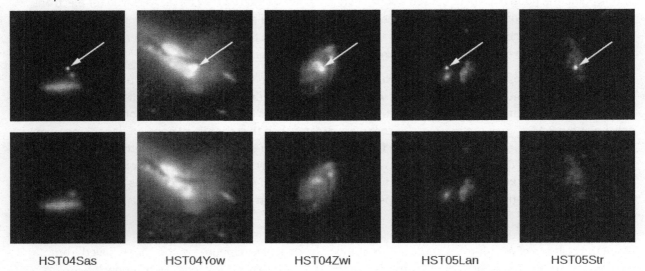

| HST04Sas | HST04Yow | HST04Zwi | HST05Lan | HST05Str |

Figure 23.7 Five Supernova Explosions in Other Galaxies. The arrows in the top row of images point to the supernovae. The bottom row shows the host galaxies before or after the stars exploded. Each of these supernovae exploded between 3.5 and 10 billion years ago. Note that the supernovae when they first explode can be as bright as an entire galaxy. (credit: modification of work by NASA, ESA, and A. Riess (STScI))

Table 23.1 summarizes the discussion so far about what happens to stars and substellar objects of different initial masses at the ends of their lives. Like so much of our scientific understanding, this list represents a progress report: it is the best we can do with our present models and observations. The mass limits corresponding to various outcomes may change somewhat as models are improved. There is much we do not yet understand about the details of what happens when stars die.

The Ultimate Fate of Stars and Substellar Objects with Different Masses

Initial Mass (Mass of Sun = 1)[1]	Final State at the End of Its Life
< 0.01	Planet
0.01 to 0.08	Brown dwarf
0.08 to 0.25	White dwarf made mostly of helium
0.25 to 8	White dwarf made mostly of carbon and oxygen
8 to 10	White dwarf made of oxygen, neon, and magnesium
10 to 40	Supernova explosion that leaves a neutron star
> 40	Supernova explosion that leaves a black hole

Table 23.1

The Supernova Giveth and the Supernova Taketh Away

After the supernova explosion, the life of a massive star comes to an end. But the death of each massive star is an important event in the history of its galaxy. The elements built up by fusion during the star's life are now "recycled" into space by the explosion, making them available to enrich the gas and dust that form new stars and planets. Because these heavy elements ejected by supernovae are critical for the formation of planets and the origin of life, it's fair to say that without mass loss from supernovae and planetary nebulae, neither the authors nor the readers of this book would exist.[2]

But the supernova explosion has one more creative contribution to make, one we alluded to in Stars from Adolescence to Old Age when we asked where the atoms in your jewelry came from. The supernova explosion produces a flood of energetic neutrons that barrel through the expanding material. These neutrons can be absorbed by iron and other nuclei where they can turn into protons. Thus, they can build up elements that are more massive than iron, possibly including such terrestrial favorites as gold, silver and uranium. Supernovae (and, as we will shortly see, the explosive mergers of neutron stars) are the only candidates we have for places where such heavier atoms can be made. Next time you wear some gold jewelry (or give some to your sweetheart), bear in mind that those gold atoms were forged long ago in these kinds of celestial explosions!

When supernovae explode, these elements (as well as the ones the star made during more stable times) are ejected into the existing gas between the stars and mixed with it. Thus, supernovae play a crucial role in enriching their galaxy with heavier elements, allowing, among other things, the chemical elements that make up earthlike planets and the building blocks of life to become more common as time goes on (Figure 23.8).

---

1 Stars in the mass ranges 0.25–8 and 8–10 may later produce a type of supernova different from the one we have discussed so far. These are discussed in The Evolution of Binary Star Systems.
2 To tell the full story, there is another way to make and recycle these heavier elements which we will discuss in The Milky Way Galaxy. That is the merger of two neutron stars, which also causes a huge explosion.

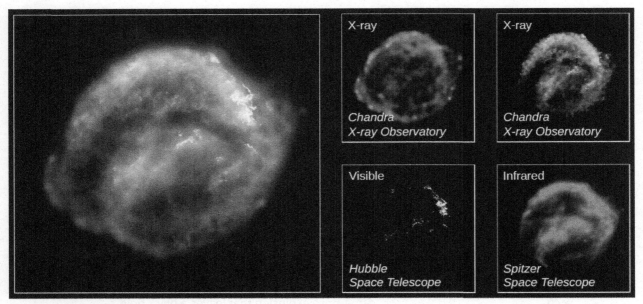

Figure 23.8 Kepler Supernova Remant. This image shows the expanding remains of a supernova explosion, which was first seen about 400 years ago by sky watchers, including the famous astronomer Johannes Kepler. The bubble-shaped shroud of gas and dust is now 14 light-years wide and is expanding at 2,000 kilometers per second (4 million miles per hour). The remnant emits energy at wavelengths from X-rays (shown in blue and green) to visible light (yellow) and into the infrared (red). The expanding shell is rich in iron, which was produced in the star that exploded. The main image combines the individual single-color images seen at the bottom into one multi-wavelength picture. (credit: modification of work by NASA, ESA, R. Sankrit and W. Blair (Johns Hopkins University))

Supernovae are also thought to be the source of many of the high-energy *cosmic ray* particles discussed in Cosmic Rays. Trapped by the magnetic field of the Galaxy, the particles from exploded stars continue to circulate around the vast spiral of the Milky Way. Scientists speculate that high-speed cosmic rays hitting the genetic material of Earth organisms over billions of years may have contributed to the steady *mutations*—subtle changes in the genetic code—that drive the evolution of life on our planet. In all the ways we have mentioned, supernovae have played a part in the development of new generations of stars, planets, and life.

But supernovae also have a dark side. Suppose a life form has the misfortune to develop around a star that happens to lie near a massive star destined to become a supernova. Such life forms may find themselves snuffed out when the harsh radiation and high-energy particles from the neighboring star's explosion reach their world. If, as some astronomers speculate, life can develop on many planets around long-lived (lower-mass) stars, then the suitability of that life's *own star* and planet may not be all that matters for its long-term evolution and survival. Life may well have formed around a number of pleasantly stable stars only to be wiped out because a massive nearby star suddenly went supernova. Just as children born in a war zone may find themselves the unjust victims of their violent neighborhood, life too close to a star that goes supernova may fall prey to having been born in the wrong place at the wrong time.

What is a safe distance to be from a supernova explosion? A lot depends on the violence of the particular explosion, what type of supernova it is (see The Evolution of Binary Star Systems), and what level of destruction we are willing to accept. Calculations suggest that a supernova less than 50 light-years away from us would certainly end all life on Earth, and that even one 100 light-years away would have drastic consequences for the radiation levels here. One minor extinction of sea creatures about 2 million years ago on Earth may actually have been caused by a supernova at a distance of about 120 light-years.

The good news is that there are at present no massive stars that promise to become supernovae within 50 light-years of the Sun. (This is in part because the kinds of massive stars that become supernovae are overall quite rare.) The massive star closest to us, Spica (in the constellation of Virgo), is about 260 light-years away, probably a safe distance, even if it were to explode as a supernova in the near future.

EXAMPLE 23.1

Extreme Gravity

In this section, you were introduced to some very dense objects. How would those objects' gravity affect you? Recall that the force of gravity, F, between two bodies is calculated as

$$F = \frac{GM_1 M_2}{R^2}$$

where G is the gravitational constant, 6.67×10^{-11} Nm²/kg², M_1 and M_2 are the masses of the two bodies, and R is their separation. Also, from Newton's second law,

$$F = M \times a$$

where a is the acceleration of a body with mass M.

So let's consider the situation of a mass—say, you—standing on a body, such as Earth or a white dwarf (where we assume you will be wearing a heat-proof space suit). You are M_1 and the body you are standing on is M_2. The distance between you and the center of gravity of the body on which you stand is its radius, R. The force exerted on you is

$$F = M_1 \times a = GM_1 M_2/R^2$$

Solving for a, the acceleration of gravity on that world, we get

$$g = \frac{(G \times M)}{R^2}$$

Note that we have replaced the general symbol for acceleration, a, with the symbol scientists use for the acceleration of gravity, g.

Say that a particular white dwarf has the mass of the Sun (2×10^{30} kg) but the radius of Earth (6.4×10^6 m). What is the acceleration of gravity at the surface of the white dwarf?

Solution

The acceleration of gravity at the surface of the white dwarf is

$$g(\text{white dwarf}) = \frac{(G \times M_{\text{Sun}})}{R_{\text{Earth}}{}^2} = \frac{\left(6.67 \times 10^{-11}\ \text{m}^2/\text{kg s}^2 \times 2 \times 10^{30}\ \text{kg}\right)}{\left(6.4 \times 10^6\ \text{m}\right)^2} = 3.26 \times 10^6\ \text{m/s}^2$$

Compare this to g on the surface of Earth, which is 9.8 m/s².

Check Your Learning

What is the acceleration of gravity at the surface if the white dwarf has the twice the mass of the Sun and is only half the radius of Earth?

Answer:

$$g(\text{white dwarf}) = \frac{(G \times 2M_{\text{Sun}})}{(0.5R_{\text{Earth}})^2} = \frac{\left(6.67 \times 10^{-11}\ \text{m}^2/\text{kg s}^2 \times 4 \times 10^{30}\ \text{kg}\right)}{(3.2 \times 10^6)^2} = 2.61 \times 10^7\ \text{m/s}^2$$

23.3 | Supernova Observations

Learning Objectives

By the end of this section, you will be able to:

> Describe the observed features of SN 1987A both before and after the supernova
> Explain how observations of various parts of the SN 1987A event helped confirm theories about supernovae

Supernovae were discovered long before astronomers realized that these spectacular cataclysms mark the death of stars (see Making Connections: Supernovae in History). The word *nova* means "new" in Latin; before telescopes, when a star too dim to be seen with the unaided eye suddenly flared up in a brilliant explosion, observers concluded it must be a brand-new star. Twentieth-century astronomers reclassified the explosions with the greatest luminosity as *supernovae*.

From historical records of such explosions, from studies of the remnants of supernovae in our Galaxy, and from analyses of supernovae in other galaxies, we estimate that, on average, one supernova explosion occurs somewhere in the Milky Way Galaxy every 25 to 100 years. Unfortunately, however, no supernova explosion has been observable in our Galaxy since the invention of the telescope. Either we have been exceptionally unlucky or, more likely, recent explosions have taken place in parts of the Galaxy where interstellar dust blocks light from reaching us.

MAKING CONNECTIONS

Supernovae in History

Although many supernova explosions in our own Galaxy have gone unnoticed, a few were so spectacular that they were clearly seen and recorded by sky watchers and historians at the time. We can use these records, going back two millennia, to help us pinpoint where the exploding stars were and thus where to look for their remnants today.

The most dramatic supernova was observed in the year 1006. It appeared in May as a brilliant point of light visible during the daytime, perhaps 100 times brighter than the planet Venus. It was bright enough to cast shadows on the ground during the night and was recorded with awe and fear by observers all over Europe and Asia. No one had seen anything like it before; Chinese astronomers, noting that it was a temporary spectacle, called it a "guest star."

Astronomers David Clark and Richard Stephenson have scoured records from around the world to find more than 20 reports of the 1006 supernova (SN 1006) (Figure 23.9). This has allowed them to determine with some accuracy where in the sky the explosion occurred. They place it in the modern constellation of Lupus; at roughly the position they have determined, we find a supernova remnant, now quite faint. From the way its filaments are expanding, it indeed appears to be about 1000 years old.

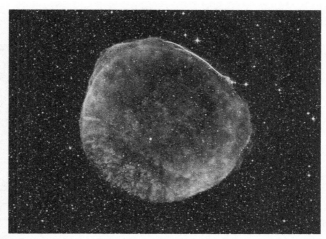

Figure 23.9 Supernova 1006 Remnant. This composite view of SN 1006 from the Chandra X-Ray Observatory shows the X-rays coming from the remnant in blue, visible light in white-yellow, and radio emission in red. (credit: modification of work by NASA, ESA, Zolt Levay(STScI))

Another guest star, now known as SN 1054, was clearly recorded in Chinese records in July 1054. The remnant of that star is one of the most famous and best-studied objects in the sky, called the Crab Nebula (Figure 23.14). It is a marvelously complex object, which has been key to understanding the death of massive stars. When its explosion was first seen, we estimate that it was about as bright as the planet Jupiter: nowhere near as dazzling as the 1006 event but still quite dramatic to anyone who kept track of objects in the sky. Another fainter supernova was seen in 1181.

The next supernova became visible in November 1572 and, being brighter than the planet Venus, was quickly spotted by a number of observers, including the young Tycho Brahe (see Orbits and Gravity). His careful measurements of the star over a year and a half showed that it was not a comet or something in Earth's atmosphere since it did not move relative to the stars. He correctly deduced that it must be a phenomenon belonging to the realm of the stars, not of the solar system. The remnant of Tycho's Supernova (as it is now called) can still be detected in many different bands of the electromagnetic spectrum.

Not to be outdone, Johannes Kepler, Tycho Brahe's scientific heir, found his own supernova in 1604, now known as Kepler's Supernova (Figure 23.8). Fainter than Tycho's, it nevertheless remained visible for about a year. Kepler wrote a book about his observations that was read by many with an interest in the heavens, including Galileo.

No supernova has been spotted in our Galaxy for the past 300 years. Since the explosion of a visible supernova is a chance event, there is no way to say when the next one might occur. Around the world, dozens of professional and amateur astronomers keep a sharp lookout for "new" stars that appear overnight, hoping to be the first to spot the next guest star in our sky and make a little history themselves.

At their maximum brightness, the most luminous supernovae have about 10 billion times the luminosity of the Sun. For a brief time, a supernova may outshine the entire galaxy in which it appears. After maximum brightness, the star's light fades and disappears from telescopic visibility within a few months or years. At the time of their outbursts, supernovae eject material at typical velocities of 10,000 kilometers per second (and speeds twice that have been observed). A speed of 20,000 kilometers per second corresponds to about 45 million miles per hour, truly an indication of great cosmic violence.

Supernovae are classified according to the appearance of their spectra, but in this chapter, we will focus on the two main causes of supernovae. Type Ia supernovae are ignited when a lot of material is dumped on degenerate white dwarfs (Figure 23.10); these supernovae will be discussed later in this chapter. For now, we will continue our story about the death of massive stars and focus on type II supernovae, which are produced when the core of a massive star collapses.

SN 2014J January 31, 2014

Figure 23.10 Supernova 2014J. This image of supernova 2014J, located in Messier 82 (M82), which is also known as the Cigar galaxy, was taken by the Hubble Space Telescope and is superposed on a mosaic image of the galaxy also taken with Hubble. The supernova event is indicated by the box and the inset. This explosion was produced by a type Ia supernova, which is theorized to be triggered in binary systems consisting of a white dwarf and another star—and could be a second white dwarf, a star like our Sun, or a giant star. This type of supernova will be discussed later in this chapter. At a distance of approximately 11.5 million light-years from Earth, this is the closest supernova of type Ia discovered in the past few decades. In the image, you can see reddish plumes of hydrogen coming from the central region of the galaxy, where a considerable number of young stars are being born. (credit: modification of work by NASA, ESA, A. Goobar (Stockholm University), and the Hubble Heritage Team (STScI/AURA))

Supernova 1987A

Our most detailed information about what happens when a type II supernova occurs comes from an event that was observed in 1987. Before dawn on February 24, Ian Shelton, a Canadian astronomer working at an observatory in Chile, pulled a photographic plate from the developer. Two nights earlier, he had begun a survey of the Large Magellanic Cloud, a small galaxy that is one of the Milky Way's nearest neighbors in space. Where he expected to see only faint stars, he saw a large bright spot. Concerned that his photograph was flawed, Shelton went outside to look at the Large Magellanic Cloud . . . and saw that a new object had indeed appeared in the sky (see Figure 23.11). He soon realized that he had discovered a supernova, one that could be seen with the unaided eye even though it was about 160,000 light-years away.

Figure 23.11 Hubble Space Telescope Image of SN 1987A. The supernova remnant with its inner and outer red rings of material is located in the Large Magellanic Cloud. This image is a composite of several images taken in 1994, 1996, and 1997—about a decade after supernova 1987A was first observed. (credit: modification of work by the Hubble Heritage Team (AURA/STScI/NASA/ESA))

Now known as SN 1987A, since it was the first supernova discovered in 1987, this brilliant newcomer to the southern sky gave astronomers their first opportunity to study the death of a relatively nearby star with modern instruments. It was also the first time astronomers had observed a star *before* it became a supernova. The star that blew up had been included in earlier surveys of the Large Magellanic Cloud, and as a result, we know the star was a blue supergiant just before the explosion.

By combining theory and observations at many different wavelengths, astronomers have reconstructed the life story of the star that became SN 1987A. Formed about 10 million years ago, it originally had a mass of about 20 M_{Sun}. For 90% of its life, it lived quietly on the main sequence, converting hydrogen into helium. At this time, its luminosity was about 60,000 times that of the Sun (L_{Sun}), and its spectral type was O. When the hydrogen in the center of the star was exhausted, the core contracted and ultimately became hot enough to fuse helium. By this time, the star was a red supergiant, emitting about 100,000 times more energy than the Sun. While in this stage, the star lost some of its mass.

This lost material has actually been detected by observations with the Hubble Space Telescope (Figure 23.12). The gas driven out into space by the subsequent supernova explosion is currently colliding with the material the star left behind when it was a red giant. As the two collide, we see a glowing ring.

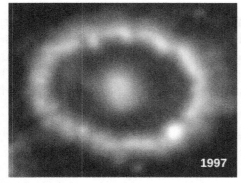
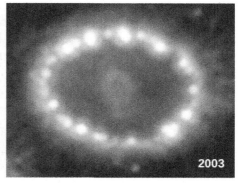

Figure 23.12 Ring around Supernova 1987A. These two images show a ring of gas expelled by a red giant star about 30,000 years before the star exploded and was observed as Supernova 1987A. The supernova, which has been artificially dimmed, is located at the

center of the ring. The left-hand image was taken in 1997 and the right-hand image in 2003. Note that the number of bright spots has increased from 1 to more than 15 over this time interval. These spots occur where high-speed gas ejected by the supernova and moving at millions of miles per hour has reached the ring and blasted into it. The collision has heated the gas in the ring and caused it to glow more brightly. The fact that we see individual spots suggests that material ejected by the supernova is first hitting narrow, inward-projecting columns of gas in the clumpy ring. The hot spots are the first signs of a dramatic and violent collision between the new and old material that will continue over the next few years. By studying these bright spots, astronomers can determine the composition of the ring and hence learn about the nuclear processes that build heavy elements inside massive stars. (credit: modification of work by NASA, P. Challis, R. Kirshner (Harvard-Smithsonian Center for Astrophysics) and B. Sugerman (STScI))

Helium fusion lasted only about 1 million years. When the helium was exhausted at the center of the star, the core contracted again, the radius of the surface also decreased, and the star became a blue supergiant with a luminosity still about equal to 100,000 L_{Sun}. This is what it still looked like on the outside when, after brief periods of further fusion, it reached the iron crisis we discussed earlier and exploded.

Some key stages of evolution of the star that became SN 1987A, including the ones following helium exhaustion, are listed in Table 23.2. While we don't expect you to remember these numbers, note the patterns in the table: each stage of evolution happens more quickly than the preceding one, the temperature and pressure in the core increase, and progressively heavier elements are the source of fusion energy. Once iron was created, the collapse began. It was a catastrophic collapse, lasting only a few tenths of a second; the speed of infall in the outer portion of the iron core reached 70,000 kilometers per second, about one-fourth the speed of light.

Evolution of the Star That Exploded as SN 1987A

Phase	Central Temperature (K)	Central Density (g/cm³)	Time Spent in This Phase
Hydrogen fusion	40×10^6	5	8×10^6 years
Helium fusion	190×10^6	970	10^6 years
Carbon fusion	870×10^6	170,000	2000 years
Neon fusion	1.6×10^9	3.0×10^6	6 months
Oxygen fusion	2.0×10^9	5.6×10^6	1 year
Silicon fusion	3.3×10^9	4.3×10^7	Days
Core collapse	200×10^9	2×10^{14}	Tenths of a second

Table 23.2

In the meantime, as the core was experiencing its last catastrophe, the outer shells of neon, oxygen, carbon, helium, and hydrogen in the star did not yet know about the collapse. Information about the physical movement of different layers travels through a star at the speed of sound and cannot reach the surface in the few tenths of a second required for the core collapse to occur. Thus, the surface layers of our star hung briefly suspended, much like a cartoon character who dashes off the edge of a cliff and hangs momentarily in space before realizing that he is no longer held up by anything.

The collapse of the core continued until the densities rose to several times that of an atomic nucleus. The resistance to further collapse then became so great that the core rebounded. Infalling material ran into the "brick wall" of the rebounding core and was thrown outward with a great shock wave. Neutrinos poured out of the core, helping the shock wave blow the star apart. The shock reached the surface of the star a few hours later, and the star began to brighten into the supernova Ian Shelton observed in 1987.

The Synthesis of Heavy Elements

The variations in the brightness of SN 1987A in the days and months after its discovery, which are shown in Figure 23.13, helped confirm our ideas about heavy element production. In a single day, the star soared in brightness by a factor of about 1000 and became just visible without a telescope. The star then continued to increase slowly in brightness until it was about the same apparent brightness as the stars in the Little Dipper. Up until about day 40 after the outburst, the energy being radiated away was produced by the explosion itself. But then SN 1987A did not continue to fade away, as we might have expected the light from the explosion to do. Instead, SN 1987A remained bright as energy from newly created radioactive elements came into play.

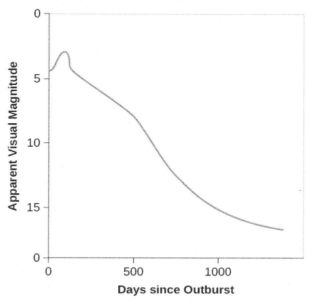

Figure 23.13 Change in the Brightness of SN 1987A over Time. Note how the rate of decline of the supernova's light slowed between days 40 and 500. During this time, the brightness was mainly due to the energy emitted by newly formed (and quickly decaying) radioactive elements. Remember that magnitudes are a backward measure of brightness: the larger the magnitude, the dimmer the object looks.

One of the elements formed in a supernova explosion is radioactive nickel, with an atomic mass of 56 (that is, the total number of protons plus neutrons in its nucleus is 56). Nickel-56 is unstable and changes spontaneously (with a half-life of about 6 days) to cobalt-56. (Recall that a half-life is the time it takes for half the nuclei in a sample to undergo radioactive decay.) Cobalt-56 in turn decays with a half-life of about 77 days to iron-56, which is stable. Energetic gamma rays are emitted when these radioactive nuclei decay. Those gamma rays then serve as a new source of energy for the expanding layers of the supernova. The gamma rays are absorbed in the overlying gas and re-emitted at visible wavelengths, keeping the remains of the star bright.

As you can see in Figure 23.13, astronomers did observe brightening due to radioactive nuclei in the first few months following the supernova's outburst and then saw the extra light die away as more and more of the radioactive nuclei decayed to stable iron. The gamma-ray heating was responsible for virtually all of the radiation detected from SN 1987A after day 40. Some gamma rays also escaped directly without being absorbed. These were detected by Earth-orbiting telescopes at the wavelengths expected for the decay of radioactive nickel and cobalt, clearly confirming our understanding that new elements were indeed formed in the crucible of the supernova.

Neutrinos from SN 1987A

If there had been any human observers in the Large Magellanic Cloud about 160,000 years ago, the explosion we call SN 1987A would have been a brilliant spectacle in their skies. Yet we know that less than 1/10 of 1% of the energy of the explosion appeared as visible light. About 1% of the energy was required to destroy the star, and the rest was carried away by neutrinos. The overall energy in these neutrinos was truly astounding. In the initial second of the event, as we noted earlier in our general discussion of supernovae, their total luminosity exceeded the luminosity of all the stars in over a billion galaxies. And the supernova generated this energy in a volume less than 50 kilometers in diameter! Supernovae are one of the most violent events in the universe, and their *light* turns out to be only the tip of the iceberg in revealing how much energy they produce.

In 1987, the neutrinos from SN 1987A were detected by two instruments—which might be called "neutrino telescopes"—almost a full day before Shelton's observations. (This is because the neutrinos get out of the exploding star more easily than light does, and also because you don't need to wait until nightfall to catch a "glimpse" of them.) Both neutrino telescopes, one in a deep mine in Japan and the other under Lake Erie, consist of several thousand tons of purified water surrounded by several hundred light-sensitive detectors. Incoming neutrinos interact with the water to produce positrons and electrons, which move rapidly through the water and emit deep blue light.

Altogether, 19 neutrinos were detected. Since the neutrino telescopes were in the Northern Hemisphere and the supernova occurred in the Southern Hemisphere, the detected neutrinos had already passed through Earth and were on their way back out into space when they were captured.

Only a few neutrinos were detected because the probability that they will interact with ordinary matter is very, very low. It is estimated that the supernova actually released 10^{58} neutrinos. A tiny fraction of these, about 30 billion, eventually passed through each square centimeter of Earth's surface. About a million people actually experienced a neutrino interaction within their bodies as a result of the supernova. This interaction happened to only a single nucleus in each person and thus had absolutely no biological effect; it went completely unnoticed by everyone concerned.

Since the neutrinos come directly from the heart of the supernova, their energies provided a measure of the temperature of the core as the star was exploding. The central temperature was about 200 billion K, a stunning figure to which no earthly analog can bring much meaning. With neutrino telescopes, we are peering into the final moment in the life stories of massive stars and observing conditions beyond all human experience. Yet we are also seeing the unmistakable hints of our own origins.

23.4 | Pulsars and the Discovery of Neutron Stars

Learning Objectives

By the end of this section, you will be able to:

> Explain the research method that led to the discovery of neutron stars, located hundreds or thousands of light-years away

> Describe the features of a neutron star that allow it to be detected as a pulsar

> List the observational evidence that links pulsars and neutron stars to supernovae

After a type II supernova explosion fades away, all that is left behind is either a neutron star or something even more strange, a black hole. We will describe the properties of black holes in Black Holes and Curved Spacetime, but for now, we want to examine how the neutron stars we discussed earlier might become observable.

Neutron stars are the densest objects in the universe; the force of gravity at their surface is 10^{11} times greater than what we experience at Earth's surface. The interior of a neutron star is composed of about 95% neutrons, with a small number of protons and electrons mixed in. In effect, a neutron star is a giant atomic nucleus, with a mass about 10^{57} times the mass of a proton. Its diameter is more like the size of a small town or an asteroid than a star. (Table 23.3 compares the properties of neutron stars and white dwarfs.) Because it is so small, a neutron star probably strikes you as the object least likely to be observed from thousands of light-years away. Yet neutron stars do manage to signal their presence across vast gulfs of space.

Properties of a Typical White Dwarf and a Neutron Star

Property	White Dwarf	Neutron Star
Mass (Sun = 1)	0.6 (always <1.4)	Always >1.4 and <3
Radius	7000 km	10 km
Density	8×10^5 g/cm^3	10^{14} g/cm^3

Table 23.3

The Discovery of Neutron Stars

In 1967, Jocelyn Bell, a research student at Cambridge University, was studying distant radio sources with a special detector that had been designed and built by her advisor Antony Hewish to find rapid variations in radio signals. The project computers spewed out reams of paper showing where the telescope had surveyed the sky, and it was the job of Hewish's graduate students to go through it all, searching for interesting phenomena. In September 1967, Bell discovered what she called "a bit of scruff"—a strange radio signal unlike anything seen before.

What Bell had found, in the constellation of Vulpecula, was a source of rapid, sharp, intense, and extremely regular pulses of radio radiation. Like the regular ticking of a clock, the pulses arrived precisely every 1.33728 seconds. Such exactness first led the scientists to speculate that perhaps they had found signals from an intelligent civilization. Radio astronomers even half-jokingly dubbed the source "LGM" for "little green men." Soon, however, three similar sources were discovered in widely separated directions in the sky.

When it became apparent that this type of radio source was fairly common, astronomers concluded that they were highly unlikely to be signals from other civilizations. By today, nearly 3000 such sources have been discovered; they are now called **pulsars**, short for "pulsating radio sources."

The pulse periods of different pulsars range from a little longer than 1/1000 of a second to nearly 10 seconds. At first, the pulsars seemed particularly mysterious because nothing could be seen at their location on visible-light photographs. But then a pulsar was discovered right in the center of the Crab Nebula, a cloud of gas produced by SN 1054, a supernova that was recorded by the Chinese in 1054 (Figure 23.14). The energy from the Crab Nebula pulsar arrives in sharp bursts that occur 30 times each second—with a regularity that would be the envy of a Swiss watchmaker. In addition to pulses of radio energy, we can observe pulses of visible light and X-rays from the Crab Nebula. The fact that the pulsar was just in the region of the supernova remnant where we expect the leftover neutron star to be immediately alerted astronomers that pulsars might be connected with these elusive "corpses" of massive stars.

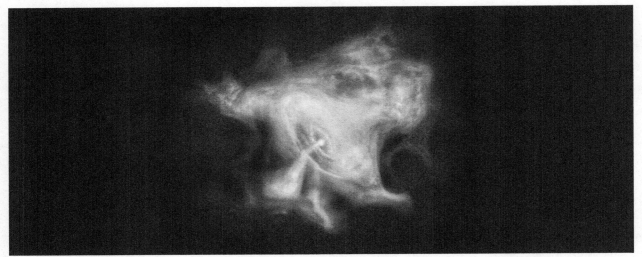

Figure 23.14 Crab Nebula. This image shows X-ray emissions from the Crab Nebula, which is about 6500 light-years away. The pulsar is the bright spot at the center of the concentric rings. Data taken over about a year show that particles stream away from the inner ring at about half the speed of light. The jet that is perpendicular to this ring is a stream of matter and antimatter electrons also moving at half the speed of light. (credit: modification of work by NASA/CXC/SAO)

The Crab Nebula is a fascinating object. The whole nebula glows with radiation at many wavelengths, and its overall energy output is more than 100,000 times that of the Sun—not a bad trick for the remnant of a supernova that exploded almost a thousand years ago. Astronomers soon began to look for a connection between the pulsar and the large energy output of the surrounding nebula.

LINK TO LEARNING

Read the transcript of an interesting interview (https://openstax.org/l/30jocbellint) with Jocelyn Bell (Burnell) to learn about her life and work (this is part of a project at the American Institute of Physics to record interviews with pathbreaking scientists while they are still alive).

A Spinning Lighthouse Model

By applying a combination of theory and observation, astronomers eventually concluded that pulsars must be *spinning neutron stars*. According to this model, a neutron star is something like a lighthouse on a rocky coast (Figure 23.15). To warn ships in all directions and yet not cost too much to operate, the light in a modern lighthouse turns, sweeping its beam across the dark sea. From the vantage point of a ship, you see a pulse of light each time the beam points in your direction. In the same way, radiation from a small region on a neutron star sweeps across the oceans of space, giving us a pulse of radiation each time the beam points toward Earth.

Figure 23.15 Lighthouse. A lighthouse in California warns ships on the ocean not to approach too close to the dangerous shoreline. The lighted section at the top rotates so that its beam can cover all directions. (credit: Anita Ritenour)

Neutron stars are ideal candidates for such a job because the collapse has made them so small that they can turn very rapidly. Recall the principle of the *conservation of angular momentum* from Newton's Great Synthesis: if an object gets smaller, it can spin more rapidly. Even if the parent star was rotating very slowly when it was on the main sequence, its rotation had to speed up as it collapsed to form a neutron star. With a diameter of only 10 to 20 kilometers, a neutron star can complete one full spin in only a fraction of a second. This is just the sort of time period we observe between pulsar pulses.

Any magnetic field that existed in the original star will be highly compressed when the core collapses to a neutron star. At the surface of the neutron star, in the outer layer consisting of ordinary matter (and not just pure neutrons), protons and electrons are caught up in this spinning field and accelerated nearly to the speed of light. In only two places—the north and south magnetic poles—can the trapped particles escape the strong hold of the magnetic field (Figure 23.16). The same effect can be seen (in reverse) on Earth, where charged particles from space are *kept out* by our planet's magnetic field everywhere except near the poles. As a result, Earth's auroras (caused when charged particles hit the atmosphere at high speed) are seen mainly near the poles.

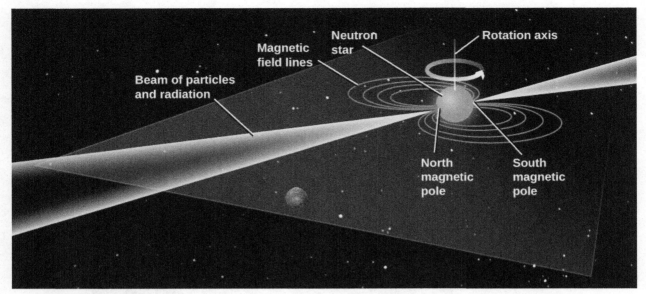

Figure 23.16 Model of a Pulsar. A diagram showing how beams of radiation at the magnetic poles of a neutron star can give rise to pulses of emission as the star rotates. As each beam sweeps over Earth, like a lighthouse beam sweeping over a distant ship, we see a short pulse of radiation. This model requires that the magnetic poles be located in different places from the rotation poles. (credit "stars": modification of work by Tony Hisgett)

Note that in a neutron star, the magnetic north and south poles do not have to be anywhere close to the north and south poles defined by the star's rotation. In the same way, we discussed in the chapter on The Giant Planets that the magnetic poles on the planets Uranus and Neptune are not lined up with the poles of the planet's spin. Figure 23.16 shows the poles of the magnetic field perpendicular to the poles of rotation, but the two kinds of poles could make any angle.

In fact, the misalignment of the rotational axis with the magnetic axis plays a crucial role in the generation of the observed pulses in this model. At the two magnetic poles, the particles from the neutron star are focused into a narrow beam and come streaming out of the whirling magnetic region at enormous speeds. They emit energy over a broad range of the electromagnetic spectrum. The radiation itself is also confined to a narrow beam, which explains why the pulsar acts like a lighthouse. As the rotation carries first one and then the other magnetic pole of the star into our view, we see a pulse of radiation each time.

Tests of the Model

This explanation of pulsars in terms of beams of radiation from highly magnetic and rapidly spinning neutron stars is a very clever idea. But what evidence do we have that it is the correct model? First, we can measure the masses of some pulsars, and they do turn out be in the range of 1.4 to 1.8 times that of the Sun—just what theorists predict for neutron stars. The masses are found using Kepler's law for those few pulsars that are members of binary star systems.

But there is an even-better confirming argument, which brings us back to the Crab Nebula and its vast energy output. When the high-energy charged particles from the neutron star pulsar hit the slower-moving material from the supernova, they energize this material and cause it to "glow" at many different wavelengths—just what we observe from the Crab Nebula. The pulsar beams are a power source that "light up" the nebula long after the initial explosion of the star that made it.

Who "pays the bills" for all the energy we see coming out of a remnant like the Crab Nebula? After all, when energy emerges from one place, it must be depleted in another. The ultimate energy source in our model is the rotation of the neutron star, which propels charged particles outward and spins its magnetic field at enormous speeds. As its rotational energy is used to excite the Crab Nebula year after year, the pulsar inside the nebula slows down. As it slows, the pulses come a little less often; more time elapses before the slower neutron star brings its beam back around.

Several decades of careful observations have now shown that the Crab Nebula pulsar is not a perfectly regular clock as we originally thought: instead, it is gradually slowing down. Having measured how much the pulsar is slowing down, we can calculate how much rotation energy the neutron star is losing. Remember that it is very densely packed and spins amazingly quickly. Even a tiny slowing down can mean an immense loss of energy.

To the satisfaction of astronomers, the rotational energy lost by the pulsar turns out to be the same as the amount of energy emerging from the nebula surrounding it. In other words, the slowing down of a rotating neutron star can explain precisely why the Crab Nebula is glowing with the amount of energy we observe.

The Evolution of Pulsars

From observations of the pulsars discovered so far, astronomers have concluded that one new pulsar is born somewhere in the Galaxy every 25 to 100 years, the same rate at which supernovae are estimated to occur. Calculations suggest that the typical lifetime of a pulsar is about 10 million years; after that, the neutron star no longer rotates fast enough to produce significant beams of particles and energy, and is no longer observable. We estimate that there are about 100 million neutron stars in our Galaxy, most of them rotating too slowly to come to our notice.

The Crab pulsar is rather young (only about 960 years old) and has a short period, whereas other, older pulsars have already slowed to longer periods. Pulsars thousands of years old have lost too much energy to emit appreciably in the visible and X-ray wavelengths, and they are observed only as radio pulsars; their periods are a second or longer.

There is one other reason we can see only a fraction of the pulsars in the Galaxy. Consider our lighthouse model again. On Earth, all ships approach on the same plane—the surface of the ocean—so the lighthouse can be built to sweep its beam over that surface. But in space, objects can be anywhere in three dimensions. As a given pulsar's beam sweeps over a circle in space, there is absolutely no guarantee that this circle will include the direction of Earth. In fact, if you think about it, many more circles in space will *not* include Earth than will include it. Thus, we estimate that we are unable to observe a large number of neutron stars because their pulsar beams miss us entirely.

At the same time, it turns out that only a few of the pulsars discovered so far are embedded in the visible clouds of gas that mark the remnant of a supernova. This might at first seem mysterious, since we know that supernovae give rise to neutron stars and we should expect each pulsar to have begun its life in a supernova explosion. But the lifetime of a pulsar turns out to be about 100 times longer than the length of time required for the expanding gas of a supernova remnant to disperse into interstellar space. Thus, most pulsars are found with no other trace left of the explosion that produced them.

In addition, some pulsars are ejected by a supernova explosion that is not the same in all directions. If the supernova explosion is stronger on one side, it can kick the pulsar entirely out of the supernova remnant (some astronomers call this "getting a birth kick"). We know such kicks happen because we see a number of young supernova remnants in nearby galaxies where the pulsar is to one side of the remnant and racing away at several hundred miles per second (Figure 23.17).

Figure 23.17 Speeding Pulsar. This intriguing image (which combines X-ray, visible, and radio observations) shows the jet trailing behind a pulsar (at bottom right, lined up between the two bright stars). With a length of 37 light-years, the jet trail (seen in purple) is the longest ever observed from an object in the Milky Way. (There is also a mysterious shorter, comet-like tail that is almost perpendicular to the purple jet.) Moving at a speed between 2.5 and 5 million miles per hour, the pulsar is traveling away from the core of the supernova remnant where it originated. (credit: X-ray: NASA/CXC/ISDC/L.Pavan et al, Radio: CSIRO/ATNF/ATCA Optical: 2MASS/UMass/IPAC-Caltech/NASA/NSF)

MAKING CONNECTIONS

Touched by a Neutron Star

On December 27, 2004, Earth was bathed with a stream of X-ray and gamma-ray radiation from a neutron star known as SGR 1806-20. What made this event so remarkable was that, despite the distance of the source, its tidal wave of radiation had measurable effects on Earth's atmosphere. The apparent brightness of this gamma-ray flare was greater than any historical star explosion.

The primary effect of the radiation was on a layer high in Earth's atmosphere called the *ionosphere*. At night, the ionosphere is normally at a height of about 85 kilometers, but during the day, energy from the Sun ionizes more molecules and lowers the boundary of the ionosphere to a height of about 60 kilometers. The pulse of X-ray and gamma-ray radiation produced about the same level of ionization as the daytime Sun. It also caused some sensitive satellites above the atmosphere to shut down their electronics.

Measurements by telescopes in space indicate that SGR 1806-20 was a special type of fast-spinning neutron star called a *magnetar*. Astronomers Robert Duncan and Christopher Thomson gave them this name because their magnetic fields are stronger than that of any other type of astronomical source—in this case, about 800 trillion times stronger than the magnetic field of Earth.

A magnetar is thought to consist of a superdense core of neutrons surrounded by a rigid crust of atoms about a mile deep with a surface made of iron. The magnetar's field is so strong that it creates huge stresses inside that can sometimes crack open the hard crust, causing a starquarke. The vibrating crust produces an enormous blast of radiation. An astronaut 0.1 light-year from this particular magnetar would have received a fatal does from the blast in less than a second.

Fortunately, we were far enough away from magnetar SGR 1806-20 to be safe. Could a magnetar ever present a real danger to Earth? To produce enough energy to disrupt the ozone layer, a magnetar would have to be located within the cloud of comets that surround the solar system, and we know no magnetars are that close. Nevertheless, it is a fascinating discovery that events on distant star corpses can have measurable effects on Earth.

The Evolution of Binary Star Systems

Learning Objectives

By the end of this section, you will be able to:

> Describe the kind of binary star system that leads to a nova event
> Describe the type of binary star system that leads to a type Ia supernovae event
> Indicate how type Ia supernovae differ from type II supernovae

The discussion of the life stories of stars presented so far has suffered from a bias—what we might call "single-star chauvinism." Because the human race developed around a star that goes through life alone, we tend to think of most stars in isolation. But as we saw in The Stars: A Celestial Census, it now appears that as many as half of all stars may develop in *binary* systems—those in which two stars are born in each other's gravitational embrace and go through life orbiting a common center of mass.

For these stars, the presence of a close-by companion can have a profound influence on their evolution. Under the right circumstances, stars can exchange material, especially during the stages when one of them swells up into a giant or supergiant, or has a strong wind. When this happens and the companion stars are sufficiently close, material can flow from one star to another, decreasing the mass of the donor and increasing the mass of the recipient. Such *mass transfer* can be especially dramatic when the recipient is a stellar remnant such as a white dwarf or a neutron star. While the detailed story of how such binary stars evolve is beyond the scope of our book, we do want to mention a few examples of how the stages of evolution described in this chapter may change when there are two stars in a system.

White Dwarf Explosions: The Mild Kind

Let's consider the following system of two stars: one has become a white dwarf and the other is gradually transferring material onto it. As fresh hydrogen from the outer layers of its companion accumulates on the surface of the hot white dwarf, it begins to build up a layer of hydrogen. As more and more hydrogen accumulates and heats up on the surface of the degenerate star, the new layer eventually reaches a temperature that causes fusion to begin in a sudden, explosive way, blasting much of the new material away.

In this way, the white dwarf quickly (but only briefly) becomes quite bright, hundreds or thousands of times its previous luminosity. To observers before the invention of the telescope, it seemed that a new star suddenly appeared, and they called it a **nova**.[3] Novae fade away in a few months to a few years.

Hundreds of novae have been observed, each occurring in a binary star system and each later showing a shell of expelled material. A number of stars have more than one nova episode, as more material from its neighboring star accumulates on the white dwarf and the whole process repeats. As long as the episodes do not increase the mass of the white dwarf beyond the Chandrasekhar limit (by transferring too much mass too quickly), the dense white dwarf itself remains pretty much unaffected by the explosions on its surface.

White Dwarf Explosions: The Violent Kind

If a white dwarf accumulates matter from a companion star at a much faster rate, it can be pushed over the Chandrasekhar limit. The evolution of such a binary system is shown in Figure 23.18. When its mass approaches the Chandrasekhar mass limit (exceeds 1.4 M_{Sun}), such an object can no longer support itself as a white dwarf, and it begins to contract. As it does so, it heats up, and new nuclear reactions can begin in the

---

3 We now know that this historical terminology is quite misleading since novae do not originate from new stars. In fact, quite to the contrary, novae originate from white dwarfs, which are actually the endpoint of stellar evolution for low-mass stars. But since the system of two stars was too faint to be visible to the naked eye, it did seem to people, before telescopes were invented, that a star had appeared where nothing had been visible.

degenerate core. The star "simmers" for the next century or so, building up internal temperature. This simmering phase ends in less than a second, when an enormous amount of fusion (especially of carbon) takes place all at once, resulting in an explosion. The fusion energy produced during the final explosion is so great that it completely destroys the white dwarf. Gases are blown out into space at velocities of about 10,000 kilometers per second, and afterward, no trace of the white dwarf remains.

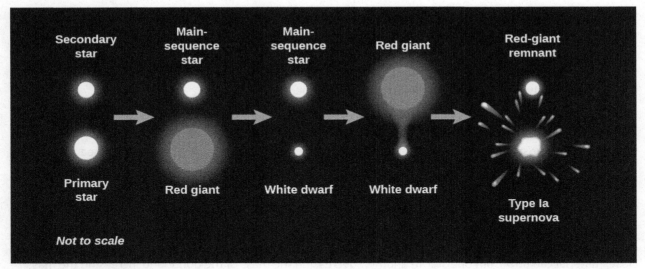

Figure 23.18 Evolution of a Binary System. The more massive star evolves first to become a red giant and then a white dwarf. The white dwarf then begins to attract material from its companion, which in turn evolves to become a red giant. Eventually, the white dwarf acquires so much mass that it is pushed over the Chandrasekhar limit and becomes a type Ia supernova.

Such an explosion is also called a supernova, since, like the destruction of a high-mass star, it produces a huge amount of energy in a very short time. However, unlike the explosion of a high-mass star, which can leave behind a neutron star or black hole remnant, the white dwarf is completely destroyed in the process, leaving behind no remnant. We call these white dwarf explosions type Ia supernovae.

We distinguish type I supernovae from those of supernovae of type II originating from the death of massive stars discussed earlier by the absence of hydrogen in their observed spectra. Hydrogen is the most common element in the universe and is a major component of massive, evolved stars. However, as we learned earlier, hydrogen is absent from the white dwarf remnant, which is primarily composed of carbon and oxygen for masses comparable to the Chandrasekhar mass limit.

The "a" subdesignation of type Ia supernovae further refers to the presence of strong silicon absorption lines, which are absent from supernovae originating from the collapse of massive stars. Silicon is one of the products that results from the fusion of carbon and oxygen, which bears out the scenario we described above—that there is a sudden onset of the fusion of the carbon (and oxygen) of which the white dwarf was made.

Observational evidence now strongly indicates that SN 1006, Tycho's Supernova, and Kepler's Supernova (see Supernovae in History) were all type Ia supernovae. For instance, in contrast to the case of SN 1054, which yielded the spinning pulsar in the Crab Nebula, none of these historical supernovae shows any evidence of stellar remnants that have survived their explosions. Perhaps even more puzzling is that, so far, astronomers have not been able to identify the companion star feeding the white dwarf in any of these historical supernovae.

Consequently, in order to address the mystery of the absent companion stars and other outstanding puzzles, astronomers have recently begun to investigate alternative mechanisms of generating type Ia supernovae. All proposed mechanisms rely upon white dwarfs composed of carbon and oxygen, which are needed to meet the observed absence of hydrogen in the type Ia spectrum. And because any isolated white dwarf below the Chandrasekhar mass is stable, all proposed mechanisms invoke a binary companion to explode the white dwarf. The leading alternative mechanism scientists believe creates a type Ia supernova is the merger of two white dwarf stars in a binary system. The two white dwarfs may have unstable orbits, such that over time, they would slowly move closer together until they merge. If their combined mass is greater than the Chandrasekhar limit, the result could also be a type Ia supernova explosion.

LINK TO LEARNING

You can watch a short video (https://openstax.org/l/30supernovavid) about Supernova SN 2014J, a type Ia supernova discovered in the Messier 82 (M82) galaxy on January 21, 2014, as well as see brief animations of the two mechanisms by which such a supernova could form.

Type Ia supernovae are of great interest to astronomers in other areas of research. This type of supernova is brighter than supernovae produced by the collapse of a massive star. Thus, type Ia supernovae can be seen at very large distances, and they are found in all types of galaxies. The energy output from most type Ia supernovae is consistent, with little variation in their maximum luminosities, or in how their light output initially increases and then slowly decreases over time. These properties make type Ia supernovae extremely valuable "standard bulbs" for astronomers looking out at great distances—well beyond the limits of our own Galaxy. You'll learn more about their use in measuring distances to other galaxies in The Big Bang.

In contrast, type II supernovae are about 5 times less luminous than type Ia supernovae and are only seen in galaxies that have recent, massive star formation. Type II supernovae are also less consistent in their energy output during the explosion and can have a range a peak luminosity values.

Neutron Stars with Companions

Now let's look at an even-more mismatched pair of stars in action. It is possible that, under the right circumstances, a binary system can even survive the explosion of one of its members as a type II supernova. In that case, an ordinary star can eventually share a system with a neutron star. If material is then transferred from the "living" star to its "dead" (and highly compressed) companion, this material will be pulled in by the strong gravity of the neutron star. Such infalling gas will be compressed and heated to incredible temperatures. It will quickly become so hot that it will experience an explosive burst of fusion. The energies involved are so great that we would expect much of the radiation from the burst to emerge as X-rays. And indeed, high-energy observatories above Earth's atmosphere (see Astronomical Instruments) have recorded many objects that undergo just these types of X-ray *bursts*.

If the neutron star and its companion are positioned the right way, a significant amount of material can be transferred to the neutron star and can set it spinning faster (as spin energy is also transferred). The radius of the neutron star would also decrease as more mass was added. Astronomers have found pulsars in binary systems that are spinning at a rate of more than 500 times per second! (These are sometimes called **millisecond pulsars** since the pulses are separated by a few thousandths of a second.)

Such a rapid spin could not have come from the birth of the neutron star; it must have been externally caused. (Recall that the Crab Nebula pulsar, one of the youngest pulsars known, was spinning "only" 30 times per second.) Indeed, some of the fast pulsars are observed to be part of binary systems, while others may be alone only because they have "fully consumed" their former partner stars through the mass transfer process. (These have sometimes been called "black widow pulsars.")

LINK TO LEARNING

View this short video (https://openstax.org/l/30scotronvid) to see Dr. Scott Ransom, of the National Radio Astronomy Observatory, explain how millisecond pulsars come about, with some nice animations.

And if you thought that a neutron star interacting with a "normal" star was unusual, there are also binary systems that consist of two neutron stars. One such system has the stars in very close orbits to one another, so much that they continually alter each other's orbit. Another binary neutron star system includes two pulsars that are orbiting each other every 2 hours and 25 minutes. As we discussed earlier, pulsars radiate away their energy, and these two pulsars are slowly moving toward one another, such that in about 85 million years, they will actually merge (see Gravitational Wave Astronomy for our first observations of such a merger).

We have now reached the end of our description of the final stages of stars, yet one piece of the story remains to be filled in. We saw that stars whose core masses are less than 1.4 M_{Sun} at the time they run out of fuel end their lives as white dwarfs. Dying stars with core masses between 1.4 and about 3 M_{Sun} become neutron stars. But there are stars whose core masses are greater than 3 M_{Sun} when they exhaust their fuel supplies. What becomes of them? The truly bizarre result of the death of such massive stellar cores (called a black hole) is the subject of our next chapter. But first, we will look at an astronomical mystery that turned out to be related to the deaths of stars and was solved through clever sleuthing and a combination of observation and theory.

23.6 | The Mystery of the Gamma–Ray Bursts

Learning Objectives

By the end of this section, you will be able to:

> Give a brief history of how gamma-ray bursts were discovered and what instruments made the discovery possible
> Explain why astronomers think that gamma-ray bursts beam their energy rather than it radiating uniformly in all directions
> Describe how the radiation from a gamma-ray burst and its afterglow is produced
> Explain how short-duration gamma-ray bursts differ from longer ones, and describe the process that makes short-duration gamma-ray bursts
> Explain why gamma-ray bursts may help us understand the early universe

Everybody loves a good mystery, and astronomers are no exception. The mystery we will discuss in this section was first discovered in the mid-1960s, not via astronomical research, but as a result of a search for the tell-tale signs of nuclear weapon explosions. The US Defense Department launched a series of *Vela* satellites to make sure that no country was violating a treaty that banned the detonation of nuclear weapons in space.

Since nuclear explosions produce the most energetic form of electromagnetic waves called *gamma rays* (see Radiation and Spectra), the *Vela* satellites contained detectors to search for this type of radiation. The satellites did not detect any confirmed events from human activities, but they did—to everyone's surprise—detect short bursts of gamma rays coming from random directions in the sky. News of the discovery was first published in 1973; however, the origin of the bursts remained a mystery. No one knew what produced the brief flashes of gamma rays or how far away the sources were.

From a Few Bursts to Thousands

With the launch of the Compton Gamma-Ray Observatory by NASA in 1991, astronomers began to identify many more bursts and to learn more about them (Figure 23.19). Approximately once per day, the NASA satellite detected a flash of gamma rays somewhere in the sky that lasted from a fraction of a second to several hundred seconds. Before the Compton measurements, astronomers had expected that the most likely place for the bursts to come from was the main disk of our own (pancake-shaped) Galaxy. If this had been the case, however, more bursts would have been seen in the crowded plane of the Milky Way than above or below it. Instead, the sources of the bursts were distributed *isotropically*; that is, they could appear anywhere in the sky with no preference for one region over another. Almost never did a second burst come from the same location.

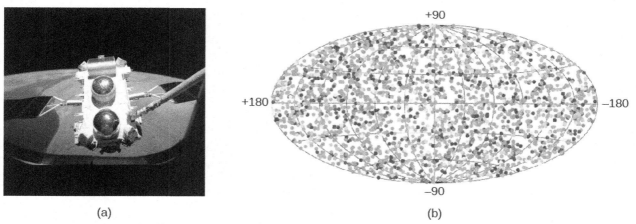

(a)   **(b)**

Figure 23.19 Compton Detects Gamma-Ray Bursts. (a) In 1991, the Compton Gamma-Ray Observatory was deployed by the Space Shuttle Atlantis. Weighing more than 16 tons, it was one of the largest scientific payloads ever launched into space. (b) This map of gamma-ray burst positions measured by the Compton Gamma-Ray Observatory shows the isotropic (same in all directions), uniform distribution of bursts on the sky. The map is oriented so that the disk of the Milky Way would stretch across the center line (or equator) of the oval. Note that the bursts show no preference at all for the plane of the Milky Way, as many other types of objects in the sky do. Colors indicate the total energy in the burst: red dots indicate long-duration, bright bursts; blue and purple dots show short, weaker bursts. (credit a: modification of work by NASA; credit b: modification of work by NASA/GSFC)

LINK TO LEARNING

To get a good visual sense of the degree to which the bursts come from all over the sky, watch this short animated NASA video (https://openstax.org/l/30swiftnasavid) showing the location of the first 500 bursts found by the later *Swift* satellite.

For several years, astronomers actively debated whether the burst sources were relatively nearby or very far away—the two possibilities for bursts that are isotropically distributed. Nearby locations might include the cloud of comets that surrounds the solar system or the halo of our Galaxy, which is large and spherical, and also surrounds us in all directions. If, on the other hand, the bursts occurred at very large distances, they could come from faraway galaxies, which are also distributed uniformly in all directions.

Both the very local and the very distant hypotheses required something strange to be going on. If the bursts were coming from the cold outer reaches of our own solar system or from the halo of our Galaxy, then astronomers had to hypothesize some new kind of physical process that could produce unpredictable flashes of high-energy gamma rays in these otherwise-quiet regions of space. And if the bursts came from galaxies millions or billions of light-years away, then they must be extremely powerful to be observable at such large distances; indeed they had to be the among the biggest explosions in the universe.

The First Afterglows

The problem with trying to figure out the source of the gamma-ray bursts was that our instruments for detecting gamma rays could not pinpoint the exact place in the sky where the burst was happening. Early gamma-ray telescopes did not have sufficient *resolution.* This was frustrating because astronomers suspected that if they *could* pinpoint the exact position of one of these rapid bursts, then they would be able to identify a counterpart (such as a star or galaxy) at other wavelengths and learn much more about the burst, including where it came from. This would, however, require either major improvements in gamma-ray detector technology to provide better resolution or detection of the burst at some other wavelength. In the end, both techniques played a role.

The breakthrough came with the launch of the Italian Dutch *BeppoSAX* satellite in 1996. *BeppoSAX* included a new type of gamma-ray telescope capable of identifying the position of a source much more accurately than previous instruments, to within a few minutes of arc on the sky. By itself, however, it was still not sophisticated enough to determine the exact source of the gamma-ray burst. After all, a box a few minutes of arc on a side could still contain many stars or other celestial objects.

However, the angular resolution of *BeppoSAX* was good enough to tell astronomers where to point other, more precise telescopes in the hopes of detecting longer-lived electromagnetic emission from the bursts at other wavelengths. Detection of a burst at visible-light or radio wavelengths could provide a position accurate to a few *seconds* of arc and allow the position to be pinpointed to an individual star or galaxy. *BeppoSAX* carried its own X-ray telescope onboard the spacecraft to look for such a counterpart, and astronomers using visible-light and radio facilities on the ground were eager to search those wavelengths as well.

Two crucial *BeppoSAX* burst observations in 1997 helped to resolve the mystery of the gamma-ray bursts. The first burst came in February from the direction of the constellation Orion. Within 8 hours, astronomers working with the satellite had identified the position of the burst, and reoriented the spacecraft to focus *BeppoSAX*'s X-ray detector on the source. To their excitement, they detected a slowly fading X-ray source 8 hours after the event—the first successful detection of an afterglow from a gamma-ray burst. This provided an even-better location of the burst (accurate to about 40 seconds of arc), which was then distributed to astronomers across the world to try to detect it at even longer wavelengths.

That very night, the 4.2-meter William Herschel Telescope on the Canary Islands found a fading visible-light source at the same position as the X-ray afterglow, confirming that such an afterglow could be detected in visible light as well. Eventually, the afterglow faded away, but left behind at the location of the original gamma-ray burst was a faint, fuzzy source right where the fading point of light had been—a distant galaxy (Figure 23.20). This was the first piece of evidence that gamma-ray bursts were indeed very energetic objects from very far away. However, it also remained possible that the burst source was much closer to us and just happened to align with a more distant galaxy, so this one observation alone was not a conclusive demonstration of the extragalactic origin of gamma-ray bursts.

Figure 23.20 Gamma-Ray Burst. This false-color Hubble Space Telescope image, taken in September 1997, shows the fading afterglow of the gamma-ray burst of February 28, 1997 and the host galaxy in which the burst originated. The left view shows the region of the burst. The enlargement shows the burst source and what appears to be its host galaxy. Note that the gamma-ray

On May 8 of the same year, a burst came from the direction of the constellation Camelopardalis. In a coordinated international effort, *BeppoSAX* again fixed a reasonably precise position, and almost immediately a telescope on Kitt Peak in Arizona was able to catch the visible-light afterglow. Within 2 days, the largest telescope in the world (the Keck in Hawaii) collected enough light to record a spectrum of the burst. The May gamma-ray burst afterglow spectrum showed absorption features from a fuzzy object that was 4 billion light-years from the Sun, meaning that the location of the burst had to be at least this far away—and possibly even farther. (How astronomers can get the distance of such an object from the Doppler shift in the spectrum is something we will discuss in Galaxies.) What that spectrum showed was clear evidence that the gamma-ray burst had taken place in a distant galaxy.

Networking to Catch More Bursts

After initial observations showed that the precise locations and afterglows of gamma-ray bursts could be found, astronomers set up a system to catch and pinpoint bursts on a regular basis. But to respond as quickly as needed to obtain usable results, astronomers realized that they needed to rely on automated systems rather than human observers happening to be in the right place at the right time.

Now, when an orbiting high-energy telescope discovers a burst, its rough location is immediately transmitted to a *Gamma-Ray Coordinates Network* based at NASA's Goddard Space Flight Center, alerting observers on the ground within a few seconds to look for the visible-light afterglow.

The first major success with this system was achieved by a team of astronomers from the University of Michigan, Lawrence Livermore National Laboratory, and Los Alamos National Laboratories, who designed an automated device they called the *Robotic Optical Transient Search Experiment* (*ROTSE*), which detected a very bright visible-light counterpart in 1999. At peak, the burst was almost as bright as Neptune—despite a distance (measured later by spectra from larger telescopes) of 9 billion light-years.

More recently, astronomers have been able to take this a step further, using wide-field-of-view telescopes to stare at large fractions of the sky in the hope that a gamma-ray burst will occur at the right place and time, and be recorded by the telescope's camera. These wide-field telescopes are not sensitive to faint sources, but ROTSE showed that gamma-ray burst afterglows could sometimes be very bright.

Astronomers' hopes were vindicated in March 2008, when an extremely bright gamma-ray burst occurred and its light was captured by two wide-field camera systems in Chile: the Polish "Pi of the Sky" and the Russian-Italian TORTORA [Telescopio Ottimizzato per la Ricerca dei Transienti Ottici Rapidi (Italian for Telescope Optimized for the Research of Rapid Optical Transients)] (see Figure 23.21). According to the data taken by these telescopes, for a period of about 30 seconds, the light from the gamma-ray burst was bright enough that it could have been seen by the unaided eye had a person been looking in the right place at the right time. Adding to our amazement, later observations by larger telescopes demonstrated that the burst occurred at a distance of 8 billion light-years from Earth!

Figure 23.21 Gamma-Ray Burst Observed in March 2008. The extremely luminous afterglow of GRB 080319B was imaged by the Swift Observatory in X-rays (left) and visible light/ultraviolet (right). (credit: modification of work by NASA/Swift/Stefan Immler, et al.)

To Beam or Not to Beam

The enormous distances to these events meant they had to have been astoundingly energetic to appear as bright as they were across such an enormous distance. In fact, they required so much energy that it posed a problem for gamma-ray burst models: if the source was radiating energy in all directions, then the energy released in gamma rays alone during a bright burst (such as the 1999 or 2008 events) would have been equivalent to the energy produced if the entire mass of a Sun-like star were suddenly converted into pure radiation.

For a source to produce this much energy this quickly (in a burst) is a real challenge. Even if the star producing the gamma-ray burst was much more massive than the Sun (as is probably the case), there is no known means of converting so much mass into radiation within a matter of seconds. However, there is one way to reduce the power required of the "mechanism" that makes gamma-ray bursts. So far, our discussion has assumed that the source of the gamma rays gives off the same amount of energy in all directions, like an incandescent light bulb.

But as we discuss in Pulsars and the Discovery of Neutron Stars, not all sources of radiation in the universe are like this. Some produce thin beams of radiation that are concentrated into only one or two directions. A laser pointer and a lighthouse on the ocean are examples of such beamed sources on Earth (Figure 23.22). If, when a burst occurs, the gamma rays come out in only one or two narrow beams, then our estimates of the luminosity of the source can be reduced, and the bursts may be easier to explain. In that case, however, the beam has to point toward Earth for us to be able to see the burst. This, in turn, would imply that for every burst we see from Earth, there are probably many others that we never detect because their beams point in other directions.

Figure 23.22 Burst That Is Beamed. This artist's conception shows an illustration of one kind of gamma-ray burst. The collapse of the core of a massive star into a black hole has produced two bright beams of light originating from the star's poles, which an observer pointed along one of these axes would see as a gamma-ray burst. The hot blue stars and gas clouds in the vicinity are meant to show that the event happened in an active star-forming region. (credit: NASA/Swift/Mary Pat Hrybyk-Keith and John Jones)

Long-Duration Gamma-Ray Bursts: Exploding Stars

After identifying and following large numbers of gamma-ray bursts, astronomers began to piece together clues about what kind of event is thought to be responsible for producing the gamma-ray burst. Or, rather, what kind of *events*, because there are at least two distinct types of gamma-ray bursts. The two—like the different types of supernovae—are produced in completely different ways.

Observationally, the crucial distinction is how long the burst lasts. Astronomers now divide gamma-ray bursts into two categories: short-duration ones (defined as lasting less than 2 seconds, but typically a fraction of a second) and long-duration ones (defined as lasting more than 2 seconds, but typically about a minute).

All of the examples we have discussed so far concern the long-duration gamma-ray bursts. These constitute most of the gamma-ray bursts that our satellites detect, and they are also brighter and easier to pinpoint. Many hundreds of long-duration gamma-ray bursts, and the properties of the galaxies in which they occurred, have now been studied in detail. Long-duration gamma-ray bursts are universally observed to come from distant galaxies that are still actively making stars. They are usually found to be located in regions of the galaxy with strong star-formation activity (such as spiral arms). Recall that the more massive a star is, the less time it spends in each stage of its life. This suggests that the bursts come from a young and short-lived, and therefore massive type of star.

Furthermore, in several cases when a burst has occurred in a galaxy relatively close to Earth (within a few billion light-years), it has been possible to search for a supernova at the same position—and in nearly all of these cases, astronomers have found evidence of a supernova of type Ic going off. A type Ic is a particular type of supernova, which we did not discuss in the earlier parts of this chapter; these are produced by a massive star that has been stripped of its outer hydrogen layer. However, only a tiny fraction of type Ic supernovae produce gamma-ray bursts.

Why would a massive star with its outer layers missing sometimes produce a gamma-ray burst at the same time that it explodes as a supernova? The explanation astronomers have in mind for the extra energy is the collapse of the star's core to form a spinning, magnetic black hole or neutron star. Because the star corpse is both magnetic and spinning rapidly, its sudden collapse is complex and can produce swirling jets of particles and powerful beams of radiation—just like in a quasar or active galactic nucleus (objects you will learn about

Active Galaxies, Quasars, and Supermassive Black Holes), but on a much faster timescale. A small amount of the infalling mass is ejected in a narrow beam, moving at speeds close to that of light. Collisions among the particles in the beam can produce intense bursts of energy that we see as a gamma-ray burst.

Within a few minutes, the expanding blast from the fireball plows into the interstellar matter in the dying star's neighborhood. This matter might have been ejected from the star itself at earlier stages in its evolution. Alternatively, it could be the gas out of which the massive star and its neighbors formed.

As the high-speed particles from the blast are slowed, they transfer their energy to the surrounding matter in the form of a shock wave. That shocked material emits radiation at longer wavelengths. This accounts for the afterglow of X-rays, visible light, and radio waves—the glow comes at longer and longer wavelengths as the blast continues to lose energy.

Short-Duration Gamma-Ray Bursts: Colliding Stellar Corpses

What about the shorter gamma-ray bursts? The gamma-ray emission from these events lasts less than 2 seconds, and in some cases may last only milliseconds—an amazingly short time. Such a timescale is difficult to achieve if they are produced in the same way as long-duration gamma-ray bursts, since the collapse of the stellar interior onto the black hole should take at least a few seconds.

Astronomers looked fruitlessly for afterglows from short-duration gamma-ray bursts found by *BeppoSAX* and other satellites. Evidently, the afterglows fade away too quickly. Fast-responding visible-light telescopes like ROTSE were not helpful either: no matter how fast these telescopes responded, the bursts were not bright enough at visible wavelengths to be detected by these small telescopes.

Once again, it took a new satellite to clear up the mystery. In this case, it was the *Swift Gamma-Ray Burst Satellite*, launched in 2004 by a collaboration between NASA and the Italian and UK space agencies (Figure 23.23). The design of *Swift* is similar to that of *BeppoSAX*. However, *Swift* is much more agile and flexible: after a gamma-ray burst occurs, the X-ray and UV telescopes can be repointed automatically within a few minutes (rather than a few hours). Thus, astronomers can observe the afterglow much earlier, when it is expected to be much brighter. Furthermore, the X-ray telescope is far more sensitive and can provide positions that are 30 times more precise than those provided by *BeppoSAX*, allowing bursts to be identified even without visible-light or radio observations.

Figure 23.23 Artist's Illustration of *Swift*. The US/UK/Italian spacecraft *Swift* contains on-board gamma-ray, X-ray, and ultraviolet detectors, and has the ability to automatically reorient itself to a gamma-ray burst detected by the gamma-ray instrument. Since its launch in 2005, *Swift* has detected and observed over a thousand bursts, including dozens of short-duration bursts. (credit: NASA, Spectrum Astro)

On May 9, 2005, *Swift* detected a flash of gamma rays lasting 0.13 seconds in duration, originating from the constellation Coma Berenices. Remarkably, the galaxy at the X-ray position looked completely different from any galaxy in which a long-duration burst had been seen to occur. The afterglow originated from the halo of a giant elliptical galaxy 2.7 billion light-years away, with no signs of any young, massive stars in its spectrum. Furthermore, no supernova was ever detected after the burst, despite extensive searching.

What could produce a burst less than a second long, originating from a region with no star formation? The leading model involves the *merger* of two compact stellar corpses: two neutron stars, or perhaps a neutron star and a black hole. Since many stars come in binary or multiple systems, it's possible to have systems where two such star corpses orbit one another. According to general relativity (which will be discussed in Black Holes and Curved Spacetime), the orbits of a binary star system composed of such objects should slowly decay with time, eventually (after millions or billions of years) causing the two objects to slam together in a violent but brief explosion. Because the decay of the binary orbit is so slow, we would expect more of these mergers to occur in old galaxies in which star formation has long since stopped.

LINK TO LEARNING

To learn more about the merger of two neutron stars and how they can produce a burst that lasts less than a second, check out this computer simulation (https://openstax.org/l/30comsimneustr) by NASA.

While it was impossible to be sure of this model based on only a single event (it is possible this burst actually came from a background galaxy and lined up with the giant elliptical only by chance), several dozen more short-duration gamma-ray bursts have since been located by *Swift*, many of which also originate from galaxies with very low star-formation rates. This has given astronomers greater confidence that this model is the correct one. Still, to be fully convinced, astronomers are searching for a "smoking gun" signature for the merger of two ultra-dense stellar remnants.

Astronomers identified two observations that would provide more direct evidence. Theoretical calculations indicate that when two neutron stars collide there will be a very special kind of explosion; neutrons stripped from the neutron stars during the violent final phase of the merger will fuse together into heavy elements and then release heat due to radioactivity, producing a short-lived but red supernova sometimes called a *kilonova*. (The term is used because it is about a thousand times brighter than an ordinary nova, but not quite as "super" as a traditional supernova.) Hubble observations of one short-duration gamma-ray burst in 2013 showed suggestive evidence of such a signature, but needed to be confirmed by future observations.

The second "smoking gun" is the detection of *gravitational waves.* As will be discussed in Black Holes and Curved Spacetime, gravitational waves are ripples in the fabric of spacetime that general relativity predicts should be produced by the acceleration of extremely massive and dense objects—such as two neutron stars or black holes spiraling toward each other and colliding. The construction of instruments to detect gravitational waves is very challenging technically, and gravitational wave astronomy became feasible only in 2015. The first few detected gravitational wave events were produced by mergers of black holes. In 2017, however, gravitational waves were observed from a source that was coincident in time and space with a gamma-ray burst. The source consisted of two objects with the masses of neutron stars. A red supernova was also observed at this location, and the ejected material was rich in heavy elements. This observation not only confirms the theory of the origin of short gamma-ray bursts, but also is a spectacular demonstration of the validity of Einstein's theory of general relativity.

Probing the Universe with Gamma-Ray Bursts

The story of how astronomers came to explain the origin of the different kinds of bursts is a good example of how the scientific process sometimes resembles good detective work. While the mystery of short-duration gamma-ray bursts is still being unraveled, the focus of studies for long-duration gamma-ray bursts has begun to change from understanding the origin of the bursts themselves (which is now fairly well-established) to using them as tools to understand the broader universe.

The reason that long-duration gamma-ray bursts are useful has to do with their extreme luminosities, if only for a short time. In fact, long-duration gamma-ray bursts are so bright that they could easily be seen at distances that correspond to a few hundred million years after the expansion of the universe began, which is when theorists think that the first generation of stars formed. Some theories predict that the first stars are likely to be massive and complete their evolution in only a million years or so. If this turns out to be the case, then gamma-ray bursts (which signal the death of some of these stars) may provide us with the best way of probing the universe when stars and galaxies first began to form.

So far, the most distant gamma-ray burst found (on April 29, 2009) was in a galaxy with a redshift that corresponds to a remarkable 13.2 billion light years—meaning it happened only 600 million years after the Big Bang itself. This is comparable to the earliest and most distant galaxies found by the Hubble Space Telescope. It is not quite old enough to expect that it formed from the first generation of stars, but its appearance at this distance still gives us useful information about the production of stars in the early universe. Astronomers continue to scan the skies, looking for even more distant events signaling the deaths of stars from even further back in time.

🔑 Key Terms

Chandrasekhar limit   the upper limit to the mass of a white dwarf (equals 1.4 times the mass of the Sun)

degenerate gas   a gas that resists further compression because no two electrons can be in the same place at the same time doing the same thing (Pauli exclusion principle)

millisecond pulsar   a pulsar that rotates so quickly that it can give off hundreds of pulses per second (and its period is therefore measured in milliseconds)

neutron star   a compact object of extremely high density composed almost entirely of neutrons

nova   the cataclysmic explosion produced in a binary system, temporarily increasing its luminosity by hundreds to thousands of times

pulsar   a variable radio source of small physical size that emits very rapid radio pulses in very regular periods that range from fractions of a second to several seconds; now understood to be a rotating, magnetic neutron star that is energetic enough to produce a detectable beam of radiation and particles

type II supernova   a stellar explosion produced at the endpoint of the evolution of stars whose mass exceeds roughly 10 times the mass of the Sun

📖 Summary

23.1 The Death of Low-Mass Stars

During the course of their evolution, stars shed their outer layers and lose a significant fraction of their initial mass. Stars with masses of 8 M_{Sun} or less can lose enough mass to become white dwarfs, which have masses less than the Chandrasekhar limit (about 1.4 M_{Sun}). The pressure exerted by degenerate electrons keeps white dwarfs from contracting to still-smaller diameters. Eventually, white dwarfs cool off to become black dwarfs, stellar remnants made mainly of carbon, oxygen, and neon.

23.2 Evolution of Massive Stars: An Explosive Finish

In a massive star, hydrogen fusion in the core is followed by several other fusion reactions involving heavier elements. Just before it exhausts all sources of energy, a massive star has an iron core surrounded by shells of silicon, sulfur, oxygen, neon, carbon, helium, and hydrogen. The fusion of iron requires energy (rather than releasing it). If the mass of a star's iron core exceeds the Chandrasekhar limit (but is less than 3 M_{Sun}), the core collapses until its density exceeds that of an atomic nucleus, forming a neutron star with a typical diameter of 20 kilometers. The core rebounds and transfers energy outward, blowing off the outer layers of the star in a type II supernova explosion.

23.3 Supernova Observations

A supernova occurs on average once every 25 to 100 years in the Milky Way Galaxy. Despite the odds, no supernova in our Galaxy has been observed from Earth since the invention of the telescope. However, one nearby supernova (SN 1987A) has been observed in a neighboring galaxy, the Large Magellanic Cloud. The star that evolved to become SN 1987A began its life as a blue supergiant, evolved to become a red supergiant, and returned to being a blue supergiant at the time it exploded. Studies of SN 1987A have detected neutrinos from the core collapse and confirmed theoretical calculations of what happens during such explosions, including the formation of elements beyond iron. Supernovae are a main source of high-energy cosmic rays and can be dangerous for any living organisms in nearby star systems.

23.4 Pulsars and the Discovery of Neutron Stars

At least some supernovae leave behind a highly magnetic, rapidly rotating neutron star, which can be observed as a pulsar if its beam of escaping particles and focused radiation is pointing toward us. Pulsars emit rapid pulses of radiation at regular intervals; their periods are in the range of 0.001 to 10 seconds. The rotating neutron star acts like a lighthouse, sweeping its beam in a circle and giving us a pulse of radiation when the beam sweeps over Earth. As pulsars age, they lose energy, their rotations slow, and their periods increase.

23.5 The Evolution of Binary Star Systems

When a white dwarf or neutron star is a member of a close binary star system, its companion star can transfer mass to it. Material falling *gradually* onto a white dwarf can explode in a sudden burst of fusion and make a nova. If material falls *rapidly* onto a white dwarf, it can push it over the Chandrasekhar limit and cause it to explode completely as a type Ia supernova. Another possible mechanism for a type Ia supernova is the merger of two white dwarfs. Material falling onto a neutron star can cause powerful bursts of X-ray radiation. Transfer of material and angular momentum can speed up the rotation of pulsars until their periods are just a few thousandths of a second.

23.6 The Mystery of the Gamma-Ray Bursts

Gamma-ray bursts last from a fraction of a second to a few minutes. They come from all directions and are now known to be associated with very distant objects. The energy is most likely beamed, and, for the ones we can detect, Earth lies in the direction of the beam. Long-duration bursts (lasting more than a few seconds) come from massive stars with their outer hydrogen layers missing that explode as supernovae. Short-duration bursts are believed to be mergers of stellar corpses (neutron stars or black holes).

 For Further Exploration

Articles

Death of Stars

Hillebrandt, W., et al. "How To Blow Up a Star." *Scientific American* (October 2006): 42. On supernova mechanisms.

Irion, R. "Pursuing the Most Extreme Stars." *Astronomy* (January 1999): 48. On pulsars.

Kalirai, J. "New Light on Our Sun's Fate." *Astronomy* (February 2014): 44. What will happen to stars like our Sun between the main sequence and the white dwarf stages.

Kirshner, R. "Supernova 1987A: The First Ten Years." *Sky & Telescope* (February 1997): 35.

Maurer, S. "Taking the Pulse of Neutron Stars." *Sky & Telescope* (August 2001): 32. Review of recent ideas and observations of pulsars.

Zimmerman, R. "Into the Maelstrom." *Astronomy* (November 1998): 44. About the Crab Nebula.

Gamma-Ray Bursts

Fox, D. & Racusin, J. "The Brightest Burst." *Sky & Telescope* (January 2009): 34. Nice summary of the brightest burst observed so far, and what we have learned from it.

Nadis, S. "Do Cosmic Flashes Reveal Secrets of the Infant Universe?" *Astronomy* (June 2008): 34. On different types of gamma-ray bursts and what we can learn from them.

Naeye, R. "Dissecting the Bursts of Doom." *Sky & Telescope* (August 2006): 30. Excellent review of gamma-ray bursts—how we discovered them, what they might be, and what they can be used for in probing the universe.

Zimmerman, R. "Speed Matters." *Astronomy* (May 2000): 36. On the quick-alert networks for finding afterglows.

Zimmerman, R. "Witness to Cosmic Collisions." *Astronomy* (July 2006): 44. On the Swift mission and what it is teaching astronomers about gamma-ray bursts.

Websites

Death of Stars

Crab Nebula: http://chandra.harvard.edu/xray_sources/crab/crab.html (http://chandra.harvard.edu/xray_sources/crab/crab.html). A short, colorfully written introduction to the history and science involving the best-known supernova remant.

Introduction to Neutron Stars: https://www.astro.umd.edu/~miller/nstar.html (https://www.astro.umd.edu/~miller/nstar.html). Coleman Miller of the University of Maryland maintains this site, which goes from easy to hard as you get into it, but it has lots of good information about corpses of massive stars.

Introduction to Pulsars (by Maryam Hobbs at the Australia National Telescope Facility): http://www.atnf.csiro.au/outreach/education/everyone/pulsars/index.html (http://www.atnf.csiro.au/outreach/education/everyone/pulsars/index.html).

Magnetars, Soft Gamma Repeaters, and Very Strong Magnetic Fields: http://solomon.as.utexas.edu/magnetar.html (http://solomon.as.utexas.edu/magnetar.html). Robert Duncan, one of the originators of the idea of magnetars, assembled this site some years ago.

Gamma-Ray Bursts

Brief Intro to Gamma-Ray Bursts (from PBS' *Seeing in the Dark*): http://www.pbs.org/seeinginthedark/astronomy-topics/gamma-ray-bursts.html (http://www.pbs.org/seeinginthedark/astronomy-topics/gamma-ray-bursts.html).

Discovery of Gamma-Ray Bursts: http://science.nasa.gov/science-news/science-at-nasa/1997/ast19sep97_2/ (http://science.nasa.gov/science-news/science-at-nasa/1997/ast19sep97_2/).

Missions to Detect and Learn More about Gamma-Ray Bursts:

- Fermi Space Telescope: https://www.nasa.gov/content/fermi/overview (https://www.nasa.gov/content/fermi/overview).
- *INTEGRAL* Spacecraft: http://www.esa.int/Science_Exploration/Space_Science/Integral_overview (http://www.esa.int/Science_Exploration/Space_Science/Integral_overview).
- *SWIFT* Spacecraft: https://swift.gsfc.nasa.gov/about_swift/ (https://swift.gsfc.nasa.gov/about_swift/).

Videos

Death of Stars

BBC interview with Antony Hewish: http://www.bbc.co.uk/archive/scientists/10608.shtml (http://www.bbc.co.uk/archive/scientists/10608.shtml). (40:54).

Black Widow Pulsars: The Vengeful Corpses of Stars: https://www.youtube.com/watch?v=Fn-3G_N0hy4 (https://www.youtube.com/watch?v=Fn-3G_N0hy4). A public talk in the Silicon Valley Astronomy Lecture Series by Dr. Roger Romani (Stanford University) (1:01:47).

Hubblecast 64: It all ends with a bang!: http://www.spacetelescope.org/videos/hubblecast64a/ (http://www.spacetelescope.org/videos/hubblecast64a/). HubbleCast Program introducing Supernovae with Dr. Joe Liske (9:48).

Space Movie Reveals Shocking Secrets of the Crab Pulsar: http://hubblesite.org/newscenter/archive/releases/2002/24/video/c/ (http://hubblesite.org/newscenter/archive/releases/2002/24/video/c/). A sequence of Hubble and Chandra Space Telescope images of the central regions of the Crab Nebula have been assembled into a very brief movie accompanied by animation showing how the pulsar affects its environment; it comes with some useful background material (40:06).

Gamma-Ray Bursts

Gamma-Ray Bursts: The Biggest Explosions Since the Big Bang!: https://www.youtube.com/watch?v=ePo_EdgV764 (https://www.youtube.com/watch?v=ePo_EdgV764). Edo Berge in a popular-level lecture at Harvard (58:50).

Gamma-Ray Bursts: Flashes in the Sky: https://www.youtube.com/watch?v=23EhcAP3O8Q (https://www.youtube.com/watch?v=23EhcAP3O8Q). American Museum of Natural History Science Bulletin on the *Swift* satellite (5:59).

Overview Animation of Gamma-Ray Burst: http://news.psu.edu/video/296729/2013/11/27/overview-animation-gamma-ray-burst (http://news.psu.edu/video/296729/2013/11/27/overview-animation-gamma-ray-burst). Brief Animation of what causes a long-duration gamma-ray burst (0:55).

Collaborative Group Activities

A. Someone in your group uses a large telescope to observe an expanding shell of gas. Discuss what measurements you could make to determine whether you have discovered a planetary nebula or the remnant of a supernova explosion.

B. The star Sirius (the brightest star in our northern skies) has a white-dwarf companion. Sirius has a mass of about 2 M_{Sun} and is still on the main sequence, while its companion is already a star corpse. Remember that a white dwarf can't have a mass greater than 1.4 M_{Sun}. Assuming that the two stars formed at the same time, your group should discuss how Sirius could have a white-dwarf companion. Hint: Was the initial mass of the white-dwarf star larger or smaller than that of Sirius?

C. Discuss with your group what people today would do if a brilliant star suddenly became visible during the daytime? What kind of fear and superstition might result from a supernova that was really bright in our skies? Have your group invent some headlines that the tabloid newspapers and the less responsible web news outlets would feature.

D. Suppose a supernova exploded only 40 light-years from Earth. Have your group discuss what effects there may be on Earth when the radiation reaches us and later when the particles reach us. Would there be any way to protect people from the supernova effects?

E. When pulsars were discovered, the astronomers involved with the discovery talked about finding "little green men." If you had been in their shoes, what tests would you have performed to see whether such a pulsating source of radio waves was natural or the result of an alien intelligence? Today, several groups around the world are actively searching for possible radio signals from intelligent civilizations. How might you expect such signals to differ from pulsar signals?

F. Your little brother, who has not had the benefit of an astronomy course, reads about white dwarfs and neutron stars in a magazine and decides it would be fun to go near them or even try to land on them. Is this a good idea for future tourism? Have your group make a list of reasons it would not be safe for children (or adults) to go near a white dwarf and a neutron star.

G. A lot of astronomers' time and many instruments have been devoted to figuring out the nature of gamma-ray bursts. Does your group share the excitement that astronomers feel about these mysterious high-energy events? What are some reasons that people outside of astronomy might care about learning about gamma-ray bursts?

⎘ Exercises

Review Questions

1. How does a white dwarf differ from a neutron star? How does each form? What keeps each from collapsing under its own weight?

2. Describe the evolution of a star with a mass like that of the Sun, from the main-sequence phase of its evolution until it becomes a white dwarf.

3. Describe the evolution of a massive star (say, 20 times the mass of the Sun) up to the point at which it becomes a supernova. How does the evolution of a massive star differ from that of the Sun? Why?

4. How do the two types of supernovae discussed in this chapter differ? What kind of star gives rise to each type?

5. A star begins its life with a mass of 5 M_{Sun} but ends its life as a white dwarf with a mass of 0.8 M_{Sun}. List the stages in the star's life during which it most likely lost some of the mass it started with. How did mass loss occur in each stage?

6. If the formation of a neutron star leads to a supernova explosion, explain why only three of the hundreds of known pulsars are found in supernova remnants.

7. How can the Crab Nebula shine with the energy of something like 100,000 Suns when the star that formed the nebula exploded almost 1000 years ago? Who "pays the bills" for much of the radiation we see coming from the nebula?

8. How is a nova different from a type Ia supernova? How does it differ from a type II supernova?

9. Apart from the masses, how are binary systems with a neutron star different from binary systems with a white dwarf?

10. What observations from SN 1987A helped confirm theories about supernovae?

11. Describe the evolution of a white dwarf over time, in particular how the luminosity, temperature, and radius change.

12. Describe the evolution of a pulsar over time, in particular how the rotation and pulse signal changes over time.

13. How would a white dwarf that formed from a star that had an initial mass of 1 M_{Sun} be different from a white dwarf that formed from a star that had an initial mass of 9 M_{Sun}?

14. What do astronomers think are the causes of longer-duration gamma-ray bursts and shorter-duration gamma-ray bursts?

15. How did astronomers finally solve the mystery of what gamma-ray bursts were? What instruments were required to find the solution?

Thought Questions

16. Arrange the following stars in order of their evolution:
A. A star with no nuclear reactions going on in the core, which is made primarily of carbon and oxygen.
B. A star of uniform composition from center to surface; it contains hydrogen but has no nuclear reactions going on in the core.
C. A star that is fusing hydrogen to form helium in its core.
D. A star that is fusing helium to carbon in the core and hydrogen to helium in a shell around the core.
E. A star that has no nuclear reactions going on in the core but is fusing hydrogen to form helium in a shell around the core.

17. Would you expect to find any white dwarfs in the Orion Nebula? (See The Birth of Stars and the Discovery of Planets outside the Solar System to remind yourself of its characteristics.) Why or why not?

18. Suppose no stars more massive than about 2 M_{Sun} had ever formed. Would life as we know it have been able to develop? Why or why not?

19. Would you be more likely to observe a type II supernova (the explosion of a massive star) in a globular cluster or in an open cluster? Why?

20. Astronomers believe there are something like 100 million neutron stars in the Galaxy, yet we have only found about 2000 pulsars in the Milky Way. Give several reasons these numbers are so different. Explain each reason.

21. Would you expect to observe every supernova in our own Galaxy? Why or why not?

22. The Large Magellanic Cloud has about one-tenth the number of stars found in our own Galaxy. Suppose the mix of high- and low-mass stars is exactly the same in both galaxies. Approximately how often does a supernova occur in the Large Magellanic Cloud?

23. Look at the list of the nearest stars in Appendix I. Would you expect any of these to become supernovae? Why or why not?

24. If most stars become white dwarfs at the ends of their lives and the formation of white dwarfs is accompanied by the production of a planetary nebula, why are there more white dwarfs than planetary nebulae in the Galaxy?

25. If a 3 and 8 M_{Sun} star formed together in a binary system, which star would:
A. Evolve off the main sequence first?
B. Form a carbon- and oxygen-rich white dwarf?
C. Be the location for a nova explosion?

26. You have discovered two star clusters. The first cluster contains mainly main-sequence stars, along with some red giant stars and a few white dwarfs. The second cluster also contains mainly main-sequence stars, along with some red giant stars, and a few neutron stars—but no white dwarf stars. What are the relative ages of the clusters? How did you determine your answer?

27. A supernova remnant was recently discovered and found to be approximately 150 years old. Provide possible reasons that this supernova explosion escaped detection.

28. Based upon the evolution of stars, place the following elements in order of least to most common in the Galaxy: gold, carbon, neon. What aspects of stellar evolution formed the basis for how you ordered the elements?

29. What observations or types of telescopes would you use to distinguish a binary system that includes a main-sequence star and a white dwarf star from one containing a main-sequence star and a neutron star?

30. How would the spectra of a type II supernova be different from a type Ia supernova? Hint: Consider the characteristics of the objects that are their source.

Figuring for Yourself

31. The ring around SN 1987A (Figure 23.12) initially became illuminated when energetic photons from the supernova interacted with the material in the ring. The radius of the ring is approximately 0.75 light-year from the supernova location. How long after the supernova did the ring become illuminated?

32. What is the acceleration of gravity (*g*) at the surface of the Sun? (See Appendix E for the Sun's key characteristics.) How much greater is this than *g* at the surface of Earth? Calculate what you would weigh on the surface of the Sun. Your weight would be your Earth weight multiplied by the ratio of the acceleration of gravity on the Sun to the acceleration of gravity on Earth. (Okay, we know that the Sun does not have a solid surface to stand on and that you would be vaporized if you were at the Sun's photosphere. Humor us for the sake of doing these calculations.)

33. What is the escape velocity from the Sun? How much greater is it than the escape velocity from Earth?

34. What is the average density of the Sun? How does it compare to the average density of Earth?

35. Say that a particular white dwarf has the mass of the Sun but the radius of Earth. What is the acceleration of gravity at the surface of the white dwarf? How much greater is this than *g* at the surface of Earth? What would you weigh at the surface of the white dwarf (again granting us the dubious notion that you could survive there)?

36. What is the escape velocity from the white dwarf in Exercise 23.35? How much greater is it than the escape velocity from Earth?

37. What is the average density of the white dwarf in Exercise 23.35? How does it compare to the average density of Earth?

38. Now take a neutron star that has twice the mass of the Sun but a radius of 10 km. What is the acceleration of gravity at the surface of the neutron star? How much greater is this than *g* at the surface of Earth? What would you weigh at the surface of the neutron star (provided you could somehow not become a puddle of protoplasm)?

39. What is the escape velocity from the neutron star in Exercise 23.38? How much greater is it than the escape velocity from Earth?

40. What is the average density of the neutron star in Exercise 23.38? How does it compare to the average density of Earth?

41. One way to calculate the radius of a star is to use its luminosity and temperature and assume that the star radiates approximately like a blackbody. Astronomers have measured the characteristics of central stars of planetary nebulae and have found that a typical central star is 16 times as luminous and 20 times as hot (about 110,000 K) as the Sun. Find the radius in terms of the Sun's. How does this radius compare with that of a typical white dwarf?

42. According to a model described in the text, a neutron star has a radius of about 10 km. Assume that the pulses occur once per rotation. According to Einstein's theory of relatively, nothing can move faster than the speed of light. Check to make sure that this pulsar model does not violate relativity. Calculate the rotation speed of the Crab Nebula pulsar at its equator, given its period of 0.033 s. (Remember that distance equals velocity × time and that the circumference of a circle is given by $2\pi R$).

43. Do the same calculations as in Exercise 23.42 but for a pulsar that rotates 1000 times per second.

44. If the Sun were replaced by a white dwarf with a surface temperature of 10,000 K and a radius equal to Earth's, how would its luminosity compare to that of the Sun?

45. A supernova can eject material at a velocity of 10,000 km/s. How long would it take a supernova remnant to expand to a radius of 1 AU? How long would it take to expand to a radius of 1 light-years? Assume that the expansion velocity remains constant and use the relationship: $\text{expansion time} = \frac{\text{distance}}{\text{expansion velocity}}$.

46. A supernova remnant was observed in 2007 to be expanding at a velocity of 14,000 km/s and had a radius of 6.5 light-years. Assuming a constant expansion velocity, in what year did this supernova occur?

47. The ring around SN 1987A (Figure 23.12) started interacting with material propelled by the shockwave from the supernova beginning in 1997 (10 years after the explosion). The radius of the ring is approximately 0.75 light-year from the supernova location. How fast is the supernova material moving, assume a constant rate of motion in km/s?

48. Before the star that became SN 1987A exploded, it evolved from a red supergiant to a blue supergiant while remaining at the same luminosity. As a red supergiant, its surface temperature would have been approximately 4000 K, while as a blue supergiant, its surface temperature was 16,000 K. How much did the radius change as it evolved from a red to a blue supergiant?

49. What is the radius of the progenitor star that became SN 1987A? Its luminosity was 100,000 times that of the Sun, and it had a surface temperature of 16,000 K.

50. What is the acceleration of gravity at the surface of the star that became SN 1987A? How does this *g* compare to that at the surface of Earth? The mass was 20 times that of the Sun and the radius was 41 times that of the Sun.

51. What was the escape velocity from the surface of the SN 1987A progenitor star? How much greater is it than the escape velocity from Earth? The mass was 20 times that of the Sun and the radius was 41 times that of the Sun.

52. What was the average density of the star that became SN 1987A? How does it compare to the average density of Earth? The mass was 20 times that of the Sun and the radius was 41 times that of the Sun.

53. If the pulsar shown in Figure 23.16 is rotating 100 times per second, how many pulses would be detected in one minute? The two beams are located along the pulsar's equator, which is aligned with Earth.

ILLUSTRATION

24

Black Holes and Curved Spacetime

Figure 24.1 Stellar Mass Black Hole. On the left, a visible-light image shows a region of the sky in the constellation of Cygnus; the red box marks the position of the X-ray source Cygnus X-1. It is an example of a black hole created when a massive star collapses at the end of its life. Cygnus X-1 is in a binary star system, and the artist's illustration on the right shows the black hole pulling material away from a massive blue companion star. This material forms a disk (shown in red and orange) that rotates around the black hole before falling into it or being redirected away from the black hole in the form of powerful jets. The material in the disk (before it falls into the black hole) is so hot that it glows with X-rays, explaining why this object is an X-ray source. (credit left: modification of work by DSS; credit right: modification of work by NASA/CXC/M.Weiss)

Chapter Outline

24.1 Introducing General Relativity

24.2 Spacetime and Gravity

24.3 Tests of General Relativity

24.4 Time in General Relativity

24.5 Black Holes

24.6 Evidence for Black Holes

24.7 Gravitational Wave Astronomy

 Thinking Ahead

For most of the twentieth century, black holes seemed the stuff of science fiction, portrayed either as monster vacuum cleaners consuming all the matter around them or as tunnels from one universe to another. But the truth about black holes is almost stranger than fiction. As we continue our voyage into the universe, we will discover that black holes are the key to explaining many mysterious and remarkable objects—including collapsed stars and the active centers of giant galaxies.

24.1 Introducing General Relativity

Learning Objectives

By the end of this section, you will be able to:

> Discuss some of the key ideas of the theory of general relativity
> Recognize that one's experiences of gravity and acceleration are interchangeable and indistinguishable
> Distinguish between Newtonian ideas of gravity and Einsteinian ideas of gravity
> Recognize why the theory of general relativity is necessary for understanding the nature of black holes

Most stars end their lives as white dwarfs or neutron stars. When a *very* massive star collapses at the end of its life, however, not even the mutual repulsion between densely packed neutrons can support the core against its own weight. If the remaining mass of the star's core is more than about three times that of the Sun (M_{Sun}), our theories predict that *no known force can stop it from collapsing forever!* Gravity simply overwhelms all other forces and crushes the core until it occupies an infinitely small volume. A star in which this occurs may become one of the strangest objects ever predicted by theory—a black hole.

To understand what a black hole is like and how it influences its surroundings, we need a theory that can describe the action of gravity under such extreme circumstances. To date, our best theory of gravity is the **general theory of relativity**, which was put forward in 1916 by Albert Einstein.

General relativity was one of the major intellectual achievements of the twentieth century; if it were music, we would compare it to the great symphonies of Beethoven or Mahler. Until recently, however, scientists had little need for a better theory of gravity; Isaac Newton's ideas that led to his law of universal gravitation (see Orbits and Gravity) are perfectly sufficient for most of the objects we deal with in everyday life. In the past half century, however, general relativity has become more than just a beautiful idea; it is now essential in understanding pulsars, quasars (which will be discussed in Active Galaxies, Quasars, and Supermassive Black Holes), and many other astronomical objects and events, including the black holes we will discuss here.

We should perhaps mention that this is the point in an astronomy course when many students start to feel a little nervous (and perhaps wish they had taken botany or some other earthbound course to satisfy the science requirement). This is because in popular culture, Einstein has become a symbol for mathematical brilliance that is simply beyond the reach of most people (Figure 24.2).

Figure 24.2 Albert Einstein (1879–1955). This famous scientist, seen here younger than in the usual photos, has become a symbol for high intellect in popular culture. (credit: NASA)

So, when we wrote that the theory of general relativity was Einstein's work, you may have worried just a bit, convinced that anything Einstein did must be beyond your understanding. This popular view is unfortunate and mistaken. Although the detailed calculations of general relativity do involve a good deal of higher mathematics, the basic ideas are not difficult to understand (and are, in fact, almost poetic in the way they give us a new perspective on the world). Moreover, general relativity goes beyond Newton's famous "inverse-square" law of gravity; it helps *explain* how matter interacts with other matter in space and time. This explanatory power is one of the requirements that any successful scientific theory must meet.

The Principle of Equivalence

The fundamental insight that led to the formulation of the general theory of relativity starts with a very simple thought: if you were able to jump off a high building and fall freely, you would not feel your own weight. In this chapter, we will describe how Einstein built on this idea to reach sweeping conclusions about the very fabric of space and time itself. He called it the "happiest thought of my life."

Einstein himself pointed out an everyday example that illustrates this effect (see Figure 24.3). Notice how your weight seems to be reduced in a high-speed elevator when it accelerates from a stop to a rapid descent. Similarly, your weight seems to increase in an elevator that starts to move quickly upward. This effect is not just a feeling you have: if you stood on a scale in such an elevator, you could measure your weight changing (you can actually perform this experiment in some science museums).

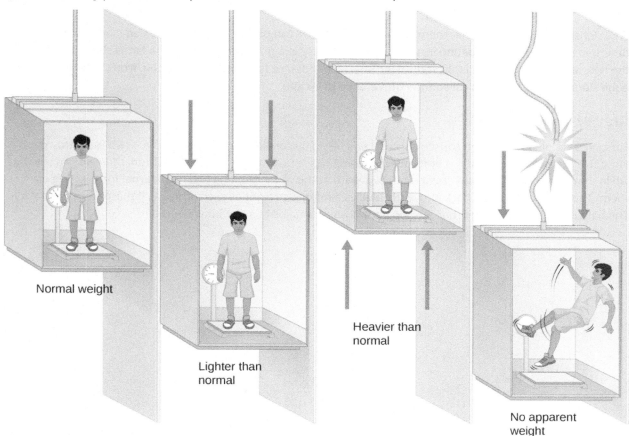

Figure 24.3 Your Weight in an Elevator. In an elevator at rest, you feel your normal weight. In an elevator that accelerates as it descends, you would feel lighter than normal. In an elevator that accelerates as it ascends, you would feel heavier than normal. If an evil villain cut the elevator cable, you would feel weightless as you fell to your doom.

In a *freely falling* elevator, with no air friction, you would lose your weight altogether. We generally don't like to cut the cables holding elevators to try this experiment, but near-weightlessness can be achieved by taking an airplane to high altitude and then dropping rapidly for a while. This is how NASA trains its astronauts for the experience of free fall in space; the scenes of weightlessness in the 1995 movie *Apollo 13* were filmed in the same way. (Moviemakers have since devised other methods using underwater filming, wire stunts, and computer graphics to create the appearance of weightlessness seen in such movies as *Gravity* and *The Martian*.)

Another way to state Einstein's idea is this: suppose we have a spaceship that contains a windowless laboratory equipped with all the tools needed to perform scientific experiments. Now, imagine that an astronomer wakes up after a long night celebrating some scientific breakthrough and finds herself sealed into this laboratory. She has no idea how it happened but notices that she is weightless. This could be because she and the laboratory are far away from any source of gravity, and both are either at rest or moving at some steady speed through space (in which case she has plenty of time to wake up). But it could also be because she and the laboratory are falling freely toward a planet like Earth (in which case she might first want to check her distance from the surface before making coffee).

What Einstein postulated is that there is no experiment she can perform inside the sealed laboratory to determine whether she is floating in space or falling freely in a gravitational field.[1] As far as she is concerned, the two situations are completely *equivalent*. This idea that free fall is indistinguishable from, and hence equivalent to, zero gravity is called the **equivalence principle**.

Gravity or Acceleration?

Einstein's simple idea has big consequences. Let's begin by considering what happens if two foolhardy people jump from opposite banks into a bottomless chasm (Figure 24.4). If we ignore air friction, then we can say that while they freely fall, they both accelerate downward at the same rate and feel no external force acting on them. They can throw a ball back and forth, always aiming it straight at each other, as if there were no gravity. The ball falls at the same rate that they do, so it always remains in a line between them.

---

1   Strictly speaking, this is true only if the laboratory is infinitesimally small. Different locations in a real laboratory that is falling freely due to gravity cannot all be at identical distances from the object(s) responsible for producing the gravitational force. In this case, objects in different locations will experience slightly different accelerations. But this point does not invalidate the principle of equivalence that Einstein derived from this line of thinking.

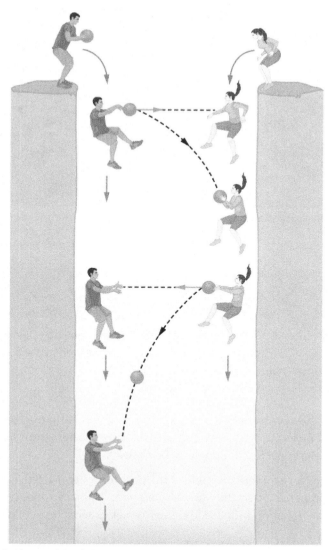

Figure 24.4 Free Fall. Two people play catch as they descend into a bottomless abyss. Since the people and ball all fall at the same speed, it appears to them that they can play catch by throwing the ball in a straight line between them. Within their frame of reference, there appears to be no gravity.

Such a game of catch is very different on the surface of Earth. Everyone who grows up feeling gravity knows that a ball, once thrown, falls to the ground. Thus, in order to play catch with someone, you must aim the ball upward so that it follows an arc—rising and then falling as it moves forward—until it is caught at the other end.

Now suppose we isolate our falling people and ball inside a large box that is falling with them. No one inside the box is aware of any gravitational force. If they let go of the ball, it doesn't fall to the bottom of the box or anywhere else but merely stays there or moves in a straight line, depending on whether it is given any motion.

Astronauts in the International Space Station (ISS) that is orbiting Earth live in an environment just like that of the people sealed in a freely falling box (Figure 24.5). The orbiting ISS is actually "falling" freely around Earth. While in free fall, the astronauts live in a strange world where there seems to be no gravitational force. One can give a wrench a shove, and it moves at constant speed across the orbiting laboratory. A pencil set in midair remains there as if no force were acting on it.

Figure 24.5 Astronauts aboard the Space Shuttle. Shane Kimbrough and Sandra Magnus are shown aboard the Endeavour in 2008 with various fruit floating freely. Because the shuttle is in free fall as it orbits Earth, everything—including astronauts—stays put or moves uniformly relative to the walls of the spacecraft. This free-falling state produces a lack of apparent gravity inside the spacecraft. (credit: NASA)

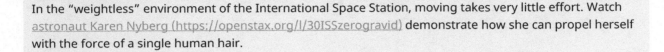

LINK TO LEARNING

In the "weightless" environment of the International Space Station, moving takes very little effort. Watch astronaut Karen Nyberg (https://openstax.org/l/30ISSzerogravid) demonstrate how she can propel herself with the force of a single human hair.

Appearances are misleading, however. There is a force in this situation. Both the ISS and the astronauts continually fall around Earth, pulled by its gravity. But since all fall together—shuttle, astronauts, wrench, and pencil—inside the ISS all gravitational forces appear to be absent.

Thus, the orbiting ISS provides an excellent example of the principle of equivalence—how local effects of gravity can be completely compensated by the right acceleration. To the astronauts, falling around Earth creates the same effects as being far off in space, remote from all gravitational influences.

The Paths of Light and Matter

Einstein postulated that the equivalence principle is a fundamental fact of nature, and that there is *no* experiment inside any spacecraft by which an astronaut can ever distinguish between being weightless in remote space and being in free fall near a planet like Earth. This would apply to experiments done with beams of light as well. But the minute we use light in our experiments, we are led to some very disturbing conclusions—and it is these conclusions that lead us to general relativity and a new view of gravity.

It seems apparent to us, from everyday observations, that beams of light travel in straight lines. Imagine that a spaceship is moving through empty space far from any gravity. Send a laser beam from the back of the ship to the front, and it will travel in a nice straight line and land on the front wall exactly opposite the point from which it left the rear wall. If the equivalence principle really applies universally, then this same experiment performed in free fall around Earth should give us the same result.

Now imagine that the astronauts again shine a beam of light along the length of their ship. But, as shown in Figure 24.6, this time the orbiting space station falls a bit between the time the light leaves the back wall and the time it hits the front wall. (The amount of the fall is grossly exaggerated in Figure 24.6 to illustrate the effect.) Therefore, if the beam of light follows a straight line but the ship's path curves downward, then the light should strike the front wall at a point higher than the point from which it left.

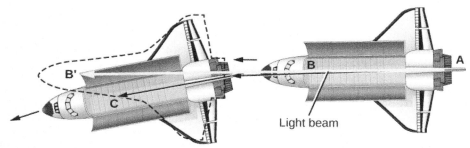

Figure 24.6 Curved Light Path. In a spaceship moving to the left (in this figure) in its orbit about a planet, light is beamed from the rear, A, toward the front, B. Meanwhile, the ship is falling out of its straight path (exaggerated here). We might therefore expect the light to strike at B', above the target in the ship. Instead, the light follows a curved path and strikes at C. In order for the principle of equivalence to be correct, gravity must be able to curve the path of a light beam just as it curves the path of the spaceship.

However, this would violate the principle of equivalence—the two experiments would give different results. We are thus faced with giving up one of our two assumptions. Either the principle of equivalence is not correct, or light does not always travel in straight lines. Instead of dropping what probably seemed at the time like a ridiculous idea, Einstein worked out what happens if light sometimes does *not* follow a straight path.

Let's suppose the principle of equivalence is right. Then the light beam must arrive directly opposite the point from which it started in the ship. The light, like the ball thrown back and forth, *must fall with the ship* that is in orbit around Earth (see Figure 24.6). This would make its path curve downward, like the path of the ball, and thus the light would hit the front wall exactly opposite the spot from which it came.

Thinking this over, you might well conclude that it doesn't seem like such a big problem: why *can't* light fall the way balls do? But, as discussed in Radiation and Spectra, light is profoundly different from balls. Balls have mass, while light does not.

Here is where Einstein's intuition and genius allowed him to make a profound leap. He gave physical meaning to the strange result of our thought experiment. Einstein suggested that the light curves down to meet the front of the shuttle because Earth's gravity actually bends the *fabric of space and time*. This radical idea—which we will explain next—keeps the behavior of light the same in both empty space and free fall, but it changes some of our most basic and cherished ideas about space and time. The reason we take Einstein's suggestion seriously is that, as we will see, experiments now clearly show his intuitive leap was correct.

24.2 | Spacetime and Gravity

Learning Objectives

By the end of this section, you will be able to:

> Describe Einstein's view of gravity as the warping of spacetime in the presence of massive objects
> Understand that Newton's concept of the gravitational force between two massive objects and Einstein's concept of warped spacetime are different explanations for the same observed accelerations of one massive object in the presence of another massive object

Is light actually bent from its straight-line path by the mass of Earth? How can light, which has no mass, be affected by gravity? Einstein preferred to think that it is *space and time* that are affected by the presence of a large mass; light beams, and everything else that travels through space and time, then find their paths affected. Light always follows the shortest path—but that path may not always be straight. This idea is true for human travel on the curved surface of planet Earth, as well. Say you want to fly from Chicago to Rome. Since an airplane can't go through the solid body of the Earth, the shortest distance is not a straight line but the arc of a *great circle*.

Linkages: Mass, Space, and Time

To show what Einstein's insight really means, let's first consider how we locate an event in space and time. For example, imagine you have to describe to worried school officials the fire that broke out in your room when your roommate tried cooking shish kebabs in the fireplace. You explain that your dorm is at 6400 College Avenue, a street that runs in the left-right direction on a map of your town; you are on the fifth floor, which tells where you are in the up-down direction; and you are the sixth room back from the elevator, which tells where you are in the forward-backward direction. Then you explain that the fire broke out at 6:23 p.m. (but was soon brought under control), which specifies the event in time. *Any* event in the universe, whether nearby or far away, can be pinpointed using the three dimensions of space and the one dimension of time.

Newton considered space and time to be completely independent, and that continued to be the accepted view until the beginning of the twentieth century. But Einstein showed that there is an intimate connection between space and time, and that only by considering the two together—in what we call **spacetime**—can we build up a correct picture of the physical world. We examine spacetime a bit more closely in the next subsection.

The gist of Einstein's general theory is that the presence of matter curves or warps the fabric of spacetime. This curving of spacetime is identified with gravity. When something else—a beam of light, an electron, or the starship *Enterprise*—enters such a region of distorted spacetime, its path will be different from what it would have been in the absence of the matter. As American physicist John Wheeler summarized it: "Matter tells spacetime how to curve; spacetime tells matter how to move."

The amount of distortion in spacetime depends on the mass of material that is involved and on how concentrated and compact it is. Terrestrial objects, such as the book you are reading, have far too little mass to introduce any significant distortion. Newton's view of gravity is just fine for building bridges, skyscrapers, or amusement park rides. General relativity does, however, have some practical applications. The GPS (Global Positioning System) in every smartphone can tell you where you are within 5 to 10 meters only because the effects of general and special relativity on the GPS satellites in orbit around the Earth are taken into account.

Unlike a book or your roommate, stars produce measurable distortions in spacetime. A white dwarf, with its stronger surface gravity, produces more distortion just above its surface than does a red giant with the same mass. So, you see, we *are* eventually going to talk about collapsing stars again, but not before discussing Einstein's ideas (and the evidence for them) in more detail.

Spacetime Examples

How can we understand the distortion of spacetime by the presence of some (significant) amount of mass? Let's try the following analogy. You may have seen maps of New York City that squeeze the full three dimensions of this towering metropolis onto a flat sheet of paper and still have enough information so tourists will not get lost. Let's do something similar with diagrams of spacetime.

Figure 24.7, for example, shows the progress of a motorist driving east on a stretch of road in Kansas where the countryside is absolutely flat. Since our motorist is traveling only in the east-west direction and the terrain is flat, we can ignore the other two dimensions of space. The amount of time elapsed since he left home is shown on the *y*-axis, and the distance traveled eastward is shown on the *x*-axis. From A to B he drove at a uniform speed; unfortunately, it was too fast a uniform speed and a police car spotted him. From B to C he stopped to receive his ticket and made no progress through space, only through time. From C to D he drove more slowly because the police car was behind him.

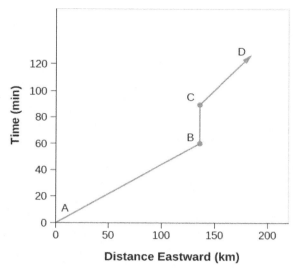

Figure 24.7 Spacetime Diagram. This diagram shows the progress of a motorist traveling east across the flat Kansas landscape. Distance traveled is plotted along the horizontal axis. The time elapsed since the motorist left the starting point is plotted along the vertical axis.

Now let's try illustrating the distortions of spacetime in two dimensions. In this case, we will (in our imaginations) use a rubber sheet that can stretch or warp if we put objects on it.

Let's imagine stretching our rubber sheet taut on four posts. To complete the analogy, we need something that normally travels in a straight line (as light does). Suppose we have an extremely intelligent ant—a friend of the comic book superhero Ant-Man, perhaps—that has been trained to walk in a straight line.

We begin with just the rubber sheet and the ant, simulating empty space with no mass in it. We put the ant on one side of the sheet and it walks in a beautiful straight line over to the other side (Figure 24.8). We next put a small grain of sand on the rubber sheet. The sand does distort the sheet a tiny bit, but this is not a distortion that we or the ant can measure. If we send the ant so it goes close to, but not on top of, the sand grain, it has little trouble continuing to walk in a straight line.

Now we grab something with a little more mass—say, a small pebble. It bends or distorts the sheet just a bit around its position. If we send the ant into this region, it finds its path slightly altered by the distortion of the sheet. The distortion is not large, but if we follow the ant's path carefully, we notice it deviating slightly from a straight line.

The effect gets more noticeable as we increase the mass of the object that we put on the sheet. Let's say we now use a massive paperweight. Such a heavy object distorts or warps the rubber sheet very effectively, putting a good sag in it. From our point of view, we can see that the sheet near the paperweight is no longer straight.

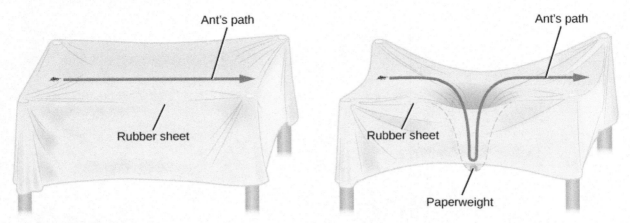

Figure 24.8 Three-Dimensional Analogy for Spacetime. On a flat rubber sheet, a trained ant has no trouble walking in a straight line. When a massive object creates a big depression in the sheet, the ant, which must walk where the sheet takes it, finds its path changed (warped) dramatically.

Now let's again send the ant on a journey that takes it close to, but not on top of, the paperweight. Far away from the paperweight, the ant has no trouble doing its walk, which looks straight to us. As it nears the paperweight, however, the ant is forced down into the sag. It must then climb up the other side before it can return to walking on an undistorted part of the sheet. All this while, the ant is following the shortest path it can, but through no fault of its own (after all, ants can't fly, so it has to stay on the sheet) this path is curved by the distortion of the sheet itself.

In the same way, according to Einstein's theory, light always follows the shortest path through spacetime. But the mass associated with large concentrations of matter distorts spacetime, and the shortest, most direct paths are no longer straight lines, but curves.

How large does a mass have to be before we can measure a change in the path followed by light? In 1916, when Einstein first proposed his theory, no distortion had been detected at the surface of Earth (so Earth might have played the role of the grain of sand in our analogy). Something with a mass like our Sun's was necessary to detect the effect Einstein was describing (we will discuss how this effect was measured using the Sun in the next section).

The paperweight in our analogy might be a white dwarf or a neutron star. The distortion of spacetime is greater near the surfaces of these compact, massive objects than near the surface of the Sun. And when, to return to the situation described at the beginning of the chapter, a star core with more than three times the mass of the Sun collapses forever, the distortions of spacetime very close to it can become truly mind-boggling.

24.3 | Tests of General Relativity

Learning Objectives

By the end of this section, you will be able to:

> Describe unusual motion of Mercury around the Sun and explain how general relativity explains the observed behavior

> Provide examples of evidence for light rays being bent by massive objects, as predicted by general relativity's theory about the warping of spacetime

What Einstein proposed was nothing less than a major revolution in our understanding of space and time. It was a new theory of gravity, in which mass determines the curvature of spacetime and that curvature, in turn, controls how objects move. Like all new ideas in science, no matter who advances them, Einstein's theory had to be tested by comparing its predictions against the experimental evidence. This was quite a challenge because the effects of the new theory were apparent only when the mass was quite large. (For smaller masses, it required measuring techniques that would not become available until decades later.)

When the distorting mass is small, the predictions of general relativity must agree with those resulting from Newton's law of universal gravitation, which, after all, has served us admirably in our technology and in guiding space probes to the other planets. In familiar territory, therefore, the differences between the predictions of the two models are subtle and difficult to detect. Nevertheless, Einstein was able to demonstrate one proof of his theory that could be found in existing data and to suggest another one that would be tested just a few years later.

The Motion of Mercury

Of the planets in our solar system, Mercury orbits closest to the Sun and is thus most affected by the distortion of spacetime produced by the Sun's mass. Einstein wondered if the distortion might produce a noticeable difference in the motion of Mercury that was not predicted by Newton's law. It turned out that the difference was subtle, but it was definitely there. Most importantly, it had already been measured.

Mercury has a highly elliptical orbit, so that it is only about two-thirds as far from the Sun at perihelion as it is at aphelion. (These terms were defined in the chapter on Orbits and Gravity.) The gravitational effects (perturbations) of the other planets on Mercury produce a calculable advance of Mercury's perihelion. What this means is that each successive perihelion occurs in a slightly different direction as seen from the Sun (Figure 24.9).

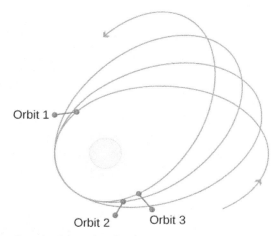

Figure 24.9 Mercury's Wobble. The major axis of the orbit of a planet, such as Mercury, rotates in space slightly because of various perturbations. In Mercury's case, the amount of rotation (or orbital precession) is a bit larger than can be accounted for by the gravitational forces exerted by other planets; this difference is precisely explained by the general theory of relativity. Mercury, being the planet closest to the Sun, has its orbit most affected by the warping of spacetime near the Sun. The change from orbit to orbit has been significantly exaggerated on this diagram.

According to Newtonian gravitation, the gravitational forces exerted by the planets will cause Mercury's perihelion to advance by about 531 seconds of arc (arcsec) per century. In the nineteenth century, however, it was observed that the actual advance is 574 arcsec per century. The discrepancy was first pointed out in 1859 by Urbain Le Verrier, the codiscoverer of Neptune. Just as discrepancies in the motion of Uranus allowed astronomers to discover the presence of Neptune, so it was thought that the discrepancy in the motion of Mercury could mean the presence of an undiscovered inner planet. Astronomers searched for this planet near the Sun, even giving it a name: Vulcan, after the Roman god of fire. (The name would later be used for the home planet of a famous character on a popular television show about future space travel.)

But no planet has ever been found nearer to the Sun than Mercury, and the discrepancy was still bothering astronomers when Einstein was doing his calculations. General relativity, however, predicts that due to the curvature of spacetime around the Sun, the perihelion of Mercury should advance slightly more than is predicted by Newtonian gravity. The result is to make the major axis of Mercury's orbit rotate slowly in space because of the Sun's gravity alone. The prediction of general relativity is that the direction of perihelion should change by an additional 43 arcsec per century. This is remarkably close to the observed discrepancy, and it gave Einstein a lot of confidence as he advanced his theory. The relativistic advance of perihelion was later also observed in the orbits of several asteroids that come close to the Sun.

Deflection of Starlight

Einstein's second test was something that had not been observed before and would thus provide an excellent confirmation of his theory. Since spacetime is more curved in regions where the gravitational field is strong, we would expect light passing very near the Sun to appear to follow a curved path (Figure 24.10), just like that of the ant in our analogy. Einstein calculated from general relativity theory that starlight just grazing the Sun's surface should be deflected by an angle of 1.75 arcsec. Could such a deflection be observed?

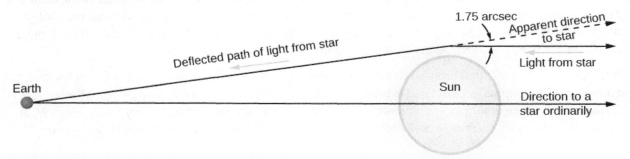

Figure 24.10 Curvature of Light Paths near the Sun. Starlight passing near the Sun is deflected slightly by the "warping" of spacetime. (This deflection of starlight is one small example of a phenomenon called gravitational lensing, which we'll discuss in more detail in The Evolution and Distribution of Galaxies.) Before passing by the Sun, the light from the star was traveling parallel to the bottom edge of the figure. When it passed near the Sun, the path was altered slightly. When we see the light, we assume the light beam has been traveling in a straight path throughout its journey, and so we measure the position of the star to be slightly different from its true position. If we were to observe the star at another time, when the Sun is not in the way, we would measure its true position.

We encounter a small "technical problem" when we try to photograph starlight coming very close to the Sun: the Sun is an outrageously bright source of starlight itself. But during a total solar eclipse, much of the Sun's light is blocked out, allowing the stars near the Sun to be photographed. In a paper published during World War I, Einstein (writing in a German journal) suggested that photographic observations during an eclipse could reveal the deflection of light passing near the Sun.

The technique involves taking a photograph of the stars six months prior to the eclipse and measuring the position of all the stars accurately. Then the same stars are photographed during the eclipse. This is when the starlight has to travel to us by skirting the Sun and moving through measurably warped spacetime. As seen from Earth, the stars closest to the Sun will seem to be "out of place"—slightly away from their regular positions as measured when the Sun is not nearby.

A single copy of that paper, passed through neutral Holland, reached the British astronomer Arthur S. Eddington, who noted that the next suitable eclipse was on May 29, 1919. The British organized two expeditions to observe it: one on the island of Príncipe, off the coast of West Africa, and the other in Sobral, in northern Brazil. Despite some problems with the weather, both expeditions obtained successful photographs. The stars seen near the Sun were indeed displaced, and to the accuracy of the measurements, which was about 20%, the shifts were consistent with the predictions of general relativity. More modern experiments with radio waves traveling close to the Sun have confirmed that the actual displacements are within 1% of what general relativity predicts.

The confirmation of the theory by the eclipse expeditions in 1919 was a triumph that made Einstein a world celebrity.

24.4 | Time in General Relativity

Learning Objectives

By the end of this section, you will be able to:

> Describe how Einsteinian gravity slows clocks and can decrease a light wave's frequency of oscillation

> Recognize that the gravitational decrease in a light wave's frequency is compensated by an increase in the light wave's wavelength—the so-called gravitational redshift—so that the light continues to travel at constant speed

General relativity theory makes various predictions about the behavior of space and time. One of these predictions, put in everyday terms, is that *the stronger the gravity, the slower the pace of time*. Such a statement goes very much counter to our intuitive sense of time as a flow that we all share. Time has always seemed the most democratic of concepts: all of us, regardless of wealth or status, appear to move together from the cradle to the grave in the great current of time.

But Einstein argued that it only seems this way to us because all humans so far have lived and died in the gravitational environment of Earth. We have had no chance to test the idea that the pace of time might depend on the strength of gravity, because we have not experienced radically different gravities. Moreover, the differences in the flow of time are extremely small until truly large masses are involved. Nevertheless, Einstein's prediction has now been tested, both on Earth and in space.

The Tests of Time

An ingenious experiment in 1959 used the most accurate atomic clock known to compare time measurements on the ground floor and the top floor of the physics building at Harvard University. For a clock, the experimenters used the frequency (the number of cycles per second) of gamma rays emitted by radioactive cobalt. Einstein's theory predicts that such a cobalt clock on the ground floor, being a bit closer to Earth's center of gravity, should run very slightly slower than the same clock on the top floor. This is precisely what the experiments observed. Later, atomic clocks were taken up in high-flying aircraft and even on one of the Gemini space flights. In each case, the clocks farther from Earth ran a bit faster. While in 1959 it didn't matter much if the clock at the top of the building ran faster than the clock in the basement, today that effect is highly relevant. Every smartphone or device that synchronizes with a GPS must correct for this (as we will see in the next section) since the clocks on satellites will run faster than clocks on Earth.

The effect is more pronounced if the gravity involved is the Sun's and not Earth's. If stronger gravity slows the pace of time, then it will take longer for a light or radio wave that passes very near the edge of the Sun to reach Earth than we would expect on the basis of Newton's law of gravity. (It takes longer because spacetime is curved in the vicinity of the Sun.) The smaller the distance between the ray of light and the edge of the Sun at closest approach, the longer will be the delay in the arrival time.

In November 1976, when the two Viking spacecraft were operating on the surface of Mars, the planet went behind the Sun as seen from Earth (Figure 24.11). Scientists had preprogrammed Viking to send a radio wave toward Earth that would go extremely close to the outer regions of the Sun. According to general relativity, there would be a delay because the radio wave would be passing through a region where time ran more slowly. The experiment was able to confirm Einstein's theory to within 0.1%.

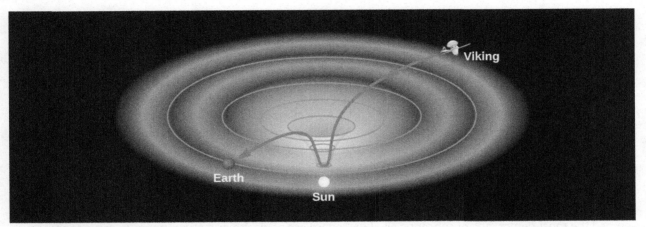

Figure 24.11 Time Delays for Radio Waves near the Sun. Radio signals from the Viking lander on Mars were delayed when they passed near the Sun, where spacetime is curved relatively strongly. In this picture, spacetime is pictured as a two-dimensional rubber sheet.

Gravitational Redshift

What does it mean to say that time runs more slowly? When light emerges from a region of strong gravity where time slows down, the light experiences a change in its frequency and wavelength. To understand what happens, let's recall that a wave of light is a repeating phenomenon—crest follows crest with great regularity. In this sense, each light wave is a little clock, keeping time with its wave cycle. If stronger gravity slows down the pace of time (relative to an outside observer), then the rate at which crest follows crest must be correspondingly slower—that is, the waves become less *frequent*.

To maintain constant light speed (the key postulate in Einstein's theories of special and general relativity), the lower frequency must be compensated by a longer wavelength. This kind of increase in wavelength (when caused by the motion of the source) is what we called a *redshift* in Radiation and Spectra. Here, because it is gravity and not motion that produces the longer wavelengths, we call the effect a **gravitational redshift**.

The advent of space-age technology made it possible to measure gravitational redshift with very high accuracy. In the mid-1970s, a hydrogen *maser*, a device akin to a laser that produces a microwave radio signal at a particular wavelength, was carried by a rocket to an altitude of 10,000 kilometers. Instruments on the ground were used to compare the frequency of the signal emitted by the rocket-borne maser with that from a similar maser on Earth. The experiment showed that the stronger gravitational field at Earth's surface really did slow the flow of time relative to that measured by the maser in the rocket. The observed effect matched the predictions of general relativity to within a few parts in 100,000.

These are only a few examples of tests that have confirmed the predictions of general relativity. Today, general relativity is accepted as our best description of gravity and is used by astronomers and physicists to understand the behavior of the centers of galaxies, the beginning of the universe, and the subject with which we began this chapter—the death of truly massive stars.

Relativity: A Practical Application

By now you may be asking: why should I be bothered with relativity? Can't I live my life perfectly well without it? The answer is you can't. Every time a pilot lands an airplane or you use a GPS to determine where you are on a drive or hike in the back country, you (or at least your GPS-enabled device) must take the effects of both general and special relativity into account.

GPS relies on an array of 24 satellites orbiting the Earth, and at least 4 of them are visible from any spot on Earth. Each satellite carries a precise atomic clock. Your GPS receiver detects the signals from those satellites that are overhead and calculates your position based on the time that it has taken those signals to reach you. Suppose you want to know where you are within 50 feet (GPS devices can actually do much better than this).

Since it takes only 50 billionths of a second for light to travel 50 feet, the clocks on the satellites must be synchronized to at least this accuracy—and relativistic effects must therefore be taken into account.

The clocks on the satellites are orbiting Earth at a speed of 14,000 kilometers per hour and are moving much faster than clocks on the surface of Earth. According to Einstein's theory of relativity, the clocks on the satellites are ticking more slowly than Earth-based clocks by about 7 millionths of a second per day. (We have not discussed the *special* theory of relativity, which deals with changes when objects move very fast, so you'll have to take our word for this part.)

The orbits of the satellites are 20,000 kilometers above Earth, where gravity is about four times weaker than at Earth's surface. General relativity says that the orbiting clocks should tick about 45 millionths of a second faster than they would on Earth. The net effect is that the time on a satellite clock advances by about 38 microseconds per day. If these relativistic effects were not taken into account, navigational errors would start to add up and positions would be off by about 7 miles in only a single day.

24.5 | Black Holes

Learning Objectives

By the end of this section, you will be able to:

> Explain the event horizon surrounding a black hole
> Discuss why the popular notion of black holes as great sucking monsters that can ingest material at great distances from them is erroneous
> Use the concept of warped spacetime near a black hole to track what happens to any object that might fall into a black hole
> Recognize why the concept of a singularity—with its infinite density and zero volume—presents major challenges to our understanding of matter

Let's now apply what we have learned about gravity and spacetime curvature to the issue we started with: the collapsing core in a very massive star. We saw that if the core's mass is greater than about 3 M_{Sun}, theory says that nothing can stop the core from collapsing forever. We will examine this situation from two perspectives: first from a pre-Einstein point of view, and then with the aid of general relativity.

Classical Collapse

Let's begin with a thought experiment. We want to know what speeds are required to escape from the gravitational pull of different objects. A rocket must be launched from the surface of Earth at a very high speed if it is to escape the pull of Earth's gravity. In fact, any object—rocket, ball, astronomy book—that is thrown into the air with a velocity less than 11 kilometers per second will soon fall back to Earth's surface. Only those objects launched with a speed greater than this *escape velocity* can get away from Earth.

The escape velocity from the surface of the Sun is higher yet—618 kilometers per second. Now imagine that we begin to compress the Sun, forcing it to shrink in diameter. Recall that the pull of gravity depends on both the mass that is pulling you and your distance from the center of gravity of that mass. If the Sun is compressed, its *mass* will remain the same, but the *distance* between a point on the Sun's surface and the center will get smaller and smaller. Thus, as we compress the star, the pull of gravity for an object on the shrinking surface will get stronger and stronger (Figure 24.12).

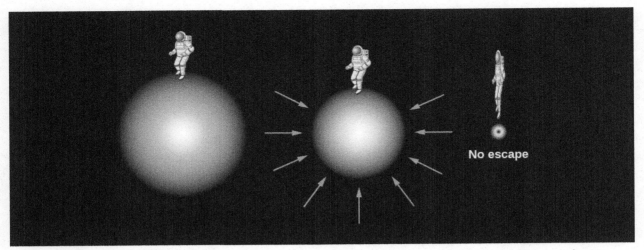

Figure 24.12 Formation of a Black Hole. At left, an imaginary astronaut floats near the surface of a massive star-core about to collapse. As the same mass falls into a smaller sphere, the gravity at its surface goes up, making it harder for anything to escape from the stellar surface. Eventually the mass collapses into so small a sphere that the escape velocity exceeds the speed of light and nothing can get away. Note that the size of the astronaut has been exaggerated. In the last picture, the astronaut is just outside the sphere we will call the event horizon and is stretched and squeezed by the strong gravity.

When the shrinking Sun reaches the diameter of a neutron star (about 20 kilometers), the velocity required to escape its gravitational pull will be about half the speed of light. Suppose we continue to compress the Sun to a smaller and smaller diameter. (We saw this can't happen to a star like our Sun in the real world because of electron degeneracy, i.e., the mutual repulsion between tightly packed electrons; this is just a quick "thought experiment" to get our bearings).

Ultimately, as the Sun shrinks, the escape velocity near the surface would exceed the speed of light. If the speed you need to get away is faster than the fastest possible speed in the universe, then nothing, not even light, is able to escape. An object with such large escape velocity emits no light, and anything that falls into it can never return.

In modern terminology, we call an object from which light cannot escape a **black hole**, a name popularized by the America scientist John Wheeler starting in the late 1960s (Figure 24.13). The idea that such objects might exist is, however, not a new one. Cambridge professor and amateur astronomer John Michell wrote a paper in 1783 about the possibility that stars with escape velocities exceeding that of light might exist. And in 1796, the French mathematician Pierre-Simon, marquis de Laplace, made similar calculations using Newton's theory of gravity; he called the resulting objects "dark bodies."

Figure 24.13 John Wheeler (1911–2008). This brilliant physicist did much pioneering work in general relativity theory and popularized the term *black hole* starting in the late 1960s. (credit: modification of work by Roy Bishop)

While these early calculations provided strong hints that something strange should be expected if very massive objects collapse under their own gravity, we really need general relativity theory to give an adequate description of what happens in such a situation.

Collapse with Relativity

General relativity tells us that gravity is really a curvature of spacetime. As gravity increases (as in the collapsing Sun of our thought experiment), the curvature gets larger and larger. Eventually, if the Sun could shrink down to a diameter of about 6 kilometers, only light beams sent out perpendicular to the surface would escape. All others would fall back onto the star (Figure 24.14). If the Sun could then shrink just a little more, even that one remaining light beam would no longer be able to escape.

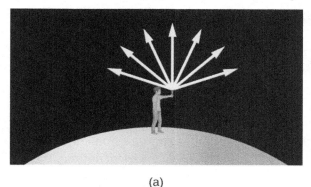
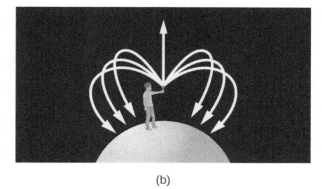

(a)   (b)

Figure 24.14 Light Paths near a Massive Object. Suppose a person could stand on the surface of a normal star with a flashlight. The light leaving the flashlight travels in a straight line no matter where the flashlight is pointed. Now consider what happens if the star collapses so that it is just a little larger than a black hole. All the light paths, except the one straight up, curve back to the surface. When the star shrinks inside the event horizon and becomes a black hole, even a beam directed straight up returns.

Keep in mind that gravity is not pulling on the light. The concentration of matter has curved spacetime, and light (like the trained ant of our earlier example) is "doing its best" to go in a straight line, yet is now confronted with a world in which straight lines that used to go outward have become curved paths that lead back in. The collapsing star is a *black hole* in this view, because the very concept of "out" has no geometrical meaning. The star has become trapped in its own little pocket of spacetime, from which there is no escape.

The star's geometry cuts off communication with the rest of the universe at precisely the moment when, in our earlier picture, the escape velocity becomes equal to the speed of light. The size of the star at this moment defines a surface that we call the **event horizon**. It's a wonderfully descriptive name: just as objects that sink below our horizon cannot be seen on Earth, so anything happening inside the event horizon can no longer interact with the rest of the universe.

Imagine a future spacecraft foolish enough to land on the surface of a massive star just as it begins to collapse in the way we have been describing. Perhaps the captain is asleep at the gravity meter, and before the crew can say "Albert Einstein," they have collapsed with the star inside the event horizon. Frantically, they send an escape pod straight outward. But paths outward twist around to become paths inward, and the pod turns around and falls toward the center of the black hole. They send a radio message to their loved ones, bidding good-bye. But radio waves, like light, must travel through spacetime, and curved spacetime allows nothing to get out. Their final message remains unheard. Events inside the event horizon can never again affect events outside it.

The characteristics of an event horizon were first worked out by astronomer and mathematician Karl Schwarzschild (Figure 24.15). A member of the German army in World War I, he died in 1916 of an illness he contracted while doing artillery shell calculations on the Russian front. His paper on the theory of event horizons was among the last things he finished as he was dying; it was the first exact solution to Einstein's equations of general relativity. The radius of the event horizon is called the *Schwarzschild radius* in his memory.

Figure 24.15 Karl Schwarzschild (1873–1916). This German scientist was the first to demonstrate mathematically that a black hole is possible and to determine the size of a nonrotating black hole's event horizon.

The event horizon is the boundary of the black hole; calculations show that it does not get smaller once the whole star has collapsed inside it. It is the region that separates the things trapped inside it from the rest of the universe. Anything coming from the outside is also trapped once it comes inside the event horizon. The horizon's size turns out to depend only on the mass inside it. If the Sun, with its mass of 1 M_{Sun}, were to become a black hole (fortunately, it can't—this is just a thought experiment), the Schwarzschild radius would be about 3 kilometers; thus, the entire black hole would be about one-third the size of a neutron star of that same mass. Feed the black hole some mass, and the horizon will grow—but not very much. Doubling the mass will make the black hole 6 kilometers in radius, still very tiny on the cosmic scale.

The event horizons of more massive black holes have larger radii. For example, if a globular cluster of 100,000 stars (solar masses) could collapse to a black hole, it would be 300,000 kilometers in radius, a little less than half the radius of the Sun. If the entire Galaxy could collapse to a black hole, it would be only about 10^{12} kilometers in radius—about a tenth of a light year. Smaller masses have correspondingly smaller horizons: for Earth to become a black hole, it would have to be compressed to a radius of only 1 centimeter—less than the size of a grape. A typical asteroid, if crushed to a small enough size to be a black hole, would have the dimensions of an atomic nucleus.

EXAMPLE 24.1

The Milky Way's Black Hole

The size of the event horizon of a black hole depends on the mass of the black hole. The greater the mass, the larger the radius of the event horizon. General relativity calculations show that the formula for the Schwarzschild radius (R_S) of the event horizon is

$$R_S = \frac{2GM}{c^2}$$

where c is the speed of light, G is the gravitational constant, and M is the mass of the black hole. Note that in this formula, 2, G, and c are all constant; only the mass changes from black hole to black hole.

As we will see in the chapter on The Milky Way Galaxy, astronomers have traced the paths of several stars near the center of our Galaxy and found that they seem to be orbiting an unseen object—dubbed Sgr A* (pronounced "Sagittarius A-star")—with a mass of about 4 million solar masses. What is the size of its Schwarzschild radius?

Solution

We can substitute data for G, M, and c (from Appendix E) directly into the equation:

$$R_S = \frac{2GM}{c^2} = \frac{2(6.67 \times 10^{-11}\,\text{N·m}^2/\text{kg}^2)(4 \times 10^6)(1.99 \times 10^{30}\,\text{kg})}{(3.00 \times 10^8\,\text{m/s})^2}$$

$$= 1.18 \times 10^{10}\,\text{m}$$

This distance is about one-fifth of the radius of Mercury's orbit around the Sun, yet the object contains 4 million solar masses and cannot be seen with our largest telescopes. You can see why astronomers are convinced this object is a black hole.

Check Your Learning

What would be the size of a black hole that contained only as much mass as a typical pickup truck (about 3000 kg)? (Note that something with so little mass could never actually form a black hole, but it's interesting to think about the result.)

Answer:

Substituting the data into our equation gives

$$R_S = \frac{2GM}{c^2} = \frac{2(6.67 \times 10^{-11}\,\text{N·m}^2/\text{kg}^2)(3000\,\text{kg})}{(3.00 \times 10^8\,\text{m/s})^2} = 4.4 \times 10^{-24}\,\text{m}.$$

For comparison, the size of a proton is usually considered to be about 8×10^{-16} m, which would be about ten million times larger.

A Black Hole Myth

Much of the modern folklore about black holes is misleading. One idea you may have heard is that black holes go about sucking things up with their gravity. Actually, it is only very close to a black hole that the strange effects we have been discussing come into play. The gravitational attraction far away from a black hole is the same as that of the star that collapsed to form it.

Remember that the gravity of any star some distance away acts as if all its mass were concentrated at a point in the center, which we call the center of gravity. For real stars, we merely *imagine* that all mass is concentrated there; for black holes, all the mass *really is* concentrated at a point in the center.

So, if you are a star or distant planet orbiting around a star that becomes a black hole, your orbit may not be significantly affected by the collapse of the star (although it may be affected by any mass loss that precedes the collapse). If, on the other hand, you venture close to the event horizon, it would be very hard for you to resist the "pull" of the warped spacetime near the black hole. You have to get really close to the black hole to experience any significant effect.

If another star or a spaceship were to pass one or two solar radii from a black hole, Newton's laws would be adequate to describe what would happen to it. Only very near the event horizon of a black hole is the gravitation so strong that Newton's laws break down. The black hole remnant of a massive star coming into our neighborhood would be far, far safer to us than its earlier incarnation as a brilliant, hot star.

MAKING CONNECTIONS

Gravity and Time Machines

Time machines are one of the favorite devices of science fiction. Such a device would allow you to move through time at a different pace or in a different direction from everyone else. General relativity suggests that it is possible, in theory, to construct a time machine using gravity that could take you into the future.

Let's imagine a place where gravity is terribly strong, such as near a black hole. General relativity predicts that the stronger the gravity, the slower the pace of time (as seen by a distant observer). So, imagine a future astronaut, with a fast and strongly built spaceship, who volunteers to go on a mission to such a high-gravity environment. The astronaut leaves in the year 2222, just after graduating from college at age 22. She takes, let's say, exactly 10 years to get to the black hole. Once there, she orbits some distance from it, taking care not to get pulled in.

She is now in a high-gravity realm where time passes much more slowly than it does on Earth. This isn't just an effect on the mechanism of her clocks—*time itself* is running slowly. That means that every way she has of measuring time will give the same slowed-down reading when compared to time passing on Earth. Her heart will beat more slowly, her hair will grow more slowly, her antique wristwatch will tick more slowly, and so on. She is not aware of this slowing down because all her readings of time, whether made by her own bodily functions or with mechanical equipment, are measuring the same—slower—time. Meanwhile, back on Earth, time passes as it always does.

Our astronaut now emerges from the region of the black hole, her mission of exploration finished, and returns to Earth. Before leaving, she carefully notes that (according to her timepieces) she spent about 2 weeks around the black hole. She then takes exactly 10 years to return to Earth. Her calculations tell her that since she was 22 when she left the Earth, she will be 42 plus 2 weeks when she returns. So, the year on Earth, she figures, should be 2242, and her classmates should now be approaching their midlife crises.

But our astronaut should have paid more attention in her astronomy class! Because time slowed down near the black hole, much less time passed for her than for the people on Earth. While her clocks measured 2 weeks spent near the black hole, more than 2000 weeks (depending on how close she got) could well have passed on Earth. That's equal to 40 years, meaning her classmates will be senior citizens in their 80s when she (a mere 42-year-old) returns. On Earth it will be not 2242, but 2282—and she will say that she has arrived *in the future*.

Is this scenario real? Well, it has a few practical challenges: we don't think any black holes are close enough for us to reach in 10 years, and we don't think any spaceship or human can survive near a black hole. But the key point about the slowing down of time is a natural consequence of Einstein's general theory of relativity, and we saw that its predictions have been confirmed by experiment after experiment.

Such developments in the understanding of science also become inspiration for science fiction writers. Recently, the film *Interstellar* featured the protagonist traveling close to a massive black hole; the resulting delay in his aging relative to his earthbound family is a key part of the plot.

Science fiction novels, such as *Gateway* by Frederik Pohl and *A World out of Time* by Larry Niven, also make use of the slowing down of time near black holes as major turning points in the story. For a list of science fiction stories based on good astronomy, you can go to http://bit.ly/astroscifi (http://bit.ly/astroscifi).

A Trip into a Black Hole

The fact that scientists cannot see inside black holes has not kept them from trying to calculate what they are like. One of the first things these calculations showed was that the formation of a black hole obliterates nearly all information about the star that collapsed to form it. Physicists like to say "black holes have no hair," meaning that nothing sticks out of a black hole to give us clues about what kind of star produced it or what material has fallen inside. The only information a black hole can reveal about itself is its mass, its spin (rotation), and whether it has any electrical charge.

What happens to the collapsing star-core that made the black hole? Our best calculations predict that the material will continue to collapse under its own weight, forming an infinitely *squozen* point—a place of zero volume and infinite density—to which we give the name **singularity**. At the singularity, spacetime ceases to exist. The laws of physics as we know them break down. We do not yet have the physical understanding or the mathematical tools to describe the singularity itself, or even if singularities actually occur. From the outside, however, the entire structure of a basic black hole (one that is not rotating) can be described as a singularity surrounded by an event horizon. Compared to humans, black holes are really very simple objects.

Scientists have also calculated what would happen if an astronaut were to fall into a black hole. Let's take up an observing position a long, safe distance away from the event horizon and watch this astronaut fall toward it. At first he falls away from us, moving ever faster, just as though he were approaching any massive star. However, as he nears the event horizon of the black hole, things change. The strong gravitational field around the black hole will make his clocks run more slowly, when seen from our outside perspective.

If, as he approaches the event horizon, he sends out a signal once per second according to his clock, we will see the spacing between his signals grow longer and longer until it becomes infinitely long when he reaches the event horizon. (Recalling our discussion of gravitational redshift, we could say that if the infalling astronaut uses a blue light to send his signals every second, we will see the light get redder and redder until its wavelength is nearly infinite.) As the spacing between clock ticks approaches infinity, it will appear to us that the astronaut is slowly coming to a stop, frozen in time at the event horizon.

In the same way, all matter falling into a black hole will also appear to an outside observer to stop at the event horizon, frozen in place and taking an infinite time to fall through it. But don't think that matter falling into a black hole will therefore be easily visible at the event horizon. The tremendous redshift will make it very difficult to observe any radiation from the "frozen" victims of the black hole.

This, however, is only how we, located far away from the black hole, see things. To the astronaut, his time goes at its normal rate and he falls right on through the event horizon into the black hole. (Remember, this horizon is not a physical barrier, but only a region in space where the curvature of spacetime makes escape impossible.)

You may have trouble with the idea that you (watching from far away) and the astronaut (falling in) have such different ideas about what has happened. This is the reason Einstein's ideas about space and time are called theories of *relativity*. What each observer measures about the world depends on (is relative to) his or her frame of reference. The observer in strong gravity measures time and space differently from the one sitting in weaker gravity. When Einstein proposed these ideas, many scientists also had difficulty with the idea that two such different views of the same event could be correct, each in its own "world," and they tried to find a mistake in the calculations. There were no mistakes: we and the astronaut really would see him fall into a black hole very differently.

For the astronaut, there is no turning back. Once inside the event horizon, the astronaut, along with any signals from his radio transmitter, will remain hidden forever from the universe outside. He will, however, not have a long time (from his perspective) to feel sorry for himself as he approaches the black hole. Suppose he is falling feet first. The force of gravity that the singularity exerts on his feet is greater than on his head, so he will be stretched slightly. Because the singularity is a point, the left side of his body will be pulled slightly toward the right, and the right slightly toward the left, bringing each side closer to the singularity. The astronaut will therefore be slightly squeezed in one direction and stretched in the other. Some scientists like to call this process of stretching and narrowing *spaghettification*. The point at which the astronaut becomes so stretched that he perishes depends on the size of the black hole. For black holes with masses billions of times the mass of the Sun, such as those found at the centers of galaxies, the spaghettification becomes significant only after the astronaut passes through the event horizon. For black holes with masses of a few solar masses, the astronaut will be stretched and ripped apart even before he reaches the event horizon.

Earth exerts similar *tidal forces* on an astronaut performing a spacewalk. In the case of Earth, the tidal forces are so small that they pose no threat to the health and safety of the astronaut. Not so in the case of a black hole. Sooner or later, as the astronaut approaches the black hole, the tidal forces will become so great that the astronaut will be ripped apart, eventually reduced to a collection of individual atoms that will continue their inexorable fall into the singularity.

LINK TO LEARNING

From the previous discussion, you will probably agree that jumping into a black hole is definitely a once-in-a-lifetime experience! You can see an engaging explanation (https://openstax.org/l/30ndegtystidfor) of death by black hole by Neil deGrasse Tyson, where he explains the effect of tidal forces on the human body until it dies by spaghettification.

An overview of black holes is given in this Discovery Channel video (https://openstax.org/l/30dischatidfor) excerpt.

24.6 | Evidence for Black Holes

Learning Objectives

By the end of this section, you will be able to:
> Describe what to look for when seeking and confirming the presence of a stellar black hole
> Explain how a black hole is inherently black yet can be associated with luminous matter
> Differentiate between stellar black holes and the black holes in the centers of galaxies

Theory tells us what black holes are like. But do they actually exist? And how do we go about looking for something that is many light years away, only about a few dozen kilometers across (if a stellar black hole), and completely black? It turns out that the trick is not to look for the black hole itself but instead to look for what it does to a nearby companion star.

As we saw, when very massive stars collapse, they leave behind their gravitational influence. What if a member of a double-star system becomes a black hole, and its companion manages to survive the death of the massive star? While the black hole disappears from our view, we may be able to deduce its presence from the things it does to its companion.

Requirements for a Black Hole

So, here is a prescription for finding a black hole: start by looking for a star whose motion (determined from the Doppler shift of its spectral lines) shows it to be a member of a binary star system. If both stars are visible, neither can be a black hole, so focus your attention on just those systems where only one star of the pair is visible, even with our most sensitive telescopes.

Being invisible is not enough, however, because a relatively faint star might be hard to see next to the glare of a brilliant companion or if it is shrouded by dust. And even if the star really is invisible, it could be a neutron star. Therefore, we must also have evidence that the unseen star has a mass too high to be a neutron star and that it is a collapsed object—an extremely small stellar remnant.

We can use Kepler's law (see Orbits and Gravity) and our knowledge of the visible star to measure the mass of the invisible member of the pair. If the mass is greater than about 3 M_{Sun}, then we are likely seeing (or, more precisely, not seeing) a black hole—as long as we can make sure the object really is a collapsed star.

If matter falls toward a compact object of high gravity, the material is accelerated to high speed. Near the event horizon of a black hole, matter is moving at velocities that approach the speed of light. As the atoms whirl chaotically toward the event horizon, they rub against each other; internal friction can heat them to temperatures of 100 million K or more. Such hot matter emits radiation in the form of flickering X-rays. The last part of our prescription, then, is to look for a source of X-rays associated with the binary system. Since X-rays do not penetrate Earth's atmosphere, such sources must be found using X-ray telescopes in space.

In our example, the infalling gas that produces the X-ray emission comes from the black hole's companion star. As we saw in The Death of Stars, stars in close binary systems can exchange mass, especially as one of the members expands into a red giant. Suppose that one star in a double-star system has evolved to a black hole and that the second star begins to expand. If the two stars are not too far apart, the outer layers of the expanding star may reach the point where the black hole exerts more gravitational force on them than do the inner layers of the red giant to which the atmosphere belongs. The outer atmosphere then passes through the point of no return between the stars and falls toward the black hole.

The mutual revolution of the giant star and the black hole causes the material falling toward the black hole to spiral around it rather than flow directly into it. The infalling gas whirls around the black hole in a pancake of matter called an **accretion disk**. It is within the inner part of this disk that matter is revolving about the black hole so fast that internal friction heats it up to X-ray–emitting temperatures (see Figure 24.1).

Another way to form an accretion disk in a binary star system is to have a powerful stellar wind come from the black hole's companion. Such winds are a characteristic of several stages in a star's life. Some of the ejected gas in the wind will then flow close enough to the black hole to be captured by it into the disk (Figure 24.16).

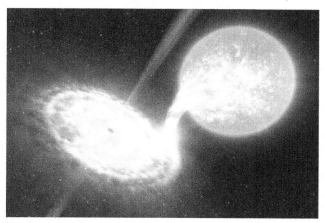

Figure 24.16 Binary Black Hole. This artist's rendition shows a black hole and star (red). As matter streams from the star, it forms a disk around the black hole. Some of the swirling material close to the black hole is pushed outward perpendicular to the disk in two narrow jets. (credit: modification of work by ESO/L. Calçada)

We should point out that, as often happens, the measurements we have been discussing are not quite as simple as they are described in introductory textbooks. In real life, Kepler's law allows us to calculate only the combined mass of the two stars in the binary system. We must learn more about the visible star of the pair and its history to ascertain the distance to the binary pair, the true size of the visible star's orbit, and how the orbit of the two stars is tilted toward Earth, something we can rarely measure. And neutron stars can also have accretion disks that produce X-rays, so astronomers must study the properties of these X-rays carefully when trying to determine what kind of object is at the center of the disk. Nevertheless, a number of systems that clearly contain black holes have now been found.

The Discovery of Stellar-Mass Black Holes

Because X-rays are such important tracers of black holes that are having some of their stellar companions for lunch, the search for black holes had to await the launch of sophisticated X-ray telescopes into space. These instruments must have the resolution to locate the X-ray sources accurately and thereby enable us to match them to the positions of binary star systems.

The first black hole binary system to be discovered is called Cygnus X-1 (see Figure 24.1). The visible star in this binary system is spectral type O. Measurements of the Doppler shifts of the O star's spectral lines show that it has an unseen companion. The X-rays flickering from it strongly indicate that the companion is a small collapsed object. The mass of the invisible collapsed companion is about 15 times that of the Sun. The companion is therefore too massive to be either a white dwarf or a neutron star.

A number of other binary systems also meet all the conditions for containing a black hole. Table 24.1 lists the characteristics of some of the best examples.

Some Black Hole Candidates in Binary Star Systems

Name/Catalog Designation[2]	Companion Star Spectral Type	Orbital Period (days)	Black Hole Mass Estimates (M_{Sun})
LMC X-1	O giant	3.9	10.9
Cygnus X-1	O supergiant	5.6	15
XTE J1819.3-254 (V4641 Sgr)	B giant	2.8	6–7
LMC X-3	B main sequence	1.7	7
4U1543-475 (IL Lup)	A main sequence	1.1	9
GRO J1655-40 (V1033 Sco)	F subgiant	2.6	7
GRS 1915+105	K giant	33.5	14
GS202+1338 (V404 Cyg)	K giant	6.5	12
XTE J1550-564	K giant	1.5	11
A0620-00 (V616 Mon)	K main sequence	0.33	9–13

Table 24.1

Name/Catalog Designation[2]	Companion Star Spectral Type	Orbital Period (days)	Black Hole Mass Estimates (M_{Sun})
H1705-250 (Nova Oph 1977)	K main sequence	0.52	5–7
GRS1124-683 (Nova Mus 1991)	K main sequence	0.43	7
GS2000+25 (QZ Vul)	K main sequence	0.35	5–10
GRS1009-45 (Nova Vel 1993)	K dwarf	0.29	8–9
XTE J1118+480	K dwarf	0.17	7
XTE J1859+226	K dwarf	0.38	5.4
GRO J0422+32	M dwarf	0.21	4

Table 24.1

Feeding a Black Hole

After an isolated star, or even one in a binary star system, becomes a black hole, it probably won't be able to grow much larger. Out in the suburban regions of the Milky Way Galaxy where we live (see The Milky Way Galaxy), stars and star systems are much too far apart for other stars to provide "food" to a hungry black hole. After all, material must approach very close to the event horizon before the gravity is any different from that of the star before it became the black hole.

But, as will see, the central regions of galaxies are quite different from their outer parts. Here, stars and raw material can be quite crowded together, and they can interact much more frequently with each other. Therefore, black holes in the centers of galaxies may have a much better opportunity to find mass close enough to their event horizons to pull in. Black holes are not particular about what they "eat": they are happy to consume other stars, asteroids, gas, dust, and even other black holes. (If two black holes merge, you just get a black hole with more mass and a larger event horizon.)

As a result, black holes in crowded regions can grow, eventually swallowing thousands or even millions of times the mass of the Sun. Ground-based observations have provided compelling evidence that there is a black hole in the center of our own Galaxy with a mass of about 4 million times the mass of the Sun (we'll discuss this further in the chapter on The Milky Way Galaxy). Observations with the Hubble Space Telescope have shown dramatic evidence for the existence of black holes in the centers of many other galaxies. These black holes can contain more than a billion solar masses. The feeding frenzy of such supermassive black holes may be responsible for some of the most energetic phenomena in the universe (see Active Galaxies, Quasars, and Supermassive Black Holes). And evidence from more recent X-ray observations is also starting to indicate the existence of "middle-weight" black holes, whose masses are dozens to thousands of times the mass of the Sun. The crowded inner regions of the globular clusters we described in Stars from Adolescence to Old Age may be just the right breeding grounds for such intermediate-mass black holes.

Over the past decades, many observations, especially with the Hubble Space Telescope and with X-ray

---

2   As you can tell, there is no standard way of naming these candidates. The chain of numbers is the location of the source in right ascension and declination (the longitude and latitude system of the sky); some of the letters preceding the numbers refer to objects (e.g., LMC) and constellations (e.g., Cygnus), while other letters refer to the satellite that discovered the candidate—A for Ariel, G for Ginga, and so on. The notations in parentheses are those used by astronomers who study binary star system or novae.

satellites, have been made that can be explained only if black holes really do exist. Furthermore, the observational tests of Einstein's general theory of relativity have convinced even the most skeptical scientists that his picture of warped or curved spacetime is indeed our best description of the effects of gravity near these black holes.

24.7 | Gravitational Wave Astronomy

Learning Objectives

By the end of this section, you will be able to:

> Describe what a gravitational wave is, what can produce it, and how fast it propagates
> Understand the basic mechanisms used to detect gravitational waves

Another part of Einstein's ideas about gravity can be tested as a way of checking the theory that underlies black holes. According to general relativity, the geometry of spacetime depends on where matter is located. Any rearrangement of matter—say, from a sphere to a sausage shape—creates a disturbance in spacetime. This disturbance is called a **gravitational wave**, and relativity predicts that it should spread outward at the speed of light. The big problem with trying to study such waves is that they are tremendously weaker than electromagnetic waves and correspondingly difficult to detect.

Proof from a Pulsar

We've had indirect evidence for some time that gravitational waves exist. In 1974, astronomers Joseph Taylor and Russell Hulse discovered a pulsar (with the designation PSR1913+16) orbiting another neutron star. Pulled by the powerful gravity of its companion, the pulsar is moving at about one-tenth the speed of light in its orbit.

According to general relativity, this system of stellar corpses should be radiating energy in the form of gravitational waves at a high enough rate to cause the pulsar and its companion to spiral closer together. If this is correct, then the orbital period should decrease (according to Kepler's third law) by one ten-millionth of a second per orbit. Continuing observations showed that the period is decreasing by precisely this amount. Such a loss of energy in the system can be due only to the radiation of gravitational waves, thus confirming their existence. Taylor and Hulse shared the 1993 Nobel Prize in physics for this work.

Direct Observations

Although such an indirect proof convinced physicists that gravitational waves exist, it is even more satisfying to detect the waves directly. What we need are phenomena that are powerful enough to produce gravitational waves with amplitudes large enough that we can measure them. Theoretical calculations suggest some of the most likely events that would give a burst of gravitational waves strong enough that our equipment on Earth could measure it:

- the coalescence of two neutron stars in a binary system that spiral together until they merge
- the swallowing of a neutron star by a black hole
- the coalescence (merger) of two black holes
- the implosion of a really massive star to form a neutron star or a black hole
- the first "shudder" when space and time came into existence and the universe began

For the last four decades, scientists have been developing an audacious experiment to try to detect gravitational waves from a source on this list. The US experiment, which was built with collaborators from the UK, Germany, Australia and other countries, is named LIGO (Laser Interferometer Gravitational-Wave Observatory). LIGO currently has two observing stations, one in Louisiana and the other in the state of Washington. The effects of gravitational waves are so small that confirmation of their detection will require simultaneous measurements by two widely separated facilities. Local events that might cause small motions within the observing stations and mimic gravitational waves—such as small earthquakes, ocean tides, and even traffic—should affect the two sites differently.

Each of the LIGO stations consists of two 4-kilometer-long, 1.2-meter-diameter vacuum pipes arranged in an L-shape. A test mass with a mirror on it is suspended by wire at each of the four ends of the pipes. Ultra-stable laser light is reflected from the mirrors and travels back and forth along the vacuum pipes (Figure 24.17). If gravitational waves pass through the LIGO instrument, then, according to Einstein's theory, the waves will affect local spacetime—they will alternately stretch and shrink the distance the laser light must travel between the mirrors ever so slightly. When one arm of the instrument gets longer, the other will get shorter, and vice versa.

Figure 24.17 Gravitational Wave Telescope. An aerial view of the LIGO facility at Livingston, Louisiana. Extending to the upper left and far right of the image are the 4-kilometer-long detectors. (credit: modification of work by Caltech/MIT/LIGO Laboratory)

The challenge of this experiment lies in that phrase "ever so slightly." In fact, to detect a gravitational wave, the change in the distance to the mirror must be measured with an accuracy of *one ten-thousandth the diameter of a proton*. In 1972, Rainer Weiss of MIT wrote a paper suggesting how this seemingly impossible task might be accomplished.

A great deal of new technology had to be developed, and work on the laboratory, with funding from the National Science Foundation, began in 1979. A full-scale prototype to demonstrate the technology was built and operated from 2002 to 2010, but the prototype was not expected to have the sensitivity required to actually detect gravitational waves from an astronomical source. Advanced LIGO, built to be more precise with the improved technology developed in the prototype, went into operation in 2015—and almost immediately detected gravitational waves.

What LIGO found was gravitational waves produced in the final fraction of a second of the merger of two black holes (Figure 24.18). The black holes had masses of 20 and 36 times the mass of the Sun, and the merger took place 1.3 billion years ago—the gravitational waves occurred so far away that it has taken that long for them, traveling at the speed of light, to reach us.

In the cataclysm of the merger, about three times the mass of the Sun was converted to energy (recall $E = mc^2$). During the tiny fraction of a second for the merger to take place, this event produced power about 10 times the power produced by all the stars in the entire visible universe—but the power was all in the form of gravitational waves and hence was invisible to our instruments, except to LIGO. The event was recorded in Louisiana about 7 milliseconds before the detection in Washington—just the right distance given the speed at which gravitational waves travel—and indicates that the source was located somewhere in the southern hemisphere sky. Unfortunately, the merger of two black holes is not expected to produce any light, so this is the only observation we have of the event.

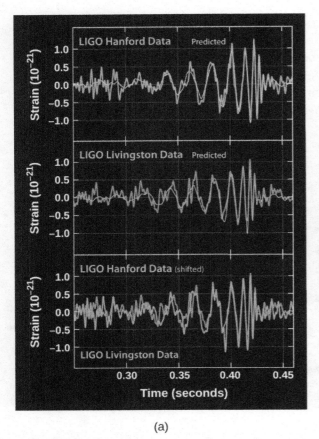

(a)   (b)

Figure 24.18 Signal Produced by a Gravitational Wave. (a) The top panel shows the signal measured at Hanford, Washington; the middle panel shows the signal measured at Livingston, Louisiana. The smoother thin curve in each panel shows the predicted signal, based on Einstein's general theory of relativity, produced by the merger of two black holes. The bottom panel shows a superposition of the waves detected at the two LIGO observatories. Note the remarkable agreement of the two independent observations and of the observations with theory. (b) The painting shows an artist's impression of two massive black holes spiraling inward toward an eventual merger. (credit a, b: modification of work by SXS)

This detection by LIGO (and another one of a different black hole merger a few months later) opened a whole new window on the universe. One of the experimenters compared the beginning of gravitational wave astronomy to the era when silent films were replaced by movies with sound (comparing the vibration of spacetime during the passing of a gravitational wave to the vibrations that sound makes).

By the end of 2021, the LIGO-Virgo collaboration (Virgo is a gravitational wave detector operated by the European Gravitational Observatory in Italy) published a catalog that now included some 90 events. Most were mergers of two black holes, and most involved black holes with a range of masses detected only by gravitational waves (see Figure 24.19). In the most extreme merger, black holes with masses of 86 and 65 times the mass of the Sun merged to form a black hole with a mass of about 142 times the mass the Sun, and released energy equivalent to 9 times the mass of our Sun (remember $E = mc^2$)—an enormous amount.

Astronomers are not yet sure how black holes in this unexpected mass range form. Bear in mind that the kind of black holes in binary star systems that we discussed in Evidence for Black Holes (and listed in Table 24.1) have masses ranging from 4 to 15 times the mass of the Sun.

To be sure, a few mergers did involve black holes with stellar masses comparable to those of black holes in X-ray binary systems. In one case, for example, the merging black holes had masses of 14 and 8 times the mass of the Sun.

While astronomers can learn about the masses of objects involved in gravitational wave events, the challenge is to locate the event in the sky precisely. A single gravitational wave detector cannot determine accurately the direction to a gravitational wave source. Four comparable detectors operating simultaneously are required to localize a source of gravitational waves *in every location* in the sky. The first observing run with four detectors, including the two LIGO detectors, Virgo, and KAGRA in Japan, is scheduled to begin in the summer of 2022. LIGO India will be a fifth, thereby enhancing the probability that at least four detectors will be operational simultaneously. Experience has shown that the LIGO and Virgo detectors are down about 25 percent of the time because these complex systems are difficult to run.

A three-observatory network does provide a sharp location for events that occur in about half of all possible locations on the sky.

In late 2017, data from the LIGO and Virgo detectors provided an accurate position for what analysis showed was the spiraling together of two neutron stars with masses of 1.1 to 1.6 times the mass of the Sun (see The Death of Stars).

With an accurate location known, follow up observations with ground-based telescopes detected electromagnetic emission from a gravitational wave event for the first time. The observations showed that this source was located in NGC 4993, a galaxy at a distance of about 130 million light-years in the direction of the constellation Hydra. The Fermi satellite detected a flash of gamma rays at the same time and in the same direction, which confirms the long-standing hypothesis that mergers of neutron stars are progenitors of short gamma-ray bursts (see The Mystery of Gamma-Ray Bursts). Spectra showed that the merger ejected material with a mass of about 6 percent of the mass of the Sun at a speed of one-tenth the speed of light.

This material is rich in heavy elements, just as the theory of kilonovas (see The Mystery of Gamma-Ray Bursts) predicted. First estimates suggest that the merger produced about 200 Earth masses of gold, and around 500 Earth masses of platinum. This makes clear that neutron star mergers are a significant source of heavy elements. More such mergers are being found (see Figure 24.19) and they will improve estimates of the frequency at which neutron star mergers occur; it may well turn out that the vast majority of heavy elements have been created in such cataclysms.

No electromagnetic observations have been detected from the mergers of two black holes.

In June 2021, scientists from LIGO and Virgo announced the first detection of mergers between black holes and neutron stars, another of the really energetic events that we listed as possible sources for a detectable burst of gravitational waves. Again, no electromagnetic waves from the two events were observed or expected, demonstrating the importance of using what we are now calling "multi-messenger astronomy" to understand the universe fully.

Observing the merger of black holes via gravitational waves means that we can now test Einstein's general theory of relativity where its effects are very strong—close to black holes—and not weak, as they are near Earth. One remarkable result from these detections is that the observed signals so closely match the theoretical predictions made using Einstein's theory. Once again, Einstein's revolutionary idea is found to be the correct description of nature.

Because of the scientific significance of the observations of gravitational waves, three of the LIGO project leaders—Rainer Weiss of MIT, and Kip Thorne and Barry Barish of Caltech—were awarded the Nobel Prize in 2017.

Ground-based gravitational-wave detectors can detect mergers of black holes with masses up to about 100 times the mass of the Sun. Astronomers would now like to look for the merger of distant supermassive black holes (see Supermassive Black Holes: What Quasars Really Are) with masses of thousands to millions of times larger, which might have occurred when the first generation of stars formed, only a few hundred million years after the Big Bang. The gravitational waves emitted by mergers of supermassive black holes are so long that it is necessary to go to space to build an observatory large enough to detect them.

ESA (the European Space Agency), with contributions from NASA, is planning to launch a facility named LISA (Laser Interferometer Space Antenna) in 2034 to search for mergers of black holes with masses thousands to millions of times larger than the mass of the Sun. The experiment will consist of three spacecraft arranged in an equilateral triangle with sides 2.5 million km long, flying along an Earth-like heliocentric orbit. A test mass floats free inside each spacecraft, effectively in free-fall, while the spacecraft around it absorbs the effects of light pressure, solar wind particles, and anything that might perturb its orbit. Lasers will be used to measure very accurately the distances between the test masses. Changes in distance will then signal the passing of gravitational waves.

We should end by acknowledging that the ideas discussed in this chapter may seem strange and overwhelming, especially the first time you read them. The consequences of the general theory of relatively take some getting used to. But they make the universe more bizarre—and interesting—than you probably thought before you took this course.

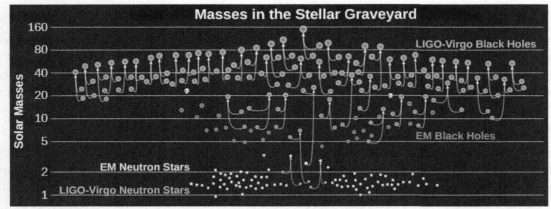

Figure 24.19 Some Black Hole Merger Events Observed with Gravitational Waves. The vertical axis shows the masses for some known black holes and neutron stars. The light blue dots represent black hole mergers seen with LIGO-Virgo, with the individual black holes and the resulting merged black hole connected with an arrow. Note that, as a result of these mergers, we are starting to observe some black holes with masses so large that new ideas may be required to explain them. The purple dots represent black holes seen with electro-magnetic radiation; these have significantly smaller masses. The orange dots represent neutron star mergers seen through the gravitational waves they emit. The yellow dots are neutron stars seen with electro-magnetic radiation. (credit: courtesy Caltech/MIT/LIGO Laboratory; LIGO-Virgo/Frank Elavsky, Aaron Geller/Northwestern)

🔑 Key Terms

accretion disk   the disk of gas and dust found orbiting newborn stars, as well as compact stellar remnants such as white dwarfs, neutron stars, and black holes when they are in binary systems and are sufficiently close to their binary companions to draw off material

black hole   a region in spacetime where gravity is so strong that nothing—not even light—can escape

equivalence principle   concept that a gravitational force and a suitable acceleration are indistinguishable within a sufficiently local environment

event horizon   a boundary in spacetime such that events inside the boundary can have no effect on the world outside it—that is, the boundary of the region around a black hole where the curvature of spacetime no longer provides any way out

general theory of relativity   Einstein's theory relating gravity and the structure (geometry) of space and time

gravitational redshift   an increase in wavelength of an electromagnetic wave (light) when propagating from or near a massive object

gravitational wave   a disturbance in the curvature of spacetime caused by changes in how matter is distributed; gravitational waves propagate at (or near) the speed of light.

singularity   the point of zero volume and infinite density to which any object that becomes a black hole must collapse, according to the theory of general relativity

spacetime   system of one time and three space coordinates, with respect to which the time and place of an event can be specified

📄 Summary

24.1 Introducing General Relativity

Einstein proposed the equivalence principle as the foundation of the theory of general relativity. According to this principle, there is no way that anyone or any experiment in a sealed environment can distinguish between free fall and the absence of gravity.

24.2 Spacetime and Gravity

By considering the consequences of the equivalence principle, Einstein concluded that we live in a curved spacetime. The distribution of matter determines the curvature of spacetime; other objects (and even light) entering a region of spacetime must follow its curvature. Light must change its path near a massive object not because light is bent by gravity, but because spacetime is.

24.3 Tests of General Relativity

In weak gravitational fields, the predictions of general relativity agree with the predictions of Newton's law of gravity. However, in the stronger gravity of the Sun, general relativity makes predictions that differ from Newtonian physics and can be tested. For example, general relativity predicts that light or radio waves will be deflected when they pass near the Sun, and that the position where Mercury is at perihelion would change by 43 arcsec per century even if there were no other planets in the solar system to perturb its orbit. These predictions have been verified by observation.

24.4 Time in General Relativity

General relativity predicts that the stronger the gravity, the more slowly time must run. Experiments on Earth and with spacecraft have confirmed this prediction with remarkable accuracy. When light or other radiation emerges from a compact smaller remnant, such as a white dwarf or neutron star, it shows a gravitational redshift due to the slowing of time.

24.5 Black Holes

Theory suggests that stars with stellar cores more massive than three times the mass of the Sun at the time they exhaust their nuclear fuel will collapse to become black holes. The surface surrounding a black hole, where the escape velocity equals the speed of light, is called the event horizon, and the radius of the surface is called the Schwarzschild radius. Nothing, not even light, can escape through the event horizon from the black hole. At its center, each black hole is thought to have a singularity, a point of infinite density and zero volume. Matter falling into a black hole appears, as viewed by an outside observer, to freeze in position at the event horizon. However, if we were riding on the infalling matter, we would pass through the event horizon. As we approach the singularity, the tidal forces would tear our bodies apart even before we reach the singularity.

24.6 Evidence for Black Holes

The best evidence of stellar-mass black holes comes from binary star systems in which (1) one star of the pair is not visible, (2) the flickering X-ray emission is characteristic of an accretion disk around a compact object, and (3) the orbit and characteristics of the visible star indicate that the mass of its invisible companion is greater than 3 M_{Sun}. A number of systems with these characteristics have been found. Black holes with masses of millions to billions of solar masses are found in the centers of large galaxies.

24.7 Gravitational Wave Astronomy

General relativity predicts that the rearrangement of matter in space should produce gravitational waves. The existence of such waves was first confirmed in observations of a pulsar in orbit around another neutron star whose orbits were spiraling closer and losing energy in the form of gravitational waves. In 2015, LIGO found gravitational waves directly by detecting the signal produced by the merger of two stellar-mass black holes, opening a new window on the universe. Since then, many other gravitational wave signals have been found, signaling the mergers of both black holes and neutron stars.

 For Further Exploration

Articles

Black Holes

Charles, P. & Wagner, R. "Black Holes in Binary Stars: Weighing the Evidence." *Sky & Telescope* (May 1996): 38. Excellent review of how we find stellar-mass black holes.

Gezari, S. "Star-Shredding Black Holes." *Sky & Telescope* (June 2013): 16. When black holes and stars collide.

Jayawardhana, R. "Beyond Black." *Astronomy* (June 2002): 28. On finding evidence of the existence of event horizons and thus black holes.

Nadis, S. "Black Holes: Seeing the Unseeable." *Astronomy* (April 2007): 26. A brief history of the black hole idea and an introduction to potential new ways to observe them.

Psallis, D. & Sheperd, D. "The Black Hole Test." *Scientific American* (September 2015): 74–79. The Event Horizon Telescope (a network of radio telescopes) will test some of the stranger predictions of general relativity for the regions near black holes. The September 2015 issue of *Scientific American* was devoted to a celebration of the 100th anniversary of the general theory of relativity.

Rees, M. "To the Edge of Space and Time." *Astronomy* (July 1998): 48. Good, quick overview.

Talcott, R. "Black Holes in our Backyard." *Astronomy* (September 2012): 44. Discussion of different kinds of black holes in the Milky Way and the 19 objects known to be black holes.

Gravitational Waves

Bartusiak, M. "Catch a Gravity Wave." *Astronomy* (October 2000): 54.

Gibbs, W. "Ripples in Spacetime." *Scientific American* (April 2002): 62.

Haynes, K., & Betz, E. "A Wrinkle in Spacetime Confirms Einstein's Gravitation." *Astronomy* (May 2016): 22. On the direct detection of gravity waves.

Sanders, G., and Beckett, D. "LIGO: An Antenna Tuned to the Songs of Gravity." *Sky & Telescope* (October 2000): 41.

Websites

Black Holes

Black Hole Encyclopedia: http://blackholes.stardate.org (http://blackholes.stardate.org). From StarDate at the University of Texas McDonald Observatory.

Black Holes: http://science.nasa.gov/astrophysics/focus-areas/black-holes (http://science.nasa.gov/astrophysics/focus-areas/black-holes). NASA overview of black holes, along with links to the most recent news and discoveries.

Introduction to Black Holes: https://www.newscientist.com/article/dn18348-introduction-black-holes/ (https://www.newscientist.com/article/dn18348-introduction-black-holes/). Basic information from New Scientist magazine.

Virtual Trips into Black Holes and Neutron Stars: http://antwrp.gsfc.nasa.gov/htmltest/rjn_bht.html (http://antwrp.gsfc.nasa.gov/htmltest/rjn_bht.html). By Robert Nemiroff at Michigan Technological University.

Gravitational Waves

Advanced LIGO: https://www.advancedligo.mit.edu (https://www.advancedligo.mit.edu). The full story on this gravitational wave observatory.

eLISA: https://www.elisascience.org (https://www.elisascience.org).

Gravitational Waves Detected, Confirming Einstein's Theory: http://www.nytimes.com/2016/02/12/science/ligo-gravitational-waves-black-holes-einstein.html (http://www.nytimes.com/2016/02/12/science/ligo-gravitational-waves-black-holes-einstein.html). *New York Times* article and videos on the discovery of gravitational waves.

Gravitational Waves Discovered from Colliding Black Holes: http://www.scientificamerican.com/article/gravitational-waves-discovered-from-colliding-black-holes1 (http://www.scientificamerican.com/article/gravitational-waves-discovered-from-colliding-black-holes1). *Scientific American* coverage of the discovery of gravitational waves (note the additional materials available in the menu at the right).

LIGO Caltech: https://www.ligo.caltech.edu (https://www.ligo.caltech.edu).

Videos

Black Holes

Black Holes: The End of Time or a New Beginning?: https://www.youtube.com/watch?v=mgtJRsdKe6Q (https://www.youtube.com/watch?v=mgtJRsdKe6Q). 2012 Silicon Valley Astronomy Lecture by Roger Blandford (1:29:52).

Death by Black Hole: https://www.youtube.com/watch?v=h1iJXOUMJpg (https://www.youtube.com/watch?v=h1iJXOUMJpg). Neil deGrasse Tyson explains spaghettification with only his hands (5:34).

Hearts of Darkness: Black Holes in Space: https://www.youtube.com/watch?v=4tiAOldypLk (https://www.youtube.com/watch?v=4tiAOldypLk). 2010 Silicon Valley Astronomy Lecture by Alex Filippenko (1:56:11).

Gravitational Waves

Journey of a Gravitational Wave: https://www.youtube.com/watch?v=FlDtXIBrAYE (https://www.youtube.com/watch?v=FlDtXIBrAYE). Introduction from LIGO Caltech (2:55).

LIGO's First Detection of Gravitational Waves: https://www.youtube.com/watch?v=gw-i_VKd6Wo (https://www.youtube.com/watch?v=gw-i_VKd6Wo). Explanation and animations from PBS Digital Studio (9:31).

Two Black Holes Merge into One: https://www.youtube.com/watch?v=I_88S8DWbcU (https://www.youtube.com/watch?v=I_88S8DWbcU). Simulation from LIGO Caltech (0:35).

What the Discovery of Gravitational Waves Means: https://www.youtube.com/watch?v=jMVAgCPYYHY (https://www.youtube.com/watch?v=jMVAgCPYYHY). TED Talk by Allan Adams (10:58).

Collaborative Group Activities

A. A computer science major takes an astronomy course like the one you are taking and becomes fascinated with black holes. Later in life, he founds his own internet company and becomes very wealthy when it goes public. He sets up a foundation to support the search for black holes in our Galaxy. Your group is the allocation committee of this foundation. How would you distribute money each year to increase the chances that more black holes will be found?

B. Suppose for a minute that stars evolve *without* losing any mass at any stage of their lives. Your group is given a list of binary star systems. Each binary contains one main-sequence star and one invisible companion. The spectral types of the main-sequence stars range from spectral type O to M. Your job is to determine whether any of the invisible companions might be black holes. Which ones are worth observing? Why? (Hint: Remember that in a binary star system, the two stars form at the same time, but the pace of their evolution depends on the mass of each star.)

C. You live in the far future, and the members of your group have been convicted (falsely) of high treason. The method of execution is to send everyone into a black hole, but you get to pick which one. Since you are doomed to die, you would at least like to see what the inside of a black hole is like—even if you can't tell anyone outside about it. Would you choose a black hole with a mass equal to that of Jupiter or one with a mass equal to that of an entire galaxy? Why? What would happen to you as you approached the event horizon in each case? (Hint: Consider the difference in force on your feet and your head as you cross over the event horizon.)

D. General relativity is one of the areas of modern astrophysics where we can clearly see the frontiers of human knowledge. We have begun to learn about black holes and warped spacetime recently and are humbled by how much we still don't know. Research in this field is supported mostly by grants from government agencies. Have your group discuss what reasons there are for our tax dollars to support such "far out" (seemingly impractical) work. Can you make a list of "far out" areas of research in past centuries that later led to practical applications? What if general relativity does not have many practical applications? Do you think a small part of society's funds should still go to exploring theories about the nature of space and time?

E. Once you all have read this chapter, work with your group to come up with a plot for a science fiction story that uses the properties of black holes.

F. Black holes seem to be fascinating not just to astronomers but to the public, and they have become part of popular culture. Searching online, have group members research examples of black holes in music, advertising, cartoons, and the movies, and then make a presentation to share the examples you found with the whole class.

G. As mentioned in the Gravity and Time Machines feature box in this chapter, the film *Interstellar* has a lot of black hole science in its plot and scenery. That's because astrophysicist Kip Thorne at Caltech had a big hand in writing the initial treatment for the movie, and later producing it. Get your group members together (be sure you have popcorn) for a viewing of the movie and then try to use your knowledge of black holes from this chapter to explain the plot. (Note that the film also uses the concept of a *wormhole,* which we don't discuss in this chapter. A wormhole is a theoretically possible way to use a large, spinning black hole to find a way to travel from one place in the universe to another without having to go through regular spacetime to get there.)

Exercises

Review Questions

1. How does the equivalence principle lead us to suspect that spacetime might be curved?

2. If general relativity offers the best description of what happens in the presence of gravity, why do physicists still make use of Newton's equations in describing gravitational forces on Earth (when building a bridge, for example)?

3. Einstein's general theory of relativity made or allowed us to make predictions about the outcome of several experiments that had not yet been carried out at the time the theory was first published. Describe three experiments that verified the predictions of the theory after Einstein proposed it.

4. If a black hole itself emits no radiation, what evidence do astronomers and physicists today have that the theory of black holes is correct?

5. What characteristics must a binary star have to be a good candidate for a black hole? Why is each of these characteristics important?

6. A student becomes so excited by the whole idea of black holes that he decides to jump into one. It has a mass 10 times the mass of our Sun. What is the trip like for him? What is it like for the rest of the class, watching from afar?

7. What is an event horizon? Does our Sun have an event horizon around it?

8. What is a gravitational wave and why was it so hard to detect?

9. What are some strong sources of gravitational waves that astronomers hope to detect in the future?

10. Suppose the amount of mass in a black hole doubles. Does the event horizon change? If so, how does it change?

Thought Questions

11. Imagine that you have built a large room around the people in Figure 24.4 and that this room is falling at exactly the same rate as they are. Galileo showed that if there is no air friction, light and heavy objects that are dropping due to gravity will fall at the same rate. Suppose that this were not true and that instead heavy objects fall faster. Also suppose that the man in Figure 24.4 is twice as massive as the woman. What would happen? Would this violate the equivalence principle?

12. A monkey hanging from a tree branch sees a hunter aiming a rifle directly at him. The monkey then sees a flash and knows that the rifle has been fired. Reacting quickly, the monkey lets go of the branch and drops so that the bullet can pass harmlessly over his head. Does this act save the monkey's life? Why or why not? (Hint: Consider the similarities between this situation and that of Exercise 24.11.)

13. Why would we not expect to detect X-rays from a disk of matter about an ordinary star?

14. Look elsewhere in this book for necessary data, and indicate what the final stage of evolution—white dwarf, neutron star, or black hole—will be for each of these kinds of stars.
    A. Spectral type-O main-sequence star
    B. Spectral type-B main-sequence star
    C. Spectral type-A main-sequence star
    D. Spectral type-G main-sequence star
    E. Spectral type-M main-sequence star

15. Which is likely to be more common in our Galaxy: white dwarfs or black holes? Why?

16. If the Sun could suddenly collapse to a black hole, how would the period of Earth's revolution about it differ from what it is now?

17. Suppose the people in Figure 24.4 are in an elevator moving upward with an acceleration equal to g, but in the opposite direction. The woman throws the ball to the man with a horizontal force. What happens to the ball?

18. You arrange to meet a friend at 5:00 p.m. on Valentine's Day on the observation deck of the Empire State Building in New York City. You arrive right on time, but your friend is not there. She arrives 5 minutes late and says the reason is that time runs faster at the top of a tall building, so she is on time but you were early. Is your friend right? Does time run slower or faster at the top of a building, as compared with its base? Is this a reasonable excuse for your friend arriving 5 minutes late?

19. You are standing on a scale in an elevator when the cable snaps, sending the elevator car into free fall. Before the automatic brakes stop your fall, you glance at the scale reading. Does the scale show your real weight? An apparent weight? Something else?

Figuring for Yourself

20. Look up G, c, and the mass of the Sun in Appendix E and calculate the radius of a black hole that has the same mass as the Sun. (Note that this is only a theoretical calculation. The Sun does not have enough mass to become a black hole.)

21. Suppose you wanted to know the size of black holes with masses that are larger or smaller than the Sun. You could go through all the steps in Exercise 24.20, wrestling with a lot of large numbers with large exponents. You could be clever, however, and evaluate all the constants in the equation once and then simply vary the mass. You could even express the mass in terms of the Sun's mass and make future calculations really easy. Show that the event horizon equation is equivalent to saying that the radius of the event horizon is equal to 3 km times the mass of the black hole in units of the Sun's mass.

22. Use the result from Exercise 24.21 to calculate the radius of a black hole with a mass equal to: the Earth, a B0-type main-sequence star, a globular cluster, and the Milky Way Galaxy. Look elsewhere in this text and the appendixes for tables that provide data on the mass of these four objects.

23. Since the force of gravity a significant distance away from the event horizon of a black hole is the same as that of an ordinary object of the same mass, Kepler's third law is valid. Suppose that Earth collapsed to the size of a golf ball. What would be the period of revolution of the Moon, orbiting at its current distance of 400,000 km? Use Kepler's third law to calculate the period of revolution of a spacecraft orbiting at a distance of 6000 km.

Figure 25.1 Milky Way Galaxy. The Milky Way rises over Square Tower, an ancestral pueblo building at Hovenweep National Monument in Utah. Many stars and dark clouds of dust combine to make a spectacular celestial sight of our home Galaxy. The location has been designated an International Dark Sky Park by the International Dark Sky Association.

Chapter Outline

25.1 The Architecture of the Galaxy
25.2 Spiral Structure
25.3 The Mass of the Galaxy
25.4 The Center of the Galaxy
25.5 Stellar Populations in the Galaxy
25.6 The Formation of the Galaxy

 ## Thinking Ahead

Today, we know that our Sun is just one of the many billions of stars that make up the huge cosmic island we call the Milky Way Galaxy. How can we "weigh" such an enormous system of stars and measure its total mass?

One of the most striking features you can see in a truly dark sky—one without light pollution—is the band of faint white light called the Milky Way, which stretches from one horizon to the other. The name comes from an ancient Greek legend that compared its faint white splash of light to a stream of spilled milk. But folktales differ from culture to culture: one East African tribe thought of the hazy band as the smoke of ancient campfires, several Native American stories tell of a path across the sky traveled by sacred animals, and in Siberia, the diffuse arc was known as the seam of the tent of the sky.

In 1610, Galileo made the first telescopic survey of the Milky Way and discovered that it is composed of a multitude of individual stars. Today, we know that the Milky Way comprises our view inward of the huge cosmic pinwheel that we call the Milky Way Galaxy and that is our home. Moreover, our Galaxy is now recognized as just one galaxy among many billions of other galaxies in the cosmos.

25.1 The Architecture of the Galaxy

Learning Objectives

By the end of this section, you will be able to:

> Explain why William and Caroline Herschel concluded that the Milky Way has a flattened structure centered on the Sun and solar system
> Describe the challenges of determining the Galaxy's structure from our vantage point within it
> Identify the main components of the Galaxy

The **Milky Way Galaxy** surrounds us, and you might think it is easy to study because it is so close. However, the very fact that we are embedded within it presents a difficult challenge. Suppose you were given the task of mapping New York City. You could do a much better job from a helicopter flying over the city than you could if you were standing in Times Square. Similarly, it would be easier to map our Galaxy if we could only get a little way outside it, but instead we are trapped inside and way out in its suburbs—far from the galactic equivalent of Times Square.

Herschel Measures the Galaxy

In 1785, William Herschel (Figure 25.2) made the first important discovery about the architecture of the Milky Way Galaxy. Using a large reflecting telescope that he had built, William and his sister Caroline counted stars in different directions of the sky. They found that most of the stars they could see lay in a flattened structure encircling the sky, and that the numbers of stars were about the same in any direction around this structure. Herschel therefore concluded that the stellar system to which the Sun belongs has the shape of a disk or wheel (he might have called it a Frisbee except Frisbees hadn't been invented yet), and that the Sun must be near the hub of the wheel (Figure 25.3).

Figure 25.2 William Herschel (1738–1822) and Caroline Herschel (1750–1848). William Herschel was a German musician who emigrated to England and took up astronomy in his spare time. He discovered the planet Uranus, built several large telescopes, and made measurements of the Sun's place in the Galaxy, the Sun's motion through space, and the comparative brightnesses of stars. This painting shows William and his sister Caroline polishing a telescope lens. (credit: modification of work by the Wellcome Library)

To understand why Herschel reached this conclusion, imagine that you are a member of a band standing in formation during halftime at a football game. If you count the band members you see in different directions and get about the same number each time, you can conclude that the band has arranged itself in a circular pattern with you at the center. Since you see no band members above you or underground, you know that the circle made by the band is much flatter than it is wide.

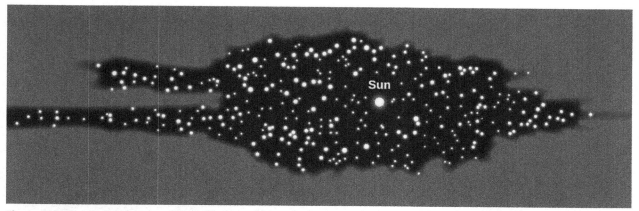

Figure 25.3 Herschel's Diagram of the Milky Way. Herschel constructed this cross section of the Galaxy by counting stars in various directions.

We now know that Herschel was right about the shape of our system, but wrong about where the Sun lies within the disk. As we saw in Between the Stars: Gas and Dust in Space, we live in a dusty Galaxy. Because interstellar dust absorbs the light from stars, Herschel could see only those stars within about 6000 light-years of the Sun. Today we know that this is a very small section of the entire 100,000-light-year-diameter disk of stars that makes up the Galaxy.

VOYAGERS IN ASTRONOMY

Harlow Shapley: Mapmaker to the Stars

Until the early 1900s, astronomers generally accepted Herschel's conclusion that the Sun is near the center of the Galaxy. The discovery of the Galaxy's true size and our actual location came about largely through the efforts of Harlow Shapley. In 1917, he was studying RR Lyrae variable stars in globular clusters. By comparing the known intrinsic luminosity of these stars to how bright they appeared, Shapley could calculate how far away they are. (Recall that it is distance that makes the stars look dimmer than they would be "up close," and that the brightness fades as the distance squared.) Knowing the distance to any star in a cluster then tells us the distance to the cluster itself.

Globular clusters can be found in regions that are free of interstellar dust and so can be seen at very large distances. When Shapley used the distances and directions of 93 globular clusters to map out their positions in space, he found that the clusters are distributed in a spherical volume, which has its center not at the Sun but at a distant point along the Milky Way in the direction of Sagittarius. Shapley then made the bold assumption, verified by many other observations since then, that the point on which the system of globular clusters is centered is also the center of the entire Galaxy (Figure 25.4).

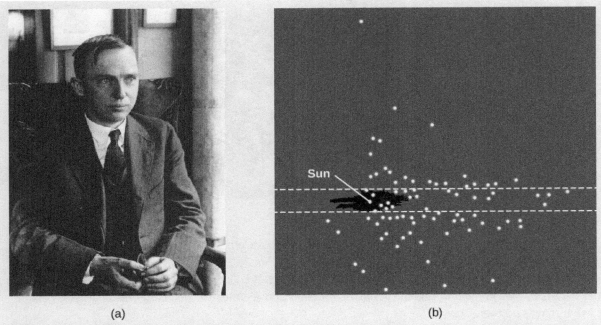

(a)   (b)

Figure 25.4 Harlow Shapley and His Diagram of the Milky Way. (a) Shapley poses for a formal portrait. (b) His diagram shows the location of globular clusters, with the position of the Sun also marked. The black area shows Herschel's old diagram, centered on the Sun, approximately to scale.

Shapley's work showed once and for all that our star has no special place in the Galaxy. We are in a nondescript region of the Milky Way, only one of 200 to 400 billion stars that circle the distant center of our Galaxy.

Born in 1885 on a farm in Missouri, Harlow Shapley at first dropped out of school with the equivalent of only a fifth-grade education. He studied at home and at age 16 got a job as a newspaper reporter covering crime stories. Frustrated by the lack of opportunities for someone who had not finished high school, Shapley went back and completed a six-year high-school program in only two years, graduating as class valedictorian.

In 1907, at age 22, he went to the University of Missouri, intent on studying journalism, but found that the school of journalism would not open for a year. Leafing through the college catalog (or so he told the story later), he chanced to see "Astronomy" among the subjects beginning with "A." Recalling his boyhood interest in the stars, he decided to study astronomy for the next year (and the rest, as the saying goes, is history).

Upon graduation Shapley received a fellowship for graduate study at Princeton and began to work with the brilliant Henry Norris Russell (see the Henry Norris Russell feature box). For his PhD thesis, Shapley made major contributions to the methods of analyzing the behavior of eclipsing binary stars. He was also able to show that cepheid variable stars are not binary systems, as some people thought at the time, but individual stars that pulsate with striking regularity.

Impressed with Shapley's work, George Ellery Hale offered him a position at the Mount Wilson Observatory, where the young man took advantage of the clear mountain air and the 60-inch reflector to do his pioneering study of variable stars in globular clusters.

Shapley subsequently accepted the directorship of the Harvard College Observatory, and over the next 30 years, he and his collaborators made contributions to many fields of astronomy, including the study of neighboring galaxies, the discovery of dwarf galaxies, a survey of the distribution of galaxies in the universe, and much more. He wrote a series of nontechnical books and articles and became known as one of the most effective popularizers of astronomy. Shapley enjoyed giving lectures around the country, including at many smaller colleges where students and faculty rarely got to interact with scientists of his caliber.

During World War II, Shapley helped rescue many scientists and their families from Eastern Europe; later, he helped found UNESCO, the United Nations Educational, Scientific, and Cultural Organization. He wrote a pamphlet called *Science from Shipboard* for men and women in the armed services who had to spend many weeks on board transport ships to Europe. And during the difficult period of the 1950s, when congressional committees began their "witch hunts" for communist sympathizers (including such liberal leaders as Shapley), he spoke out forcefully and fearlessly in defense of the freedom of thought and expression. A man of many interests, he was fascinated by the behavior of ants, and wrote scientific papers about them as well as about galaxies.

By the time he died in 1972, Shapley was acknowledged as one of the pivotal figures of modern astronomy, a "twentieth-century Copernicus" who mapped the Milky Way and showed us our place in the Galaxy.

LINK TO LEARNING

To find more information about Shapley's life and work (https://openstax.org/l/30shapbrumed), see the entry for him on the Bruce Medalists website. (This site features the winners of the Bruce Medal of the Astronomical Society of the Pacific, one of the highest honors in astronomy; the list is a who's who of some of the greatest astronomers of the last twelve decades.)

Disks and Haloes

With modern instruments, astronomers can now penetrate the "smog" of the Milky Way by studying radio and infrared emissions from distant parts of the Galaxy. Measurements at these wavelengths (as well as observations of other galaxies like ours) have given us a good idea of what the Milky Way would look like if we *could* observe it from a distance.

Figure 25.5 sketches what we would see if we could view the Galaxy face-on and edge-on. The brightest part of the Galaxy consists of a thin, circular, rotating disk of stars distributed across a region about 100,000 light-years in diameter and about 2000 light-years thick. (Given how thin the disk is, perhaps a CD is a more appropriate analogy than a wheel.) The very youngest stars, and the dust and gas from which stars form, are found typically within 100 light-years of the plane of the Milky Way Galaxy. The mass of the interstellar matter is about 15% of the mass of the stars in this disk.

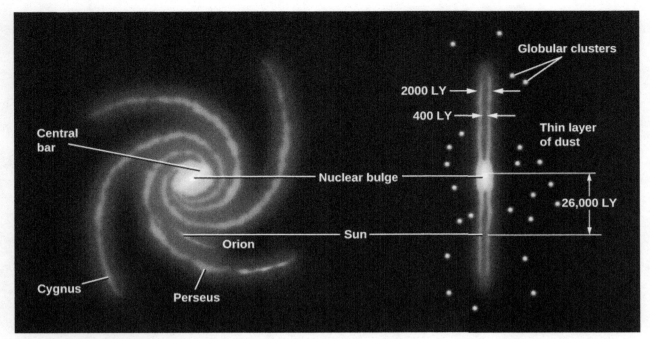

Figure 25.5 Schematic Representation of the Galaxy. The left image shows the face-on view of the spiral disk; the right image shows the view looking edge-on along the disk. The major spiral arms are labeled. The Sun is located on the inside edge of the short Orion spur.

As the diagram in Figure 25.5 shows, the stars, gas, and dust are not spread evenly throughout the disk but are concentrated into a central bar and a series of spiral arms. Recent infrared observations have confirmed that the central bar is composed mostly of old yellow-red stars. The two main spiral arms appear to connect with the ends of the bar. They are highlighted by the blue light from young hot stars. We know many other spiral galaxies that also have bar-shaped concentrations of stars in their central regions; for that reason they are called *barred spirals.* Figure 25.6 shows two other galaxies—one without a bar and one with a strong bar—to give you a basis for comparison to our own. We will describe our spiral structure in more detail shortly. The Sun is located about halfway between the center of the Galaxy and the edge of the disk and only about 70 light-years above its central plane.

(a)   (b)

Figure 25.6 Unbarred and Barred Spiral Galaxies. (a) This image shows the unbarred spiral galaxy M74. It contains a small central bulge of mostly old yellow-red stars, along with spiral arms that are highlighted with the blue light from young hot stars. (b) This image shows the strongly barred spiral galaxy NGC 1365. The bulge and the fainter bar both appear yellowish because the brightest stars in them are mostly old yellow and red giants. Two main spiral arms project from the ends of the bar. As in M74, these spiral arms are populated with blue stars and red patches of glowing gas—hallmarks of recent star formation. The Milky Way Galaxy is thought to have a barred spiral structure that is intermediate between these two examples. (credit a: modification of work by ESO/ PESSTO/S. Smartt; credit b: modification of work by ESO)

Our thin disk of young stars, gas, and dust is embedded in a thicker but more diffuse disk of older stars; this thicker disk extends about 1000 light-years above and 1000 light-years below the midplane of the thin disk and contains only about 5% as much mass as the thin disk. The stars thin out with distance from the galactic plane and don't have a sharp edge. Approximately 2/3 of the stars in the thick disk are within 1000 light-years of midplane.

Close in to the galactic center (within about 10,000 light-years), the stars are no longer confined to the disk but form a **central bulge** (or nuclear bulge). When we observe with visible light, we can glimpse the stars in the bulge only in those rare directions where there happens to be relatively little interstellar dust. The first picture that actually succeeded in showing the bulge as a whole was taken at infrared wavelengths (Figure 25.7).

Figure 25.7 Inner Part of the Milky Way Galaxy. This beautiful infrared map, showing half a billion stars, was obtained as part of the Two Micron All Sky Survey (2MASS). Because interstellar dust does not absorb infrared as strongly as visible light, this view reveals the previously hidden bulge of old stars that surrounds the center of our Galaxy, along with the Galaxy's thin disk component. (credit: modification of work by 2MASS/J. Carpenter, T. H. Jarrett, and R. Hurt)

The fact that much of the bulge is obscured by dust makes its shape difficult to determine. For a long time, astronomers assumed it was spherical. However, infrared images and other data indicate that the bulge is about two times longer than it is wide, and shaped rather like a peanut. The relationship between this elongated inner bulge and the larger bar of stars remains uncertain. At the very center of the nuclear bulge is a tremendous concentration of matter, which we will discuss later in this chapter.

In our Galaxy, the thin and thick disks and the nuclear bulge are embedded in a spherical **halo** of very old, faint stars that extends to a distance of at least 150,000 light-years from the galactic center. Most of the globular clusters are also found in this halo.

The mass in the Milky Way extends even farther out, well beyond the boundary of the luminous stars to a distance of at least 200,000 light-years from the center of the Galaxy. This invisible mass has been give the name *dark matter* because it emits no light and cannot be seen with any telescope. Its composition is unknown, and it can be detected only because of its gravitational effects on the motions of luminous matter that we can see. We know that this extensive **dark matter halo** exists because of its effects on the orbits of distant star clusters and other dwarf galaxies that are associated with the Galaxy. This mysterious halo will be a subject of the section on The Mass of the Galaxy, and the properties of dark matter will be discussed more in the chapter on The Big Bang.

Some vital statistics of the thin and thick disks and the stellar halo are given in Table 25.1, with an illustration in Figure 25.8. Note particularly how the ages of stars correlate with where they are found. As we shall see, this information holds important clues to how the Milky Way Galaxy formed.

Characteristics of the Milky Way Galaxy

Property	Thin Disk	Thick Disk	Stellar Halo (Excludes Dark Matter)
Stellar mass	$4 \times 10^{10}\ M_{Sun}$	A few percent of the thin disk mass	$10^{10}\ M_{Sun}$
Luminosity	$3 \times 10^{10}\ L_{Sun}$	A few percent of the thin disk luminosity	$8 \times 10^{8}\ L_{Sun}$
Typical age of stars	1 million to 10 billion years	11 billion years	13 billion years
Heavier-element abundance	High	Intermediate	Very low
Rotation	High	Intermediate	Very low

Table 25.1

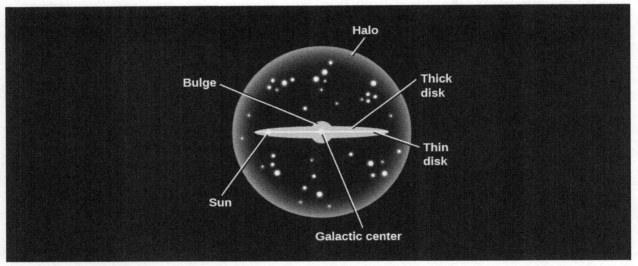

Figure 25.8 Major Parts of the Milky Way Galaxy. This schematic shows the major components of our Galaxy.

Establishing this overall picture of the Galaxy from our dust-shrouded viewpoint inside the thin disk has been one of the great achievements of modern astronomy (and one that took decades of effort by astronomers working with a wide range of telescopes). One thing that helped enormously was the discovery that our Galaxy is not unique in its characteristics. There are many other flat, spiral-shaped islands of stars, gas, and dust in the universe. For example, the Milky Way somewhat resembles the Andromeda galaxy, which, at a distance of about 2.3 million light-years, is our nearest neighboring giant spiral galaxy. Just as you can get a much better picture of yourself if someone else takes the photo from a distance away, pictures and other diagnostic observations of nearby galaxies that resemble ours have been vital to our understanding of the properties of the Milky Way.

LINK TO LEARNING

Start this video (https://openstax.org/l/30gaiastars), then click-and-drag to look in all directions from Earth in the Gaia space telescope's visualization of the stars of our galaxy.

The Milky Way viewer (https://openstax.org/l/30milkywayview) allows you to move to different locations in the Milky Way (using the Move and Zoom buttons) and view different stellar populations.

MAKING CONNECTIONS

The Milky Way Galaxy in Myth and Legend

To most of us living in the twenty-first century, the Milky Way Galaxy is an elusive sight. We must make an effort to leave our well-lit homes and streets and venture beyond our cities and suburbs into less populated environments. Once the light pollution subsides to negligible levels, the Milky Way can be readily spotted arching over the sky on clear, moonless nights. The Milky Way is especially bright in late summer and early fall in the Northern Hemisphere. Some of the best places to view the Milky Way are in our national and state parks, where residential and industrial developments have been kept to a minimum. Some of these parks host special sky-gazing events that are definitely worth checking out—especially during the two weeks surrounding the new moon, when the faint stars and Milky Way don't have to compete with the Moon's brilliance.

Go back a few centuries, and these starlit sights would have been the norm rather than the exception. Before the advent of electric or even gas lighting, people relied on short-lived fires to illuminate their homes and byways. Consequently, their night skies were typically much darker. Confronted by myriad stellar patterns and the Milky Way's gauzy band of diffuse light, people of all cultures developed myths to make sense of it all.

Some of the oldest myths relating to the Milky Way are maintained by the aboriginal Australians through their rock painting and storytelling. These legacies are thought to go back tens of thousands of years, to when the aboriginal people were being "dreamed" along with the rest of the cosmos. The Milky Way played a central role as an arbiter of the Creation. Taking the form of a great serpent, it joined with the Earth serpent to dream and thus create all the creatures on Earth.

The ancient Greeks viewed the Milky Way as a spray of milk that spilled from the breast of the goddess Hera. In this legend, Zeus had secretly placed his infant son Heracles at Hera's breast while she was asleep in order to give his half-human son immortal powers. When Hera awoke and found Heracles suckling, she pushed him away, causing her milk to spray forth into the cosmos (Figure 25.9).

The dynastic Chinese regarded the Milky Way as a "silver river" that was made to separate two star-crossed lovers. To the east of the Milky Way, Zhi Nu, the weaving maiden, was identified with the bright star Vega in the constellation of Lyra the Harp. To the west of the Milky Way, her lover Niu Lang, the cowherd, was associated with the star Altair in the constellation of Aquila the Eagle. They had been exiled on opposite sides of the Milky Way by Zhi Nu's mother, the Queen of Heaven, after she heard of their secret marriage and the birth of their two children. However, once a year, they are permitted to reunite. On the seventh day of the seventh lunar month (which typically occurs in our month of August), they would meet on a bridge over the Milky Way that thousands of magpies had made (Figure 25.9). This romantic time continues to be celebrated today as Qi Xi, meaning "Double Seventh," with couples reenacting the cosmic reunion of Zhi Nu and Niu Lang.

(a)   (b)

Figure 25.9 The Milky Way in Myth. (a) *Origin of the Milky Way* by Jacopo Tintoretto (circa 1575) illustrates the Greek myth that explains the formation of the Milky Way. (b) *The Moon of the Milky Way* by Japanese painter Tsukioka Yoshitoshi depicts the Chinese legend of Zhi Nu and Niu Lang.

To the Quechua Indians of Andean Peru, the Milky Way was seen as the celestial abode for all sorts of cosmic creatures. Arrayed along the Milky Way are myriad dark patches that they identified with partridges, llamas, a toad, a snake, a fox, and other animals. The Quechua's orientation toward the dark regions rather than the glowing band of starlight appears to be unique among all the myth makers. Likely, their access to the richly structured southern Milky Way had something to do with it.

Among Finns, Estonians, and related northern European cultures, the Milky Way is regarded as the "pathway of birds" across the night sky. Having noted that birds seasonally migrate along a north-south route, they identified this byway with the Milky Way. Recent scientific studies have shown that this myth is rooted in fact: the birds of this region use the Milky Way as a guide for their annual migrations.

Today, we regard the Milky Way as our galactic abode, where the foment of star birth and star death plays out on a grand stage, and where sundry planets have been found to be orbiting all sorts of stars. Although our perspective on the Milky Way is based on scientific investigations, we share with our forebears an affinity for telling stories of origin and transformation. In these regards, the Milky Way continues to fascinate and inspire us.

25.2 | Spiral Structure

Learning Objectives

By the end of this section, you will be able to:

> Describe the structure of the Milky Way Galaxy and how astronomers discovered it
> Compare theoretical models for the formation of spiral arms in disk galaxies

Astronomers were able to make tremendous progress in mapping the spiral structure of the Milky Way after the discovery of the 21-cm line that comes from cool hydrogen (see Between the Stars: Gas and Dust in Space). Remember that the obscuring effect of interstellar dust prevents us from seeing stars at large distances in the disk at visible wavelengths. However, radio waves of 21-cm wavelength pass right through the dust, enabling astronomers to detect hydrogen atoms throughout the Galaxy. More recent surveys of the infrared emission from stars in the disk have provided a similar dust-free perspective of our Galaxy's stellar distribution. Despite all this progress over the past fifty years, we are still just beginning to pin down the precise structure of our Galaxy.

The Arms of the Milky Way

Our radio observations of the disk's gaseous component indicate that the Galaxy has two major spiral arms that emerge from the bar and several fainter arms and shorter spurs. You can see a recently assembled map of our Galaxy's arm structure—derived from studies in the infrared—in Figure 25.10.

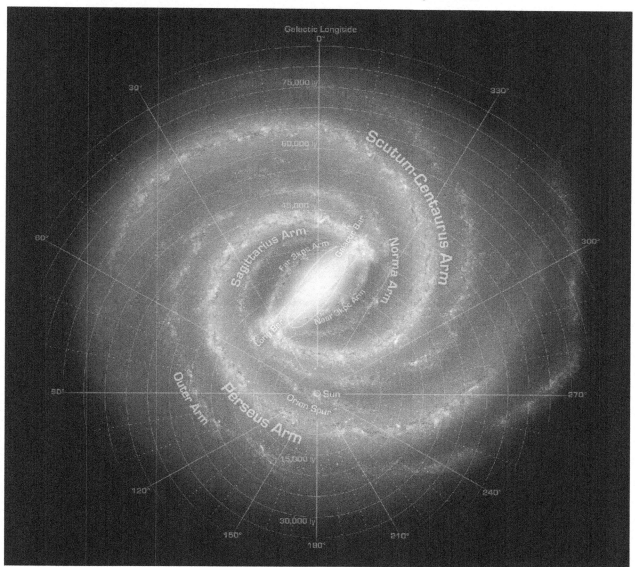

Figure 25.10 Milky Way Bar and Arms. Here, we see the Milky Way Galaxy as it would look from above. This image, assembled from data from NASA's Spitzer mission, shows that the Milky Way Galaxy has a modest bar in its central regions. Two spiral arms, Scutum-Centaurus and Perseus, emerge from the ends of the bar and wrap around the bulge. The Sagittarius and Outer arms have fewer stars than the other two arms. (credit: modification of work by NASA/JPL-Caltech/R. Hurt (SSC/Caltech))

The Sun is near the inner edge of a short arm called the Orion Spur, which is about 10,000 light-years long and contains such conspicuous features as the Cygnus Rift (the great dark nebula in the summer Milky Way) and the bright Orion Nebula. Figure 25.11 shows a few other objects that share this small section of the Galaxy with us and are easy to see. Remember, the farther away we try to look from our own arm, the more the dust in the Galaxy builds up and makes it hard to see with visible light.

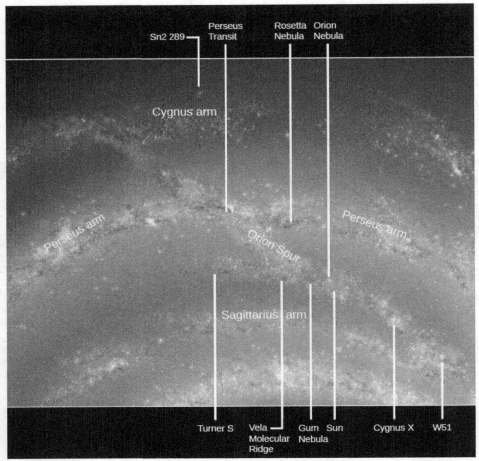

Figure 25.11 Orion Spur. The Sun is located in the Orion Spur, which is a minor spiral arm located between two other arms. In this diagram, the white lines point to some other noteworthy objects that share this feature of the Milky Way Galaxy with the Sun. (credit: modification of work by NASA/JPL-Caltech)

Formation of Spiral Structure

At the Sun's distance from its center, the Galaxy does not rotate like a solid wheel or a CD inside your player. Instead, the way individual objects turn around the center of the Galaxy is more like the solar system. Stars, as well as the clouds of gas and dust, obey Kepler's third law. Objects farther from the center take longer to complete an orbit around the Galaxy than do those closer to the center. In other words, stars (and interstellar matter) in larger orbits in the Galaxy trail behind those in smaller ones. This effect is called **differential galactic rotation**.

Differential rotation would appear to explain why so much of the material in the disk of the Milky Way is concentrated into elongated features that resemble **spiral arms**. No matter what the original distribution of the material might be, the differential rotation of the Galaxy can stretch it out into spiral features. Figure 25.12 shows the development of spiral arms from two irregular blobs of interstellar matter. Notice that as the portions of the blobs closest to the galactic center move faster, those farther out trail behind.

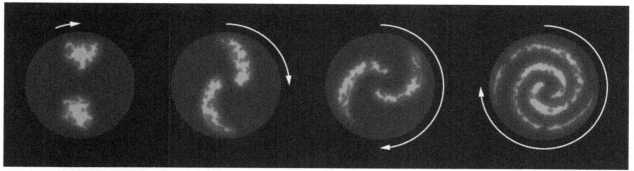

Figure 25.12 Simplified Model for the Formation of Spiral Arms. This sketch shows how spiral arms might form from irregular clouds of interstellar material stretched out by the different rotation rates throughout the Galaxy. The regions farthest from the galactic center take longer to complete their orbits and thus lag behind the inner regions. If this were the only mechanism for creating spiral arms, then over time the spiral arms would completely wind up and disappear. Since many galaxies have spiral arms, they must be long-lived, and there must be other processes at work to maintain them.

But this picture of spiral arms presents astronomers with an immediate problem. If that's all there were to the story, differential rotation—over the roughly 13-billion-year history of the Galaxy—would have wound the Galaxy's arms tighter and tighter until all semblance of spiral structure had disappeared. But did the Milky Way actually have spiral arms when it formed 13 billion years ago? And do spiral arms, once formed, last for that long a time?

With the advent of the Hubble Space Telescope, it has become possible to observe the structure of very distant galaxies and to see what they were like shortly after they began to form more than 13 billion years ago. What the observations show is that galaxies in their infancy had bright, clumpy star-forming regions, but no regular spiral structure.

Over the next few billion years, the galaxies began to "settle down." The galaxies that were to become spirals lost their massive clumps and developed a central bulge. The turbulence in these galaxies decreased, rotation began to dominate the motions of the stars and gas, and stars began to form in a much quieter disk. Smaller star-forming clumps began to form fuzzy, not-very-distinct spiral arms. Bright, well-defined spiral arms began to appear only when the galaxies were about 3.6 billion years old. Initially, there were two well-defined arms. Multi-armed structures in galaxies like we see in the Milky Way appeared only when the universe was about 8 billion years old.

We will discuss the history of galaxies in more detail in The Evolution and Distribution of Galaxies. But, even from our brief discussion, you can get the sense that the spiral structures we now observe in mature galaxies have come along later in the full story of how things develop in the universe.

Scientists have used supercomputer calculations to model the formation and evolution of the arms. These calculations follow the motions of up to 100 million "star particles" to see whether gravitational forces can cause them to form spiral structure. What these calculations show is that giant molecular clouds (which we discussed in Between the Stars: Gas and Dust in Space) have enough gravitational influence over their surroundings to initiate the formation of structures that look like spiral arms. These arms then become self-perpetuating and can survive for at least several billion years. The arms may change their brightness over time as star formation comes and goes, but they are not temporary features. The concentration of matter in the arms exerts sufficient gravitational force to keep the arms together over long periods of time.

25.3 | The Mass of the Galaxy

Learning Objectives

By the end of this section, you will be able to:

> Describe historical attempts to determine the mass of the Galaxy

> Interpret the observed rotation curve of our Galaxy to suggest the presence of dark matter whose distribution extends well beyond the Sun's orbit

When we described the sections of the Milky Way, we said that the stars are now known to be surrounded by a much larger halo of invisible matter. Let's see how this surprising discovery was made.

Kepler Helps Weigh the Galaxy

The Sun, like all the other stars in the Galaxy, orbits the center of the Milky Way. Our star's orbit is nearly circular and lies in the Galaxy's disk. The speed of the Sun in its orbit is about 200 kilometers per second, which means it takes us approximately 225 million years to go once around the center of the Galaxy. We call the period of the Sun's revolution the *galactic year*. It is a long time compared to human time scales; during the entire lifetime of Earth, only about 20 galactic years have passed. This means that we have gone only a tiny fraction of the way around the Galaxy in all the time that humans have gazed into the sky.

We can use the information about the Sun's orbit to estimate the mass of the Galaxy (just as we could "weigh" the Sun by monitoring the orbit of a planet around it—see Orbits and Gravity). Let's assume that the Sun's orbit is circular and that the Galaxy is roughly spherical, (we know the Galaxy is shaped more like a disk, but to simplify the calculation we will make this assumption, which illustrates the basic approach). Long ago, Newton showed that if you have matter distributed in the shape of a sphere, then it is simple to calculate the pull of gravity on some object just outside that sphere: you can assume that gravity acts as if all the matter were concentrated at a point in the center of the sphere. For our calculation, then, we can assume that all the mass that lies inward of the Sun's position is concentrated at the center of the Galaxy, and that the Sun orbits that point from a distance of about 26,000 light-years.

This is the sort of situation to which Kepler's third law (as modified by Newton) can be directly applied. Plugging numbers into Kepler's formula, we can calculate the sum of the masses of the Galaxy and the Sun. However, the mass of the Sun is completely trivial compared to the mass of the Galaxy. Thus, for all practical purposes, the result (about 100 billion times the mass of the Sun) is the mass of the Milky Way. More sophisticated calculations based on more sophisticated models give a similar result.

Our estimate tells us how much mass is contained in the volume inside the Sun's orbit. This is a good estimate for the total mass of the Galaxy only if hardly any mass lies outside the Sun's orbit. For many years astronomers thought this assumption was reasonable. The number of bright stars and the amount of *luminous matter* (meaning any material from which we can detect electromagnetic radiation) both drop off dramatically at distances of more than about 30,000 light-years from the galactic center. Little did we suspect how wrong our assumption was.

A Galaxy of Mostly Invisible Matter

In science, what seems to be a reasonable assumption can later turn out to be wrong (which is why we continue to do observations and experiments every chance we get). There is a lot more to the Milky Way than meets the eye (or our instruments). While there is relatively little luminous matter beyond 30,000 light-years, we now know that a lot of *invisible matter* exists at great distances from the galactic center.

We can understand how astronomers detected this invisible matter by remembering that according to Kepler's third law, objects orbiting at large distances from a massive object will move more slowly than objects that are closer to that central mass. In the case of the solar system, for example, the outer planets move more slowly in their orbits than the planets close to the Sun.

There are a few objects, including globular clusters and some nearby small satellite galaxies, that lie well outside the luminous boundary of the Milky Way. If most of the mass of our Galaxy were concentrated within the luminous region, then these very distant objects should travel around their galactic orbits at lower speeds than, for example, the Sun does.

It turns out, however, that the few objects seen at large distances from the luminous boundary of the Milky Way Galaxy are *not* moving more slowly than the Sun. There are some globular clusters and RR Lyrae stars between 30,000 and 150,000 light-years from the center of the Galaxy, and their orbital velocities are even greater than the Sun's (Figure 25.13).

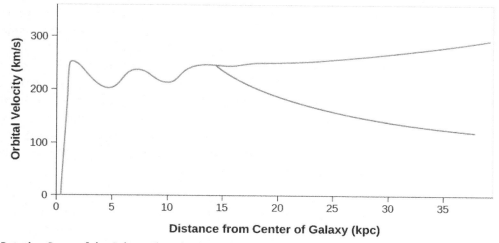

Figure 25.13 Rotation Curve of the Galaxy. The orbital speed of carbon monoxide (CO) and hydrogen (H) gas at different distances from the center of the Milky Way Galaxy is shown in red. The blue curve shows what the rotation curve would look like if all the matter in the Galaxy were located inside a radius of 50,000 light-years. Instead of going down, the speed of gas clouds farther out remains high, indicating a great deal of mass beyond the Sun's orbit. The horizontal axis shows the distance from the galactic center in kiloparsecs (where a kiloparsec equals 3,260 light-years).

What do these higher speeds mean? Kepler's third law tells us how fast objects must orbit a source of gravity if they are neither to fall in (because they move too slowly) nor to escape (because they move too fast). If the Galaxy had only the mass calculated by Kepler, then the high-speed outer objects should long ago have escaped the grip of the Milky Way. The fact that they have not done so means that our Galaxy must have more gravity than can be supplied by the luminous matter—in fact, a *lot* more gravity. The high speed of these outer objects tells us that the source of this extra gravity must extend outward from the center far beyond the Sun's orbit.

If the gravity were supplied by stars or by something else that gives off radiation, we should have spotted this additional outer material long ago. We are therefore forced to the reluctant conclusion that this matter is invisible and has, except for its gravitational pull, gone entirely undetected.

Studies of the motions of the most remote globular clusters and the small galaxies that orbit our own show that the total mass of the Galaxy is at least 2×10^{12} M_{Sun}, which is about twenty times greater than the amount of luminous matter. Moreover, the **dark matter** (as astronomers have come to call the invisible material) extends to a distance of at least 200,000 light-years from the center of the Galaxy. Observations indicate that this dark matter halo is almost but not quite spherical.

The obvious question is: what is the dark matter made of? Let's look at a list of "suspects" taken from our study of astronomy so far. Since this matter is invisible, it clearly cannot be in the form of ordinary stars. And it cannot be gas in any form (remember that there has to be a lot of it). If it were neutral hydrogen gas, its 21-cm wavelength spectral-line emission would have been detected as radio waves. If it were ionized hydrogen, it should be hot enough to emit visible radiation. If a lot of hydrogen atoms out there had combined into hydrogen molecules, these should produce dark features in the ultraviolet spectra of objects lying beyond the Galaxy, but such features have not been seen. Nor can the dark matter consist of interstellar dust, since in the required quantities, the dust would significantly obscure the light from distant galaxies.

What are our other possibilities? The dark matter cannot be a huge number of black holes (of stellar mass) or old neutron stars, since interstellar matter falling onto such objects would produce more X-rays than are observed. Also, recall that the formation of black holes and neutron stars is preceded by a substantial amount of mass loss, which scatters heavy elements into space to be incorporated into subsequent generations of stars. If the dark matter consisted of an enormous number of any of those objects, they would have blown off and recycled a lot of heavier elements over the history of the Galaxy. In that case, the young stars we observe in our Galaxy today would contain much greater abundances of heavy elements than they actually do.

Brown dwarfs and lone Jupiter-like planets have also been ruled out. First of all, there would have to be an awful lot of them to make up so much dark matter. But we have a more direct test of whether so many low-mass objects could actually be lurking out there. As we learned in Black Holes and Curved Spacetime, the general theory of relativity predicts that the path traveled by light is changed when it passes near a concentration of mass. It turns out that when the two objects appear close enough together in the sky, the mass closer to us can bend the light from farther away. With just the right alignment, the image of the more distant object also becomes significantly brighter. By looking for the temporary brightening that occurs when a dark matter object in our own Galaxy moves across the path traveled by light from stars in the Magellanic Clouds, astronomers have now shown that the dark matter cannot be made up of a lot of small objects with masses between one-millionth and one-tenth the mass of the Sun.

What's left? One possibility is that the dark matter is composed of exotic subatomic particles of a type not yet detected on Earth. Very sophisticated (and difficult) experiments are now under way to look for such particles. Stay tuned to see whether anything like that turns up.

We should add that the problem of dark matter is by no means confined to the Milky Way. Observations show that dark matter must also be present in other galaxies (whose outer regions also orbit too fast "for their own good"—they also have flat rotation curves). As we will see, dark matter even exists in great clusters of galaxies whose members are now known to move around under the influence of far more gravity than can be accounted for by luminous matter alone.

Stop a moment and consider how astounding the conclusion we have reached really is. Perhaps as much as 95% of the mass in our Galaxy (and many other galaxies) is not only invisible, but we do not even know what it is made of. The stars and raw material we can observe may be merely the tip of the cosmic iceberg; underlying it all may be other matter, perhaps familiar, perhaps startlingly new. Understanding the nature of this dark matter is one of the great challenges of astronomy today; you will learn more about this in A Universe of (Mostly) Dark Matter and Dark Energy.

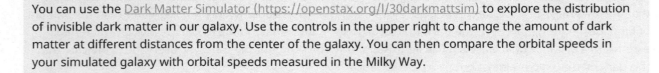

LINK TO LEARNING

You can use the Dark Matter Simulator (https://openstax.org/l/30darkmattsim) to explore the distribution of invisible dark matter in our galaxy. Use the controls in the upper right to change the amount of dark matter at different distances from the center of the galaxy. You can then compare the orbital speeds in your simulated galaxy with orbital speeds measured in the Milky Way.

25.4 | The Center of the Galaxy

Learning Objectives

By the end of this section, you will be able to:

> Describe the radio and X-ray observations that indicate energetic phenomena are occurring at the galactic center

> Explain what has been revealed by high-resolution near-infrared imaging of the galactic center

> Discuss how these near-infrared images, when combined with Kepler's third law of motion, can be used to derive the mass of the central gravitating object

At the beginning of this chapter, we hinted that the core of our Galaxy contains a large concentration of mass. In fact, we now have evidence that the very center contains a black hole with a mass equivalent to 4.6 million Suns and that all this mass fits within a sphere that has less than the diameter of Mercury's orbit. Such monster black holes are called **supermassive black holes** by astronomers, to indicate that the mass they contain is far greater than that of the typical black hole created by the death of a single star. It is amazing that we have very convincing evidence that this black hole really does exist. After all, recall from the chapter on Black Holes and Curved Spacetime that we cannot see a black hole directly because by definition it radiates no energy. And we cannot even see into the center of the Galaxy in visible light because of absorption by the interstellar dust that lies between us and the galactic center. Light from the central region of the Galaxy is dimmed by a factor of a trillion (10^{12}) by all this dust.

Fortunately, we are not so blind at other wavelengths. Infrared and radio radiation, which have long wavelengths compared to the sizes of the interstellar dust grains, flow unimpeded past the dust particles and so reach our telescopes with hardly any dimming. In fact, the very bright radio source in the nucleus of the Galaxy, now known as Sagittarius A* (pronounced "Sagittarius A-star" and abbreviated Sgr A*), was the first cosmic radio source astronomers discovered.

A Journey toward the Center

Let's take a voyage to the mysterious heart of our Galaxy and see what's there. Figure 25.14 is a radio image of a region about 1500 light-years across, centered on Sagittarius A, a bright radio source that contains the smaller Sagittarius A*. Much of the radio emission comes from hot gas heated either by clusters of hot stars (the stars themselves do not produce radio emission and can't be seen in the image) or by supernova blast waves. Most of the hollow circles visible on the radio image are supernova remnants. The other main source of radio emission is from electrons moving at high speed in regions with strong magnetic fields. The bright thin arcs and "threads" on the figure show us where this type of emission is produced.

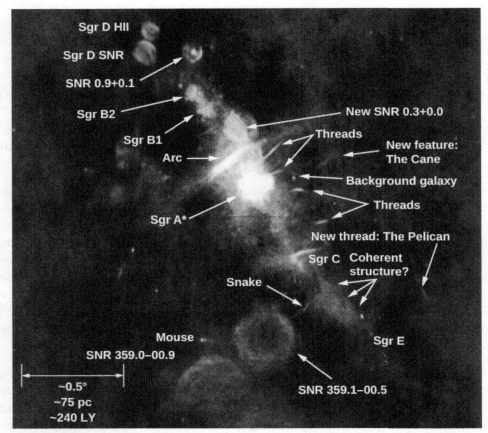

Figure 25.14 Radio Image of Galactic Center Region. This radio map of the center of the Galaxy (at a wavelength of 90 centimeters) was constructed from data obtained with the Very Large Array (VLA) of radio telescopes in Socorro, New Mexico. Brighter regions are more intense in radio waves. The galactic center is inside the region labeled Sagittarius A. Sagittarius B1 and B2 are regions of active star formation. Many filaments or threadlike features are seen, as well as a number of shells (labeled SNR), which are supernova remnants. The scale bar at the bottom left is about 240 light-years long. Notice that radio astronomers also give fanciful animal names to some of the structures, much as visible-light nebulae are sometimes given the names of animals they resemble. (credit: modification of work by N. E. Kassim, D. S. Briggs, T. J. W. Lazio, T. N. LaRosa, and J. Imamura (NRL/RSD))

Now let's focus in on the central region using a more energetic form of electromagnetic radiation. Figure 25.15 shows the X-ray emission from a smaller region 400 light-years wide and 900 light-years across centered in Sagittarius A*. Seen in this picture are hundreds of hot white dwarfs, neutron stars, and stellar black holes with accretion disks glowing with X-rays. The diffuse haze in the picture is emission from gas that lies among the stars and is at a temperature of 10 million K.

Figure 25.15 Galactic Center in X-Rays. This artificial-color mosaic of 30 images taken with the Chandra X-ray satellite shows a region 400 × 900 light-years in extent and centered on Sagittarius A*, the bright white source in the center of the picture. The X-ray-emitting point sources are white dwarfs, neutron stars, and stellar black holes. The diffuse "haze" is emission from gas at a temperature of 10 million K. This hot gas is flowing away from the center out into the rest of the Galaxy. The colors indicate X-ray energy bands: red (low energy), green (medium energy), and blue (high energy). (credit: modification of work by NASA/CXC/ UMass/

D. Wang et al.)

As we approach the center of the Galaxy, we find the supermassive black hole Sagittarius A*. There are also thousands of stars within a few lightyears of Sagittarius A*. Most of these are old, reddish main-sequence stars. But there are also about a hundred hot OB stars that must have formed within the last few million years. There is as yet no good explanation for how stars could have formed recently so close to a supermassive black hole. Perhaps they formed in a dense cluster of stars that was originally at a larger distance from the black hole and subsequently migrated closer.

There is currently no star formation at the galactic center, but there is lots of dust and molecular gas that is revolving around the black hole, along with some ionized gas streamers that are heated by the hot stars. Figure 25.16 is a radio map that shows these gas streamers.

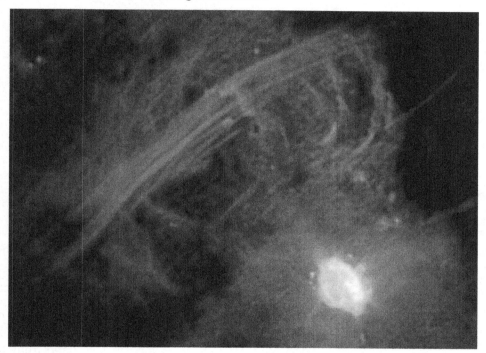

Figure 25.16 Sagittarius A. This image, taken with the Very Large Array of radio telescopes, shows the radio emission from hot, ionized gas in the center of the Milky Way. The lines slanting across the top of the image are gas streamers. Sagittarius A* is the bright spot in the lower right. (credit: modification of work by Farhad Zadeh et al. (Northwestern), VLA, NRAO)

Finding the Heart of the Galaxy

Just what is Sagittarius A*, which lies right at the center our Galaxy? To establish that there really is a black hole there, we must show that there is a very large amount of mass crammed into a very tiny volume. As we saw in Black Holes and Curved Spacetime, proving that a black hole exists is a challenge because the black hole itself emits no radiation. What astronomers must do is prove that a black hole is the only possible explanation for our observations—that a small region contains far more mass than could be accounted for by a very dense cluster of stars or something else made of ordinary matter.

To put some numbers with this discussion, the radius of the event horizon of a *galactic black hole* with a mass of about 4 million M_{Sun} would be only about 17 times the size of the Sun—the equivalent of a single red giant star. The corresponding density within this region of space would be much higher than that of any star cluster or any other ordinary astronomical object. Therefore, we must measure both the diameter of Sagittarius A* and its mass. Both radio and infrared observations are required to give us the necessary evidence.

First, let's look at how the mass can be measured. If we zero in on the inner few light-days of the Galaxy with an infrared telescope equipped with adaptive optics, we see a region crowded with individual stars (Figure 25.17). These stars have now been observed for almost two decades, and astronomers have detected their rapid orbital motions around the very center of the Galaxy.

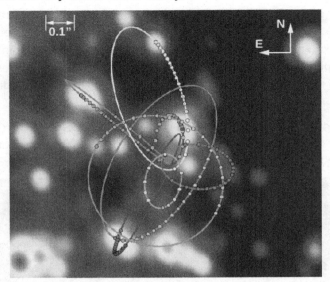

Figure 25.17 Near-Infrared View of the Galactic Center. This image shows the inner 1 arcsecond, or 0.13 light-year, at the center of the Galaxy, as observed with the giant Keck Telescope. Tracks of the orbiting stars measured from 1995 to 2014 have been added to this "snapshot." The stars are moving around the center very fast, and their tracks are all consistent with a single massive "gravitator" that resides in the very center of this image. (credit: modification of work by Andrea Ghez, UCLA Galactic Center Group, W.M. Keck Observatory Laser Team)

LINK TO LEARNING

Check out an animated version (https://openstax.org/l/30anifiginfgal) of Figure 25.17, showing the motion of the stars over the years.

If we combine observations of their periods and the size of their orbits with Kepler's third law, we can estimate the mass of the object that keeps them in their orbits. One of the stars has been observed for its full orbit of 15.6 years. Its closest approach takes it to a distance of only 124 AU or about 17 light-hours from the black hole. This orbit, when combined with observations of other stars close to the galactic center, indicates that a mass of 4.6 million M_{Sun} must be concentrated inside the orbit—that is, within 17 light-hours of the center of the Galaxy.

Even tighter limits on the size of the concentration of mass at the center of the Galaxy come from radio astronomy, which provided the first clue that a black hole might lie at the center of the Galaxy. As matter spirals inward toward the event horizon of a black hole, it is heated in a whirling *accretion disk* and produces radio radiation. (Such accretion disks were explained in Black Holes and Curved Spacetime.) Measurements of the size of the accretion disk with the Very Long Baseline Array, which provides very high spatial resolution, show that the diameter of the radio source Sagittarius A* is no larger than about 0.3 AU, or about the size of Mercury's orbit. (In light units, that's only 2.5 light-*minutes*!)

The observations thus show that 4.6 million solar masses are crammed into a volume that has a diameter that is no larger than the orbit of Mercury. If this were anything other than a supermassive black hole—low-mass stars that emit very little light or neutron stars or a very large number of small black holes— calculations show that these objects would be so densely packed that they would collapse to a single black hole within a hundred thousand years. That is a very short time compared with the age of the Galaxy, which probably began forming more than 13 billion years ago. Since it seems very unlikely that we would have caught such a complex cluster of objects just before it collapsed, the evidence for a supermassive black hole at the center of the Galaxy is convincing indeed.

Finding the Source

Where did our galactic black hole come from? The origin of supermassive black holes in galaxies like ours is currently an active field of research. One possibility is that a large cloud of gas near the center of the Milky Way collapsed directly to form a black hole. Since we find large black holes at the centers of most other large galaxies (see Active Galaxies, Quasars, and Supermassive Black Holes)—even ones that are very young—this collapse probably would have taken place when the Milky Way was just beginning to take shape. The initial mass of this black hole might have been only a few tens of solar masses. Another way it could have started is that a massive star might have exploded to leave behind a seed black hole, or a dense cluster of stars might have collapsed into a black hole.

Once a black hole exists at the center of a galaxy, it can grow over the next several billion years by devouring nearby stars and gas clouds in the crowded central regions. It can also grow by merging with other black holes.

It appears that the monster black hole at the center of our Galaxy is not finished "eating." At the present time, we observe clouds of gas and dust falling into the galactic center at the rate of about 1 M_{Sun} per thousand years. Stars are also on the black hole's menu. The density of stars near the galactic center is high enough that we would expect a star to pass near the black hole and be swallowed by it every ten thousand years or so. As this happens, some of the energy of infall is released as radiation. As a result, the center of the Galaxy might flare up and even briefly outshine all the stars in the Milky Way. Other objects might also venture too close to the black hole and be pulled in. How great a flare we observe would depend on the mass of the object falling in.

In 2013, the Chandra X-ray satellite detected a flare from the center of our Galaxy that was 400 times brighter than the usual output from Sagittarius A*. A year later, a second flare, only half as bright, was also detected. This is much less energy than swallowing a whole star would produce. There are two theories to account for the flares. First, an asteroid might have ventured too close to the black hole and been heated to a very high temperature before being swallowed up. Alternatively, the flares might have involved interactions of the magnetic fields near the galactic center in a process similar to the one described for solar flares (see The Sun: A Garden-Variety Star). Astronomers continue to monitor the galactic center area for flares or other activity. Although the monster in the center of the Galaxy is not close enough to us to represent any danger, we still want to keep our eyes on it.

VOYAGERS IN ASTRONOMY

Andrea Ghez

A lover of puzzles, Andrea Ghez has been pursuing one of the greatest mysteries in astronomy: what strange entity lurks within the center of our Milky Way Galaxy?

Figure 25.18 Andrea Ghez. Research by Ghez and her team has helped shape our understanding of supermassive black holes. (credit: modification of work by John D. and Catherine T. MacArthur Foundation)

As a child living in Chicago during the late 1960s, Andrea Ghez (Figure 25.18) was fascinated by the Apollo Moon landings. But she was also drawn to ballet and to solving all sorts of puzzles. By high school, she had lost the ballet bug in favor of competing in field hockey, playing the flute, and digging deeper into academics. Her undergraduate years at MIT were punctuated by a number of changes in her major—from mathematics to chemistry, mechanical engineering, aerospace engineering, and finally physics—where she felt her options were most open. As a physics major, she became involved in astronomical research under the guidance of one of her instructors. Once she got to do some actual observing at Kitt Peak National Observatory in Arizona, and later at Cerro Tololo Inter-American Observatory in Chile, Ghez had found her calling.

Pursuing her graduate studies at Caltech, she stuck with physics but oriented her efforts toward observational astrophysics, an area where Caltech had access to cutting-edge facilities. Though initially attracted to studying the black holes that were suspected of dwelling inside most massive galaxies, Ghez ended up spending most of her graduate study and later postdoctoral research at the University of Arizona studying stars in formation. By taking very high-resolution (detailed) imaging of regions where new stars are born, she discovered that most stars form as members of binary systems. As technologies advanced, she was able to track the orbits danced by these stellar pairings and thereby could ascertain their respective masses.

Now an astronomy professor at UCLA, Ghez has since used similar high-resolution imaging techniques to study the orbits of stars in the innermost core of the Milky Way. These orbits take years to delineate, so Ghez and her science team have logged more than 20 years of taking super-resolution infrared images with the giant Keck telescopes in Hawaii. Based on the resulting stellar orbits, the UCLA Galactic Center Group has settled (as we saw) on a gravitational solution that requires the presence of a supermassive black hole with a mass equivalent to 4.6 million Suns—all nestled within a space smaller than that occupied by our solar system. Ghez's achievements have been recognized with one of the "genius" awards given by the MacArthur Foundation. More recently, her team discovered glowing clouds of warm ionized gas that co-orbit with the stars but may be more vulnerable to the disruptive effects of the central black hole. By monitoring these clouds, the team hopes to better understand the evolution of supermassive black holes and their immediate environs. They also hope to test Einstein's theory of general relativity by carefully scrutinizing the orbits of stars that career closest to the intensely gravitating black hole.

In 2020, Ghez received the Nobel Prize in Physics for her work on the black hole at the center of the Galaxy. You can see her explain her work in non-technical language in this video (https://openstax.org/l/ghez). Besides her pioneering work as an astronomer, Ghez competes as a master swimmer, enjoys family life as a mother of two children, and actively encourages other women to pursue scientific careers.

25.5 | Stellar Populations in the Galaxy

Learning Objectives

By the end of this section, you will be able to:

> Distinguish between population I and population II stars according to their locations, motions, heavy-element abundances, and ages

> Explain why the oldest stars in the Galaxy are poor in elements heavier than hydrogen and helium, while stars like the Sun and even younger stars are typically richer in these heavy elements

In the first section of his chapter, we described the thin disk, thick disk, and stellar halo. Look back at Table 25.1 and note some of the patterns. Young stars lie in the thin disk, are rich in metals, and orbit the Galaxy's center at high speed. The stars in the halo are old, have low abundances of elements heavier than hydrogen and helium, and have highly elliptical orbits randomly oriented in direction (see Figure 25.19). Halo stars can plunge through the disk and central bulge, but they spend most of their time far above or below the plane of the Galaxy. The stars in the thick disk are intermediate between these two extremes. Let's first see why age and heavier-element abundance are correlated and then see what these correlations tell us about the origin of our Galaxy.

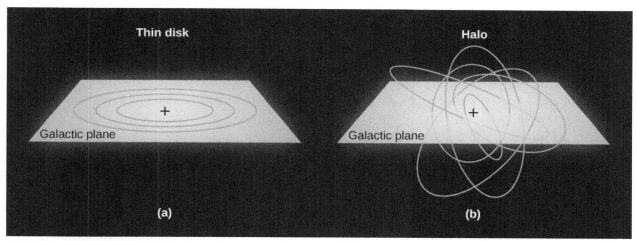

Figure 25.19 How Objects Orbit the Galaxy. (a) In this image, you see stars in the thin disk of our Galaxy in nearly circular orbits. (b) In this image, you see the motion of stars in the Galaxy's halo in randomly oriented and elliptical orbits.

Two Kinds of Stars

The discovery that there are two different kinds of stars was first made by Walter Baade during World War II. As a German national, Baade was not allowed to do war research as many other U.S.-based scientists were doing, so he was able to make regular use of the Mount Wilson telescopes in southern California. His observations were aided by the darker skies that resulted from the wartime blackout of Los Angeles.

Among the things a large telescope and dark skies enabled Baade to examine carefully were *other* galaxies—neighbors of our Milky Way Galaxy. We will discuss other galaxies in the next chapter (Galaxies), but for now we will just mention that the nearest Galaxy that resembles our own (with a similar disk and spiral structure) is often called the Andromeda galaxy, after the constellation in which we find it.

Baade was impressed by the similarity of the mainly reddish stars in the Andromeda galaxy's nuclear bulge to those in our Galaxy's globular clusters and the halo. He also noted the difference in color between all these and the bluer stars found in the spiral arms near the Sun (Figure 25.20). On this basis, he called the bright blue stars in the spiral arms **population I** and all the stars in the halo and globular clusters **population II**.

Figure 25.20 Andromeda Galaxy (M31). This neighboring spiral looks similar to our own Galaxy in that it is a disk galaxy with a central bulge. Note the bulge of older, yellowish stars in the center, the bluer and younger stars in the outer regions, and the dust in the disk that blocks some of the light from the bulge. (credit: Adam Evans)

We now know that the populations differ not only in their locations in the Galaxy, but also in their chemical composition, age, and orbital motions around the center of the Galaxy. Population I stars are found only in the disk and follow nearly circular orbits around the galactic center. Examples are bright supergiant stars, main-sequence stars of high luminosity (spectral classes O and B), which are concentrated in the spiral arms, and members of young open star clusters. Interstellar matter and molecular clouds are found in the same places as population I stars.

Population II stars show no correlation with the location of the spiral arms. These objects are found throughout the Galaxy. Some are in the disk, but many others follow eccentric elliptical orbits that carry them high above the galactic disk into the halo. Examples include stars surrounded by planetary nebulae and RR Lyrae variable stars. The stars in globular clusters, found almost entirely in the Galaxy's halo, are also classified as population II.

Today, we know much more about stellar evolution than astronomers did in the 1940s, and we can determine the ages of stars. Population I includes stars with a wide range of ages. While some are as old as 10 billion years, others are still forming today. For example, the Sun, which is about 5 billion years old, is a population I star. But so are the massive young stars in the Orion Nebula that have formed in the last few million years. Population II, on the other hand, consists entirely of old stars that formed very early in the history of the Galaxy; typical ages are 11 to 13 billion years.

We also now have good determinations of the compositions of stars. These are based on analyses of the stars' detailed spectra. Nearly all stars appear to be composed mostly of hydrogen and helium, but their abundances of the heavier elements differ. In the Sun and other population I stars, the heavy elements (those heavier than hydrogen and helium) account for 1–4% of the total stellar mass. Population II stars in the outer galactic halo and in globular clusters have much lower abundances of the heavy elements—often less than one-hundredth the concentrations found in the Sun and in rare cases even lower. The oldest population II star discovered to date has less than one ten-millionth as much iron as the Sun, for example.

As we discussed in earlier chapters, heavy elements are created deep within the interiors of stars. They are added to the Galaxy's reserves of raw material when stars die, and their material is recycled into new generations of stars. Thus, as time goes on, stars are born with larger and larger supplies of heavy elements. Population II stars formed when the abundance of elements heavier than hydrogen and helium was low. Population I stars formed later, after mass lost by dying members of the first generations of stars had seeded the interstellar medium with elements heavier than hydrogen and helium. Some are still forming now, when further generations have added to the supply of heavier elements available to new stars.

The Real World

With rare exceptions, we should never trust any theory that divides the world into just two categories. While they can provide a starting point for hypotheses and experiments, they are often oversimplifications that need refinement a research continue. The idea of two populations helped organize our initial thoughts about the Galaxy, but we now know it cannot explain everything we observe. Even the different structures of the Galaxy—disk, halo, central bulge—are not so cleanly separated in terms of their locations, ages, and the heavy element content of the stars within them.

The exact definition of the Galaxy's disk depends on what objects we use to define it, and, as we saw earlier, it has no sharp boundary. The hottest young stars and their associated gas and dust clouds are mostly in a region about 200 light-years thick. Older stars define a thicker disk that is about 2000 light-years thick. Halo stars spend most of their time high above or below the disk but pass through it on their highly elliptical orbits and so are sometimes found relatively near the Sun.

The highest density of stars is found in the central bulge, that bar-shaped inner region of the Galaxy. There are a few hot, young stars in the bulge, but most of the bulge stars are more than 10 billion years old. Yet unlike the halo stars of similar age, the abundance of heavy elements in the bulge stars is about the same as in the Sun. Why would that be?

Astronomers think that star formation in the crowded nuclear bulge occurred very rapidly just after the Milky Way Galaxy formed. After a few million years, the first generation of massive and short-lived stars then expelled heavy elements in supernova explosions and thereby enriched subsequent generations of stars. Thus, even stars that formed in the bulge more than 10 billion years ago started with a good supply of heavy elements.

Exactly the opposite occurred in the Small Magellanic Cloud, a small galaxy near the Milky Way, visible from Earth's Southern Hemisphere. Even the youngest stars in this galaxy are deficient in heavy elements. We think this is because the little galaxy is not especially crowded, and star formation has occurred quite slowly. As a result there have been, so far, relatively few supernova explosions. Smaller galaxies also have more trouble holding onto the gas expelled by supernova explosions in order to recycle it. Low-mass galaxies exert only a modest gravitational force, and the high-speed gas ejected by supernovae can easily escape from them.

Which elements a star is endowed with thus depends not only on when the star formed in the history of its galaxy, but also on how many stars in its part of the galaxy had already completed their lives by the time the star is ready to form.

25.6 | The Formation of the Galaxy

Learning Objectives

By the end of this section, you will be able to:

> Describe the roles played by the collapse of a single cloud and mergers with other galaxies in building the Milky Way Galaxy we see today

> Provide examples of globular clusters and satellite galaxies affected by the Milky Way's strong gravity.

Information about stellar populations holds vital clues to how our Galaxy was built up over time. The flattened disk shape of the Galaxy suggests that it formed through a process similar to the one that leads to the formation of a protostar (see The Birth of Stars and the Discovery of Planets outside the Solar System).

Building on this idea, astronomers first developed models that assumed the Galaxy formed from a single rotating cloud. But, as we shall see, this turns out to be only part of the story.

The Protogalactic Cloud and the Monolithic Collapse Model

Because the oldest stars—those in the halo and in globular clusters—are distributed in a sphere centered on the nucleus of the Galaxy, it makes sense to assume that the *protogalactic* cloud that gave birth to our Galaxy was roughly spherical. The oldest stars in the halo have ages of 12 to 13 billion years, so we estimate that the formation of the Galaxy began about that long ago. (See the chapter on The Big Bang for other evidence that galaxies in general began forming a little more than 13 billion years ago.) Then, just as in the case of star formation, the protogalactic cloud collapsed and formed a thin rotating disk. Stars born before the cloud collapsed did not participate in the collapse, but have continued to orbit in the halo to the present day (Figure 25.21).

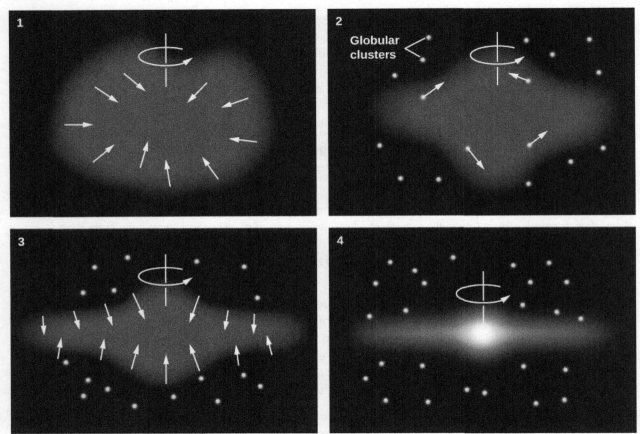

Figure 25.21 Monolithic Collapse Model for the Formation of the Galaxy. According to this model, the Milky Way Galaxy initially formed from a rotating cloud of gas that collapsed due to gravity. Halo stars and globular clusters either formed prior to the collapse or were formed elsewhere. Stars in the disk formed later, when the gas from which they were made was already "contaminated" with heavy elements produced in earlier generations of stars.

Gravitational forces caused the gas in the thin disk to fragment into clouds or clumps with masses like those of star clusters. These individual clouds then fragmented further to form stars. Since the oldest stars in the disk are nearly as old as the youngest stars in the halo, the collapse must have been rapid (astronomically speaking), requiring perhaps no more than a few hundred million years.

Collision Victims and the Multiple Merger Model

In past decades, astronomers have learned that the evolution of the Galaxy has not been quite as peaceful as this monolithic collapse model suggests. In 1994, astronomers discovered a small new galaxy in the direction of the constellation of Sagittarius. The Sagittarius dwarf galaxy is currently about 70,000 light-years away from

Earth and 50,000 light-years from the center of the Galaxy. It is the closest galaxy known (Figure 25.22). It is very elongated, and its shape indicates that it is being torn apart by our Galaxy's gravitational tides—just as Comet Shoemaker-Levy 9 was torn apart when it passed too close to Jupiter in 1992.

The Sagittarius galaxy is much smaller than the Milky Way and is about 10,000 times less massive than our Galaxy. All of the stars in the Sagittarius dwarf galaxy seem destined to end up in the bulge and halo of the Milky Way. But don't sound the funeral bells for the little galaxy quite yet; the ingestion of the Sagittarius dwarf will take another 100 million years or so, and the stars themselves will survive.

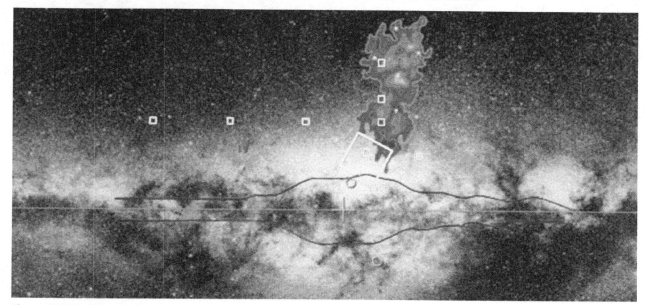

Figure 25.22 Sagittarius Dwarf. In 1994, British astronomers discovered a galaxy in the constellation of Sagittarius, located only about 50,000 light-years from the center of the Milky Way and falling into our Galaxy. This image covers a region approximately 70° × 50° and combines a black-and-white view of the disk of our Galaxy with a red contour map showing the brightness of the dwarf galaxy. The dwarf galaxy lies on the other side of the galactic center from us. The white stars in the red region mark the locations of several globular clusters contained within the Sagittarius dwarf galaxy. The cross marks the galactic center. The horizontal line corresponds to the galactic plane. The blue outline on either side of the galactic plane corresponds to the infrared image in Figure 25.7. The boxes mark regions where detailed studies of individual stars led to the discovery of this galaxy. (credit: modification of work by R. Ibata (UBC), R. Wyse (JHU), R. Sword (IoA))

Since that discovery, evidence has been found for many more close encounters between our Galaxy and other neighbor galaxies. When a small galaxy ventures too close, the force of gravity exerted by our Galaxy tugs harder on the near side than on the far side. The net effect is that the stars that originally belonged to the small galaxy are spread out into a long stream that orbits through the halo of the Milky Way (Figure 25.23).

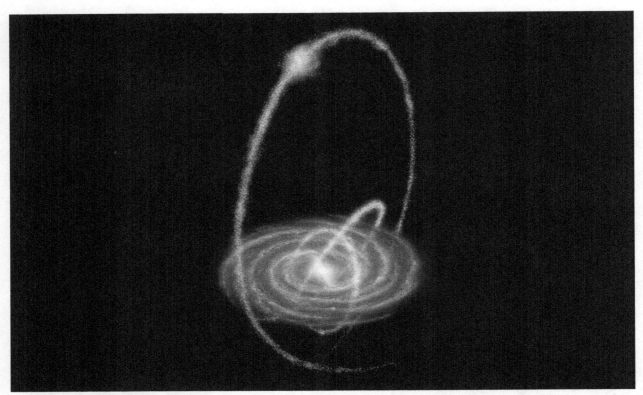

Figure 25.23 Streams in the Galactic Halo. When a small galaxy is swallowed by the Milky Way, its member stars are stripped away and form streams of stars in the galactic halo. This image is based on calculations of what some of these tidal streams might look like if the Milky Way swallowed 50 dwarf galaxies over the past 10 billion years. (credit: modification of work by NASA/JPL-Caltech/R. Hurt (SSC/Caltech))

Such a tidal stream can maintain its identity for billions of years. To date, astronomers have now identified streams originating from 12 small galaxies that ventured too close to the much larger Milky Way. Six more streams are associated with globular clusters. It has been suggested that large globular clusters, like Omega Centauri, are actually dense nuclei of cannibalized dwarf galaxies. The globular cluster M54 is now thought to be the nucleus of the Sagittarius dwarf we discussed earlier, which is currently merging with the Milky Way (Figure 25.24). The stars in the outer regions of such galaxies are stripped off by the gravitational pull of the Milky Way, but the central dense regions may survive.

Figure 25.24 Globular Cluster M54. This beautiful Hubble Space Telescope image shows the globular cluster that is now believed to be the nucleus of the Sagittarius Dwarf Galaxy. (credit: ESA/Hubble & NASA)

Calculations indicate that the Galaxy's thick disk may be a product of one or more such collisions with other galaxies. Accretion of a satellite galaxy would stir up the orbits of the stars and gas clouds originally in the thin disk and cause them to move higher above and below the mid-plane of the Galaxy. Meanwhile, the Galaxy's stars would add to the fluffed-up mix. If such a collision happened about 10 billion years ago, then any gas in the two galaxies that had not yet formed into stars would have had plenty of time to settle back down into the thin disk. The gas could then have begun forming subsequent generations of population I stars. This timing is also consistent with the typical ages of stars in the thick disk.

The Milky Way has more collisions in store. An example is the Canis Major dwarf galaxy, which has a mass of about 1% of the mass of the Milky Way. Already long tidal tails have been stripped from this galaxy, which have wrapped themselves around the Milky Way three times. Several of the globular clusters found in the Milky Way may also have come from the Canis Major dwarf, which is expected to merge gradually with the Milky Way over about the next billion years.

In about 4 billion years, the Milky Way itself will be swallowed up, since it and the Andromeda galaxy are on a collision course. Our computer models show that after a complex interaction, the two will merge to form a larger, more rounded galaxy (Figure 25.25).

Figure 25.25 Collision of the Milky Way with Andromeda. In about 3 billion years, the Milky Way Galaxy and Andromeda Galaxy will begin a long process of colliding, separating, and then coming back together to form an elliptical galaxy. The whole interaction will take 3 to 4 billion years. These computer-simulated images show the following sequence: (1) In 3.75 billion years, Andromeda has approached the Milky Way. (2) New star formation fills the sky 3.85 billion years from now. (3) Star formation continues at 3.9 billion years. (4) The galaxy shapes change as they interact, with Andromeda being stretched and our Galaxy becoming warped, about 4 billion years from now. (5) In 5.1 billion years, the cores of the two galaxies are bright lobes. (6) In 7 billion years, the merged galaxies form a huge elliptical galaxy whose brightness fills the night sky. This artist's illustrations show events from a vantage point 25,000 light-years from the center of the Milky Way. However, we should mention that the Sun may not be at that distance throughout the sequence of events, as the collision readjusts the orbits of many stars within each galaxy. (credit: NASA; ESA; Z. Levay, R. van der Marel, STScl; T. Hallas, and A. Mellinger)

We are thus coming to realize that "environmental influences" (and not just a galaxy's original characteristics) play an important role in determining the properties and development of our Galaxy. In future chapters we will see that collisions and mergers are a major factor in the evolution of many other galaxies as well.

🔑 Key Terms

central bulge   (or nuclear bulge) the central (round) part of the Milky Way or a similar galaxy

dark matter   nonluminous mass, whose presence can be inferred only because of its gravitational influence on luminous matter; the composition of the dark matter is not known

dark matter halo   the mass in the Milky Way that extends well beyond the boundary of the luminous stars to a distance of at least 200,000 light-years from the center of the Galaxy; although we deduce its existence from its gravity, the composition of this matter remains a mystery

differential galactic rotation   the idea that different parts of the Galaxy turn at different rates, since the parts of the Galaxy follow Kepler's third law: more distant objects take longer to complete one full orbit around the center of the Galaxy

halo   the outermost extent of our Galaxy (or another galaxy), containing a sparse distribution of stars and globular clusters in a more or less spherical distribution

Milky Way Galaxy   the band of light encircling the sky, which is due to the many stars and diffuse nebulae lying near the plane of the Milky Way Galaxy

population I star   a star containing heavy elements; typically young and found in the disk

population II star   a star with very low abundance of heavy elements; found throughout the Galaxy

spiral arm   a spiral-shaped region, characterized by relatively dense interstellar material and young stars, that is observed in the disks of spiral galaxies

supermassive black hole   the object in the center of most large galaxies that is so massive and compact that light cannot escape from it; the Milky Way's supermassive black hole contains 4.6 millions of Suns' worth of mass

📖 Summary

25.1 The Architecture of the Galaxy

The Milky Way Galaxy consists of a thin disk containing dust, gas, and young and old stars; a spherical halo containing populations of very old stars, including RR Lyrae variable stars and globular star clusters; a thick, more diffuse disk with stars that have properties intermediate between those in the thin disk and the halo; a peanut-shaped nuclear bulge of mostly old stars around the center; and a supermassive black hole at the very center. The Sun is located roughly halfway out of the Milky Way, about 26,000 light-years from the center.

25.2 Spiral Structure

The gaseous distribution in the Galaxy's disk has two main spiral arms that emerge from the ends of the central bar, along with several fainter arms and short spurs; the Sun is located in one of those spurs. Measurements show that the Galaxy does not rotate as a solid body, but instead its stars and gas follow differential rotation, such that the material closer to the galactic center completes its orbit more quickly. Observations show that galaxies like the Milky Way take several billion years after they began to form to develop spiral structure.

25.3 The Mass of the Galaxy

The Sun revolves completely around the galactic center in about 225 million years (a galactic year). The mass of the Galaxy can be determined by measuring the orbital velocities of stars and interstellar matter. The total mass of the Galaxy is about 2×10^{12} M_{Sun}. As much as 95% of this mass consists of dark matter that emits no electromagnetic radiation and can be detected only because of the gravitational force it exerts on visible stars and interstellar matter. This dark matter is located mostly in the Galaxy's halo; its nature is not well understood at present.

25.4 The Center of the Galaxy

A supermassive black hole is located at the center of the Galaxy. Measurements of the velocities of stars located within a few light-days of the center show that the mass inside their orbits around the center is about 4.6 million M_{Sun}. Radio observations show that this mass is concentrated in a volume with a diameter similar to that of Mercury's orbit. The density of this matter concentration exceeds that of the densest known star clusters by a factor of nearly a million. The only known object with such a high density and total mass is a black hole.

25.5 Stellar Populations in the Galaxy

We can roughly divide the stars in the Galaxy into two categories. Old stars with few heavy elements are referred to as population II stars and are found in the halo and in globular clusters. Population I stars contain more heavy elements than globular cluster and halo stars, are typically younger and found in the disk, and are especially concentrated in the spiral arms. The Sun is a member of population I. Population I stars formed after previous generations of stars had produced heavy elements and ejected them into the interstellar medium. The bulge stars, most of which are more than 10 billion years old, have unusually high amounts of heavy elements, presumably because there were many massive first-generation stars in this dense region, and these quickly seeded the next generations of stars with heavier elements.

25.6 The Formation of the Galaxy

The Galaxy began forming a little more than 13 billion years ago. Models suggest that the stars in the halo and globular clusters formed first, while the Galaxy was spherical. The gas, somewhat enriched in heavy elements by the first generation of stars, then collapsed from a spherical distribution to a rotating disk-shaped distribution. Stars are still forming today from the gas and dust that remain in the disk. Star formation occurs most rapidly in the spiral arms, where the density of interstellar matter is highest. The Galaxy captured (and still is capturing) additional stars and globular clusters from small galaxies that ventured too close to the Milky Way. In 3 to 4 billion years, the Galaxy will begin to collide with the Andromeda galaxy, and after about 7 billion years, the two galaxies will merge to form a giant elliptical galaxy.

 For Further Exploration

Articles

Blitz, L. "The Dark Side of the Milky Way." *Scientific American* (October 2011): 36–43. How we find dark matter and what it tells us about our Galaxy, its warped disk, and its satellite galaxies.

Dvorak, J. "Journey to the Heart of the Milky Way." *Astronomy* (February 2008): 28. Measuring nearby stars to determine the properties of the black hole at the center.

Gallagher, J., Wyse, R., & Benjamin, R. "The New Milky Way." *Astronomy* (September 2011): 26. Highlights all aspects of the Milky Way based on recent observations.

Goldstein, A. "Finding our Place in the Milky Way." *Astronomy* (August 2015): 50. On the history of observations that pinpointed the Sun's location in the Galaxy.

Haggard, D., & Bower, G. "In the Heart of the Milky Way." *Sky & Telescope* (February 2016): 16. On observations of the Galaxy's nucleus and the supermassive black hole and magnetar there.

Ibata, R., & Gibson, B. "The Ghosts of Galaxies Past." *Scientific American* (April 2007): 40. About star streams in the Galaxy that are evidence of past mergers and collisions.

Irion, R. "A Crushing End for Our Galaxy." *Science* (January 7, 2000): 62. On the role of mergers in the evolution of the Milky Way.

Irion, R. "Homing in on Black Holes." *Smithsonian* (April 2008). On how astronomers probe the large black hole at the center of the Milky Way Galaxy.

Kruesi, L. "How We Mapped the Milky Way." *Astronomy* (October 2009): 28.

Kruesi, L. "What Lurks in the Monstrous Heart of the Milky Way?" *Astronomy* (October 2015): 30. On the center of the Galaxy and the black hole there.

Laughlin, G., & Adams, F. "Celebrating the Galactic Millennium." *Astronomy* (November 2001): 39. The long-term future of the Milky Way in the next 90 billion years.

Loeb, A., & Cox, T.J. "Our Galaxy's Date with Destruction." *Astronomy* (June 2008): 28. Describes the upcoming merger of Milky Way and Andromeda.

Szpir, M. "Passing the Bar Exam." *Astronomy* (March 1999): 46. On evidence that our Galaxy is a barred spiral.

Tanner, A. "A Trip to the Galactic Center." *Sky & Telescope* (April 2003): 44. Nice introduction, with observations pointing to the presence of a black hole.

Trimble, V., & Parker, S. "Meet the Milky Way." *Sky & Telescope* (January 1995): 26. Overview of our Galaxy.

Wakker, B., & Richter, P. "Our Growing, Breathing Galaxy." *Scientific American* (January 2004): 38. Evidence that our Galaxy is still being built up by the addition of gas and smaller neighbors.

Waller, W. "Redesigning the Milky Way." *Sky & Telescope* (September 2004): 50. On recent multi-wavelength surveys of the Galaxy.

Whitt, K. "The Milky Way from the Inside." *Astronomy* (November 2001): 58. Fantastic panorama image of the Galaxy, with finder charts and explanations.

Websites

International Dark Sky Sanctuaries: http://darksky.org/idsp/sanctuaries/ (http://darksky.org/idsp/sanctuaries/). A listing of dark-sky sanctuaries, parks, and reserves.

Multiwavelength Milky Way: https://asd.gsfc.nasa.gov/archive/mwmw/mmw_edu.html (https://asd.gsfc.nasa.gov/archive/mwmw/mmw_edu.html). This NASA site shows the plane of our Galaxy in a variety of wavelength bands, and includes background material and other resources.

Shapley-Curtis Debate in 1920: https://apod.nasa.gov/debate/debate20.html (https://apod.nasa.gov/debate/debate20.html). In 1920, astronomers Harlow Shapley and Heber Curtis engaged in a historic debate about how large our Galaxy was and whether other galaxies existed. Here you can find historical and educational material about the debate.

UCLA Galactic Center Group: http://www.galacticcenter.astro.ucla.edu/ (http://www.galacticcenter.astro.ucla.edu/). Learn more about the work of Andrea Ghez and colleagues on the central region of the Milky Way Galaxy.

Videos

Crash of the Titans: http://www.spacetelescope.org/videos/hubblecast55a/ (http://www.spacetelescope.org/videos/hubblecast55a/). This Hubblecast from 2012 features Jay Anderson and Roeland van der Marel explaining how Andromeda will collide with the Milky Way in the distant future (5:07).

Diner at the Center of the Galaxy: https://www.youtube.com/watch?v=UP7ig8Gxftw (https://www.youtube.com/watch?v=UP7ig8Gxftw). A short discussion from NASA ScienceCast of NuSTAR observations of flares from our Galaxy's central black hole (3:23).

Hunt for a Supermassive Black Hole: https://www.ted.com/talks/andrea_ghez_the_hunt_for_a_supermassive_black_hole (https://www.ted.com/talks/andrea_ghez_the_hunt_for_a_supermassive_black_hole). 2009 TED talk by Andrea Ghez on searching for supermassive black holes, particularly the one at the center of the Milky Way (16:19).

Journey to the Galactic Center: https://www.youtube.com/watch?v=36xZsgZ0oSo (https://www.youtube.com/

watch?v=36xZsgZ0oSo). A brief silent trip into the cluster of stars near the galactic center showing their motions around the center (3:00).

The Nobel Prize Lecture by Dr. Andrea Ghez in 2020: https://www.youtube.com/watch?v=wGw6_CdvGKM (https://www.youtube.com/watch?v=wGw6_CdvGKM).

Collaborative Group Activities

A. You are captured by space aliens, who take you inside a complex cloud of interstellar gas, dust, and a few newly formed stars. To escape, you need to make a map of the cloud. Luckily, the aliens have a complete astronomical observatory with equipment for measuring all the bands of the electromagnetic spectrum. Using what you have learned in this chapter, have your group discuss what kinds of maps you would make of the cloud to plot your most effective escape route.

B. The diagram that Herschel made of the Milky Way has a very irregular outer boundary (see Figure 25.3). Can your group think of a reason for this? How did Herschel construct his map?

C. Suppose that for your final exam in this course, your group is assigned telescope time to observe a star selected for you by your professor. The professor tells you the position of the star in the sky (its right ascension and declination) but nothing else. You can make any observations you wish. How would you go about determining whether the star is a member of population I or population II?

D. The existence of dark matter comes as a great surprise, and its nature remains a mystery today. Someday astronomers will know a lot more about it (you can learn more about current findings in The Evolution and Distribution of Galaxies). Can your group make a list of earlier astronomical observations that began as a surprise and mystery, but wound up (with more observations) as well-understood parts of introductory textbooks?

E. Physicist Gregory Benford has written a series of science fiction novels that take place near the center of the Milky Way Galaxy in the far future. Suppose your group were writing such a story. Make a list of ways that the environment near the galactic center differs from the environment in the "galactic suburbs," where the Sun is located. Would life as we know it have an easier or harder time surviving on planets that orbit stars near the center (and why)?

F. These days, in most urban areas, city lights completely swamp the faint light of the Milky Way in our skies. Have each member of your group survey 5 to 10 friends or relatives (you could spread out on campus to investigate or use social media or the phone), explaining what the Milky Way is and then asking if they have seen it. Also ask their age. Report back to your group and discuss your reactions to the survey. Is there any relationship between a person's age and whether they have seen the Milky Way? How important is it that many kids growing up on Earth today never (or rarely) get to see our home Galaxy in the sky?

Exercises

Review Questions

1. Explain why we see the Milky Way as a faint band of light stretching across the sky.

2. Explain where in a spiral galaxy you would expect to find globular clusters, molecular clouds, and atomic hydrogen.

3. Describe several characteristics that distinguish population I stars from population II stars.

4. Briefly describe the main parts of our Galaxy.

5. Describe the evidence indicating that a black hole may be at the center of our Galaxy.

6. Explain why the abundances of heavy elements in stars correlate with their positions in the Galaxy.

7. What will be the long-term future of our Galaxy?

Thought Questions

8. Suppose the Milky Way was a band of light extending only halfway around the sky (that is, in a semicircle). What, then, would you conclude about the Sun's location in the Galaxy? Give your reasoning.

9. Suppose somebody proposed that rather than invoking dark matter to explain the increased orbital velocities of stars beyond the Sun's orbit, the problem could be solved by assuming that the Milky Way's central black hole was much more massive. Does simply increasing the assumed mass of the Milky Way's central supermassive black hole correctly resolve the issue of unexpectedly high orbital velocities in the Galaxy? Why or why not?

10. The globular clusters revolve around the Galaxy in highly elliptical orbits. Where would you expect the clusters to spend most of their time? (Think of Kepler's laws.) At any given time, would you expect most globular clusters to be moving at high or low speeds with respect to the center of the Galaxy? Why?

11. Shapley used the positions of globular clusters to determine the location of the galactic center. Could he have used open clusters? Why or why not?

12. Consider the following five kinds of objects: open cluster, giant molecular cloud, globular cluster, group of O and B stars, and planetary nebulae.
A. Which occur only in spiral arms?
B. Which occur only in the parts of the Galaxy other than the spiral arms?
C. Which are thought to be very young?
D. Which are thought to be very old?
E. Which have the hottest stars?

13. The dwarf galaxy in Sagittarius is the one closest to the Milky Way, yet it was discovered only in 1994. Can you think of a reason it was not discovered earlier? (Hint: Think about what else is in its constellation.)

14. Suppose three stars lie in the disk of the Galaxy at distances of 20,000 light-years, 25,000 light-years, and 30,000 light-years from the galactic center, and suppose that right now all three are lined up in such a way that it is possible to draw a straight line through them and on to the center of the Galaxy. How will the relative positions of these three stars change with time? Assume that their orbits are all circular and lie in the plane of the disk.

15. Why does star formation occur primarily in the disk of the Galaxy?

16. Where in the Galaxy would you expect to find Type II supernovae, which are the explosions of massive stars that go through their lives very quickly? Where would you expect to find Type I supernovae, which involve the explosions of white dwarfs?

17. Suppose that stars evolved without losing mass—that once matter was incorporated into a star, it remained there forever. How would the appearance of the Galaxy be different from what it is now? Would there be population I and population II stars? What other differences would there be?

Figuring for Yourself

18. Assume that the Sun orbits the center of the Galaxy at a speed of 220 km/s and a distance of 26,000 light-years from the center.
A. Calculate the circumference of the Sun's orbit, assuming it to be approximately circular. (Remember that the circumference of a circle is given by $2\pi R$, where R is the radius of the circle. Be sure to use consistent units. The conversion from light-years to km/s can be found in an online calculator or appendix, or you can calculate it for yourself: the speed of light is 300,000 km/s, and you can determine the number of seconds in a year.)
B. Calculate the Sun's period, the "galactic year." Again, be careful with the units. Does it agree with the number we gave above?

19. The Sun orbits the center of the Galaxy in 225 million years at a distance of 26,000 light-years. Given that $a^3 = (M_1 + M_2) \times P^2$, where a is the semimajor axis and P is the orbital period, what is the mass of the Galaxy within the Sun's orbit?

20. Suppose the Sun orbited a little farther out, but the mass of the Galaxy inside its orbit remained the same as we calculated in Exercise 25.19. What would be its period at a distance of 30,000 light-years?

21. We have said that the Galaxy rotates differentially; that is, stars in the inner parts complete a full 360° orbit around the center of the Galaxy more rapidly than stars farther out. Use Kepler's third law and the mass we derived in Exercise 25.19 to calculate the period of a star that is only 5000 light-years from the center. Now do the same calculation for a globular cluster at a distance of 50,000 light-years. Suppose the Sun, this star, and the globular cluster all fall on a straight line through the center of the Galaxy. Where will they be relative to each other after the Sun completes one full journey around the center of the Galaxy? (Assume that all the mass in the Galaxy is concentrated at its center.)

22. If our solar system is 4.6 billion years old, how many galactic years has planet Earth been around?

23. Suppose the average mass of a star in the Galaxy is one-third of a solar mass. Use the value for the mass of the Galaxy that we calculated in Exercise 25.19, and estimate how many stars are in the Milky Way. Give some reasons it is reasonable to assume that the mass of an average star is less than the mass of the Sun.

24. The first clue that the Galaxy contains a lot of dark matter was the observation that the orbital velocities of stars did not decreases with increasing distance from the center of the Galaxy. Construct a rotation curve for the solar system by using the orbital velocities of the planets, which can be found in Appendix F. How does this curve differ from the rotation curve for the Galaxy? What does it tell you about where most of the mass in the solar system is concentrated?

25. The best evidence for a black hole at the center of the Galaxy also comes from the application of Kepler's third law. Suppose a star at a distance of 20 light-hours from the center of the Galaxy has an orbital speed of 6200 km/s. How much mass must be located inside its orbit?

26. The next step in deciding whether the object in Exercise 25.25 is a black hole is to estimate the density of this mass. Assume that all of the mass is spread uniformly throughout a sphere with a radius of 20 light-hours. What is the density in kg/km³? (Remember that the volume of a sphere is given by $V = \frac{4}{3}\pi R^3$.) Explain why the density might be even higher than the value you have calculated. How does this density compare with that of the Sun or other objects we have talked about in this book?

27. Suppose the Sagittarius dwarf galaxy merges completely with the Milky Way and adds 150,000 stars to it. Estimate the percentage change in the mass of the Milky Way. Will this be enough mass to affect the orbit of the Sun around the galactic center? Assume that all of the Sagittarius galaxy's stars end up in the nuclear bulge of the Milky Way Galaxy and explain your answer.

Figure 26.1 Spiral Galaxy. NGC 6946 is a spiral galaxy also known as the "Fireworks galaxy." It is at a distance of about 18 million light-years, in the direction of the constellations Cepheus and Cygnus. It was discovered by William Herschel in 1798. This galaxy is about one-third the size of the Milky Way. Note on the left how the colors of the galaxy change from the yellowish light of old stars in the center to the blue color of hot, young stars and the reddish glow of hydrogen clouds in the spiral arms. As the image shows, this galaxy is rich in dust and gas, and new stars are still being born here. In the right-hand image, the x-rays coming from this galaxy are shown in purple, which has been added to other colors showing visible light. (Credit left: modification of work by NASA, ESA, STScI, R. Gendler, and the Subaru Telescope (NAOJ); credit right: modification of work by X-ray: NASA/CXC/MSSL/R.Soria et al, Optical: AURA/Gemini OBs)

Chapter Outline

26.1 The Discovery of Galaxies
26.2 Types of Galaxies
26.3 Properties of Galaxies
26.4 The Extragalactic Distance Scale
26.5 The Expanding Universe

 ## Thinking Ahead

In the last chapter, we explored our own Galaxy. But is it the only one? If there are others, are they like the Milky Way? How far away are they? Can we see them? As we shall learn, some galaxies turn out to be so far away that it has taken billions of years for their light to reach us. These remote galaxies can tell us what the universe was like when it was young.

In this chapter, we start our exploration of the vast realm of galaxies. Like tourists from a small town making their first visit to the great cities of the world, we will be awed by the beauty and variety of the galaxies. And yet, we will recognize that much of what we see is not so different from our experiences at home, and we will be impressed by how much we can learn by looking at structures built long ago.

We begin our voyage with a guide to the properties of galaxies, much as a tourist begins with a guidebook to the main features of the cities on the itinerary. In later chapters, we will look more carefully at the past history of galaxies, how they have changed over time, and how they acquired their many different forms. First, we'll begin our voyage through the galaxies with the question: is our Galaxy the only one?

26.1 | The Discovery of Galaxies

Learning Objectives

By the end of this section, you will be able to:

> Describe the discoveries that confirmed the existence of galaxies that lie far beyond the Milky Way Galaxy
> Explain why galaxies used to be called nebulae and why we don't include them in that category any more

Growing up at a time when the Hubble Space Telescope orbits above our heads and giant telescopes are springing up on the great mountaintops of the world, you may be surprised to learn that we were not sure about the existence of other galaxies for a very long time. The very idea that other galaxies exist used to be controversial. Even into the 1920s, many astronomers thought the Milky Way encompassed *all* that exists in the universe. The evidence found in 1924 that meant our Galaxy is not alone was one of the great scientific discoveries of the twentieth century.

It was not that scientists weren't asking questions. They questioned the composition and structure of the universe as early as the eighteenth century. However, with the telescopes available in earlier centuries, galaxies looked like small fuzzy patches of light that were difficult to distinguish from the star clusters and gas-and-dust clouds that are part of our own Galaxy. All objects that were not sharp points of light were given the same name, *nebulae*, the Latin word for "clouds." Because their precise shapes were often hard to make out and no techniques had yet been devised for measuring their distances, the nature of the nebulae was the subject of much debate.

As early as the eighteenth century, the philosopher Immanuel Kant (1724–1804) suggested that some of the nebulae might be distant systems of stars (other Milky Ways), but the evidence to support this suggestion was beyond the capabilities of the telescopes of that time.

Other Galaxies

By the early twentieth century, some nebulae had been correctly identified as star clusters, and others (such as the Orion Nebula) as gaseous nebulae. Most nebulae, however, looked faint and indistinct, even with the best telescopes, and their distances remained unknown. (For more on how such nebulae are named, by the way, see the feature box on Naming the Nebulae in the chapter on interstellar matter.) If these nebulae were nearby, with distances comparable to those of observable stars, they were most likely clouds of gas or groups of stars within our Galaxy. If, on the other hand, they were remote, far beyond the edge of the Galaxy, they could be other star systems containing billions of stars.

To determine what the nebulae are, astronomers had to find a way of measuring the distances to at least some of them. When the 2.5-meter (100-inch) telescope on Mount Wilson in Southern California went into operation, astronomers finally had the large telescope they needed to settle the controversy.

Working with the 2.5-meter telescope, Edwin Hubble was able to resolve individual stars in several of the brighter spiral-shaped nebulae, including M31, the great spiral in Andromeda (Figure 26.2). Among these stars, he discovered some faint variable stars that—when he analyzed their light curves—turned out to be cepheids. Here were reliable indicators that Hubble could use to measure the distances to the nebulae using the technique pioneered by Henrietta Leavitt (see the chapter on Celestial Distances). After painstaking work, he estimated that the Andromeda galaxy was about 900,000 light-years away from us. At that enormous distance, it had to be a separate galaxy of stars located well outside the boundaries of the Milky Way. Today, we know the Andromeda galaxy is actually slightly more than twice as distant as Hubble's first estimate, but his conclusion about its true nature remains unchanged.

Figure 26.2 Andromeda Galaxy. Also known by its catalog number M31, the Andromeda galaxy is a large spiral galaxy very similar in appearance to, and slightly larger than, our own Galaxy. At a distance of about 2.5 million light-years, Andromeda is the spiral galaxy that is nearest to our own in space. Here, it is seen with two of its satellite galaxies, M32 (top) and M110 (bottom). (credit: Adam Evans)

No one in human history had ever measured a distance so great. When Hubble's paper on the distances to nebulae was read before a meeting of the American Astronomical Society on the first day of 1925, the entire room erupted in a standing ovation. A new era had begun in the study of the universe, and a new scientific field—extragalactic astronomy—had just been born.

VOYAGERS IN ASTRONOMY

Edwin Hubble: Expanding the Universe

The son of a Missouri insurance agent, Edwin Hubble (Figure 26.3) graduated from high school at age 16. He excelled in sports, winning letters in track and basketball at the University of Chicago, where he studied both science and languages. Both his father and grandfather wanted him to study law, however, and he gave in to family pressure. He received a prestigious Rhodes scholarship to Oxford University in England, where he studied law with only middling enthusiasm. Returning to be the United States, he spent a year teaching high school physics and Spanish as well as coaching basketball, while trying to determine his life's direction.

Figure 26.3 Edwin Hubble (1889–1953). Edwin Hubble established some of the most important ideas in the study of galaxies.

The pull of astronomy eventually proved too strong to resist, and so Hubble went back to the University of Chicago for graduate work. Just as he was about to finish his degree and accept an offer to work at the soon-to-be-completed 2.5-meter telescope, the United States entered World War I, and Hubble enlisted as an officer. Although the war had ended by the time he arrived in Europe, he received more officer's training abroad and enjoyed a brief time of further astronomical study at Cambridge before being sent home.

In 1919, at age 30, he joined the staff at Mount Wilson and began working with the world's largest telescope. Ripened by experience, energetic, disciplined, and a skillful observer, Hubble soon established some of the most important ideas in modern astronomy. He showed that other galaxies existed, classified them on the basis of their shapes, found a pattern to their motion (and thus put the notion of an expanding universe on a firm observational footing), and began a lifelong program to study the distribution of galaxies in the universe. Although a few others had glimpsed pieces of the puzzle, it was Hubble who put it all together and showed that an understanding of the large-scale structure of the universe was feasible.

His work brought Hubble much renown and many medals, awards, and honorary degrees. As he became better known (he was the first astronomer to appear on the cover of *Time* magazine), he and his wife enjoyed and cultivated friendships with movie stars and writers in Southern California. Hubble was instrumental (if you'll pardon the pun) in the planning and building of the 5-meter telescope on Palomar Mountain, and he had begun to use it for studying galaxies when he passed away from a stroke in 1953.

When astronomers built a space telescope that would allow them to extend Hubble's work to distances he could only dream about, it seemed natural to name it in his honor. It was fitting that observations with the Hubble Space Telescope (and his foundational work on expansion of the universe) contributed to the 2011 Nobel Prize in Physics, given for the discovery that the expansion of the universe is accelerating (a topic we will expand upon in the chapter on The Big Bang).

Types of Galaxies

Learning Objectives

By the end of this section, you will be able to:

> Describe the properties and features of elliptical, spiral, and irregular galaxies
> Explain what may cause a galaxy's appearance to change over time

Having established the existence of other galaxies, Hubble and others began to observe them more closely—noting their shapes, their contents, and as many other properties as they could measure. This was a daunting task in the 1920s when obtaining a single photograph or spectrum of a galaxy could take a full night of tireless observing. Today, larger telescopes and electronic detectors have made this task less difficult, although observing the most distant galaxies (those that show us the universe in its earliest phases) still requires enormous effort.

The first step in trying to understand a new type of object is often simply to describe it. Remember, the first step in understanding stellar spectra was simply to sort them according to appearance (see Analyzing Starlight). As it turns out, the biggest and most luminous galaxies come in one of two basic shapes: either they are flatter and have spiral arms, like our own Galaxy, or they appear to be elliptical (blimp- or cigar-shaped). Many smaller galaxies, in contrast, have an irregular shape.

Spiral Galaxies

Our own Galaxy and the Andromeda galaxy are typical, large **spiral galaxies** (see Figure 26.2). They consist of a central bulge, a halo, a disk, and spiral arms. Interstellar material is usually spread throughout the disks of spiral galaxies. Bright emission nebulae and hot, young stars are present, especially in the spiral arms, showing that new star formation is still occurring. The disks are often dusty, which is especially noticeable in those systems that we view almost edge on (Figure 26.4).

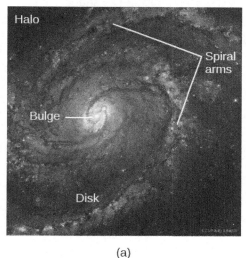

(a)   (b)

Figure 26.4 Spiral Galaxies. (a) The spiral arms of M100, shown here, are bluer than the rest of the galaxy, indicating young, high-mass stars and star-forming regions. (b) We view this spiral galaxy, NGC 4565, almost exactly edge on, and from this angle, we can see the dust in the plane of the galaxy; it appears dark because it absorbs the light from the stars in the galaxy. (credit a: modification of work by Hubble Legacy Archive, NASA, ESA, and Judy Schmidt; credit b: modification of work by "Jschulman555"/ Wikimedia)

In galaxies that we see face on, the bright stars and emission nebulae make the arms of spirals stand out like those of a pinwheel on the fourth of July. Open star clusters can be seen in the arms of nearer spirals, and globular clusters are often visible in their halos. Spiral galaxies contain a mixture of young and old stars, just as the Milky Way does. All spirals rotate, and the direction of their spin is such that the arms appear to trail much like the wake of a boat.

About two-thirds of the nearby spiral galaxies have boxy or peanut-shaped bars of stars running through their centers (Figure 26.5). Showing great originality, astronomers call these galaxies barred spirals.

Figure 26.5 Barred Spiral Galaxy. NGC 1300, shown here, is a barred spiral galaxy. Note that the spiral arms begin at the ends of the bar. (credit: NASA, ESA, and the Hubble Heritage Team(STScI/AURA))

As we noted in The Milky Way Galaxy chapter, our Galaxy has a modest bar too (see Figure 25.10). The spiral arms usually begin from the ends of the bar. The fact that bars are so common suggests that they are long lived; it may be that most spiral galaxies form a bar at some point during their evolution.

In both barred and unbarred spiral galaxies, we observe a range of different shapes. At one extreme, the central bulge is large and luminous, the arms are faint and tightly coiled, and bright emission nebulae and supergiant stars are inconspicuous. Hubble, who developed a system of classifying galaxies by shape, gave these galaxies the designation Sa. Galaxies at this extreme may have no clear spiral arm structure, resulting in a lens-like appearance (they are sometimes referred to as lenticular galaxies). These galaxies seem to share as many properties with elliptical galaxies as they do with spiral galaxies

At the other extreme, the central bulge is small and the arms are loosely wound. In these Sc galaxies, luminous stars and emission nebulae are very prominent. Our Galaxy and the Andromeda galaxy are both intermediate between the two extremes. Photographs of spiral galaxies, illustrating the different types, are shown in Figure 26.6, along with elliptical galaxies for comparison.

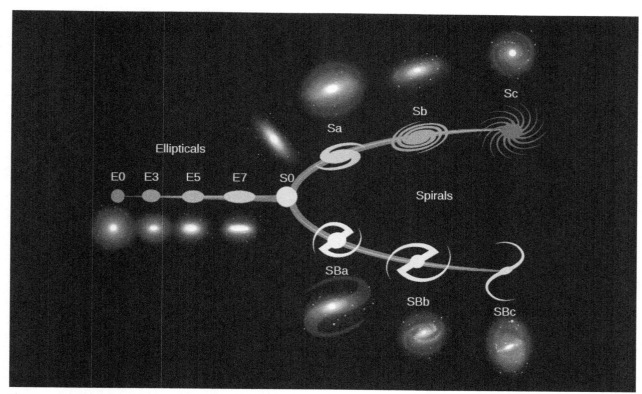

Figure 26.6 Hubble Classification of Galaxies. This figure shows Edwin Hubble's original classification of galaxies. Elliptical galaxies are on the left. On the right, you can see the basic spiral shapes illustrated, alongside images of actual barred and unbarred spirals. (credit: modification of work by NASA, ESA)

The luminous parts of spiral galaxies appear to range in diameter from about 20,000 to more than 100,000 light-years. Recent studies have found that there is probably a large amount of galactic material that extends well beyond the apparent edge of galaxies. This material appears to be thin, cold gas that is difficult to detect in most observations.

From the observational data available, the masses of the visible portions of spiral galaxies are estimated to range from 1 billion to 1 trillion Suns (10^9 to 10^{12} M_{Sun}). The total luminosities of most spirals fall in the range of 100 million to 100 billion times the luminosity of our Sun (10^8 to 10^{11} L_{Sun}). Our Galaxy and M31 are relatively large and massive, as spirals go. There is also considerable dark matter in and around the galaxies, just as there is in the Milky Way; we deduce its presence from how fast stars in the outer parts of the Galaxy are moving in their orbits.

Elliptical Galaxies

Elliptical galaxies consist almost entirely of old stars and have shapes that are spheres or ellipsoids (somewhat squashed spheres) (Figure 26.7). They contain no trace of spiral arms. Their light is dominated by older reddish stars (the population II stars discussed in The Milky Way Galaxy). In the larger nearby ellipticals, many globular clusters can be identified. Dust and emission nebulae are not conspicuous in elliptical galaxies, but many do contain a small amount of interstellar matter.

| (a) | (b) |

Figure 26.7 Elliptical Galaxies. (a) ESO 325-G004 is a giant elliptical galaxy. Other elliptical galaxies can be seen around the edges of this image. (b) This elliptical galaxy probably originated from the collision of two spiral galaxies. (credit a: modification of work by NASA, ESA, and The Hubble Heritage Team (STScI/AURA); credit b: modification of work by ESA/Hubble, NASA)

Elliptical galaxies show various degrees of flattening, ranging from systems that are approximately spherical to those that approach the flatness of spirals. The rare giant ellipticals (for example, ESO 325-G004 in Figure 26.7) reach luminosities of 10^{11} L_{Sun}. The mass in a giant elliptical can be as large as 10^{13} M_{Sun}. The diameters of these large galaxies extend over several hundred thousand light-years and are considerably larger than the largest spirals. Although individual stars orbit the center of an elliptical galaxy, the orbits are not all in the same direction, as occurs in spirals. Therefore, ellipticals don't appear to rotate in a systematic way, making it difficult to estimate how much dark matter they contain.

We find that elliptical galaxies range all the way from the giants, just described, to dwarfs, which may be the most common kind of galaxy. *Dwarf ellipticals* (sometimes called dwarf spheroidals) escaped our notice for a long time because they are very faint and difficult to see. An example of a dwarf elliptical is the Leo I Dwarf Spheroidal galaxy shown in Figure 26.8. The luminosity of this typical dwarf is about equal to that of the brightest globular clusters.

Intermediate between the giant and dwarf elliptical galaxies are systems such as M32 and M110, the two companions of the Andromeda galaxy. While they are often referred to as dwarf ellipticals, these galaxies are significantly larger than galaxies such as Leo I.

Figure 26.8 Dwarf Elliptical Galaxy. M32, a dwarf elliptical galaxy and one of the companions to the giant Andromeda galaxy M31. M32 is a dwarf by galactic standards, as it is only 2400 light-years across. (credit: NOAO/AURA/NSF)

Irregular Galaxies

Hubble classified galaxies that do not have the regular shapes associated with the categories we just described into the catchall bin of an **irregular galaxy**, and we continue to use his term. Typically, irregular galaxies have lower masses and luminosities than spiral galaxies. Irregular galaxies often appear disorganized, and many are undergoing relatively intense star formation activity. They contain both young population I stars and old population II stars.

The two best-known irregular galaxies are the Large Magellanic Cloud and Small Magellanic Cloud (Figure 26.9), which are at a distance of a little more than 160,000 light-years away and are among our nearest extragalactic neighbors. Their names reflect the fact that Ferdinand Magellan and his crew, making their round-the-world journey, were the first European travelers to notice them. Although not visible from the United States and Europe, these two systems are prominent from the Southern Hemisphere, where they look like wispy clouds in the night sky. Since they are only about one-tenth as distant as the Andromeda galaxy, they present an excellent opportunity for astronomers to study nebulae, star clusters, variable stars, and other key objects in the setting of another galaxy. For example, the Large Magellanic Cloud contains the 30 Doradus complex (also known as the Tarantula Nebula), one of the largest and most luminous groups of supergiant stars known in any galaxy.

Figure 26.9 4-Meter Telescope at Cerro Tololo Inter-American Observatory Silhouetted against the Southern Sky. The Milky Way is seen to the right of the dome, and the Large and Small Magellanic Clouds are seen to the left. (credit: Roger Smith/NOAO/AURA/NSF)

The Small Magellanic Cloud is considerably less massive than the Large Magellanic Cloud, and it is six times longer than it is wide. This narrow wisp of material points directly toward our Galaxy like an arrow. The Small Magellanic Cloud was most likely contorted into its current shape through gravitational interactions with the Milky Way. A large trail of debris from this interaction between the Milky Way and the Small Magellanic Cloud has been strewn across the sky and is seen as a series of gas clouds moving at abnormally high velocity, known as the Magellanic Stream. We will see that this kind of interaction between galaxies will help explain the irregular shapes of this whole category of small galaxies,

LINK TO LEARNING

View this beautiful album showcasing the different types of galaxies (https://openstax.org/l/30galaxphohubb) that have been photographed by the Hubble Space Telescope.

Galaxy Evolution

Encouraged by the success of the H-R diagram for stars (see Analyzing Starlight), astronomers studying galaxies hoped to find some sort of comparable scheme, where differences in appearance could be tied to different evolutionary stages in the life of galaxies. Wouldn't it be nice if every elliptical galaxy evolved into a spiral, for example, just as every main-sequence star evolves into a red giant? Several simple ideas of this kind were tried, some by Hubble himself, but none stood the test of time (and observation).

Because no simple scheme for evolving one type of galaxy into another could be found, astronomers then tended to the opposite point of view. For a while, most astronomers thought that all galaxies formed very early in the history of the universe and that the differences between them had to do with the rate of star formation. Ellipticals were those galaxies in which all the interstellar matter was converted rapidly into stars. Spirals were galaxies in which star formation occurred slowly over the entire lifetime of the galaxy. This idea turned out to be too simple as well.

Today, we understand that at least some galaxies have changed types over the billions of years since the universe began. As we shall see in later chapters, collisions and mergers between galaxies may dramatically change spiral galaxies into elliptical galaxies. Even isolated spirals (with no neighbor galaxies in sight) can change their appearance over time. As they consume their gas, the rate of star formation will slow down, and the spiral arms will gradually become less conspicuous. Over long periods, spirals therefore begin to look more like the galaxies at the middle of Figure 26.6 (which astronomers refer to as S0 types).

Over the past several decades, the study of how galaxies evolve over the lifetime of the universe has become one of the most active fields of astronomical research. We will discuss the evolution of galaxies in more detail in The Evolution and Distribution of Galaxies, but let's first see in a little more detail just what different galaxies are like.

26.3 | Properties of Galaxies

Learning Objectives

By the end of this section, you will be able to:
> Describe the methods through which astronomers can estimate the mass of a galaxy
> Characterize each type of galaxy by its mass-to-light ratio

The technique for deriving the masses of galaxies is basically the same as that used to estimate the mass of the Sun, the stars, and our own Galaxy. We measure how fast objects in the outer regions of the galaxy are orbiting the center, and then we use this information along with Kepler's third law to calculate how much mass is inside that orbit.

Masses of Galaxies

Astronomers can measure the rotation speed in spiral galaxies by obtaining spectra of either stars or gas, and looking for wavelength shifts produced by the Doppler effect. Remember that the faster something is moving toward or away from us, the greater the shift of the lines in its spectrum. Kepler's law, together with such observations of the part of the Andromeda galaxy that is bright in visible light, for example, show it to have a galactic mass of about 4×10^{11} M_{Sun} (enough material to make 400 billion stars like the Sun).

The total mass of the Andromeda galaxy is greater than this, however, because we have not included the mass of the material that lies beyond its visible edge. Fortunately, there is a handful of objects—such as isolated stars, star clusters, and satellite galaxies—beyond the visible edge that allows astronomers to estimate how much additional matter is hidden out there. Recent studies show that the amount of dark matter beyond the visible edge of Andromeda may be as large as the mass of the bright portion of the galaxy. Indeed, using Kepler's third law and the velocities of its satellite galaxies, the Andromeda galaxy is estimated to have a mass closer to 1.4×10^{12} M_{Sun}. The mass of the Milky Way Galaxy is estimated to be 8.5×10^{11} M_{Sun}, and so our Milky Way is turning out to be somewhat smaller than Andromeda.

Elliptical galaxies do not rotate in a systematic way, so we cannot determine a rotational velocity; therefore, we must use a slightly different technique to measure their mass. Their stars are still orbiting the galactic center, but not in the organized way that characterizes spirals. Since elliptical galaxies contain stars that are billions of years old, we can assume that the galaxies themselves are not flying apart. Therefore, if we can measure the various speeds with which the stars are moving in their orbits around the center of the galaxy, we can calculate how much mass the galaxy must contain in order to hold the stars within it.

In practice, the spectrum of a galaxy is a composite of the spectra of its many stars, whose different motions produce different Doppler shifts (some red, some blue). The result is that the lines we observe from the entire galaxy contain the combination of many Doppler shifts. When some stars provide blueshifts and others provide redshifts, they create a wider or broader absorption or emission feature than would the same lines in a hypothetical galaxy in which the stars had no orbital motion. Astronomers call this phenomenon line broadening. The amount by which each line broadens indicates the range of speeds at which the stars are moving with respect to the center of the galaxy. The range of speeds depends, in turn, on the force of gravity that holds the stars within the galaxies. With information about the speeds, it is possible to calculate the mass of an elliptical galaxy.

Table 26.1 summarizes the range of masses (and other properties) of the various types of galaxies. Interestingly enough, the most and least massive galaxies are ellipticals. On average, irregular galaxies have less mass than spirals.

Characteristics of the Different Types of Galaxies

Characteristic	Spirals	Ellipticals	Irregulars
Mass (M_{Sun})	10^9 to 10^{12}	10^5 to 10^{13}	10^8 to 10^{11}
Diameter (thousands of light-years)	15 to 150	3 to >700	3 to 30
Luminosity (L_{Sun})	10^8 to 10^{11}	10^6 to 10^{11}	10^7 to 2×10^9
Populations of stars	Old and young	Old	Old and young
Interstellar matter	Gas and dust	Almost no dust; little gas	Much gas; some have little dust, some much dust

Table 26.1

Characteristic	Spirals	Ellipticals	Irregulars
Mass-to-light ratio in the visible part	2 to 10	10 to 20	1 to 10
Mass-to-light ratio for total galaxy	100	100	?

Table 26.1

Mass-to-Light Ratio

A useful way of characterizing a galaxy is by noting the ratio of its mass (in units of the Sun's mass) to its light output (in units of the Sun's luminosity). This single number tells us roughly what kind of stars make up most of the luminous population of the galaxy, and it also tells us whether a lot of dark matter is present. For stars like the Sun, the **mass-to-light ratio** is 1 by our definition.

Galaxies are not, of course, composed entirely of stars that are identical to the Sun. The overwhelming majority of stars are less massive and less luminous than the Sun, and usually these stars contribute most of the mass of a system without accounting for very much light. The mass-to-light ratio for low-mass stars is greater than 1 (you can verify this using the data in Table 18.3). Therefore, a galaxy's mass-to-light ratio is also generally greater than 1, with the exact value depending on the ratio of high-mass stars to low-mass stars.

Galaxies in which star formation is still occurring have many massive stars, and their mass-to-light ratios are usually in the range of 1 to 10. Galaxies consisting mostly of an older stellar population, such as ellipticals, in which the massive stars have already completed their evolution and have ceased to shine, have mass-to-light ratios of 10 to 20.

But these figures refer only to the inner, conspicuous parts of galaxies (Figure 26.10). In The Milky Way Galaxy and above, we discussed the evidence for dark matter in the outer regions of our own Galaxy, extending much farther from the galactic center than do the bright stars and gas. Recent measurements of the rotation speeds of the outer parts of nearby galaxies, such as the Andromeda galaxy we discussed earlier, suggest that they too have extended distributions of dark matter around the visible disk of stars and dust. This largely invisible matter adds to the mass of the galaxy while contributing nothing to its luminosity, thus increasing the mass-to-light ratio. If dark invisible matter is present in a galaxy, its mass-to-light ratio can be as high as 100. The two different mass-to-light ratios measured for various types of galaxies are given in Table 26.1.

Figure 26.10 M101, the Pinwheel Galaxy. This galaxy is a face-on spiral at a distance of 21 million light-years. M101 is almost twice the diameter of the Milky Way, and it contains at least 1 trillion stars. (credit: NASA, ESA, K. Kuntz (Johns Hopkins University), F. Bresolin (University of Hawaii), J. Trauger (Jet Propulsion Lab), J. Mould (NOAO), Y.-H. Chu (University of Illinois, Urbana), and STScI)

These measurements of other galaxies support the conclusion already reached from studies of the rotation of our own Galaxy—namely, that most of the material in the universe cannot at present be observed directly in any part of the electromagnetic spectrum. An understanding of the properties and distribution of this invisible matter is crucial to our understanding of galaxies. It's becoming clearer and clearer that, through the gravitational force it exerts, dark matter plays a dominant role in galaxy formation and early evolution. There is an interesting parallel here between our time and the time during which Edwin Hubble was receiving his training in astronomy. By 1920, many scientists were aware that astronomy stood on the brink of important breakthroughs—if only the nature and behavior of the nebulae could be settled with better observations. In the same way, many astronomers today feel we may be closing in on a far more sophisticated understanding of the large-scale structure of the universe—if only we can learn more about the nature and properties of dark matter. If you follow astronomy articles in the news (as we hope you will), you should be hearing more about dark matter in the years to come.

26.4 | The Extragalactic Distance Scale

Learning Objectives

By the end of this section, you will be able to:

> Describe the use of variable stars to estimate distances to galaxies
> Explain how standard bulbs and the Tully-Fisher relation can be used to estimate distances to galaxies

To determine many of the properties of a galaxy, such as its luminosity or size, we must first know how far away it is. If we know the distance to a galaxy, we can convert how bright the galaxy appears to us in the sky into its true luminosity because we know the precise way light is dimmed by distance. (The same galaxy 10 times farther away, for example, would look 100 times dimmer.) But the measurement of galaxy distances is one of the most difficult problems in modern astronomy: all galaxies are far away, and most are so distant that we cannot even make out individual stars in them.

For decades after Hubble's initial work, the techniques used to measure galaxy distances were relatively inaccurate, and different astronomers derived distances that differed by as much as a factor of two. (Imagine if the distance between your home or dorm and your astronomy class were this uncertain; it would be difficult to make sure you got to class on time.) In the past few decades, however, astronomers have devised new techniques for measuring distances to galaxies; most importantly, all of them give the same answer to within an accuracy of about 10%. As we will see, this means we may finally be able to make reliable estimates of the size of the universe.

Variable Stars

Before astronomers could measure distances to other galaxies, they first had to establish the scale of cosmic distances using objects in our own Galaxy. We described the chain of these distance methods in Celestial Distances (and we recommend that you review that chapter if it has been a while since you've read it). Astronomers were especially delighted when they discovered that they could measure distances using certain kinds of intrinsically luminous *variable stars*, such as cepheids, which can be seen at very large distances (Figure 26.11).

After the variables in nearby galaxies had been used to make distance measurements for a few decades, Walter Baade showed that there were actually two kinds of cepheids and that astronomers had been unwittingly mixing them up. As a result, in the early 1950s, the distances to all of the galaxies had to be increased by about a factor of two. We mention this because we want you to bear in mind, as you read on, that science is always a study in progress. Our first tentative steps in such difficult investigations are always subject to future revision as our techniques become more reliable.

The amount of work involved in finding cepheids and measuring their periods can be enormous. Hubble, for example, obtained 350 long-exposure photographs of the Andromeda galaxy over a period of 18 years and was able to identify only 40 cepheids. Even though cepheids are fairly luminous stars, they can be detected in only about 30 of the nearest galaxies with the world's largest ground-based telescopes.

As mentioned in Celestial Distances, one of the main projects carried out during the first years of operation of the Hubble Space Telescope was the measurement of cepheids in more distant galaxies to improve the accuracy of the extragalactic distance scale. Recently, astronomers working with the Hubble Space Telescope have extended such measurements out to 108 million light-years—a triumph of technology and determination.

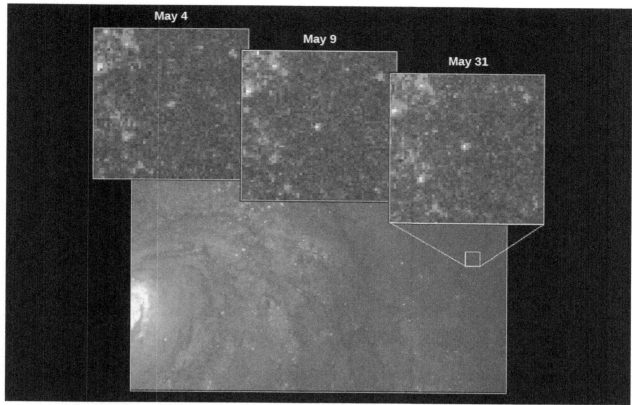

Figure 26.11 Cepheid Variable Star. In 1994, using the Hubble Space Telescope, astronomers were able to make out an individual cepheid variable star in the galaxy M100 and measure its distance to be 56 million light-years. The insets show the star on three different nights; you can see that its brightness is indeed variable. (credit: modification of work by Wendy L. Freedman, Observatories of the Carnegie Institution of Washington, and NASA/ESA)

Nevertheless, we can only use cepheids to measure distances within a small fraction of the universe of galaxies. After all, to use this method, we must be able to resolve single stars and follow their subtle variations. Beyond a certain distance, even our finest space telescopes cannot help us do this. Fortunately, there are other ways to measure the distances to galaxies.

Standard Bulbs

We discussed in Celestial Distances the great frustration that astronomers felt when they realized that the stars in general were not standard *bulbs*. If every light bulb in a huge auditorium is a standard 100-watt bulb, then bulbs that look brighter to us must be closer, whereas those that look dimmer must be farther away. If every star were a standard luminosity (or wattage), then we could similarly "read off" their distances based on how bright they appear to us. Alas, as we have learned, neither stars nor galaxies come in one standard-issue luminosity. Nonetheless, astronomers have been searching for objects out there that do act in some way like a standard bulb—that have the same intrinsic (built-in) brightness wherever they are.

A number of suggestions have been made for what sorts of objects might be effective standard bulbs, including the brightest supergiant stars, planetary nebulae (which give off a lot of ultraviolet radiation), and the average globular cluster in a galaxy. One object turns out to be particularly useful: the **type Ia supernova**. These supernovae involve the explosion of a white dwarf in a binary system (see The Evolution of Binary Star Systems) Observations show that supernovae of this type all reach nearly the same luminosity (about 4.5×10^9 L_{Sun}) at maximum light. With such tremendous luminosities, these supernovae have been detected out to a distance of more than 8 billion light-years and are therefore especially attractive to astronomers as a way of determining distances on a large scale (Figure 26.12).

Figure 26.12 Type Ia Supernova. The bright object at the bottom left of center is a type Ia supernova near its peak intensity. The supernova easily outshines its host galaxy. This extreme increase and luminosity help astronomers use Ia supernova as standard bulbs. (credit: NASA, ESA, A. Riess (STScI))

Several other kinds of standard bulbs visible over great distances have also been suggested, including the overall brightness of, for example, giant ellipticals and the brightest member of a galaxy cluster. Type Ia supernovae, however, have proved to be the most accurate standard bulbs, and they can be seen in more distant galaxies than the other types of calibrators. As we will see in the chapter on The Big Bang, observations of this type of supernova have profoundly changed our understanding of the evolution of the universe.

Other Measuring Techniques

Another technique for measuring galactic distances makes use of an interesting relationship noticed in the late 1970s by Brent Tully of the University of Hawaii and Richard Fisher of the National Radio Astronomy Observatory. They discovered that the luminosity of a spiral galaxy is related to its rotational velocity (how fast it spins). Why would this be true?

The more mass a galaxy has, the faster the objects in its outer regions must orbit. A more massive galaxy has more stars in it and is thus more luminous (ignoring dark matter for a moment). Thinking back to our discussion from the previous section, we can say that if the mass-to-light ratios for various spiral galaxies are pretty similar, then we can estimate the luminosity of a spiral galaxy by measuring its mass, and we can estimate its mass by measuring its rotational velocity.

Tully and Fisher used the 21-cm line of cold hydrogen gas to determine how rapidly material in spiral galaxies is orbiting their centers (you can review our discussion of the 21-cm line in Between the Stars: Gas and Dust in Space). Since 21-cm radiation from stationary atoms comes in a nice narrow line, the width of the 21-cm line produced by a whole rotating galaxy tells us the range of orbital velocities of the galaxy's hydrogen gas. The broader the line, the faster the gas is orbiting in the galaxy, and the more massive and luminous the galaxy turns out to be.

It is somewhat surprising that this technique works, since much of the mass associated with galaxies is dark matter, which does not contribute at all to the luminosity but does affect the rotation speed. There is also no obvious reason why the mass-to-light ratio should be similar for all spiral galaxies. Nevertheless, observations of nearer galaxies (where we have other ways of measuring distance) show that measuring the rotational velocity of a galaxy provides an accurate estimate of its intrinsic luminosity. Once we know how luminous the galaxy really is, we can compare the luminosity to the apparent brightness and use the difference to calculate its distance.

While the Tully-Fisher relation works well, it is limited—we can only use it to determine the distance to a spiral galaxy. There are other methods that can be used to estimate the distance to an elliptical galaxy; however, those methods are beyond the scope of our introductory astronomy course.

Table 26.2 lists the type of galaxy for which each of the distance techniques is useful, and the range of distances over which the technique can be applied.

Some Methods for Estimating Distance to Galaxies

Method	Galaxy Type	Approximate Distance Range (millions of light-years)
Planetary nebulae	All	0–70
Cepheid variables	Spiral, irregulars	0–110
Tully-Fisher relation	Spiral	0–300
Type Ia supernovae	All	0–11,000
Redshifts (Hubble's law)	All	300–13,000

Table 26.2

26.5 | The Expanding Universe

Learning Objectives

By the end of this section, you will be able to:

> Describe the discovery that galaxies getting farther apart as the universe evolves
> Explain how to use Hubble's law to determine distances to remote galaxies
> Describe models for the nature of an expanding universe
> Explain the variation in Hubble's constant

We now come to one of the most important discoveries ever made in astronomy—the fact that the universe is expanding. Before we describe how the discovery was made, we should point out that the first steps in the study of galaxies came at a time when the techniques of spectroscopy were also making great strides. Astronomers using large telescopes could record the spectrum of a faint star or galaxy on photographic plates, guiding their telescopes so they remained pointed to the same object for many hours and collected more light. The resulting spectra of galaxies contained a wealth of information about the composition of the galaxy and the velocities of these great star systems.

Slipher's Pioneering Observations

Curiously, the discovery of the expansion of the universe began with the search for Martians and other solar systems. In 1894, the controversial (and wealthy) astronomer Percival Lowell established an observatory in Flagstaff, Arizona, to study the planets and search for life in the universe. Lowell thought that the spiral nebulae might be solar systems in the process of formation. He therefore asked one of the observatory's young astronomers, Vesto M. Slipher (Figure 26.13), to photograph the spectra of some of the spiral nebulae to see if their spectral lines might show chemical compositions like those expected for newly forming planets.

Figure 26.13 Vesto M. Slipher (1875–1969). Slipher spent his entire career at the Lowell Observatory, where he discovered the large radial velocities of galaxies. (credit: Lowell Observatory)

The Lowell Observatory's major instrument was a 24-inch refracting telescope, which was not at all well suited to observations of faint spiral nebulae. With the technology available in those days, photographic plates had to be exposed for 20 to 40 hours to produce a good spectrum (in which the positions of the lines could reveal a galaxy's motion). This often meant continuing to expose the same photograph over several nights. Beginning in 1912, and making heroic efforts over a period of about 20 years, Slipher managed to photograph the spectra of more than 40 of the spiral nebulae (which would all turn out to be galaxies).

To his surprise, the spectral lines of most galaxies showed an astounding **redshift**. By "redshift" we mean that the lines in the spectra are displaced toward longer wavelengths (toward the red end of the visible spectrum). Recall from the chapter on Radiation and Spectra that a redshift is seen when the source of the waves is moving away from us. Slipher's observations showed that most spirals are racing away at huge speeds; the highest velocity he measured was 1800 kilometers per second.

Only a few spirals—such as the Andromeda and Triangulum Galaxies and M81—all of which are now known to be our close neighbors, turned out to be approaching us. All the other galaxies were moving away. Slipher first announced this discovery in 1914, years before Hubble showed that these objects were other galaxies and before anyone knew how far away they were. No one at the time quite knew what to make of this discovery.

Hubble's Law

The profound implications of Slipher's work became apparent only during the 1920s. Georges Lemaître was a Belgian priest and a trained astronomer. In 1927, he published a paper in French in an obscure Belgian journal in which he suggested that we live in an expanding universe. The title of the paper (translated into English) is "A Homogenous Universe of Constant Mass and Growing Radius Accounting for the Radial Velocity of Extragalactic Nebulae." Lemaître had discovered that Einstein's equations of relativity were consistent with an expanding universe (as had the Russian scientist Alexander Friedmann independently in 1922). Lemaître then went on to use Slipher's data to support the hypothesis that the universe actually is expanding and to estimate the rate of expansion. Initially, scientists paid little attention to this paper, perhaps because the Belgian journal was not widely available.

In the meantime, Hubble was making observations of galaxies with the 2.5-meter telescope on Mt. Wilson, which was then the world's largest. Hubble carried out the key observations in collaboration with a remarkable man, Milton Humason, who dropped out of school in the eighth grade and began his astronomical career by driving a mule train up the trail on Mount Wilson to the observatory (Figure 26.14). In those early days, supplies had to be brought up that way; even astronomers hiked up to the mountaintop for their turns at the telescope. Humason became interested in the work of the astronomers and, after marrying the daughter of the observatory's electrician, took a job as janitor there. After a time, he became a night assistant, helping the astronomers run the telescope and record data. Eventually, he made such a mark that he became a full astronomer at the observatory.

Figure 26.14 Milton Humason (1891–1972). Humason was Hubble's collaborator on the great task of observing, measuring, and classifying the characteristics of many galaxies. (credit: Caltech Archives)

By the late 1920s, Humason was collaborating with Hubble by photographing the spectra of faint galaxies with the 2.5-meter telescope. (By then, there was no question that the spiral nebulae were in fact galaxies.) Hubble had found ways to improve the accuracy of the estimates of distances to spiral galaxies, and he was able to measure much fainter and more distant galaxies than Slipher could observe with his much-smaller telescope. When Hubble laid his own distance estimates next to measurements of the recession velocities (the speed with which the galaxies were moving away), he found something stunning: there was a relationship between distance and velocity for galaxies. *The more distant the galaxy, the faster it was receding from us.*

In 1931, Hubble and Humason jointly published the seminal paper where they compared distances and velocities of remote galaxies moving away from us at speeds as high as 20,000 kilometers per second and were able to show that the recession velocities of galaxies are directly proportional to their distances from us (Figure 26.15), just as Lemaître had suggested.

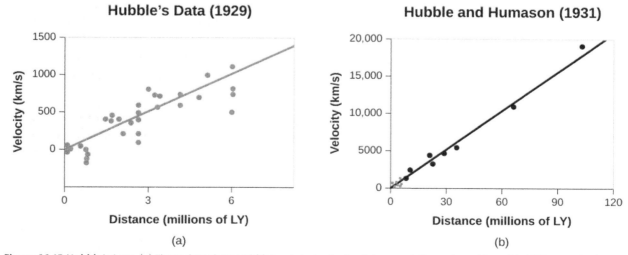

Figure 26.15 Hubble's Law. (a) These data show Hubble's original velocity-distance relation, adapted from his 1929 paper in the *Proceedings of the National Academy of Sciences*. (b) These data show Hubble and Humason's velocity-distance relation, adapted from their 1931 paper in *The Astrophysical Journal*. The red dots at the lower left are the points in the diagram in the 1929 paper.

Comparison of the two graphs shows how rapidly the determination of galactic distances and redshifts progressed in the 2 years between these publications.

We now know that this relationship holds for every galaxy except a few of the nearest ones. Nearly all of the galaxies that are approaching us turn out to be part of the Milky Way's own group of galaxies, which have their own individual motions, just as birds flying in a group may fly in slightly different directions at slightly different speeds even though the entire flock travels through space together.

Written as a formula, the relationship between velocity and distance is

$$v = H \times d$$

where *v* is the velocity, *d* is the distance, and *H* is a number called the **Hubble constant**. This equation is now known as **Hubble's law**.[1]

ASTRONOMY BASICS

Constants of Proportionality

Mathematical relationships such as Hubble's law are pretty common in life. To take a simple example, suppose your college or university hires you to call rich alumni and ask for donations. You are paid $2.50 for each call; the more calls you can squeeze in between studying astronomy and other courses, the more money you take home. We can set up a formula that connects *p*, your pay, and *n*, the number of calls

$$p = A \times n$$

where *A* is the alumni constant, with a value of $2.50. If you make 20 calls, you will earn $2.50 times 20, or $50.

Suppose your boss forgets to tell you what you will get paid for each call. You can calculate the alumni constant that governs your pay by keeping track of how many calls you make and noting your gross pay each week. If you make 100 calls the first week and are paid $250, you can deduce that the constant is $2.50 (in units of dollars per call). Hubble, of course, had no "boss" to tell him what his constant would be—he *had* to calculate its value from the measurements of distance and velocity.

Astronomers express the value of Hubble's constant in units that relate to how they measure speed and distance for galaxies. In this book, we will use kilometers per second per million light-years as that unit. For many years, estimates of the value of the Hubble constant have been in the range of 15 to 30 kilometers per second per million light-years The most recent work appears to be converging on a value near 22 kilometers per second per million light-years If *H* is 22 kilometers per second per million light-years, a galaxy moves away from us at a speed of 22 kilometers per second for every million light-years of its distance. As an example, a galaxy 100 million light-years away is moving away from us at a speed of 2200 kilometers per second.

Hubble's law tells us something fundamental about the universe. Since all but the nearest galaxies appear to be in motion away from us, with the most distant ones moving the fastest, we must be living in an expanding universe. We will explore the implications of this idea shortly, as well as in the final chapters of this text. For now, we will just say that Hubble's observation underlies all our theories about the origin and evolution of the universe.

Hubble's Law and Distances

The regularity expressed in Hubble's law has a built-in bonus: it gives us a new way to determine the distances to remote galaxies. First, we must reliably establish Hubble's constant by measuring both the distance and the

---

1   In 2020, the International Astronomical Union suggested that it would be more fair to call it the Hubble-Lemaître law. In telling the history in this textbook, we have acknowledged the role Lemaître played, and we urge our readers to keep his contributions in mind as they read on.

velocity of many galaxies in many directions to be sure Hubble's law is truly a universal property of galaxies. But once we have calculated the value of this constant and are satisfied that it applies everywhere, much more of the universe opens up for distance determination. Basically, if we can obtain a spectrum of a galaxy, we can immediately tell how far away it is.

The procedure works like this. We use the spectrum to measure the speed with which the galaxy is moving away from us. If we then put this speed and the Hubble constant into Hubble's law equation, we can solve for the distance.

EXAMPLE 26.1

Hubble's Law

Hubble's law ($v = H \times d$) allows us to calculate the distance to any galaxy. Here is how we use it in practice.

We have measured Hubble's constant to be 22 km/s per million light-years. This means that if a galaxy is 1 million light-years farther away, it will move away 22 km/s faster. So, if we find a galaxy that is moving away at 18,000 km/s, what does Hubble's law tells us about the distance to the galaxy?

Solution

$$d = \frac{v}{H} = \frac{18,000 \text{ km/s}}{\frac{22 \text{ km/s}}{1 \text{ million light-years}}} = \frac{18,000}{22} \times \frac{1 \text{ million light-years}}{1} = 818 \text{ million light-years}$$

Note how we handled the units here: the km/s in the numerator and denominator cancel, and the factor of million light-years in the denominator of the constant must be divided correctly before we get our distance of 818 million light-years.

Check Your Learning

Using 22 km/s/million light-years for Hubble's constant, what recessional velocity do we expect to find if we observe a galaxy at 500 million light-years?

Answer:

$$v = d \times H = 500 \text{ million light-years} \times \frac{22 \text{ km/s}}{1 \text{ million light-years}} = 11,000 \text{ km/s}$$

Variation of Hubble's Constant

The use of redshift is potentially a very important technique for determining distances because as we have seen, most of our methods for determining galaxy distances are limited to approximately the nearest few hundred million light-years (and they have large uncertainties at these distances). The use of Hubble's law as a distance indicator requires only a spectrum of a galaxy and a measurement of the Doppler shift, and with large telescopes and modern spectrographs, spectra can be taken of extremely faint galaxies.

But, as is often the case in science, things are not so simple. This technique works if, and only if, the Hubble constant has been truly constant throughout the entire life of the universe. When we observe galaxies billions of light-years away, we are seeing them as they were billions of years ago. What if the Hubble "constant" was different billions of years ago? Before 1998, astronomers thought that, although the universe is expanding, the expansion should be slowing down, or decelerating, because the overall gravitational pull of all matter in the universe would have a dominant, measureable effect. If the expansion is decelerating, then the Hubble constant should be decreasing over time.

The discovery that type Ia supernovae are standard bulbs gave astronomers the tool they needed to observe extremely distant galaxies and measure the rate of expansion billions of years ago. The results were completely unexpected. It turns out that the expansion of the universe is *accelerating* over time! What makes this result so astounding is that there is no way that existing physical theories can account for this observation. While a decelerating universe could easily be explained by gravity, there was no force or property in the universe known to astronomers that could account for the acceleration. In The Big Bang chapter, we will look in more detail at the observations that led to this totally unexpected result and explore its implications for the ultimate fate of the universe.

In any case, if the Hubble constant is not really a constant when we look over large spans of space and time, then the calculation of galaxy distances using the Hubble constant won't be accurate. As we shall see in the chapter on The Big Bang, the accurate calculation of distances requires a model for how the Hubble constant has changed over time. The farther away a galaxy is (and the longer ago we are seeing it), the more important it is to include the effects of the change in the Hubble constant. For galaxies within a few billion light-years, however, the assumption that the Hubble constant is indeed constant gives good estimates of distance.

Models for an Expanding Universe

At first, thinking about Hubble's law and being a fan of the work of Copernicus and Harlow Shapley, you might be shocked. Are all the galaxies really moving *away from us*? Is there, after all, something special about our position in the universe? Worry not; the fact that galaxies are receding from us and that more distant galaxies are moving away more rapidly than nearby ones shows only that the universe is expanding uniformly.

A uniformly expanding universe is one that is expanding at the same rate everywhere. In such a universe, we and all other observers, no matter where they are located, must observe a proportionality between the velocities and distances of equivalently remote galaxies. (Here, we are ignoring the fact that the Hubble constant is not constant over all time, but if at any given time in the evolution of the universe the Hubble constant has the same value everywhere, this argument still works.)

To see why, first imagine a ruler made of stretchable rubber, with the usual lines marked off at each centimeter. Now suppose someone with strong arms grabs each end of the ruler and slowly stretches it so that, say, it doubles in length in 1 minute (Figure 26.16). Consider an intelligent ant sitting on the mark at 2 centimeters—a point that is not at either end nor in the middle of the ruler. He measures how fast other ants, sitting at the 4-, 7-, and 12-centimeter marks, move away from him as the ruler stretches.

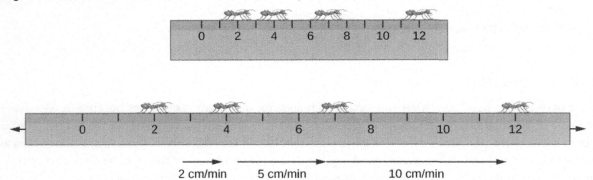

Figure 26.16 Stretching a Ruler. Ants on a stretching ruler see other ants move away from them. The speed with which another ant moves away is proportional to its distance.

The ant at 4 centimeters, originally 2 centimeters away from our ant, has doubled its distance in 1 minute; it therefore moved away at a speed of 2 centimeters per minute. The ant at the 7-centimeters mark, which was originally 5 centimeters away from our ant, is now 10 centimeters away; it thus had to move at 5 centimeters per minute. The one that started at the 12-centimeters mark, which was 10 centimeters away from the ant doing the counting, is now 20 centimeters away, meaning it must have raced away at a speed of 10 centimeters per minute. Ants at different distances move away at different speeds, and their speeds are proportional to their distances (just as Hubble's law indicates for galaxies). Yet, notice in our example that all the ruler was doing was stretching uniformly. Also, notice that none of the ants were actually moving of their own accord, it was the stretching of the ruler that moved them apart.

Now let's repeat the analysis, but put the intelligent ant on some other mark—say, on 7 or 12 centimeters. We discover that, as long as the ruler stretches uniformly, this ant also finds every other ant moving away at a speed proportional to its distance. In other words, the kind of relationship expressed by Hubble's law can be explained by a uniform stretching of the "world" of the ants. And all the ants in our simple diagram will see the other ants moving away from them as the ruler stretches.

For a three-dimensional analogy, let's look at the loaf of raisin bread in Figure 26.17. The chef has accidentally put too much yeast in the dough, and when she sets the bread out to rise, it doubles in size during the next hour, causing all the raisins to move farther apart. On the figure, we again pick a representative raisin (that is not at the edge or the center of the loaf) and show the distances from it to several others in the figure (before and after the loaf expands).

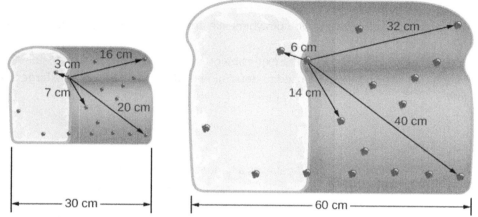

Figure 26.17 Expanding Raisin Bread. As the raisin bread rises, the raisins "see" other raisins moving away. More distant raisins move away faster in a uniformly expanding bread.

Measure the increases in distance and calculate the speeds for yourself on the raisin bread, just like we did for the ruler. You will see that, since each distance doubles during the hour, each raisin moves away from our selected raisin at a speed proportional to its distance. The same is true no matter which raisin you start with.

Our two analogies are useful for clarifying our thinking, but you must not take them literally. On both the ruler and the raisin bread, there are points that are at the end or edge. You can use these to pinpoint the middle of the ruler and the loaf. While our models of the universe have some resemblance to the properties of the ruler and the loaf, the universe has no boundaries, no edges, and no center (all mind-boggling ideas that we will discuss in a later chapter).

What is useful to notice about both the ants and the raisins is that they themselves did not "cause" their motion. It isn't as if the raisins decided to take a trip away from each other and then hopped on a hoverboard to get away. No, in both our analogies, it was the stretching of the medium (the ruler or the bread) that moved the ants or the raisins farther apart. In the same way, we will see in The Big Bang chapter that the galaxies don't have rocket motors propelling them away from each other. Instead, they are passive participants in the *expansion of space*. As space stretches, the galaxies are carried farther and farther apart much as the ants and the raisins were. (If this notion of the "stretching" of space surprises or bothers you, now would be a good time to review the information about spacetime in Black Holes and Curved Spacetime. We will discuss these ideas further as our discussion broadens from galaxies to the whole universe.)

The expansion of the universe, by the way, does not imply that the individual galaxies and clusters of galaxies themselves are expanding. Neither raisins nor the ants in our analogy grow in size as the loaf expands. Similarly, gravity holds galaxies and clusters of galaxies together, and they get farther away from each other—without themselves changing in size—as the universe expands.

🔑 Key Terms

elliptical galaxy   a galaxy whose shape is an ellipse and that contains no conspicuous interstellar material

Hubble constant   a constant of proportionality in the law relating the velocities of remote galaxies to their distances

Hubble's law   a rule that the radial velocities of remote galaxies are proportional to their distances from us

irregular galaxy   a galaxy without any clear symmetry or pattern; neither a spiral nor an elliptical galaxy

mass-to-light ratio   the ratio of the total mass of a galaxy to its total luminosity, usually expressed in units of solar mass and solar luminosity; the mass-to-light ratio gives a rough indication of the types of stars contained within a galaxy and whether or not substantial quantities of dark matter are present

redshift   when lines in the spectra are displaced toward longer wavelengths (toward the red end of the visible spectrum)

spiral galaxy   a flattened, rotating galaxy with pinwheel-like arms of interstellar material and young stars, winding out from its central bulge

type Ia supernova   a supernova formed by the explosion of a white dwarf in a binary system and reach a luminosity of about 4.5×10^9 L_{Sun}; can be used to determine distances to galaxies on a large scale

📖 Summary

26.1 The Discovery of Galaxies

Faint star clusters, clouds of glowing gas, and galaxies all appeared as faint patches of light (or nebulae) in the telescopes available at the beginning of the twentieth century. It was only when Hubble measured the distance to the Andromeda galaxy using cepheid variables with the giant 2.5-meter reflector on Mount Wilson in 1924 that the existence of other galaxies similar to the Milky Way in size and content was established.

26.2 Types of Galaxies

The majority of bright galaxies are either spirals or ellipticals. Spiral galaxies contain both old and young stars, as well as interstellar matter, and have typical masses in the range of 10^9 to 10^{12} M_{Sun}. Our own Galaxy is a large spiral. Ellipticals are spheroidal or slightly elongated systems that consist almost entirely of old stars, with very little interstellar matter. Elliptical galaxies range in size from giants, more massive than any spiral, down to dwarfs, with masses of only about 10^6 M_{Sun}. Dwarf ellipticals are probably the most common type of galaxy in the nearby universe. A small percentage of galaxies with more disorganized shapes are classified as irregulars. Galaxies may change their appearance over time due to collisions with other galaxies or by a change in the rate of star formation.

26.3 Properties of Galaxies

The masses of spiral galaxies are determined from measurements of their rates of rotation. The masses of elliptical galaxies are estimated from analyses of the motions of the stars within them. Galaxies can be characterized by their mass-to-light ratios. The luminous parts of galaxies with active star formation typically have mass-to-light ratios in the range of 1 to 10; the luminous parts of elliptical galaxies, which contain only old stars, typically have mass-to-light ratios of 10 to 20. The mass-to-light ratios of whole galaxies, including their outer regions, are as high as 100, indicating the presence of a great deal of dark matter.

26.4 The Extragalactic Distance Scale

Astronomers determine the distances to galaxies using a variety of methods, including the period-luminosity relationship for cepheid variables; objects such as type Ia supernovae, which appear to be standard bulbs; and the Tully-Fisher relation, which connects the line broadening of 21-cm radiation to the luminosity of spiral galaxies. Each method has limitations in terms of its precision, the kinds of galaxies with which it can be used, and the range of distances over which it can be applied.

26.5 The Expanding Universe

The universe is expanding. Observations show that the spectral lines of distant galaxies are redshifted, and that their recession velocities are proportional to their distances from us, a relationship known as Hubble's law. The rate of recession, called the Hubble constant, is approximately 22 kilometers per second per million light-years. We are not at the center of this expansion: an observer in any other galaxy would see the same pattern of expansion that we do. The expansion described by Hubble's law is best understood as a stretching of space.

 For Further Exploration

Articles

Andrews, B. "What Are Galaxies Trying to Tell Us?" *Astronomy* (February 2011): 24. Introduction to our understanding of the shapes and evolution of different types of galaxies.

Bothun, G. "Beyond the Hubble Sequence." *Sky & Telescope* (May 2000): 36. History and updating of Hubble's classification scheme.

Christianson, G. "Mastering the Universe." *Astronomy* (February 1999): 60. Brief introduction to Hubble's life and work.

Dalcanton, J. "The Overlooked Galaxies." *Sky & Telescope* (April 1998): 28. On low-brightness galaxies, which have been easy to miss.

Freedman, W. "The Expansion Rate and Size of the Universe." *Scientific American* (November 1992): 76.

Hodge, P. "The Extragalactic Distance Scale: Agreement at Last?" *Sky & Telescope* (October 1993): 16.

Jones, B. "The Legacy of Edwin Hubble." *Astronomy* (December 1989): 38.

Kaufmann, G. and van den Bosch, F. "The Life Cycle of Galaxies." *Scientific American* (June 2002): 46. On galaxy evolution and how it leads to the different types of galaxies.

Martin, P. and Friedli, D. "At the Hearts of Barred Galaxies." *Sky & Telescope* (March 1999): 32. On barred spirals.

Osterbrock, D. "Edwin Hubble and the Expanding Universe." *Scientific American* (July 1993): 84.

Russell, D. "Island Universes from Wright to Hubble." *Sky & Telescope* (January 1999) 56. A history of our discovery of galaxies.

Smith, R. "The Great Debate Revisited." *Sky & Telescope* (January 1983): 28. On the Shapley-Curtis debate concerning the extent of the Milky Way and the existence of other galaxies.

Websites

ABC's of Distance: http://www.astro.ucla.edu/~wright/distance.htm (http://www.astro.ucla.edu/~wright/distance.htm). A concise summary by astronomer Ned Wright of all the different methods we use to get distances in astronomy.

Cosmic Times 1929: http://cosmictimes.gsfc.nasa.gov/online_edition/1929Cosmic/index.html (http://cosmictimes.gsfc.nasa.gov/online_edition/1929Cosmic/index.html). NASA project explaining Hubble's work and surrounding discoveries as if you were reading newspaper articles.

Edwin Hubble: Biography: https://www.nasa.gov/content/about-story-edwin-hubble (https://www.nasa.gov/content/about-story-edwin-hubble). Concise biography from the people at the Hubble Space Telescope.

Edwin Hubble: http://apod.nasa.gov/diamond_jubilee/d_1996/sandage_hubble.html (http://apod.nasa.gov/diamond_jubilee/d_1996/sandage_hubble.html). An article on the life and work of Hubble by his student and successor, Allan Sandage. A bit technical in places, but giving a real picture of the man and the science.

International Astronomical Union resolution to call Hubble Law the Hubble-Lemaître Law (with background information): https://www.iau.org/news/pressreleases/detail/iau1812/ (https://www.iau.org/news/pressreleases/detail/iau1812/).

NASA Science: Introduction to Galaxies: http://science.nasa.gov/astrophysics/focus-areas/what-are-galaxies/ (http://science.nasa.gov/astrophysics/focus-areas/what-are-galaxies/). A brief overview with links to other pages, and recent Hubble Space Telescope discoveries.

National Optical Astronomy Observatories Gallery of Galaxies: https://www.noao.edu/image_gallery/galaxies.html (https://www.noao.edu/image_gallery/galaxies.html). A collection of images and information about galaxies and galaxy groups of different types. Another impressive archive can be found at the European Southern Observatory site: https://www.eso.org/public/images/archive/category/galaxies/ (https://www.eso.org/public/images/archive/category/galaxies/).

Sloan Digital Sky Survey: Introduction to Galaxies: http://skyserver.sdss.org/dr1/en/astro/galaxies/galaxies.asp (http://skyserver.sdss.org/dr1/en/astro/galaxies/galaxies.asp). Another brief overview.

Universe Expansion: https://hubblesite.org/contents/news-releases/1999/news-1999-19.html (https://hubblesite.org/contents/news-releases/1999/news-1999-19.html). The background material here provides a nice chronology of how we discovered and measured the expansion of the universe.

Videos

Edwin Hubble (Hubblecast Episode 89): http://www.spacetelescope.org/videos/hubblecast89a/ (http://www.spacetelescope.org/videos/hubblecast89a/). (5:59).

Galaxies: https://www.youtube.com/watch?v=I82ADyJC7wE (https://www.youtube.com/watch?v=I82ADyJC7wE). An introduction.

Hubble's Views of the Deep Universe: https://www.youtube.com/watch?v=argR2U15w-M (https://www.youtube.com/watch?v=argR2U15w-M). A 2015 public talk by Brandon Lawton of the Space Telescope Science Institute about galaxies and beyond (1:26:20).

Collaborative Group Activities

A. Throughout much of the last century, the 100-inch telescope on Mt. Wilson (completed in 1917) and the 200-inch telescope on Palomar Mountain (completed in 1948) were the only ones large enough to obtain spectra of faint galaxies. Only a handful of astronomers (all male—since, until the 1960s, women were not given time on these two telescopes) were allowed to use these facilities, and in general the observers did not compete with each other but worked on different problems. Now there are many other telescopes, and several different groups do often work on the same problem. For example, two different groups have independently developed the techniques for using supernovae to determine the distances to galaxies at high redshifts. Which approach do you think is better for the field of astronomy? Which is more cost effective? Why?

B. A distant relative, whom you invite to dinner so you can share all the exciting things you have learned in your astronomy class, says he does not believe that other galaxies are made up of stars. You come back to your group and ask them to help you respond. What kinds of measurements would you make to show that other galaxies are composed of stars?

C. Look at Figure 26.1 with your group. What does the difference in color between the spiral arms and the bulge of Andromeda tell you about the difference in the types of stars that populate these two regions of the galaxy? Which side of the galaxy is closer to us? Why?

D. What is your reaction to reading about the discovery of the expanding universe? Discuss how the members of the group feel about a universe "in motion." Einstein was not comfortable with the notion of a universe that had some overall movement to it, instead of being at rest. He put a kind of "fudge factor" into his equations of general relativity for the universe as a whole to keep it from moving (although later, hearing about Hubble and Humason's work, he called it "the greatest blunder" he ever made). Do you share Einstein's original sense that this is not the kind of universe you feel comfortable with? What do you think could have caused space to be expanding?

E. In science fiction, characters sometimes talk about visiting other galaxies. Discuss with your group how realistic this idea is. Even if we had fast spaceships (traveling close to the speed of light, the speed limit of the universe) how likely are we to be able to reach another galaxy? Why?

F. Despite his son's fascination with astronomy in college, Edwin Hubble's father did not want him to go into astronomy as a profession. He really wanted his son to be a lawyer and pushed him hard to learn the law when he won a fellowship to study abroad. Hubble eventually defied his father and went into astronomy, becoming, as you learned in this chapter, one of the most important astronomers of all time. His dad didn't live to see his son's remarkable achievements. Do you think he would have reconciled himself to his son's career choice if he had? Do you or does anyone in your group or among your friends have to face a choice between the passion in your heart and what others want you to do? Discuss how people in college today are dealing with such choices.

Exercises

Review Questions

1. Describe the main distinguishing features of spiral, elliptical, and irregular galaxies.

2. Why did it take so long for the existence of other galaxies to be established?

3. Explain what the mass-to-light ratio is and why it is smaller in spiral galaxies with regions of star formation than in elliptical galaxies.

4. If we now realize dwarf ellipticals are the most common type of galaxy, why did they escape our notice for so long?

5. What are the two best ways to measure the distance to a nearby spiral galaxy, and how would it be measured?

6. What are the two best ways to measure the distance to a distant, isolated spiral galaxy, and how would it be measured?

7. Why is Hubble's law considered one of the most important discoveries in the history of astronomy?

8. What does it mean to say that the universe is expanding? What is expanding? For example, is your astronomy classroom expanding? Is the solar system? Why or why not?

9. Was Hubble's original estimate of the distance to the Andromeda galaxy correct? Explain.

10. Does an elliptical galaxy rotate like a spiral galaxy? Explain.

11. Why does the disk of a spiral galaxy appear dark when viewed edge on?

12. What causes the largest mass-to-light ratio: gas and dust, dark matter, or stars that have burnt out?

13. What is the most useful standard bulb method for determining distances to galaxies?

14. When comparing two isolated spiral galaxies that have the same apparent brightness, but rotate at different rates, what can you say about their relative luminosity?

15. If all distant galaxies are expanding away from us, does this mean we're at the center of the universe?

16. Is the Hubble constant actually constant?

Thought Questions

17. Where might the gas and dust (if any) in an elliptical galaxy come from?

18. Why can we not determine distances to galaxies by the same method used to measure the parallaxes of stars?

19. Which is redder—a spiral galaxy or an elliptical galaxy?

20. Suppose the stars in an elliptical galaxy all formed within a few million years shortly after the universe began. Suppose these stars have a range of masses, just as the stars in our own galaxy do. How would the color of the elliptical change over the next several billion years? How would its luminosity change? Why?

21. Starting with the determination of the size of Earth, outline a sequence of steps necessary to obtain the distance to a remote cluster of galaxies. (Hint: Review the chapter on Celestial Distances.)

22. Suppose the Milky Way Galaxy were truly isolated and that no other galaxies existed within 100 million light-years. Suppose that galaxies were observed in larger numbers at distances greater than 100 million light-years. Why would it be more difficult to determine accurate distances to those galaxies than if there were also galaxies relatively close by?

23. Suppose you were Hubble and Humason, working on the distances and Doppler shifts of the galaxies. What sorts of things would you have to do to convince yourself (and others) that the relationship you were seeing between the two quantities was a real feature of the behavior of the universe? (For example, would data from two galaxies be enough to demonstrate Hubble's law? Would data from just the nearest galaxies—in what astronomers call "the Local Group"—suffice?)

24. What does it mean if one elliptical galaxy has broader spectrum lines than another elliptical galaxy?

25. Based on your analysis of galaxies in Table 26.1, is there a correlation between the population of stars and the quantity of gas or dust? Explain why this might be.

26. Can a higher mass-to-light ratio mean that there is gas and dust present in the system that is being analyzed?

Figuring for Yourself

27. According to Hubble's law, what is the recessional velocity of a galaxy that is 10^8 light-years away from us? (Assume a Hubble constant of 22 km/s per million light-years.)

28. A cluster of galaxies is observed to have a recessional velocity of 60,000 km/s. Find the distance to the cluster. (Assume a Hubble constant of 22 km/s per million light-years.)

29. Suppose we could measure the distance to a galaxy using one of the distance techniques listed in Table 26.2 and it turns out to be 200 million light-years. The galaxy's redshift tells us its recessional velocity is 5000 km/s. What is the Hubble constant?

30. Calculate the mass-to-light ratio for a globular cluster with a luminosity of 10^6 L_{Sun} and 10^5 stars. (Assume that the average mass of a star in such a cluster is 1 M_{Sun}.)

31. Calculate the mass-to-light ratio for a luminous star of 100 M_{Sun} having the luminosity of 10^6 L_{Sun}.

27

Active Galaxies, Quasars, and Supermassive Black Holes

Figure 27.1 Hubble Ultra-Deep Field. The deepest picture of the sky in visible light (left) shows huge numbers of galaxies in a tiny patch of sky, only 1/100 the area of the full Moon. In contrast, the deepest picture of the sky taken in X-rays (right) shows large numbers of point-like quasars, which astronomers have shown are supermassive black holes at the very centers of galaxies. (credit left: modification of work by NASA, ESA, H. Teplitz and M. Rafelski (IPAC/Caltech), A. Koekemoer (STScI), R. Windhorst (Arizona State University), and Z. Levay (STScI); credit right: modification of work by ESO/Mario Nonino, Piero Rosati, ESO GOODS Team)

Chapter Outline

27.1 Quasars
27.2 Supermassive Black Holes: What Quasars Really Are
27.3 Quasars as Probes of Evolution in the Universe

Thinking Ahead

During the first half of the twentieth century, astronomers viewed the universe of galaxies as a mostly peaceful place. They assumed that galaxies formed billions of years ago and then evolved slowly as the populations of stars within them formed, aged, and died. That placid picture completely changed in the last few decades of the twentieth century.

Today, astronomers can see that the universe is often shaped by violent events, including cataclysmic explosions of supernovae, collisions of whole galaxies, and the tremendous outpouring of energy as matter interacts in the environment surrounding very massive black holes. The key event that began to change our view of the universe was the discovery of a new class of objects: quasars.

27.1 | Quasars

Learning Objectives

By the end of this section, you will be able to:

> Describe how quasars were discovered
> Explain how astronomers determined that quasars are at the distances implied by their redshifts
> Justify the statement that the enormous amount of energy produced by quasars is generated in a very small volume of space

The name "quasars" started out as short for "quasi-stellar radio sources" (here "quasi-stellar" means "sort of like stars"). The discovery of radio sources that appeared point-like, just like stars, came with the use of surplus World War II radar equipment in the 1950s. Although few astronomers would have predicted it, the sky turned out to be full of strong sources of radio waves. As they improved the images that their new radio telescopes could make, scientists discovered that some radio sources were in the same location as faint blue "stars." No known type of star in our Galaxy emits such powerful radio radiation. What then were these "quasi-stellar radio sources"?

Redshifts: The Key to Quasars

The answer came when astronomers obtained visible-light spectra of two of those faint "blue stars" that were strong sources of radio waves (Figure 27.2). Spectra of these radio "stars" only deepened the mystery: they had emission lines, but astronomers at first could not identify them with any known substance. By the 1960s, astronomers had a century of experience in identifying elements and compounds in the spectra of stars. Elaborate tables had been published showing the lines that each element would produce under a wide range of conditions. A "star" with unidentifiable lines in the ordinary visible light spectrum had to be something completely new.

Figure 27.2 Typical Quasar. The arrow in this image marks the quasar known by its catalog number, PKS 1117-248. Note that nothing in this image distinguishes the quasar from an ordinary star. Its spectrum, however, shows that it is moving away from us at a speed of 36% the speed of light, or 67,000 miles per second. In contrast, the maximum speed observed for any star is only a few hundred miles per second. (credit: modification of work by WIYN Telescope, Kitt Peak National Observatory, NOAO)

In 1963 at Caltech's Palomar Observatory, Maarten Schmidt (Figure 27.3) was puzzling over the spectrum of one of the radio stars, which was named 3C 273 because it was the 273rd entry in the third Cambridge catalog of radio sources (part (b) of Figure 27.3). There were strong emission lines in the spectrum, and Schmidt recognized that they had the same spacing between them as the Balmer lines of hydrogen (see Radiation and Spectra). But the lines in 3C 273 were shifted far to the red of the wavelengths at which the Balmer lines are normally located. Indeed, these lines were at such long wavelengths that if the redshifts were attributed to the Doppler effect, 3C 273 was receding from us at a speed of 45,000 kilometers per second, or about 15% the speed of light! Since stars don't show Doppler shifts this large, no one had thought of considering high redshifts to be the cause of the strange spectra.

(a)   (b)

Figure 27.3 Quasar Pioneers and Quasar 3C 273. (a) Maarten Schmidt (left), who solved the puzzle of the quasar spectra in 1963, shares a joke in this 1987 photo with Allan Sandage, who took the first spectrum of a quasar. Sandage was also instrumental in measuring the value of Hubble's constant. (b) This is the first quasar for which a redshift was measured. The redshift showed that the light from it took about 2.5 billion years to reach us. Despite this great distance, it is still one of the quasars closest to the Milky Way Galaxy. Note also the faint streak going toward the upper left from the quasar. Some quasars, like 3C 273, eject super-fast jets of material. The jet from 3C 273 is about 200,000 light-years long. (credit a: modification of work by Andrew Fraknoi; credit b: modification of work by ESA/Hubble/NASA)

The puzzling emission lines in other star-like radio sources were then reexamined to see if they, too, might be well-known lines with large redshifts. This proved to be the case, but the other objects were found to be receding from us at even greater speeds. Their astounding speeds showed that the radio "stars" could not possibly be stars in our own Galaxy. Any true star moving at more than a few hundred kilometers per second would be able to overcome the gravitational pull of the Galaxy and completely escape from it. (As we shall see later in this chapter, astronomers eventually discovered that there was also more to these "stars" than just a point of light.)

It turns out that these high-velocity objects only look like stars because they are compact and very far away. Later, astronomers discovered objects with large redshifts that appear star-like but have no radio emission. Observations also showed that quasars were bright in the infrared and X-ray bands too, and not all these X-ray or infrared-bright quasars could be seen in either the radio or the visible-light bands of the spectrum. Today, all these objects are referred to as *quasi-stellar objects* (*QSOs*), or, as they are more popularly known, **quasars**. (The name was also soon appropriated by a manufacturer of home electronics.)

LINK TO LEARNING

Read an interview (https://openstax.org/l/30SchmidtIntv) with Maarten Schmidt on the fiftieth anniversary of his insight about the spectrum of quasars and their redshifts.

Over a million quasars have now been discovered, and spectra are available for over a hundred thousand. All these spectra show redshifts, none show blueshifts, and their redshifts can be very large. Yet in a photo they look just like stars (Figure 27.4).

Figure 27.4 Typical Quasar Imaged by the Hubble Space Telescope. One of these two bright "stars" in the middle is in our Galaxy, while the other is a quasar 9 billion light-years away. From this picture alone, there's no way to say which is which. (The quasar is the one in the center of the picture.) (credit: Charles Steidel (CIT)/NASA/ESA)

In the record-holding quasars, the first Lyman series line of hydrogen, with a laboratory wavelength of 121.5 nanometers in the ultraviolet portion of the spectrum, is shifted all the way through the visible region to the infrared. At such high redshifts, the simple formula for converting a Doppler shift to speed (Radiation and Spectra) must be modified to take into account the effects of the theory of relativity. If we apply the relativistic form of the Doppler shift formula, we find that these redshifts correspond to velocities of about 96% of the speed of light.

EXAMPLE 27.1

Recession Speed of a Quasar

The formula for the Doppler shift, which astronomers denote by the letters z, is

$$z = \frac{\Delta\lambda}{\lambda} = \frac{v}{c}$$

where λ is the wavelength emitted by a source of radiation that is not moving, $\Delta\lambda$ is the difference between that wavelength and the wavelength we measure, v is the speed with which the source moves away, and c (as usual) is the speed of light.

A line in the spectrum of a galaxy is at a wavelength of 393 nanometers (nm, or 10^{-9} m) when the source is at rest. Let's say the line is measured to be longer than this value (redshifted) by 7.86 nm. Then its redshift $z = \frac{7.86 \text{ nm}}{393 \text{ nm}} = 0.02$, so its speed away from us is 2% of the speed of light $\left(\frac{v}{c} = 0.02\right)$.

This formula is fine for galaxies that are relatively nearby and are moving away from us slowly in the expansion of the universe. But the quasars and distant galaxies we discuss in this chapter are moving away at speeds close to the speed of light. In that case, converting a Doppler shift (redshift) to a distance must include the effects of the special theory of relativity, which explains how measurements of space and time change when we see things moving at high speeds. The details of how this is done are way beyond the level of this text, but we can share with you the relativistic formula for the Doppler shift:

$$\frac{v}{c} = \frac{(z+1)^2 - 1}{(z+1)^2 + 1}$$

Let's do an example. Suppose a distant quasar has a redshift of 3. At what fraction of the speed of light is the quasar moving away?

Solution

We calculate the following:

$$\frac{v}{c} = \frac{(3+1)^2 - 1}{(3+1)^2 + 1} = \frac{16 - 1}{16 + 1} = \frac{15}{17} = 0.882$$

The quasar is thus receding from us at about 88% the speed of light.

Check Your Learning

Several lines of hydrogen absorption in the visible spectrum have rest wavelengths of 410 nm, 434 nm, 486 nm, and 656 nm. In a spectrum of a distant galaxy, these same lines are observed to have wavelengths of 492 nm, 521 nm, 583 nm, and 787 nm respectively. What is the redshift of this galaxy? What is the recession speed of this galaxy?

Answer:

Because this is the same galaxy, we could pick any one of the four wavelengths and calculate how much it has shifted. If we use a rest wavelength of 410 nm and compare it to the shifted wavelength of 492 nm, we see that

$$z = \frac{\Delta\lambda}{\lambda} = \frac{(492 \, nm - 410 \, nm)}{410 \, nm} = \frac{82 \, nm}{410 \, nm} = 0.20$$

In the classical view, this galaxy is receding at 20% of the speed of light; however, at 20% of the speed of light, relativistic effects are starting to become important. So, using the relativistic Doppler equation, we compute the true recession rate as

$$\frac{v}{c} = \frac{(z+1)^2 - 1}{(z+1)^2 + 1} = \frac{(0.2+1)^2 - 1}{(0.2+1)^2 + 1} = \frac{1.44 - 1}{1.44 + 1} = \frac{0.44}{2.44} = 0.18$$

Therefore, the actual recession speed is only 18% of the speed of light. While this may not initially seem like a big difference from the classical measurement, there is already an 11% deviation between the classical and the relativistic solutions; and at greater recession speeds, the divergence between the classical and relativistic speeds increases rapidly!

Quasars Obey the Hubble Law

The first question astronomers asked was whether quasars obeyed the Hubble law and were really at the large distances implied by their redshifts. If they did not obey the rule that large redshift means large distance, then they could be much closer, and their luminosity could be a lot less. One straightforward way to show that quasars had to obey the Hubble law was to demonstrate that they were actually part of galaxies, and that their redshift was the same as the galaxy that hosted them. Since ordinary galaxies *do* obey the Hubble law, anything within them would be subject to the same rules.

Observations with the Hubble Space Telescope provided the strongest evidence showing that quasars are located at the centers of galaxies. Hints that this is true had been obtained with ground-based telescopes, but space observations were required to make a convincing case. The reason is that quasars can outshine their entire galaxies by factors of 10 to 100 or even more. When this light passes through Earth's atmosphere, it is blurred by turbulence and drowns out the faint light from the surrounding galaxy—much as the bright headlights from an oncoming car at night make it difficult to see anything close by.

The Hubble Space Telescope, however, is not affected by atmospheric turbulence and can detect the faint glow from some of the galaxies that host quasars (Figure 27.5). Quasars have been found in the cores of both spiral and elliptical galaxies, and each quasar has the same redshift as its host galaxy. A wide range of studies with the Hubble Space Telescope now clearly demonstrate that quasars are indeed far away. If so, they must be producing a truly impressive amount of energy to be detectable as points of light that are much brighter than their galaxy. Interestingly, many quasar host galaxies are found to be involved in a collision with a second galaxy, providing, as we shall see, an important clue to the source of their prodigious energy output.

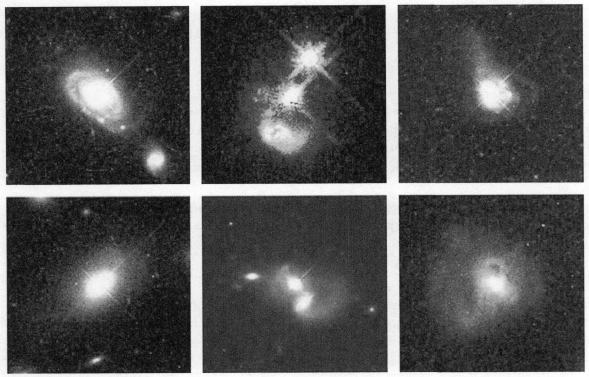

Figure 27.5 Quasar Host Galaxies. The Hubble Space Telescope reveals the much fainter "host" galaxies around quasars. The top left image shows a quasar that lies at the heart of a spiral galaxy 1.4 billion light-years from Earth. The bottom left image shows a quasar that lies at the center of an elliptical galaxy some 1.5 billion light-years from us. The middle images show remote pairs of interacting galaxies, one of which harbors a quasar. Each of the right images shows long tails of gas and dust streaming away from a galaxy that contains a quasar. Such tails are produced when one galaxy collides with another. (credit: modification of work by John Bahcall, Mike Disney, NASA)

The Size of the Energy Source

Given their large distances, quasars have to be extremely luminous to be visible to us at all—far brighter than any normal galaxy. In visible light alone, most are far more energetic than the brightest elliptical galaxies. But, as we saw, quasars also emit energy at X-ray and ultraviolet wavelengths, and some are radio sources as well. When all their radiation is added together, some QSOs have total luminosities as large as a hundred trillion Suns (10^{14} L_{Sun}), which is 10 to 100 times the brightness of luminous elliptical galaxies.

Finding a mechanism to produce the large amount of energy emitted by a quasar would be difficult under any circumstances. But there is an additional problem. When astronomers began monitoring quasars carefully, they found that some vary in luminosity on time scales of months, weeks, or even, in some cases, days. This variation is irregular and can change the brightness of a quasar by a few tens of percent in both its visible light and radio output.

Think about what such a change in luminosity means. A quasar at its dimmest is still more brilliant than any normal galaxy. Now imagine that the brightness increases by 30% in a few weeks. Whatever mechanism is responsible must be able to release new energy at rates that stagger our imaginations. The most dramatic changes in quasar brightness are equivalent to the energy released by 100,000 billion Suns. To produce this much energy we would have to convert the total mass of about ten Earths into energy every minute.

Moreover, because the fluctuations occur in such short times, the part of a quasar that is varying must be smaller than the distance light travels in the time it takes the variation to occur—typically a few months. To see why this must be so, let's consider a cluster of stars 10 light-years in diameter at a very large distance from Earth (see Figure 27.6, in which Earth is off to the right). Suppose every star in this cluster somehow brightens simultaneously and remains bright. When the light from this event arrives at Earth, we would first see the brighter light from stars on the near side; 5 years later we would see increased light from stars at the center. Ten years would pass before we detected more light from stars on the far side.

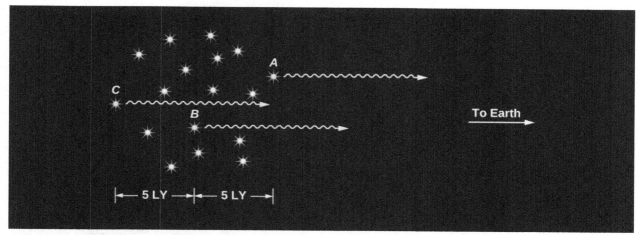

Figure 27.6 How the Size of a Source Affects the Timescale of Its Variability. This diagram shows why light variations from a large region in space appear to last for an extended period of time as viewed from Earth. Suppose all the stars in this cluster, which is 10 light-years across, brighten simultaneously and instantaneously. From Earth, star *A* will appear to brighten 5 years before star *B*, which in turn will appear to brighten 5 years earlier than star *C*. It will take 10 years for an Earth observer to get the full effect of the brightening.

Even though all stars in the cluster brightened at the same time, the fact that the cluster is 10 light-years wide means that 10 years must elapse before the increased light from every part of the cluster reaches us. From Earth we would see the cluster get brighter and brighter, as light from more and more stars began to reach us. Not until 10 years after the brightening began would we see the cluster reach maximum brightness. In other words, if an extended object suddenly flares up, it will seem to brighten over a period of time equal to the time it takes light to travel across the object from its far side.

We can apply this idea to brightness changes in quasars to estimate their diameters. Because quasars typically vary (get brighter and dimmer) over periods of a few months, the region where the energy is generated can be no larger than a few light-months across. If it were larger, it would take longer than a few months for the light from the far side to reach us.

How large is a region of a few light-months? Pluto, usually the outermost (dwarf) planet in our solar system, is about 5.5 light-hours from us, while the nearest star is 4 light-years away. Clearly a region a few light months across is tiny relative to the size of the entire Galaxy. And some quasars vary even more rapidly, which means their energy is generated in an even smaller region. Whatever mechanism powers the quasars must be able to generate more energy than that produced by an entire galaxy in a volume of space that, in some cases, is not much larger than our solar system.

Earlier Evidence

Even before the discovery of quasars, there had been hints that something very strange was going on in the centers of at least some galaxies. Back in 1918, American astronomer Heber Curtis used the large Lick Observatory telescope to photograph the galaxy Messier 87 in the constellation Virgo. On that photograph, he saw what we now call a jet coming from the center, or nucleus, of the galaxy (Figure 27.7). This jet literally and figuratively pointed to some strange activity going on in that galaxy nucleus. But he had no idea what it was.

No one else knew what to do with this space oddity either.

The random factoid that such a central jet existed lay around for a quarter century, until Carl Seyfert, a young astronomer at Mount Wilson Observatory, also in California, found half a dozen galaxies with extremely bright nuclei that were almost stellar, rather than fuzzy in appearance like most galaxy nuclei. Using spectroscopy, he found that these nuclei contain gas moving at up to two percent the speed of light. That may not sound like much, but it is 6 million miles per hour, and more than 10 times faster than the typical motions of stars in galaxies.

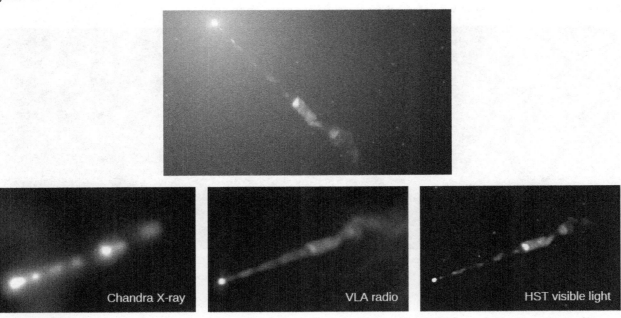

Figure 27.7 M87 Jet. Streaming out like a cosmic searchlight from the center of the galaxy, M87 is one of nature's most amazing phenomena, a huge jet of electrons and other particles traveling at nearly the speed of light. In this Hubble Space Telescope image, the blue of the jet contrasts with the yellow glow from the combined light of billions of unseen stars and yellow, point-like globular clusters that make up the galaxy (at the upper left). As we shall see later in this chapter, the jet, which is several thousand light-years long, originates in a disk of superheated gas swirling around a giant black hole at the center of M87. The light that we see is produced by electrons twisting along magnetic field lines in the jet, a process known as synchrotron radiation, which gives the jet its bluish tint. The jet in M87 can be observed in X-ray, radio, and visible light, as shown in the bottom three images. At the extreme left of each bottom image, we see the bright galactic nucleus harboring a supermassive black hole. (credit top: modification of work by NASA, The Hubble Heritage Team(STScI/AURA); credit bottom: modification of work by X-ray: H. Marshall (MIT), et al., CXC, NASA; Radio: F. Zhou, F. Owen (NRAO), J. Biretta (STScI); Optical: E. Perlman (UMBC), et al.)

After decades of study, astronomers identified many other strange objects beyond our Milky Way Galaxy; they populate a whole "zoo" of what are now called **active galaxies** or **active galactic nuclei (AGN)**. Astronomers first called them by many different names, depending on what sorts of observations discovered each category, but now we know that we are always looking at the same basic mechanism. What all these galaxies have in common is some activity in their nuclei that produces an enormous amount of energy in a very small volume of space. In the next section, we describe a model that explains all these galaxies with strong central activity—both the AGNs and the QSOs.

LINK TO LEARNING

To see a jet for yourself, check out a time-lapse video (https://openstax.org/l/30timelapsejet) of the jet ejected from NGC 3862.

27.2 Supermassive Black Holes: What Quasars Really Are

Learning Objectives

By the end of this section, you will be able to:

> Describe the characteristics common to all quasars
> Justify the claim that supermassive black holes are the source of the energy emitted by quasars (and AGNs)
> Explain how a quasar's energy is produced

In order to find a common model for quasars (and their cousins, the AGNs), let's first list the common characteristics we have been describing—and add some new ones:

- Quasars are hugely powerful, emitting more power in radiated light than all the stars in our Galaxy combined.
- Quasars are tiny, about the size of our solar system (to astronomers, that is really small!).
- Some quasars are observed to be shooting out pairs of straight jets at close to the speed of light, in a tight beam, to distances far beyond the galaxies they live in. These jets are themselves powerful sources of radio and gamma-ray radiation.
- Because quasars put out so much power from such a small region, they can't be powered by nuclear fusion the way stars are; they must use some process that is far more efficient.
- As we shall see later in this chapter, quasars were much more common when the universe was young than they are today. That means they must have been able to form in the first billion years or so after the universe began to expand.

The readers of this text are in a much better position than the astronomers who discovered quasars in the 1960s to guess what powers the quasars. That's because the key idea in solving the puzzle came from observations of the black holes we discussed in Black Holes and Curved Spacetime. The discovery of the first stellar mass black hole in the binary system Cygnus X-1 was announced in 1971, several years after the discovery of quasars. Proof that there is a black hole at the center of our own Galaxy came even later. Back when astronomers first began trying to figure out what powered quasars, black holes were simply one of the more exotic predictions of the general theory of relativity that still waited to be connected to the real world.

It was only as proof of the existence of black holes accumulated over several decades that it became clearer that only supermassive black holes could account for all the observed properties of quasars and AGNs. As we saw in The Milky Way Galaxy, our own Galaxy has a black hole in its center, and the energy is emitted from a small central region. While our black hole doesn't have the mass or energy of the quasar black holes, the mechanism that powers them is similar. The evidence now shows that most—and probably all—elliptical galaxies and all spirals with nuclear bulges have black holes at their centers. The amount of energy emitted by material near the black hole depends on two things: the mass of the black hole and the amount of matter that is falling into it.

If a black hole with a billion Suns' worth of mass inside (10^9 M_{Sun}) accretes (gathers) even a relatively modest amount of additional material—say, about 10 M_{Sun} per year—then (as we shall see) it can, in the process, produce as much energy as a thousand normal galaxies. This is enough to account for the total energy of a quasar. If the mass of the black hole is smaller than a billion solar masses or the accretion rate is low, then the amount of energy emitted can be much smaller, as it is in the case of the Milky Way.

LINK TO LEARNING

Watch a video (https://openstax.org/l/30mataccrsupblh) of an artist's impression of matter accreting around a supermassive black hole.

Observational Evidence for Black Holes

In order to prove that a black hole is present at the center of a galaxy, we must demonstrate that so much mass is crammed into so small a volume that no normal objects—massive stars or clusters of stars—could possibly account for it (just as we did for the black hole in the Milky Way). We already know from observations (discussed in Black Holes and Curved Spacetime) that an accreting black hole is surrounded by a hot *accretion disk* with gas and dust that swirl around the black hole before it falls in.

If we assume that the energy emitted by quasars is also produced by a hot accretion disk, then, as we saw in the previous section, the size of the disk must be given by the time the quasar energy takes to vary. For quasars, the emission in visible light varies on typical time scales of 5 to 2000 days, limiting the size of the disk to that many light-days.

In the X-ray band, quasars vary even more rapidly, so the light travel time argument tells us that this more energetic radiation is generated in an even smaller region. Therefore, the mass around which the accretion disk is swirling must be confined to a space that is even smaller. If the quasar mechanism involves a great deal of mass, then the only astronomical object that can confine a lot of mass into a very small space is a black hole. In a few cases, it turns out that the X-rays are emitted from a region just a few times the size of the black hole event horizon.

The next challenge, then, is to "weigh" this central mass in a quasar. In the case of our own Galaxy, we used observations of the orbits of stars very close to the galactic center, along with Kepler's third law, to estimate the mass of the central black hole (The Milky Way Galaxy). In the case of distant galaxies, we cannot measure the orbits of individual stars, but we can measure the orbital speed of the gas in the rotating accretion disk. The Hubble Space Telescope is especially well suited to this task because it is above the blurring of Earth's atmosphere and can obtain spectra very close to the bright central regions of active galaxies. The Doppler effect is then used to measure radial velocities of the orbiting material and so derive the speed with which it moves around.

One of the first galaxies to be studied with the Hubble Space Telescope is our old favorite, the giant elliptical M87. Hubble Space Telescope images showed that there is a disk of hot (10,000 K) gas swirling around the center of M87 (Figure 27.8). It was surprising to find hot gas in an elliptical galaxy because this type of galaxy is usually devoid of gas and dust. But the discovery was extremely useful for pinning down the existence of the black hole. Astronomers measured the Doppler shift of spectral lines emitted by this gas, found its speed of rotation, and then used the speed to derive the amount of mass inside the disk—applying Kepler's third law.

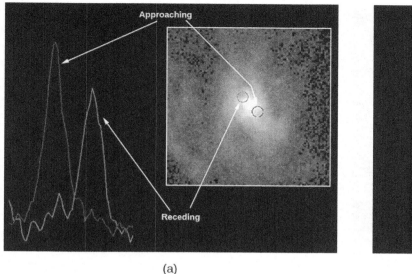

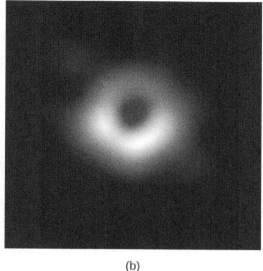

(a)   (b)

Figure 27.8 Evidence for a Black Hole at the Center of M87. (a) The disk of whirling gas at right was discovered at the center of the giant elliptical galaxy M87 with the Hubble Space Telescope. Observations made on opposite sides of the disk show that one side is approaching us (the spectral lines are blueshifted by the Doppler effect) while the other is receding (lines redshifted), a clear indication that the disk is rotating. The rotation speed is about 550 kilometers per second or 1.2 million miles per hour. Such a high rotation speed is evidence that there is a very massive black hole at the center of M87. (b) By 2019, the Event Horizon Telescope, a collaborative venture among radio telescopes spanning our planet, was able to produce the first radio image of the shadow of the M87 black hole on the accretion disk surrounding it. This was taken with radio waves, so color is arbitrary. The shadow is about 2 ½ times as big as the event horizon of the supermassive black hole. (credit (a): modification of work by Holland Ford, STScI/JHU; Richard Harms, Linda Dressel, Ajay K. Kochhar, Applied Research Corp.; Zlatan Tsvetanov, Arthur Davidsen, Gerard Kriss, Johns Hopkins; Ralph Bohlin, George Hartig, STScI; Bruce Margon, University of Washington in Seattle; NASA; credit (b): modification of work by European Southern Observatory (ESO), Event Horizon Telescope Collaboration)

Modern estimates show that there is a mass of at least 3.5 billion M_{Sun} concentrated in a tiny region at the very center of M87. So much mass in such a small volume of space must be a black hole. Let's stop for a moment and take in this figure: a single black hole that has swallowed enough material to make 3.5 billion stars like the Sun. Few astronomical measurements have ever led to so mind-boggling a result. What a strange environment the neighborhood of such a supermassive black hole must be.

Another example is shown in Figure 27.9. Here, we see a disk of dust and gas that surrounds a 300-million-M_{Sun} black hole in the center of an elliptical galaxy. (The bright spot in the center is produced by the combined light of stars that have been pulled close together by the gravitational force of the black hole.) The mass of the black hole was again derived from measurements of the rotational speed of the disk. The gas in the disk is moving around at 155 kilometers per second at a distance of only 186 light-years from its center. Given the pull of the mass at the center, we expect that the whole dust disk should be swallowed by the black hole in several billion years.

 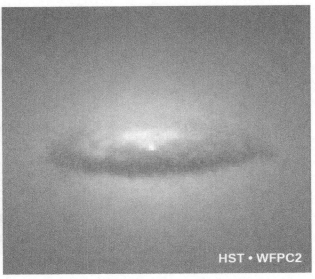

Figure 27.9 Another Galaxy with a Black-Hole Disk. The ground-based image shows an elliptical galaxy called NGC 7052 located in the constellation of Vulpecula, almost 200 million light-years from Earth. At the galaxy's center (right) is a dust disk roughly 3700 light-years in diameter. The disk rotates like a giant merry-go-round: gas in the inner part (186 light-years from the center) whirls around at a speed of 155 kilometers per second (341,000 miles per hour). From these measurements and Kepler's third law, it is possible to estimate that the disk is orbiting around a central black hole with a mass of 300 million Suns. (credit: modification of work by Roeland P. van der Marel (STScI), Frank C. van den Bosch (University of Washington), NASA)

But do we *have* to accept black holes as the only explanation of what lies at the center of these galaxies? What else could we put in such a small space other than a giant black hole? The alternative is stars. But to explain the masses in the centers of galaxies without a black hole we need to put at least a million stars in a region the size of the solar system. To fit, they would have be only 2 star diameters apart. Collisions between stars would happen all the time. And these collisions would lead to mergers of stars, and very soon the one giant star that they form would collapse into a black hole. So there is really no escape: only a black hole can fit so much mass into so small a space.

As we saw earlier, observations now show that all the galaxies with a spherical concentration of stars—either elliptical galaxies or spiral galaxies with nuclear bulges (see the chapter on Galaxies)—harbor one of these giant black holes at their centers. Among them is our neighbor spiral galaxy, the Andromeda galaxy, M31. The masses of these central black holes range from a just under a million up to at least 30 billion times the mass of the Sun. Several black holes may be even more massive, but the mass estimates have large uncertainties and need verification. We call these black holes "supermassive" to distinguish them from the much smaller black holes that form when some stars die (see The Death of Stars). So far, the most massive black holes from stars—those detected through gravitational waves detected by LIGO—have masses of about 40 solar masses.

Energy Production around a Black Hole

By now, you may be willing to entertain the idea that huge black holes lurk at the centers of active galaxies. But we still need to answer the question of how such a black hole can account for one of the most powerful sources of energy in the universe. As we saw in Black Holes and Curved Spacetime, a black hole itself can radiate no energy. Any energy we detect from it must come from material very close to the black hole, but not inside its event horizon.

In a galaxy, a central black hole (with its strong gravity) attracts matter—stars, dust, and gas—orbiting in the dense nuclear regions. This matter spirals in toward the spinning black hole and forms an accretion disk of material around it. As the material spirals ever closer to the black hole, it accelerates and becomes compressed, heating up to temperatures of millions of degrees. Such hot matter can radiate prodigious amounts of energy as it falls in toward the black hole.

To convince yourself that falling into a region with strong gravity can release a great deal of energy, imagine dropping a printed version of your astronomy textbook out the window of the ground floor of the library. It will land with a thud, and maybe give a surprised pigeon a nasty bump, but the energy released by its fall will not be very great. Now take the same book up to the fifteenth floor of a tall building and drop it from there. For anyone below, astronomy could suddenly become a deadly subject; when the book hits, it does so with a great deal of energy.

Dropping things from far away into the much stronger gravity of a black hole is much more effective in turning the energy released by infall into other forms of energy. Just as the falling book can heat up the air, shake the ground, or produce sound energy that can be heard some distance away, so the energy of material falling toward a black hole can be converted to significant amounts of electromagnetic radiation.

What a black hole has to work with is not textbooks but streams of infalling gas. If a dense blob of gas moves through a thin gas at high speed, it heats up as it slows by friction. As it slows down, kinetic (motion) energy is turned into heat energy. Just like a spaceship reentering the atmosphere (Figure 27.10), gas approaching a black hole heats up and glows where it meets other gas. But this gas, as it approaches the event horizon, reaches speeds of 10% the speed of light and more. It therefore gets far, far hotter than a spaceship, which reaches no more than about 1500 K. Indeed, gas near a supermassive black hole reaches a temperature of about 150,000 K, about 100 times hotter than a spaceship returning to Earth. It can even get so hot—millions of degrees—that it radiates X-rays.

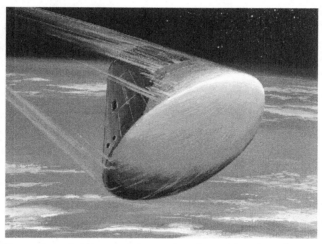

Figure 27.10 Friction in Earth's Atmosphere. In this artist's impression, the rapid motion of a spacecraft (the Apollo mission reentry capsule) through the atmosphere compresses and heats the air ahead of it, which heats the spacecraft in turn until it glows red hot. Pushing on the air slows down the spacecraft, turning the kinetic energy of the spacecraft into heat. Fast-moving gas falling into a quasar heats up in a similar way. (credit: modification of work by NASA)

The amount of energy that can be liberated this way is enormous. Einstein showed that mass and energy are interchangeable with his famous formula $E = mc^2$ (see The Sun: A Nuclear Powerhouse). A hydrogen bomb releases just 1% of that energy, as does a star. Quasars are much more efficient than that. The energy released falling to the event horizon of a black hole can easily reach 10% or, in the extreme theoretical limit, 32%, of that energy. (Unlike the hydrogen atoms in a bomb or a star, the gas falling into the black hole is not actually losing mass from its atoms to free up the energy; the energy is produced just because the gas is falling closer and closer to the black hole.) This huge energy release explains how a tiny volume like the region around a black hole can release as much power as a whole galaxy. But to radiate all that energy, instead of just falling inside the event horizon with barely a peep, the hot gas must take the time to swirl around the star in the accretion disk and emit some of its energy.

Most black holes don't show any signs of quasar emission. We call them "quiescent." But, like sleeping dragons, they can be woken up by being roused with a fresh supply of gas. Our own Milky Way black hole is currently quiescent, but it may have been a quasar just a few million years ago (Figure 27.11). Two giant bubbles that extend 25,000 light-years above and below the galactic center are emitting gamma rays. Were these produced a few million years ago when a significant amount of matter fell into the black hole at the center of the galaxy? Astronomers are still working to understand what remarkable event might have formed these enormous bubbles.

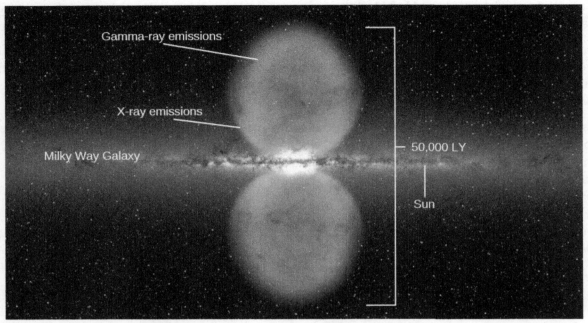

Figure 27.11 Fermi Bubbles in the Galaxy. Giant bubbles shining in gamma-ray light lie above and below the center of the Milky Way Galaxy, as seen by the Fermi satellite. (The gamma-ray and X-ray image is superimposed on a visible-light image of the inner parts of our Galaxy.) The bubbles may be evidence that the supermassive black hole at the center of our Galaxy was a quasar a few million years ago. (credit: modification of work by NASA's Goddard Space Flight Center)

The physics required to account for the exact way in which the energy of infalling material is converted to radiation near a black hole is far more complicated than our simple discussion suggests. To understand what happens in the "rough and tumble" region around a massive black hole, astronomers and physicists must resort to computer simulations (and they require supercomputers, fast machines capable of awesome numbers of calculations per second). The details of these models are beyond the scope of our book, but they support the basic description presented here.

Radio Jets

So far, our model seems to explain the central energy source in quasars and active galaxies. But, as we have seen, there is more to quasars and other active galaxies than the point-like energy source. They can also have long jets that glow with radio waves, light, and sometimes even X-rays, and that extend far beyond the limits of the parent galaxy. Can we find a way for our black hole and its accretion disk to produce these jets of energetic particles as well?

Many different observations have now traced these jets to within 3 to 30 light-years of the parent quasar or galactic nucleus. While the black hole and accretion disk are typically smaller than 1 light-year, we nevertheless presume that if the jets come this close, they probably originate in the vicinity of the black hole. Another characteristic of the jets we need to explain is that they contain matter moving close to the speed of light.

Why are energetic electrons and other particles near a supermassive black hole ejected into jets, and often into two oppositely directed jets, rather than in all directions? Again, we must use theoretical models and supercomputer simulations of what happens when a lot of material whirls inward in a crowded black hole accretion disk. Although there is no agreement on exactly how jets form, it has become clear that any material escaping from the neighborhood of the black hole has an easier time doing so *perpendicular to* the disk.

In some ways, the inner regions of black hole accretion disks resemble a baby that is just learning to eat by herself. As much food as goes into the baby's mouth can sometimes wind up being spit out in various directions. In the same way, some of the material whirling inward toward a black hole finds itself under tremendous pressure and orbiting with tremendous speed. Under such conditions, simulations show that a significant amount of material can be flung outward—not back along the disk, where more material is crowding in, but above and below the disk. If the disk is thick (as it tends to be when a lot of material falls in quickly), it can channel the outrushing material into narrow beams perpendicular to the disk (Figure 27.12).

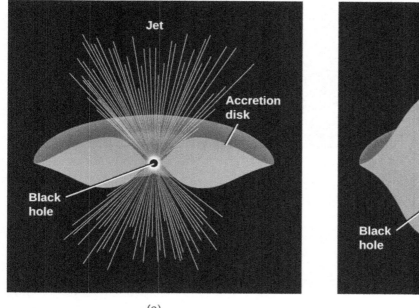
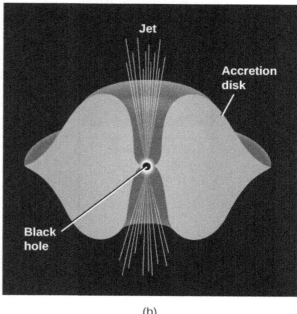

(a)   (b)

Figure 27.12 Models of Accretion Disks. These schematic drawings show what accretion disks might look like around large black holes for (a) a thin accretion disk and (b) a "fat" disk—the type needed to account for channeling the outflow of hot material into narrow jets oriented perpendicular to the disk.

Figure 27.13 shows observations of an elliptical galaxy that behaves in exactly this way. At the center of this active galaxy, there is a ring of dust and gas about 400 light-years in diameter, surrounding a 1.2-billion-M_{Sun} black hole. Radio observations show that two jets emerge in a direction perpendicular to the ring, just as the model predicts.

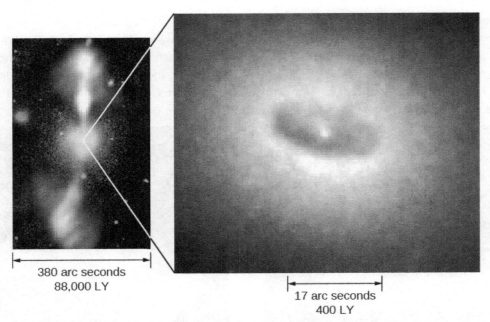

380 arc seconds
88,000 LY

17 arc seconds
400 LY

Figure 27.13 Jets and Disk in an Active Galaxy. The picture on the left shows the active elliptical galaxy NGC 4261, which is located in the Virgo Cluster at a distance of about 100 million light-years. The galaxy itself—the white circular region in the center—is shown the way it looks in visible light, while the jets are seen at radio wavelengths. A Hubble Space Telescope image of the central portion of the galaxy is shown on the right. It contains a ring of dust and gas about 800 light-years in diameter, surrounding a supermassive black hole. Note that the jets emerge from the galaxy in a direction perpendicular to the plane of the ring. (credit: modification of work by ESA/HST)

MAKING CONNECTIONS

Quasars and the Attitudes of Astronomers

The discovery of quasars in the early 1960s was the first in a series of surprises astronomers had in store. Within another decade they would find neutron stars (in the form of pulsars), the first hints of black holes (in binary X-ray sources), and even the radio echo of the Big Bang itself. Many more new discoveries lay ahead.

As Maarten Schmidt reminisced in 1988, "This had, I believe, a profound impact on the conduct of those practicing astronomy. Before the 1960s, there was much authoritarianism in the field. New ideas expressed at meetings would be instantly judged by senior astronomers and rejected if too far out." We saw a good example of this in the trouble Chandrasekhar had in finding acceptance for his ideas about the death of stars with cores greater than 1.4 M_{Sun} (see the feature box on Subrahmanyan Chandrasekhar).

"The discoveries of the 1960s," Schmidt continued, "were an embarrassment, in the sense that they were totally unexpected and could not be evaluated immediately. In reaction to these developments, an attitude has evolved where even outlandish ideas in astronomy are taken seriously. Given our lack of solid knowledge in extragalactic astronomy, this is probably to be preferred over authoritarianism."[1]

That is not to say that astronomers (being human) don't continue to have prejudices and preferences. For example, a small group of astronomers who thought that the redshifts of quasars were not connected with their distances (which was definitely a minority opinion) often felt excluded from meetings or from access to telescopes in the 1960s and 1970s. It's not so clear that they actually *were ex*cluded, as much as that they felt the very difficult pressure of knowing that most of their colleagues strongly disagreed with them. As it

---

1   M. Schmidt, "The Discovery of Quasars," in *Modern Cosmology in Retrospect*, ed. B. Bertotti et al. (Cambridge University Press, 1990).

turned out, the evidence—which must ultimately decide all scientific questions—was not on their side either.

But today, as better instruments bring solutions to some problems and starkly illuminate our ignorance about others, the entire field of astronomy seems more open to discussing unusual ideas. Of course, before any hypotheses become accepted, they must be tested—again and again—against the evidence that nature itself reveals. Still, the many strange proposals published about what dark matter might be (see The Evolution and Distribution of Galaxies) attest to the new openness that Schmidt described.

With this black hole model, we have come a long way toward understanding the quasars and active galaxies that seemed very mysterious only a few decades ago. As often happens in astronomy, a combination of better instruments (making better observations) and improved theoretical models enabled us to make significant progress on a puzzling aspect of the cosmos.

27.3 | Quasars as Probes of Evolution in the Universe

Learning Objectives

By the end of this section, you will be able to:

> Trace the rise and fall of quasars over cosmic time
> Describe some of the ways in which galaxies and black holes influence each other's growth
> Describe some ways the first black holes may have formed
> Explain why some black holes are not producing quasar emission but rather are quiescent

The quasars' brilliance and large distance make them ideal probes of the far reaches of the universe and its remote past. Recall that when first introducing quasars, we mentioned that they generally tend to be far away. When we see extremely distant objects, we are seeing them as they were long ago. Radiation from a quasar 8 billion light-years away is telling us what that quasar and its environment were like 8 billion years ago, much closer to the time that the galaxy that surrounds it first formed. Astronomers have now detected light emitted from quasars that were already formed only a few hundred million years after the universe began its expansion 13.8 billion years ago. Thus, they give us a remarkable opportunity to learn about the time when large structures were first assembling in the cosmos.

The Evolution of Quasars

Quasars provide compelling evidence that we live in an evolving universe—one that changes with time. They tell us that astronomers living billions of years ago would have seen a universe that is very different from the universe of today. Counts of the number of quasars at different redshifts (and thus at different times in the evolution of the universe) show us how dramatic these changes are (Figure 27.14). We now know that the number of quasars was greatest at the time when the universe was only 20% of its present age.

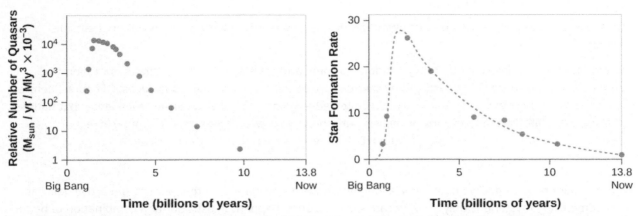

Figure 27.14 Relative Number of Quasars and Rate at Which Stars Formed as a Function of the Age of the Universe. An age of 0 on the plots corresponds to the beginning of the universe; an age of 13.8 corresponds to the present time. Both the number of quasars and the rate of star formation were at a peak when the universe was about 20% as old as it is now.

As you can see, the drop-off in the numbers of quasars as time gets nearer to the present day is quite abrupt. Observations also show that the emission from the accretion disks around the most massive black holes peaks early and then fades. The most powerful quasars are seen only at early times. In order to explain this result, we make use of our model of the energy source of the quasars—namely that quasars are black holes with enough fuel to make a brilliant accretion disk right around them.

The fact that there were more quasars long ago (far away) than there are today (nearby) could be explained if there was more material available to be accreted by black holes early in the history of the universe. You might say that the quasars were more active when their black holes had fuel for their "energy-producing engines." If that fuel was mostly consumed in the first few billion years after the universe began its expansion, then later in its life, a "hungry" black hole would have very little left with which to light up the galaxy's central regions.

In other words, if matter in the accretion disk is continually being depleted by falling into the black hole or being blown out from the galaxy in the form of jets, then a quasar can continue to radiate only as long as new gas is available to replenish the accretion disk.

In fact, there *was* more gas around to be accreted early in the history of the universe. Back then, most gas had not yet collapsed to form stars, so there was more fuel available for both the feeding of black holes and the forming of new stars. Much of that fuel was subsequently consumed in the formation of stars during the first few billion years after the universe began its expansion. Later in its life, a galaxy would have little left to feed a hungry black hole or to form more new stars. As we see from Figure 27.14, both star formation and black hole growth peaked together when the universe was about 2 billion years old. Ever since, both have been in sharp decline. We are late to the party of the galaxies and have missed some of the early excitement.

Observations of nearer galaxies (seen later in time) indicate that there is another source of fuel for the central black holes—the collision of galaxies. If two galaxies of similar mass collide and merge, or if a smaller galaxy is pulled into a larger one, then gas and dust from one may come close enough to the black hole in the other to be devoured by it and so provide the necessary fuel. Astronomers have found that collisions were also much more common early in the history of the universe than they are today. There were more small galaxies in those early times because over time, as we shall see (in The Evolution and Distribution of Galaxies), small galaxies tend to combine into larger ones. Again, this means that we would expect to see more quasars long ago (far away) than we do today (nearby)—as we in fact do.

Codependence of Black Holes and Galaxies

Once black hole masses began to be measured reliably in the late 1990s, they posed an enigma. It looked as though the mass of the central black hole depended on the mass of the galaxy. The black holes in galaxies always seem to be just 1/200 the mass of the galaxy they live in. This result is shown schematically in Figure

27.15, and some of the observations are plotted in Figure 27.16.

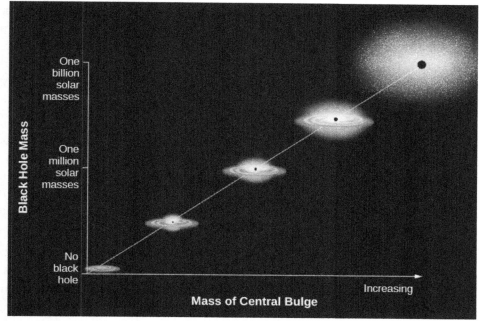

Figure 27.15 Relationship between Black Hole Mass and the Mass of the Host Galaxy. Observations show that there is a close correlation between the mass of the black hole at the center of a galaxy and the mass of the spherical distribution of stars that surrounds the black hole. That spherical distribution may be in the form of either an elliptical galaxy or the central bulge of a spiral galaxy. (credit: modification of work by K. Cordes, S. Brown (STScI))

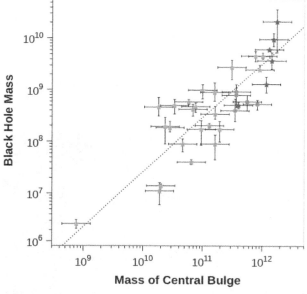

Figure 27.16 Correlation between the Mass of the Central Black Hole and the Mass Contained within the Bulge of Stars Surrounding the Black Hole, Using Data from Real Galaxies. The black hole always turns out to be about 1/200 the mass of the stars surrounding it. The horizontal and vertical bars surrounding each point show the uncertainty of the measurement. (credit: modification of work by Nicholas J. McConnell, Chung-Pei Ma, "Revisiting the Scaling Relations of Black Hole Masses and Host Galaxy Properties," *The Astrophysical Journal*, 764:184 (14 pp.), February 20, 2013.)

Somehow black hole mass and the mass of the surrounding bulge of stars are connected. But why does this correlation exist? Unfortunately, astronomers do not yet know the answer to this question. We do know, however, that the black hole can influence the rate of star formation in the galaxy, and that the properties of the surrounding galaxy can influence how fast the black hole grows. Let's see how these processes work.

How a Galaxy Can Influence a Black Hole in Its Center

Let's look first at how the surrounding galaxy might influence the growth and size of the black hole. Without large quantities of fresh "food," the surroundings of black holes glow only weakly as bits of local material spiral inward toward the black hole. So somehow large amounts of gas have to find their way to the black hole from the galaxy in order to feed the quasar and make it grow and give off the energy to be noticed. Where does this "food" for the black hole come from originally and how might it be replenished? The jury is still out, but the options are pretty clear.

One obvious source of fuel for the black hole is matter from the host galaxy itself. Galaxies start out with large amounts of interstellar gas and dust, and at least some of this interstellar matter is gradually converted into stars as the galaxy evolves. On the other hand, as stars go through their lives and die, they lose mass all the time into the space between them, thereby returning some of the gas and dust to the interstellar medium. We expect to find more gas and dust in the central regions early in a galaxy's life than later on, when much of it has been converted into stars. Any of the interstellar matter that ventures too close to the black hole may be accreted by it. This means that we would expect that the number and luminosity of quasars powered in this way would decline with time. And as we have seen, that is just what we find.

Today both *elliptical galaxies* and the *nuclear bulges of spiral galaxies* have very little raw material left to serve as a source of fuel for the black hole. And most of the giant black holes in nearby galaxies, including the one in our own Milky Way, are now dark and relatively quiet—mere shadows of their former selves. So that fits with our observations.

We should note that even if you have a quiescent supermassive black hole, a star in the area could occasionally get close to it. Then the powerful tidal forces of the black hole can pull the whole star apart into a stream of gas. This stream quickly forms an accretion disk that gives off energy in the normal way and makes the black hole region into a temporary quasar. However, the material will fall into the black hole after only a few weeks or months. The black hole then goes back into its lurking, quiescent state, until another victim wanders by.

This sort of "cannibal" event happens only once every 100,000 years or so in a typical galaxy. But we can monitor millions of galaxies in the sky, so a few of these "tidal disruption events" are found each year (Figure 27.17). However, these individual events, dramatic as they are, are too rare to account for the huge masses of the central black holes.

Figure 27.17 A Black Hole Snacks on a Star. This artist's impression shows three stages of a star (red) swinging too close to a giant black hole (black circle). The star starts off (top left) in its normal spherical shape, then begins to be pulled into a long football shape

by tides raised by the black hole (center). When the star gets closer still, the tides become stronger than the gravity holding the star together, and it breaks up into a streamer (right). Much of the star's matter forms a temporary accretion disk that lights up as a quasar for a few weeks or months. (credit: modification of work by NASA/CXC/M. Weiss)

Another source of fuel for the black hole is the collision of its host galaxy with another galaxy. Some of the brightest galaxies turn out, when a detailed picture is taken, to be pairs of colliding galaxies. And most of them have quasars inside them, not easily visible to us because they are buried by enormous amounts of dust and gas.

A collision between two cars creates quite a mess, pushing parts out of their regular place. In the same way, if two galaxies collide and merge, then gas and dust (though not so much the stars) can get pushed out of their regular orbits. Some may veer close enough to the black hole in one galaxy or the other to be devoured by it and so provide the necessary fuel to power a quasar. As we saw, galaxy collisions and mergers happened most frequently when the universe was young and probably help account for the fact that quasars were most common when the universe was only about 20% of its current age.

Collisions in today's universe are less frequent, but they do happen. Once a galaxy reaches the size of the Milky Way, most of the galaxies it merges with will be much smaller galaxies—*dwarf galaxies* (see the chapter on Galaxies). These don't disrupt the big galaxy much, but they can supply some additional gas to its black hole.

By the way, if two galaxies, each of which contains a black hole, collide, then the two black holes may merge and form an even larger black hole (Figure 27.18). In this process they will emit a burst of gravitational waves. One of the main goals of the European Space Agency's planned LISA (Laser Interferometer Space Antenna) mission is to detect the gravitational wave signals from the merging of supermassive black holes.

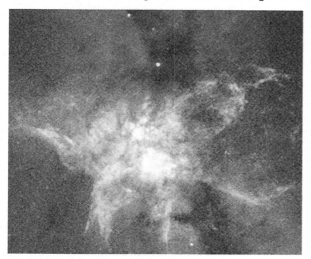 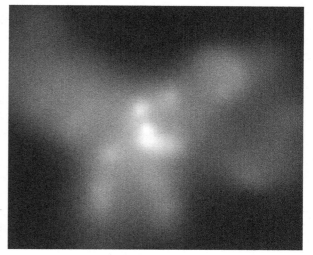

Figure 27.18 Colliding Galaxies with Two Black Holes. We compare Hubble Space Telescope visible-light (left) and Chandra X-ray (right) images of the central regions of NGC 6240, a galaxy about 400 million light-years away. It is a prime example of a galaxy in which stars are forming, evolving, and exploding at an exceptionally rapid rate due to a relatively recent merger (30 million years ago). The Chandra image shows two bright X-ray sources, each produced by hot gas surrounding a black hole. Over the course of the next few hundred million years, the two supermassive black holes, which are about 3000 light-years apart, will drift toward each other and merge to form an even larger black hole. This detection of a binary black hole supports the idea that black holes can grow to enormous masses in the centers of galaxies by merging with nearby galaxies. (credit left: modification of work by NASA/CXC/MPE/S.Komossa et al; credit right: NASA/STScI/R. P. van der Marel, J. Gerssen)

LINK TO LEARNING

Watch two galaxies collide (https://openstax.org/l/30galcolsuperma) to form a supermassive black hole.

How Does the Black Hole Influence the Formation of Stars in the Galaxy?

We have seen that the material in galaxies can influence the growth of the black hole. The black hole in turn can also influence the galaxy in which it resides. It can do so in three ways: through its jets, through winds of particles that manage to stream away from the accretion disk, and through radiation from the accretion disk. As they stream away from the black hole, all three can either promote star formation by compressing the surrounding gas and dust—or instead suppress star formation by heating the surrounding gas and shredding molecular clouds, thereby inhibiting or preventing star formation. The outflowing energy can even be enough to halt the accretion of new material and starve the black hole of fuel. Astronomers are still trying to evaluate the relative importance of these effects in determining the overall evolution of galactic bulges and the rates of star formation.

In summary, we have seen how galaxies and supermassive black holes can each influence the evolution of the other: the galaxy supplies fuel to the black hole, and the quasar can either support or suppress star formation. The balance of these processes probably helps account for the correlation between black hole and bulge masses, but there are as yet no theories that explain quantitatively and in detail why the correlation between black hole and bulge masses is as tight as it is or why the black hole mass is always about 1/200 times the mass of the bulge.

The Birth of Black Holes and Galaxies

While the connection between quasars and galaxies is increasingly clear, the biggest puzzle of all—namely, how the supermassive black holes in galaxies got started—remains unsolved. Observations show that they existed when the universe was very young. One dramatic example is the discovery of a quasar that was already shining when the universe was only 700 million years old. What does it take to create a large black hole so quickly? A related problem is that in order to eventually build black holes containing more than 2 billion solar masses, it is necessary to have giant "seed" black holes with masses at least 2000 times the mass of the Sun—and they must somehow have been created shortly after the expansion of the universe began.

Astronomers are now working actively to develop models for how these seed black holes might have formed. Theories suggest that galaxies formed from collapsing clouds of dark matter and gas. Some of the gas formed stars, but perhaps some of the gas settled to the center where it became so concentrated that it formed a black hole. If this happened, the black hole could form right away—although this requires that the gas should not be rotating very much initially.

A more likely scenario is that the gas will have some angular momentum (rotation) that will prevent direct collapse to a black hole. In that case, the very first generation of stars will form, and some of them, according to calculations, will have masses hundreds of times that of the Sun. When these stars finish burning hydrogen, just a few million years later, the supernovae they end with will create black holes a hundred or so times the mass of the Sun. These can then merge with others or accrete the rich gas supply available at these early times.

The challenge is growing these smaller black holes quickly enough to make the much larger black holes we see a few hundred million years later. It turns out to be difficult because there are limits on how fast they can accrete matter. These should make sense to you from what we discussed earlier in the chapter. If the rate of accretion becomes too high, then the energy streaming outward from the black hole's accretion disk will become so strong as to blow away the infalling matter.

What if, instead, a collapsing gas cloud doesn't form a black hole directly or break up and form a group of regular stars, but stays together and makes one fairly massive star embedded within a dense cluster of thousands of lower mass stars and large quantities of dense gas? The massive star will have a short lifetime and will soon collapse to become a black hole. It can then begin to attract the dense gas surrounding it. But calculations show that the gravitational attraction of the many nearby stars will cause the black hole to zigzag randomly within the cluster and will prevent the formation of an accretion disk. If there is no accretion disk, then matter can fall freely into the black hole from all directions. Calculations suggest that under these conditions, a black hole even as small as 10 times the mass of the Sun could grow to more than 10 billion times the mass of the Sun by the time the universe is a billion years old.

Scientists are exploring other ideas for how to form the seeds of supermassive black holes, and this remains a very active field of research. Whatever mechanism caused the rapid formation of these supermassive black holes, they do give us a way to observe the youthful universe when it was only about five percent as old as it is now.

LINK TO LEARNING

Take a look at some new results (https://openstax.org/l/30chanxrayobser) from the Chandra X-ray Observatory about the formation of supermassive black holes in the early universe.

 Key Terms

active galactic nuclei (AGN)   galaxies that are almost as luminous as quasars and share many of their properties, although to a less spectacular degree; abnormal amounts of energy are produced in their centers

active galaxies   galaxies that house active galactic nuclei

quasar   an object of very high redshift that looks like a star but is extragalactic and highly luminous; also called a quasi-stellar object, or QSO

Summary

27.1 Quasars

The first quasars discovered looked like stars but had strong radio emission. Their visible-light spectra at first seemed confusing, but then astronomers realized that they had much larger redshifts than stars. The quasar spectra obtained so far show redshifts ranging from 15% to more than 96% the speed of light. Observations with the Hubble Space Telescope show that quasars lie at the centers of galaxies and that both spirals and ellipticals can harbor quasars. The redshifts of the underlying galaxies match the redshifts of the quasars embedded in their centers, thereby proving that quasars obey the Hubble law and are at the great distances implied by their redshifts. To be noticeable at such great distances, quasars must have 10 to 100 times the luminosity of the brighter normal galaxies. Their variations show that this tremendous energy output is generated in a small volume—in some cases, in a region not much larger than our own solar system. A number of galaxies closer to us also show strong activity at their centers—activity now known to be caused by the same mechanism as the quasars.

27.2 Supermassive Black Holes: What Quasars Really Are

Both active galactic nuclei and quasars derive their energy from material falling toward, and forming a hot accretion disk around, a massive black hole. This model can account for the large amount of energy emitted and for the fact that the energy is produced in a relatively small volume of space. It can also explain why jets coming from these objects are seen in two directions: those directions are perpendicular to the accretion disk.

27.3 Quasars as Probes of Evolution in the Universe

Quasars and galaxies affect each other: the galaxy supplies fuel to the black hole, and the quasar heats and disrupts the gas clouds in the galaxy. The balance between these two processes probably helps explain why the black hole seems always to be about 1/200 the mass of the spherical bulge of stars that surrounds the black hole.

Quasars were much more common billions of years ago than they are now, and astronomers speculate that they mark an early stage in the formation of galaxies. Quasars were more likely to be active when the universe was young and fuel for their accretion disk was more available.

Quasar activity can be re-triggered by a collision between two galaxies, which provides a new source of fuel to feed the black hole.

For Further Exploration

Articles

Bartusiak, M. "A Beast in the Core." *Astronomy* (July 1998): 42. On supermassive black holes at the centers of galaxies.

Disney, M. "A New Look at Quasars." *Scientific American* (June 1998): 52.

Djorgovski, S. "Fires at Cosmic Dawn." *Astronomy* (September 1995): 36. On quasars and what we can learn from them.

Ford, H., & Tsvetanov, Z. "Massive Black Holes at the Hearts of Galaxies." *Sky & Telescope* (June 1996): 28. Nice overview.

Irion, R. "A Quasar in Every Galaxy?" *Sky & Telescope* (July 2006): 40. Discusses how supermassive black holes powering the centers of galaxies may be more common than thought.

Kormendy, J. "Why Are There so Many Black Holes?" *Astronomy* (August 2016): 26. Discussion of why supermassive black holes are so common in the universe.

Kruesi, L. "Secrets of the Brightest Objects in the Universe." *Astronomy* (July 2013): 24. Review of our current understanding of quasars and how they help us learn about black holes.

Miller, M., et al. "Supermassive Black Holes: Shaping their Surroundings." *Sky & Telescope* (April 2005): 42. Jets from black hole disks.

Nadis, S. "Exploring the Galaxy–Black Hole Connection." *Astronomy* (May 2010): 28. Overview.

Nadis, S. "Here, There, and Everywhere." *Astronomy* (February 2001): 34. On Hubble observations showing how common supermassive black holes are in galaxies.

Nadis, S. "Peering inside a Monster Galaxy." *Astronomy* (May 2014): 24. What X-ray observations tell us about the mechanism that powers the active galaxy M87.

Olson, S. "Black Hole Hunters." *Astronomy* (May 1999): 48. Profiles four astronomers who search for "hungry" black holes at the centers of active galaxies.

Peterson, B. "Solving the Quasar Puzzle." *Sky & Telescope* (September 2013): 24. A review article on how we figured out that black holes were the power source for quasars, and how we view them today.

Tucker, W., et al. "Black Hole Blowback." *Scientific American* (March 2007): 42. How supermassive black holes create giant bubbles in the intergalactic medium.

Voit, G. "The Rise and Fall of Quasars." *Sky & Telescope* (May 1999): 40. Good overview of how quasars fit into cosmic history.

Wanjek, C. "How Black Holes Helped Build the Universe." *Sky & Telescope* (January 2007): 42. On the energy and outflow from disks around supermassive black holes; nice introduction.

Websites

Monsters in Galactic Nuclei: https://chandra.as.utexas.edu/stardate.html (https://chandra.as.utexas.edu/stardate.html). An article on supermassive black holes by John Kormendy, from *StarDate* magazine.

Quasar Astronomy Forty Years On: https://www.astr.ua.edu/keel/agn/quasar40.html (https://www.astr.ua.edu/keel/agn/quasar40.html). A 2003 popular article by William Keel.

Quasars and Active Galactic Nuclei: https://www.astr.ua.edu/keel/agn/ (https://www.astr.ua.edu/keel/agn/). An annotated gallery of images showing the wide range of activity in galaxies. There is also an introduction, a glossary, and background information. Also by William Keel.

Quasars: "The Light Fantastic": https://hubblesite.org/newscenter/archive/releases/1996/35/background/ (https://hubblesite.org/newscenter/archive/releases/1996/35/background/). This brief "backgrounder" from the public information office at the HubbleSite gives a bit of the history of the discovery and understanding of quasars.

Videos

Active Galaxies: https://vimeo.com/21079798 (https://vimeo.com/21079798). Part of the *Astronomy:*

Observations and Theories series; half-hour introduction to quasars and related objects (27:28).

Black Hole Chaos: The Environments of the Most Supermassive Black Holes in the Universe: https://www.youtube.com/watch?v=hzSgU-3d8QY (https://www.youtube.com/watch?v=hzSgU-3d8QY). May 2013 lecture by Dr. Belinda Wilkes and Dr. Francesca Civano of the Center for Astrophysics in the CfA Observatory Nights Lecture Series (50:14).

Hubble and Black Holes: https://www.spacetelescope.org/videos/hubblecast43a/ (https://www.spacetelescope.org/videos/hubblecast43a/). Hubblecast on black holes and active galactic nuclei (9:10).

Monster Black Holes: https://www.youtube.com/watch?v=LN9oYjNKBm8 (https://www.youtube.com/watch?v=LN9oYjNKBm8). May 2013 lecture by Professor Chung-Pei Ma of the University of California, Berkeley; part of the Silicon Valley Astronomy Lecture Series (1:18:03).

Image of a Black Hole: https://www.eso.org/public/videos/eso1907a/ (https://www.eso.org/public/videos/eso1907a/). Video about the Event Horizon Telescope and its pioneering image of the shadow of the event horizon in the active galaxy M87.

Collaborative Group Activities

A. When quasars were first discovered and the source of their great energy was unknown, some astronomers searched for evidence that quasars are much nearer to us than their redshifts imply. (That way, they would not have to produce so much energy to look as bright as they do.) One way was to find a "mismatched pair"—a quasar and a galaxy with different redshifts that lie in very nearly the same direction in the sky. Suppose you do find one and only one galaxy with a quasar very close by, and the redshift of the quasar is six times larger than that of the galaxy. Have your group discuss whether you could then conclude that the two objects are at the same distance and that redshift is *not* a reliable indicator of distance. Why? Suppose you found three such pairs, each with different mismatched redshifts? Suppose *every* galaxy has a nearby quasar with a different redshift. How would your answer change and why?

B. Large ground-based telescopes typically can grant time to only one out of every four astronomers who apply for observing time. One prominent astronomer tried for several years to establish that the redshifts of quasars do not indicate their distances. At first, he was given time on the world's largest telescope, but eventually it became clearer that quasars were just the centers of active galaxies and that their redshifts really did indicate distance. At that point, he was denied observing time by the committee of astronomers who reviewed such proposals. Suppose your group had been the committee. What decision would you have made? Why? (In general, what criteria should astronomers have for allowing astronomers whose views completely disagree with the prevailing opinion to be able to pursue their research?)

C. Based on the information in this chapter and in Black Holes and Curved Spacetime, have your group discuss what it would be like near the event horizon of a supermassive black hole in a quasar or active galaxy. Make a list of all the reasons a trip to that region would not be good for your health. Be specific.

D. Before we understood that the energy of quasars comes from supermassive black holes, astronomers were baffled by how such small regions could give off so much energy. A variety of models were suggested, some involving new physics or pretty "far out" ideas from current physics. Can your group come up with some areas of astronomy that you have studied in this course where we don't yet have an explanation for something happening in the cosmos?

Exercises

Review Questions

1. Describe some differences between quasars and normal galaxies.

2. Describe the arguments supporting the idea that quasars are at the distances indicated by their redshifts.

3. In what ways are active galaxies like quasars but different from normal galaxies?

4. Why could the concentration of matter at the center of an active galaxy like M87 not be made of stars?

5. Describe the process by which the action of a black hole can explain the energy radiated by quasars.

6. Describe the observations that convinced astronomers that M87 is an active galaxy.

7. Why do astronomers believe that quasars represent an early stage in the evolution of galaxies?

8. Why were quasars and active galaxies not initially recognized as being "special" in some way?

9. What do we now understand to be the primary difference between normal galaxies and active galaxies?

10. What is the typical structure we observe in a quasar at radio frequencies?

11. What evidence do we have that the luminous central region of a quasar is small and compact?

Thought Questions

12. Suppose you observe a star-like object in the sky. How can you determine whether it is actually a star or a quasar?

13. Why don't any of the methods for establishing distances to galaxies, described in Galaxies (other than Hubble's law itself), work for quasars?

14. One of the early hypotheses to explain the high redshifts of quasars was that these objects had been ejected at very high speeds from other galaxies. This idea was rejected, because no quasars with large blueshifts have been found. Explain why we would expect to see quasars with both blueshifted and redshifted lines if they were ejected from nearby galaxies.

15. A friend of yours who has watched many Star Trek episodes and movies says, "I thought that black holes pulled everything into them. Why then do astronomers think that black holes can explain the great outpouring of energy from quasars?" How would you respond?

16. Could the Milky Way ever become an active galaxy? Is it likely to ever be as luminous as a quasar?

17. Why are quasars generally so much more luminous (why do they put out so much more energy) than active galaxies?

18. Suppose we detect a powerful radio source with a radio telescope. How could we determine whether or not this was a newly discovered quasar and not some nearby radio transmission?

19. A friend tries to convince you that she can easily see a quasar in her backyard telescope. Would you believe her claim?

Figuring for Yourself

20. Show that no matter how big a redshift (z) we measure, v/c will never be greater than 1. (In other words, no galaxy we observe can be moving away faster than the speed of light.)

21. If a quasar has a redshift of 3.3, at what fraction of the speed of light is it moving away from us?

22. If a quasar is moving away from us at $v/c = 0.8$, what is the measured redshift?

23. In the chapter, we discussed that the largest redshifts found so far are greater than 6. Suppose we find a quasar with a redshift of 6.1. With what fraction of the speed of light is it moving away from us?

24. Rapid variability in quasars indicates that the region in which the energy is generated must be small. You can show why this is true. Suppose, for example, that the region in which the energy is generated is a transparent sphere 1 light-year in diameter. Suppose that in 1 s this region brightens by a factor of 10 and remains bright for two years, after which it returns to its original luminosity. Draw its light curve (a graph of its brightness over time) as viewed from Earth.

25. Large redshifts move the positions of spectral lines to longer wavelengths and change what can be observed from the ground. For example, suppose a quasar has a redshift of $\frac{\Delta\lambda}{\lambda} = 4.1$. At what wavelength would you make observations in order to detect its Lyman line of hydrogen, which has a laboratory or rest wavelength of 121.6 nm? Would this line be observable with a ground-based telescope in a quasar with zero redshift? Would it be observable from the ground in a quasar with a redshift of $\frac{\Delta\lambda}{\lambda} = 4.1$?

26. Once again in this chapter, we see the use of Kepler's third law to estimate the mass of supermassive black holes. In the case of NGC 4261, this chapter supplied the result of the calculation of the mass of the black hole in NGC 4261. In order to get this answer, astronomers had to measure the velocity of particles in the ring of dust and gas that surrounds the black hole. How high were these velocities? Turn Kepler's third law around and use the information given in this chapter about the galaxy NGC 4261—the mass of the black hole at its center and the diameter of the surrounding ring of dust and gas—to calculate how long it would take a dust particle in the ring to complete a single orbit around the black hole. Assume that the only force acting on the dust particle is the gravitational force exerted by the black hole. Calculate the velocity of the dust particle in km/s.

27. In the Check Your Learning section of Example 27.1, you were told that several lines of hydrogen absorption in the visible spectrum have rest wavelengths of 410 nm, 434 nm, 486 nm, and 656 nm. In a spectrum of a distant galaxy, these same lines are observed to have wavelengths of 492 nm, 521 nm, 583 nm, and 787 nm, respectively. The example demonstrated that $z = 0.20$ for the 410 nm line. Show that you will obtain the same redshift regardless of which absorption line you measure.

28. In the Check Your Learning section of Example 27.1, the author commented that even at $z = 0.2$, there is already an 11% deviation between the relativistic and the classical solution. What is the percentage difference between the classical and relativistic results at $z = 0.1$? What is it for $z = 0.5$? What is it for $z = 1$?

29. The quasar that appears the brightest in our sky, 3C 273, is located at a distance of 2.4 billion light-years. The Sun would have to be viewed from a distance of 1300 light-years to have the same apparent magnitude as 3C 273. Using the inverse square law for light, estimate the luminosity of 3C 273 in solar units.

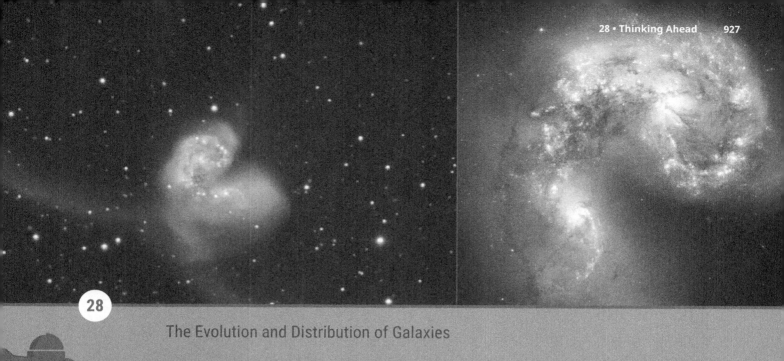

28

The Evolution and Distribution of Galaxies

Figure 28.1 Colliding Galaxies. Collisions and mergers of galaxies strongly influence their evolution. On the left is a ground-based image of two colliding galaxies (NCG 4038 and 4039), sometimes nicknamed the Antennae galaxies. The long, luminous tails are material torn out of the galaxies by tidal forces during the collision. The right image shows the inner regions of these two galaxies, as taken by the Hubble Space Telescope. The cores of the twin galaxies are the orange blobs to the lower left and upper right of the center of the image. Note the dark lanes of dust crossing in front of the bright regions. The bright pink and blue star clusters are the result of a burst of star formation stimulated by the collision. (credit left: modification of work by Bob and Bill Twardy/Adam Block/NOAO/AURA/NSF; credit right: modification of work by NASA, ESA, and the Hubble Heritage Team (STScI/AURA)-ESA/Hubble Collaboration)

Chapter Outline

28.1 Observations of Distant Galaxies
28.2 Galaxy Mergers and Active Galactic Nuclei
28.3 The Distribution of Galaxies in Space
28.4 The Challenge of Dark Matter
28.5 The Formation and Evolution of Galaxies and Structure in the Universe

 Thinking Ahead

How and when did galaxies like our Milky Way form? Which formed first: stars or galaxies? Can we see direct evidence of the changes galaxies undergo over their lifetimes? If so, what determines whether a galaxy will "grow up" to be spiral or elliptical? And what is the role of "nature versus nurture"? That is to say, how much of a galaxy's development is determined by what it looks like when it is born and how much is influenced by its environment?

Astronomers today have the tools needed to explore the universe almost back to the time it began. The huge new telescopes and sensitive detectors built in the last decades make it possible to obtain both images and spectra of galaxies so distant that their light has traveled to reach us for more than 13 billion years—more than 90% of the way back to the Big Bang: we can use the finite speed of light and the vast size of the universe as a cosmic time machine to peer back and observe how galaxies formed and evolved over time. Studying galaxies so far away in any detail is always a major challenge, largely because their distance makes them appear very faint. However, today's large telescopes on the ground and in space are finally making such a task possible.

28.1 Observations of Distant Galaxies

Learning Objectives

By the end of this section, you will be able to:

> Explain how astronomers use light to learn about distant galaxies long ago
> Discuss the evidence showing that the first stars formed when the universe was less than 10% of its current age
> Describe the major differences observed between galaxies seen in the distant, early universe and galaxies seen in the nearby universe today

Let's begin by exploring some techniques astronomers use to study how galaxies are born and change over cosmic time. Suppose you wanted to understand how adult humans got to be the way they are. If you were very dedicated and patient, you could actually observe a sample of babies from birth, following them through childhood, adolescence, and into adulthood, and making basic measurements such as their heights, weights, and the proportional sizes of different parts of their bodies to understand how they change over time.

Unfortunately, we have no such possibility for understanding how galaxies grow and change over time: in a human lifetime—or even over the entire history of human civilization—individual galaxies change hardly at all. We need other tools than just patiently observing single galaxies in order to study and understand those long, slow changes.

We do, however, have one remarkable asset in studying galactic evolution. As we have seen, the universe itself is a kind of time machine that permits us to observe remote galaxies as they were long ago. For the closest galaxies, like the Andromeda galaxy, the time the light takes to reach us is on the order of a few hundred thousand to a few million years. Typically not much changes over times that short—individual stars in the galaxy may be born or die, but the overall structure and appearance of the galaxy will remain the same. But we have observed galaxies so far away that we are seeing them as they were when the light left them more than 10 billion years ago.

By observing more distant objects, we look further back toward a time when both galaxies and the universe were young (Figure 28.2). For something similar on Earth, imagine your college doing a project where families of international students around the world are asked to send a daily newspaper from their hometown. Because there is a limited budget, the papers are sent via ordinary mail. The farther a town is from the United States, the longer it takes for the papers to get to the college, and the older the news is by the time it arrives.

Figure 28.2 Astronomical Time Travel. This true-color, long-exposure image, made during 70 orbits of Earth with the Hubble Space Telescope, shows a small area in the direction of the constellation Sculptor. The massive cluster of galaxies named Abell 2744 appears in the foreground of this image. It contains several hundred galaxies, and we are seeing them as they looked 3.5 billion years ago. The immense gravity in Abell 2744 acts as a gravitational lens (see the Astronomy Basics feature box on Gravitational Lensing later in this chapter) to warp space and brighten and magnify images of nearly 3000 distant background galaxies. The more distant galaxies (many of them quite blue) appear as they did more than 12 billion years ago, not long after the Big Bang. Blue galaxies were much more common in that earlier time than they are today. These galaxies appear blue because they are undergoing active star formation and making hot, bright blue stars. (credit: NASA, ESA, STScI)

If we can't directly detect the changes over time in individual galaxies because they happen too slowly, how then can we ever understand those changes and the origins of galaxies? The solution is to observe many galaxies at many different cosmic distances and, therefore, look-back times (how far back in time we are seeing the galaxy). If we can study a thousand very distant "baby" galaxies when the universe was 1 billion years old, and another thousand slightly closer "toddler" galaxies when it was 2 billion years old, and so on until the present 13.8-billion-year-old universe of mature "adult" galaxies near us today, then maybe we can piece together a coherent picture of how the whole ensemble of galaxies evolves over time. This allows us to reconstruct the "life story" of galaxies since the universe began, even though we can't follow a single galaxy from infancy to old age.

Fortunately, there is no shortage of galaxies to study. Hold up your pinky at arm's length: the part of the sky blocked by your fingernail contains about one million galaxies, layered farther and farther back in space and time. In fact, the sky is filled with galaxies, all of them, except for Andromeda and the Magellanic Clouds, too faint to see with the naked eye—more than 2 trillion (2000 billion) galaxies in the observable universe, each one with about 100 billion stars.

This cosmic time machine, then, lets us peer into the past to answer fundamental questions about where galaxies come from and how they got to be the way they are today. Astronomers call those galactic changes over cosmic time **evolution**, a word that recalls the work of Darwin and others on the development of life on Earth. But note that galaxy evolution refers to the changes in *individual* galaxies over time, while the kind of evolution biologists study is changes in *successive generations* of living organisms over time.

Spectra, Colors, and Shapes

Astronomy is one of the few sciences in which all measurements must be made at a distance. Geologists can take samples of the objects they are studying; chemists can conduct experiments in their laboratories to determine what a substance is made of; archeologists can use carbon dating to determine how old something is. But astronomers can't pick up and play with a star or galaxy. As we have seen throughout this book, if they want to know what galaxies are made of and how they have changed over the lifetime of the universe, they must decode the messages carried by the small number of photons that reach Earth.

Fortunately (as you have learned) electromagnetic radiation is a rich source of information. The distance to a galaxy is derived from its *redshift* (how much the lines in its spectrum are shifted to the red because of the expansion of the universe). The conversion of redshift to a distance depends on certain properties of the universe, including the value of the Hubble constant and how much mass it contains. We will describe the currently accepted model of the universe in The Big Bang. For the purposes of this chapter, it is enough to know that the current best estimate for the age of the universe is 13.8 billion years. In that case, if we see an object that emitted its light 6 billion light-years ago, we are seeing it as it was when the universe was almost 8 billion years old. If we see something that emitted its light 13 billion years ago, we are seeing it as it was when the universe was less than a billion years old. So astronomers measure a galaxy's redshift from its spectrum, use the Hubble constant plus a model of the universe to turn the redshift into a distance, and use the distance and the constant speed of light to infer how far back in time they are seeing the galaxy—the look-back time.

In addition to distance and look-back time, studies of the Doppler shifts of a galaxy's spectral lines can tell us how fast the galaxy is rotating and hence how massive it is (as explained in Galaxies). Detailed analysis of such lines can also indicate the types of stars that inhabit a galaxy and whether it contains large amounts of interstellar matter.

Unfortunately, many galaxies are so faint that collecting enough light to produce a detailed spectrum is currently impossible. Astronomers thus have to use a much rougher guide to estimate what kinds of stars inhabit the faintest galaxies—their overall colors. Look again at Figure 28.2 and notice that some of the galaxies are very blue and others are reddish-orange. Now remember that hot, luminous blue stars are very massive and have lifetimes of only a few million years. If we see a galaxy where blue colors dominate, we know that it must have many hot, luminous blue stars, and that star formation must have taken place in the few million years before the light left the galaxy. In a yellow or red galaxy, on the other hand, the young, luminous blue stars that surely were made in the galaxy's early bursts of star formation must have died already; it must contain mostly old yellow and red stars that last a long time in their main-sequence stages and thus typically formed billions of years before the light that we now see was emitted.

Another important clue to the nature of a galaxy is its shape. Spiral galaxies can be distinguished from elliptical galaxies by shape. Observations show that spiral galaxies contain young stars and large amounts of interstellar matter, while elliptical galaxies have mostly old stars and very little or no star formation. Elliptical galaxies turned most of their interstellar matter into stars many billions of years ago, while star formation has continued until the present day in spiral galaxies.

If we can count the number of galaxies of each type during each epoch of the universe, it will help us understand how the pace of star formation changes with time. As we will see later in this chapter, galaxies in the distant universe—that is, young galaxies—look very different from the older galaxies that we see nearby in the present-day universe.

The First Generation of Stars

In addition to looking at the most distant galaxies we can find, astronomers look at the oldest stars (what we might call the fossil record) of our own Galaxy to probe what happened in the early universe. Since stars are the source of nearly all the light emitted by galaxies, we can learn a lot about the evolution of galaxies by studying the stars within them. What we find is that nearly all galaxies contain at least some very old stars. For example, our own Galaxy contains globular clusters with stars that are at least 13 billion years old, and some may be even older than that. Therefore, if we count the age of the Milky Way as the age of its oldest constituents, the Milky Way must have been born at least 13 billion years ago.

As we will discuss in The Big Bang, astronomers have discovered that the universe is expanding, and have traced the expansion backward in time. In this way, they have discovered that the universe itself is only about 13.8 billion years old. Thus, it appears that at least some of the globular-cluster stars in the Milky Way must have formed less than a billion years after the expansion began.

Several other observations also establish that star formation in the cosmos began very early. Astronomers have used spectra to determine the composition of some elliptical galaxies that are so far away that the light we see left them when the universe was only half as old as it is now. Yet these ellipticals contain old red stars, which must have formed billions of years earlier still.

When we make computer models of how such galaxies evolve with time, they tell us that star formation in elliptical galaxies began less than a billion years or so after the universe started its expansion, and new stars continued to form for a few billion years. But then star formation apparently stopped. When we compare distant elliptical galaxies with ones nearby, we find that ellipticals have not changed very much since the universe reached about half its current age. We'll return to this idea later in the chapter.

Observations of the most luminous galaxies take us even further back in time. Recently, as we have already noted, astronomers have discovered a few galaxies that are so far away that the light we see now left them less than a billion years or so after the beginning (Figure 28.3). Yet the spectra of some of these galaxies already contain lines of heavy elements, including carbon, silicon, aluminum, and sulfur. These elements were not present when the universe began but had to be manufactured in the interiors of stars. This means that when the light from these galaxies was emitted, an entire generation of stars had already been born, lived out their lives, and died—spewing out the new elements made in their interiors through supernova explosions—even before the universe was a billion years old. And it wasn't just a few stars in each galaxy that got started this way. Enough had to live and die to affect the overall composition of the galaxy, in a way that we can still measure in the spectrum from far away.

Figure 28.3 Very Distant Galaxy. This image was made with the Hubble Space Telescope and shows the field around a luminous galaxy at a redshift $z = 8.68$, which corresponds to 13.2 billion light years. This means that we are seeing this galaxy as it appeared about 13.2 billion years ago. The galaxy itself is indicated by the arrow. Long exposures in the far-red and infrared wavelengths were combined to make the image, and additional infrared exposures with the Spitzer Space Telescope, which has lower spatial resolution than the Hubble (lower inset), show the redshifted light of normal stars. The very distant galaxy was detected because it has a strong emission line of hydrogen. This line is produced in regions where the formation of hot, young stars is taking place. (credit: modification of work by I. Labbé (Leiden University), NASA/ESA/JPL-Caltech)

Observations of *quasars* (galaxies whose centers contain a supermassive black hole) support this conclusion.

We can measure the abundances of heavy elements in the gas near quasar black holes (explained in Active Galaxies, Quasars, and Supermassive Black Holes). The composition of this gas in quasars that emitted their light 12.5 billion light-years ago is very similar to that of the Sun. This means that a large portion of the gas surrounding the black holes must have already been cycled through stars during the first 1.3 billion years after the expansion of the universe began. If we allow time for this cycling, then their first stars must have formed when the universe was only a few hundred million years old.

A Changing Universe of Galaxies

Back in the middle decades of the twentieth century, the observation that all galaxies contain some old stars led astronomers to the hypothesis that galaxies were born fully formed near the time when the universe began its expansion. This hypothesis was similar to suggesting that human beings were born as adults and did not have to pass through the various stages of development from infancy through the teens. If this hypothesis were correct, the most distant galaxies should have shapes and sizes very much like the galaxies we see nearby. According to this old view, galaxies, after they formed, should then change only slowly, as successive generations of stars within them formed, evolved, and died. As the interstellar matter was slowly used up and fewer new stars formed, the galaxies would gradually become dominated by fainter, older stars and look dimmer and dimmer.

Thanks to the new generation of large ground- and space-based telescopes, we now know that this picture of galaxies evolving peacefully and in isolation from one another is completely wrong. As we will see later in this chapter, galaxies in the distant universe do not look like the Milky Way and nearby galaxies such as Andromeda, and the story of their development is more complex and involves far more interaction with their neighbors.

Why were astronomers so wrong? Up until the early 1990s, the most distant normal galaxy that had been observed emitted its light 8 billion years ago. Since that time, many galaxies—and particularly the giant ellipticals, which are the most luminous and therefore the easiest to see at large distances—did evolve peacefully and slowly. But the Hubble, Spitzer, Herschel, Keck, and other powerful new telescopes that have come on line since the 1990s make it possible to pierce the 8-billion-light-year barrier. We now have detailed views of many thousands of galaxies that emitted their light much earlier (some more than 13 billion years ago—see Figure 28.3).

Much of the recent work on the evolution of galaxies has progressed by studying a few specific small regions of the sky where the Hubble, Spitzer, and ground-based telescopes have taken extremely long exposure images. This allowed astronomers to detect very faint, very distant, and therefore very *young* galaxies (Figure 28.4). Our deep space telescope images show some galaxies that are 100 times fainter than the faintest objects that can be observed spectroscopically with today's giant ground-based telescopes. This turns out to mean that we can obtain the spectra needed to determine redshifts for only the very brightest five percent of the galaxies in these images.

Figure 28.4 Hubble Ultra-Deep Field. This image is the result of an 11-day-long observation with the Hubble Space Telescope of a tiny region of sky, located toward the constellation Fornax near the south celestial pole. This is an area that has only a handful of Milky Way stars. (Since the Hubble orbits Earth every 96 minutes, the telescope returned to view the same tiny piece of sky over and over again until enough light was collected and added together to make this very long exposure.) There are about 10,000 objects in this single image, nearly all of them galaxies, each with tens or hundreds of billions of stars. We can see some pinwheel-shaped spiral galaxies, which are like the Milky Way. But we also find a large variety of peculiar-shaped galaxies that are in collision with companion galaxies. Elliptical galaxies, which contain mostly old stars, appear as reddish blobs. (credit: modification of work by NASA, ESA, H. Teplitz and M. Rafelski (IPAC/Caltech), A. Koekemoer (STScI), R. Windhorst (Arizona State University), and Z. Levay (STScI))

Although we do not have spectra for most of the faint galaxies, the Hubble Space Telescope is especially well suited to studying their *shapes* because the images taken in space are not blurred by Earth's atmosphere. To the surprise of astronomers, the distant galaxies did not fit Hubble's classification scheme at all. Remember that Hubble found that nearly all nearby galaxies could be classified into a few categories, depending on whether they were ellipticals or spirals. The distant galaxies observed by the Hubble Space Telescope look very different from present-day galaxies, without identifiable spiral arms, disks, and bulges (Figure 28.5). They also tend to be much clumpier than most galaxies today. In other words, it's becoming clear that the shapes of galaxies have changed significantly over time. In fact, we now know that the Hubble scheme works well for only the last half of the age of the universe. Before then, galaxies were much more chaotic.

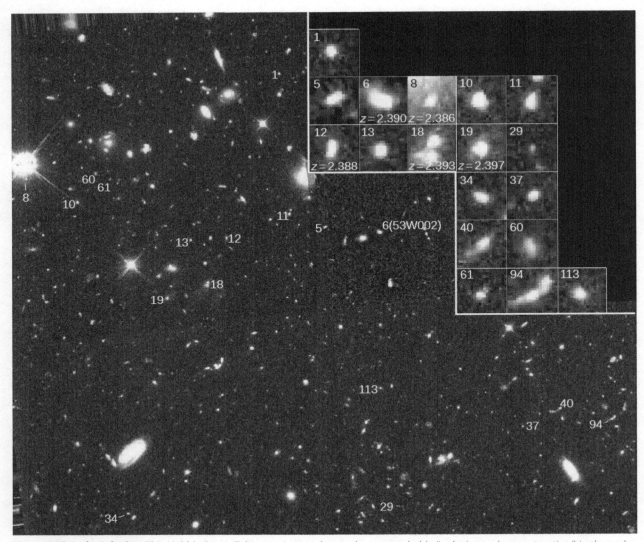

Figure 28.5 Early Galaxies. This Hubble Space Telescope image shows what are probably "galaxies under construction" in the early universe. The boxes in this color image show enlargements of 18 groups of stars smaller than galaxies as we know them. All these objects emitted their light about 11 billion years ago. They are typically only about 2,000 light-years across, which is much smaller than the Milky Way, with its diameter of 100,000 light-years. These 18 objects are found in a region only 2 million light-years across and are close enough together that they will probably collide and merge to build one or more normal galaxies. (credit: modification of work by Rogier Windhorst (Arizona State University) and NASA)

It's not just the shapes that are different. Nearly all the galaxies with red-shifts that correspond to 11 billion light-years or more—that is, galaxies that we are seeing when they were less than 3 billion years old—are extremely blue, indicating that they contain a lot of young stars and that star formation in them is occurring at a higher rate than in nearby galaxies. Observations also show that very distant galaxies are systematically smaller on average than nearby galaxies. Relatively few galaxies present before the universe was about 8 billion years old have masses greater than 10^{11} M_{Sun}. That's 1/20 the mass of the Milky Way if we include its dark matter halo. Eleven billion years ago, there were only a few galaxies with masses greater than 10^{10} M_{Sun}. What we see instead seem to be small pieces or fragments of galactic material (Figure 28.6). When we look at galaxies that emitted their light 11 to 12 billion years ago, we now believe we are seeing the *seeds* of elliptical galaxies and of the central bulges of spirals. Over time, these smaller galaxies collided and merged to build up today's large galaxies.

Bear in mind that stars that formed more than 11 billion years ago will be very old stars today. Indeed when we look nearby (at galaxies we see closer to our time), we find mostly old stars in the nuclear bulges of nearby spirals and in elliptical galaxies.

Figure 28.6 One of the Farthest, Faintest, and Smallest Galaxies Ever Seen. The small white boxes, labeled *a*, *b*, and *c*, mark the positions of three images of the same galaxy. These multiple images were produced by the massive cluster of galaxies known as Abell 2744, which is located between us and the galaxy and acts as a gravitational lens. The arrows in the enlarged insets at right point to the galaxy. Each magnified image makes the galaxy appear as much as 10 times larger and brighter than it would look without the intervening lens. This galaxy emitted the light we observe today when the universe was only about 500 million years old. When the light was emitted the galaxy was tiny—only 850 light-years across, or 500 times smaller than the Milky, and its mass was only 40 million times the mass of the Sun. Star formation is going on in this galaxy, but it appears red in the image because of its large redshift. (credit: modification of work by NASA, ESA, A. Zitrin (California Institute of Technology), and J. Lotz, M. Mountain, A. Koekemoer, and the HFF Team (STScI))

What such observations are showing us is that galaxies have grown in size as the universe has aged. Not only were galaxies smaller several billion years ago, but there were more of them; gas-rich galaxies, particularly the less luminous ones, were much more numerous then than they are today.

Those are some of the basic observations we can make of individual galaxies (and their evolution) looking back in cosmic time. Now we want to turn to the larger context. If stars are grouped into galaxies, are the galaxies also grouped in some way? In the third section of this chapter, we'll explore the largest structures known in the universe.

28.2 | Galaxy Mergers and Active Galactic Nuclei

Learning Objectives

By the end of this section, you will be able to:
> Explain how galaxies grow by merging with other galaxies and by consuming smaller galaxies (for lunch)
> Describe the effects that supermassive black holes in the centers of most galaxies have on the fate of their host galaxies

One of the conclusions astronomers have reached from studying distant galaxies is that collisions and mergers of whole galaxies play a crucial role in determining how galaxies acquired the shapes and sizes we see today. Only a few of the nearby galaxies are currently involved in collisions, but detailed studies of those tell us what to look for when we seek evidence of mergers in very distant and very faint galaxies. These in turn give us important clues about the different evolutionary paths galaxies have taken over cosmic time. Let's examine in more detail what happens when two galaxies collide.

Mergers and Cannibalism

Figure 28.1 shows a dynamic view of two galaxies that are colliding. The stars themselves in this pair of galaxies will not be affected much by this cataclysmic event. (See the Astronomy Basics feature box Why Galaxies Collide but Stars Rarely Do.) Since there is a lot of space between the stars, a direct collision between two stars is very unlikely. However, the *orbits* of many of the stars will be changed as the two galaxies move through each other, and the change in orbits can totally alter the appearance of the interacting galaxies. A gallery of interesting colliding galaxies is shown in Figure 28.7. Great rings, huge tendrils of stars and gas, and other complex structures can form in such cosmic collisions. Indeed, these strange shapes are the signposts that astronomers use to identify colliding galaxies.

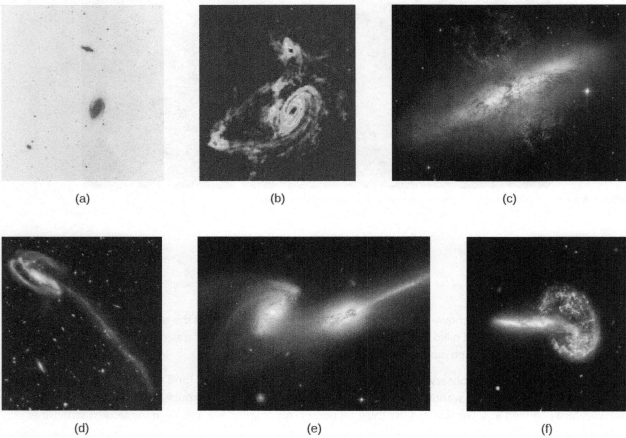

(a)   (b)   (c)

(d)   (e)   (f)

Figure 28.7 Gallery of Interacting Galaxies. (a and b) M82 (smaller galaxy at top) and M83 (spiral) are seen (a) in a black-and-white visible light image and (b) in radio waves given off by cold hydrogen gas. The hydrogen image shows that the two galaxies are wrapped in a common shroud of gas that is being tugged and stretched by the gravity of the two galaxies. (c) This close-up view by the Hubble Space Telescope shows some of the effects of this interaction on galaxy M82, including gas streaming outward (red tendrils) powered by supernovae explosions of massive stars formed in the burst of star formation that was a result of the collision. (d) Galaxy UGC 10214 ("The Tadpole") is a barred spiral galaxy 420 million light-years from the Milky Way that has been disrupted by the passage of a smaller galaxy. The interloper's gravity pulled out the long tidal tail, which is about 280,000 light-years long, and triggered bursts of star formation seen as blue clumps along the tail. (e) Galaxies NGC 4676 A and B are nicknamed "The Mice." In this Hubble Space Telescope image, you can see the long, narrow tails of stars pulled away from the galaxies by the interactions of the two spirals. (f) Arp 148 is a pair of galaxies that are caught in the act of merging to become one new galaxy. The two appear to have already passed through each other once, causing a shockwave that reformed one into a bright blue ring of star formation, like the ripples from a stone tossed into a pond. (credit a, b: modification of work by NRAO/AUI; credit c: modification of work by NASA, ESA, and The Hubble Heritage Team (STScI/AURA); credit d, e: modification of work by NASA, H. Ford (JHU), G. Illingworth (UCSC/LO), M.Clampin (STScI), G. Hartig (STScI), the ACS Science Team, and ESA; credit f: modification of work by NASA, ESA, the Hubble Heritage (STScI/AURA)-ESA/Hubble Collaboration, and A. Evans (University of Virginia, Charlottesville/NRAO/Stony Brook University))

ASTRONOMY BASICS

Why Galaxies Collide but Stars Rarely Do

Throughout this book we have emphasized the large distances between objects in space. You might therefore have been surprised to hear about collisions between galaxies. Yet (except at the very cores of galaxies) we have not worried at all about stars inside a galaxy colliding with each other. Let's see why there is a difference.

The reason is that stars are pitifully small compared to the distances between them. Let's use our Sun as an example. The Sun is about 1.4 million kilometers wide, but is separated from the closest other star by about 4 light-years, or about 38 trillion kilometers. In other words, the Sun is 27 million of its own diameters from its nearest neighbor. If the Sun were a grapefruit in New York City, the nearest star would be another grapefruit in San Francisco. This is typical of stars that are not in the nuclear bulge of a galaxy or inside star clusters. Let's contrast this with the separation of galaxies.

The visible disk of the Milky Way is about 100,000 light-years in diameter. We have three satellite galaxies that are just one or two Milky Way diameters away from us (and will probably someday collide with us). The closest major spiral is the Andromeda Galaxy (M31), about 2.4 million light-years away. If the Milky Way were a pancake at one end of a big breakfast table, M31 would be another pancake at the other end of the same table. Our nearest large galaxy neighbor is only 24 of our Galaxy's diameters from us, and it will begin to crash into the Milky Way in about 4.5 billion years.

Galaxies in rich clusters are even closer together than those in our neighborhood (see The Distribution of Galaxies in Space). Thus, the chances of galaxies colliding are far greater than the chances of stars in the disk of a galaxy colliding. And we should note that the difference between the separation of galaxies and stars also means that when galaxies do collide, their stars almost always pass right by each other like smoke passing through a screen door.

The details of galaxy collisions are complex, and the process can take hundreds of millions of years. Thus, collisions are best simulated on a computer (Figure 28.8), where astronomers can calculate the slow interactions of stars, and clouds of gas and dust, via gravity. These calculations show that if the collision is slow, the colliding galaxies may coalesce to form a single galaxy.

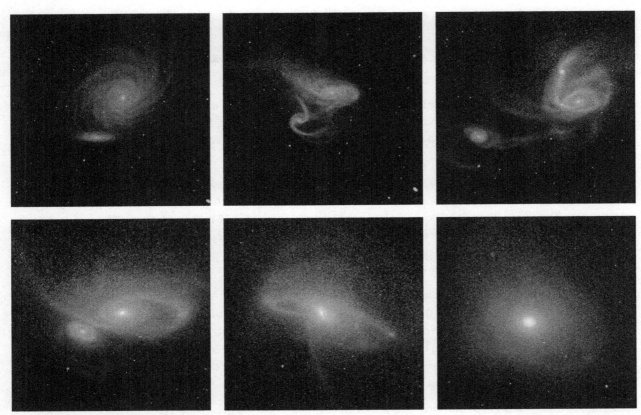

Figure 28.8 Computer Simulation of a Galaxy Collision. This computer simulation starts with two spiral galaxies merging and ends with a single elliptical galaxy. The colors show the colors of stars in the system; note the bursts of blue color as copious star formation gets triggered by the interaction. The timescale from start to finish in this sequence is about a billion years. (credit: modification of work by P. Jonsson (Harvard-Smithsonian Center for Astrophysics), G. Novak (Princeton University), and T. J. Cox (Carnegie Observatories))

When two galaxies of equal size are involved in a collision, we call such an interaction a **merger** (the term applied in the business world to two equal companies that join forces). But small galaxies can also be swallowed by larger ones—a process astronomers have called, with some relish, **galactic cannibalism** (Figure 28.9).

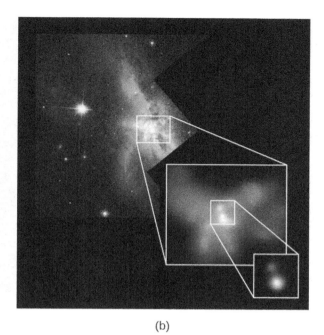

(a)   (b)

Figure 28.9 Galactic Cannibalism. (a) This Hubble image shows the eerie silhouette of dark dust clouds against the glowing nucleus of the elliptical galaxy NGC 1316. Elliptical galaxies normally contain very little dust. These clouds are probably the remnant of a small companion galaxy that was cannibalized (eaten) by NGC 1316 about 100 million years ago. (b) The highly disturbed galaxy NGC 6240, imaged by Hubble Space Telescope (background image) and Chandra X-ray Telescope (both insets) is apparently the product of a merger between two gas-rich spiral galaxies. The X-ray images show that there is not one but two nuclei, both glowing brightly in X-rays and separated by only 4000 light-years. These are likely the locations of two supermassive black holes that inhabited the cores of the two galaxies pre-merger; here they are participating in a kind of "death spiral," in which the two black holes themselves will merge to become one. (credit a: modification of work by NASA, ESA, and The Hubble Heritage Team (STScI/AURA); credit b: X-ray: NASA/CXC/MPE/S.Komossa et al.; Optical: NASA/STScI/R.P.van der Marel & J.Gerssen)

The very large elliptical galaxies we discussed in Galaxies probably form by cannibalizing a variety of smaller galaxies in their clusters. These "monster" galaxies frequently possess more than one nucleus and have probably acquired their unusually high luminosities by swallowing nearby galaxies. The multiple nuclei are the remnants of their victims (Figure 28.9). Many of the large, peculiar galaxies that we observe also owe their chaotic shapes to past interactions. Slow collisions and mergers can even transform two or more spiral galaxies into a single elliptical galaxy.

A change in shape is not all that happens when galaxies collide. If either galaxy contains interstellar matter, the collision can compress the gas and trigger an increase in the rate at which stars are being formed—by as much as a factor of 100. Astronomers call this abrupt increase in the number of stars being formed a **starburst**, and the galaxies in which the increase occurs are termed starburst galaxies (Figure 28.10). In some interacting galaxies, star formation is so intense that all the available gas is exhausted in only a few million years; the burst of star formation is clearly only a temporary phenomenon. While a starburst is going on, however, the galaxy where it is taking place becomes much brighter and much easier to detect at large distances.

Figure 28.10 Starburst Associated with Colliding Galaxies. (a) Three of the galaxies in the small group known as Stephan's Quintet are interacting gravitationally with each other (the galaxy at upper left is actually much closer than the other three and is not part of this interaction), resulting in the distorted shapes seen here. Long strings of young, massive blue stars and hundreds of star formation regions glowing in the pink light of excited hydrogen gas are also results of the interaction. The ages of the star clusters range from 2 million to 1 billion years old, suggesting that there have been several different collisions within this group of galaxies, each leading to bursts of star formation. The three interacting members of Stephan's Quintet are located at a distance of 270 million light-years. (b) Most galaxies form new stars at a fairly slow rate, but members of a rare class known as starburst galaxies blaze with extremely active star formation. The galaxy II Zw 096 is one such starburst galaxy, and this combined image using both Hubble and Spitzer Space Telescope data shows that it is forming bright clusters of new stars at a prodigious rate. The blue colors show the merging galaxies in visible light, while the red colors show infrared radiation from the dusty region where star formation is happening. This galaxy is at a distance of 500 million light-years and has a diameter of about 50,000 light-years, about half the size of the Milky Way. (credit a: modification of work by NASA, ESA, and the Hubble SM4 ERO Team; credit b: modification of work by NASA/JPL-Caltech/STScI)

When astronomers finally had the tools to examine a significant number of galaxies that emitted their light 11 to 12 billion years ago, they found that these very young galaxies often resemble nearby starburst galaxies that are involved in mergers: they also have multiple nuclei and peculiar shapes, they are usually clumpier than normal galaxies today, with multiple intense knots and lumps of bright starlight, and they have higher rates of star formation than isolated galaxies. They also contain lots of blue, young, type O and B stars, as do nearby merging galaxies.

Galaxy mergers in today's universe are rare. Only about five percent of nearby galaxies are currently involved in interactions. Interactions were much more common billions of years ago (Figure 28.11) and helped build up the "more mature" galaxies we see in our time. Clearly, interactions of galaxies have played a crucial role in their evolution.

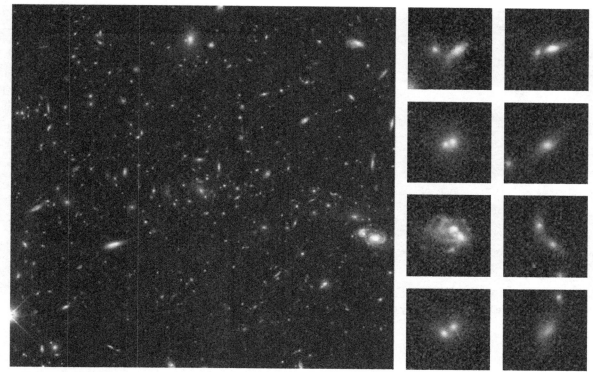

Figure 28.11 Collisions of Galaxies in a Distant Cluster. The large picture on the left shows the Hubble Space Telescope image of a cluster of galaxies at a distance of about 8 billion light-years. Among the 81 galaxies in the cluster that have been examined in some detail, 13 are the result of recent collisions of pairs of galaxies. The eight smaller images on the right are close-ups of some of the colliding galaxies. The merger process typically takes a billion years or so. (credit: modification of work by Pieter van Dokkum, Marijn Franx (University of Groningen/Leiden), ESA and NASA)

Active Galactic Nuclei and Galaxy Evolution

While galaxy mergers are huge, splashy events that completely reshape entire galaxies on scales of hundreds of thousands of light-years and can spark massive bursts of star formation, accreting black holes inside galaxies can also disturb and alter the evolution of their host galaxies. You learned in Active Galaxies, Quasars, and Supermassive Black Holes about a family of objects known as *active galactic nuclei* (AGN), all of them powered by supermassive black holes. If the black hole is surrounded by enough gas, some of the gas can fall into the black hole, getting swept up on the way into an accretion disk, a compact, swirling maelstrom perhaps only 100 AU across (the size of our solar system).

Within the disk the gas heats up until it shines brilliantly even in X-rays, often outshining the rest of the host galaxy with its billions of stars. Supermassive black holes and their accretion disks can be violent and powerful places, with some material getting sucked into the black hole but even more getting shot out along huge jets perpendicular to the disk. These powerful jets can extend far outside the starry edge of the galaxy.

AGN were much more common in the early universe, in part because frequent mergers provided a fresh gas supply for the black hole accretion disks. Examples of AGN in the nearby universe today include the one in galaxy M87 (see Figure 27.7), which sports a jet of material shooting out from its nucleus at speeds close to the speed of light, and the one in the bright galaxy NGC 5128, also known as Centaurus A (see Figure 28.12).

Figure 28.12 Composite View of the Galaxy Centaurus A. This artificially colored image was made using data from three different telescopes: submillimeter radiation with a wavelength of 870 microns is shown in orange; X-rays are seen in blue; and visible light from stars is shown in its natural color. Centaurus A has an active galactic nucleus that is powering two jets, seen in blue and orange, reaching in opposite directions far outside the galaxy's stellar disk, and inflating two huge lobes, or clouds, of hot X-ray-emitting gas. Centaurus is at a distance of 13 million light-years, making it one of the closest active galaxies we know. (credit: modification of work by ESO/WFI (Optical); MPIfR/ESO/APEX/A. Weiss et al. (Submillimeter); NASA/CXC/CfA/R.Kraft et al. (X-ray))

Many highly accelerated particles move with the jets in such galaxies. Along the way, the particles in the jets can plow into gas clouds in the interstellar medium, breaking them apart and scattering them. Since denser clouds of gas and dust are required for material to clump together to make stars, the disruption of the clouds can halt star formation in the host galaxy or cut it off before it even begins.

In this way, quasars and other kinds of AGN can play a crucial role in the evolution of their galaxies. For example, there is growing evidence that the merger of two gas-rich galaxies not only produces a huge burst of star formation, but also triggers AGN activity in the core of the new galaxy. That activity, in turn, could then slow down or shut off the burst of star formation—which could have significant implications for the apparent shape, brightness, chemical content, and stellar components of the entire galaxy. Astronomers refer to that process as *AGN feedback*, and it is apparently an important factor in the evolution of most galaxies.

28.3 | The Distribution of Galaxies in Space

Learning Objectives

By the end of this section, you will be able to:

> Explain the cosmological principle and summarize the evidence that it applies on the largest scales of the known universe
> Describe the contents of the Local Group of galaxies
> Distinguish among groups, clusters, and superclusters of galaxies
> Describe the largest structures seen in the universe, including voids

In the preceding section, we emphasized the role of mergers in shaping the evolution of galaxies. In order to collide, galaxies must be fairly close together. To estimate how often collisions occur and how they affect galaxy evolution, astronomers need to know how galaxies are distributed in space and over cosmic time. Are most of them isolated from one another or do they congregate in groups? If they congregate, how large are the groups and how and when did they form? And how, in general, are galaxies and their groups arranged in the cosmos? Are there as many in one direction of the sky as in any other, for example? How did galaxies get to be arranged the way we find them today?

Edwin Hubble found answers to some of these questions only a few years after he first showed that the spiral nebulae were galaxies and not part of our Milky Way. As he examined galaxies all over the sky, Hubble made two discoveries that turned out to be crucial for studies of the evolution of the universe.

The Cosmological Principle

Hubble made his observations with what were then the world's largest telescopes—the 100-inch and 60-inch reflectors on Mount Wilson. These telescopes have small fields of view: they can see only a small part of the heavens at a time. To photograph the entire sky with the 100-inch telescope, for example, would have taken longer than a human lifetime. So instead, Hubble sampled the sky in many regions, much as Herschel did with his star gauging (see The Architecture of the Galaxy). In the 1930s, Hubble photographed 1283 sample areas, and on each print, he carefully counted the numbers of galaxy images (Figure 28.13).

The first discovery Hubble made from his survey was that the number of galaxies visible in each area of the sky is about the same. (Strictly speaking, this is true only if the light from distant galaxies is not absorbed by dust in our own Galaxy, but Hubble made corrections for this absorption.) He also found that the numbers of galaxies increase with faintness, as we would expect if the density of galaxies is about the same at all distances from us.

To understand what we mean, imagine you are taking snapshots in a crowded stadium during a sold-out concert. The people sitting near you look big, so only a few of them will fit into a photo. But if you focus on the people sitting in seats way on the other side of the stadium, they look so small that many more will fit into your picture. If all parts of the stadium have the same seat arrangements, then as you look farther and farther away, your photo will get more and more crowded with people. In the same way, as Hubble looked at fainter and fainter galaxies, he saw more and more of them.

Figure 28.13 Hubble at Work. Edwin Hubble at the 100-inch telescope on Mount Wilson. (credit: NASA)

Hubble's findings are enormously important, for they indicate that the universe is both **isotropic** and **homogeneous**—it looks the same in all directions, and a large volume of space at any given redshift or distance is much like any other volume at that redshift. If that is so, it does not matter what section of the universe we observe (as long as it's a sizable portion): any section will look the same as any other.

Hubble's results—and many more that have followed in the nearly 100 years since then—imply not only that the universe is about the same everywhere (apart from changes with time) but also that aside from small-scale local differences, the part we can see around us is representative of the whole. The idea that the universe is the same everywhere is called the **cosmological principle** and is the starting assumption for nearly all theories that describe the entire universe (see The Big Bang).

Without the cosmological principle, we could make no progress at all in studying the universe. Suppose our own local neighborhood were unusual in some way. Then we could no more understand what the universe is like than if we were marooned on a warm south-sea island without outside communication and were trying to understand the geography of Earth. From our limited island vantage point, we could not know that some parts of the planet are covered with snow and ice, or that large continents exist with a much greater variety of terrain than that found on our island.

Hubble merely counted the numbers of galaxies in various directions without knowing how far away most of them were. With modern instruments, astronomers have measured the velocities and distances of hundreds of thousands of galaxies, and so built up a meaningful picture of the large-scale structure of the universe. In the rest of this section, we describe what we know about the distribution of galaxies, beginning with those that are nearby.

The Local Group

The region of the universe for which we have the most detailed information is, as you would expect, our own local neighborhood. It turns out that the Milky Way Galaxy is a member of a small group of galaxies called, not too imaginatively, the **Local Group**. It is spread over about 3 million light-years and contains 60 or so members. There are three large spiral galaxies (our own, the Andromeda galaxy, and M33), two intermediate ellipticals, and many dwarf ellipticals and irregular galaxies.

New members of the Local Group are still being discovered. We mentioned in The Milky Way Galaxy a dwarf galaxy only about 80,000 light-years from Earth and about 50,000 light-years from the center of the galaxy that was discovered in 1994 in the constellation of Sagittarius. (This dwarf is actually venturing too close to the much larger Milky Way and will eventually be consumed by it.)

Many of the recent discoveries have been made possible by the new generation of automated, sensitive, wide-field surveys, such as the Sloan Digital Sky Survey, that map the positions of millions of stars across most of the visible sky. By digging into the data with sophisticated computer programs, astronomers have turned up numerous tiny, faint dwarf galaxies that are all but invisible to the eye even in those deep telescopic images. These new findings may help solve a long-standing problem: the prevailing theories of how galaxies form predicted that there should be more dwarf galaxies around big galaxies like the Milky Way than had been observed—and only now do we have the tools to find these faint and tiny galaxies and begin to compare the numbers of them with theoretical predictions.

LINK TO LEARNING

You can read more about the Sloan survey (https://openstax.org/l/30sloansurvey) and its dramatic results. And check out this brief animation (https://openstax.org/l/30anifliarrgal) of a flight through the arrangement of the galaxies as revealed by the survey.

Several new dwarf galaxies have also been found near the Andromeda galaxy. Such dwarf galaxies are difficult to find because they typically contain relatively few stars, and it is hard to distinguish them from the foreground stars in our own Milky Way.

Figure 28.14 is a rough sketch showing where the brighter members of the Local Group are located. The average of the motions of all the galaxies in the Local Group indicates that its total mass is about 4×10^{12} M_{Sun}, and at least half of this mass is contained in the two giant spirals—the Andromeda galaxy and the Milky Way Galaxy. And bear in mind that a substantial amount of the mass in the Local Group is in the form of dark matter.

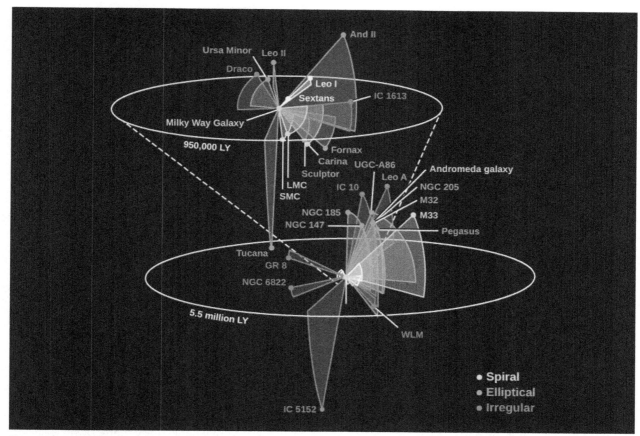

Figure 28.14 Local Group. This illustration shows some members of the Local Group of galaxies, with our Milky Way at the center. The exploded view at the top shows the region closest to the Milky Way and fits into the bigger view at the bottom as shown by the dashed lines. The three largest galaxies among the three dozen or so members of the Local Group are all spirals; the others are small irregular galaxies and dwarf ellipticals. A number of new members of the group have been found since this map was made.

Neighboring Groups and Clusters

Small galaxy groups like ours are hard to notice at larger distances. However, there are much more substantial groups called galaxy clusters that are easier to spot even many millions of light-years away. Such clusters are described as *poor* or *rich* depending on how many galaxies they contain. Rich clusters have thousands or even tens of thousands of galaxies, although many of the galaxies are quite faint and hard to detect.

The nearest moderately rich galaxy cluster is called the Virgo Cluster, after the constellation in which it is seen. It is about 50 million light-years away and contains thousands of members, of which a few are shown in Figure 28.15. The giant elliptical (and very active) galaxy M87, which you came to know and love in the chapter on Active Galaxies, Quasars, and Supermassive Black Holes, belongs to the Virgo Cluster.

Figure 28.15 Central Region of the Virgo Cluster. Virgo is the nearest rich cluster and is at a distance of about 50 million light-years. It contains hundreds of bright galaxies. In this picture you can see only the central part of the cluster, including the giant elliptical galaxy M87, just below center. Other spirals and ellipticals are visible; the two galaxies to the top right are known as "The Eyes." (credit: modification of work by Chris Mihos (Case Western Reserve University)/ESO)

A good example of a cluster that is much larger than the Virgo complex is the Coma cluster, with a diameter of at least 10 million light-years (Figure 28.16). A little over 300 million light-years distant, this cluster is centered on two giant ellipticals whose luminosities equal about 400 billion Suns each. Thousands of galaxies have been observed in Coma, but the galaxies we see are almost certainly only part of what is really there. Dwarf galaxies are too faint to be seen at the distance of Coma, but we expect they are part of this cluster just as they are part of nearer ones. If so, then Coma likely contains tens of thousands of galaxies. The total mass of this cluster is about 4×10^{15} M_{Sun} (enough mass to make 4 million billion stars like the Sun).

Let's pause here for a moment of perspective. We are now discussing numbers by which even astronomers sometimes feel overwhelmed. The Coma cluster may have 10, 20, or 30 thousand galaxies, and each galaxy has billions and billions of stars. If you were traveling at the speed of light, it would still take you more than 10 million years (longer than the history of the human species) to cross this giant swarm of galaxies. And if you lived on a planet on the outskirts of one of these galaxies, many other members of the cluster would be close enough to be noteworthy sights in your nighttime sky.

Figure 28.16 Central Region of the Coma Cluster. This combined visible-light (from the Sloan Digital Sky Survey) and infrared (from the Spitzer Space Telescope) image has been color coded so that faint dwarf galaxies are seen as green. Note the number of little green smudges on the image. The cluster is roughly 320 million light-years away from us. (credit: modification of work by NASA/JPL-Caltech/L. Jenkins (GSFC))

Really rich clusters such as Coma usually have a high concentration of galaxies near the center. We can see giant elliptical galaxies in these central regions but few, if any, spiral galaxies. The spirals that do exist generally occur on the outskirts of clusters.

We might say that ellipticals are highly "social": they are often found in groups and very much enjoy "hanging out" with other ellipticals in crowded situations. It is precisely in such crowds that collisions are most likely and, as we discussed earlier, we think that most large ellipticals are built through mergers of smaller galaxies.

Spirals, on the other hand, are more "shy": they are more likely to be found in poor clusters or on the edges of rich clusters where collisions are less likely to disrupt the spiral arms or strip out the gas needed for continued star formation.

ASTRONOMY BASICS

Gravitational Lensing

As we saw in <u>Black Holes and Curved Spacetime</u>, spacetime is more strongly curved in regions where the gravitational field is strong. Light passing very near a concentration of matter appears to follow a curved path. In the case of starlight passing close to the Sun, we measure the position of the distant star to be slightly different from its true position.

Now let's consider the case of light from a distant galaxy or quasar that passes near a concentration of matter such as a cluster of galaxies on its journey to our telescopes. According to general relativity, the light path may be bent in a variety of ways; as a result we can observe distorted and even multiple images (<u>Figure 28.17</u>).

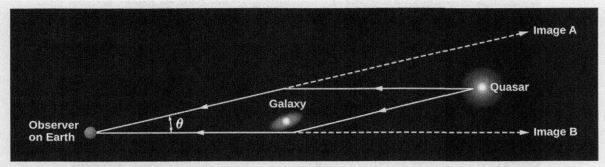

Figure 28.17 Gravitational Lensing. This drawing shows how a gravitational lens can make two images. Two light rays from a distant quasar are shown being bent while passing a foreground galaxy; they then arrive together at Earth. Although the two beams of light contain the same information, they now appear to come from two different points on the sky. This sketch is oversimplified and not to scale, but it gives a rough idea of the lensing phenomenon.

Gravitational lenses can produce not only double images, as shown in Figure 28.17, but also multiple images, arcs, or rings. The first gravitational lens discovered, in 1979, showed two images of the same distant object. Eventually, astronomers used the Hubble Space Telescope to capture remarkable images of the effects of gravitational lenses. One example is shown in Figure 28.18.

Figure 28.18 Multiple Images of a Gravitationally Lensed Supernova. Light from a supernova at a distance of 9 billion light-years passed near a galaxy in a cluster at a distance of about 5 billion light-years. In the enlarged inset view of the galaxy, the arrows point to the multiple images of the exploding star. The images are arranged around the galaxy in a cross-shaped pattern called an Einstein Cross. The blue streaks wrapping around the galaxy are the stretched images of the supernova's host spiral galaxy, which has been distorted by the warping of space. (credit: modification of work by NASA, ESA, and S. Rodney (JHU) and the FrontierSN team; T. Treu (UCLA), P. Kelly (UC Berkeley), and the GLASS team; J. Lotz (STScI) and the Frontier Fields team; M. Postman (STScI) and the CLASH team; and Z. Levay (STScI))

General relativity predicts that the light from a distant object may also be amplified by the lensing effect, thereby making otherwise invisible objects bright enough to detect. This is particularly useful for probing the earliest stages of galaxy formation, when the universe was young. Figure 28.19 shows an example of a very distant faint galaxy that we can study in detail only because its light path passes through a large concentration of massive galaxies and we now see a brighter image of it.

Figure 28.19 Distorted Images of a Distant Galaxy Produced by Gravitational Lensing in a Galaxy Cluster. The rounded outlines show the location of distinct, distorted images of the background galaxy resulting from lensing by the mass in the cluster. The image in the box at lower left is a reconstruction of what the lensed galaxy would look like in the absence of the cluster, based on a model of the cluster's mass distribution, which can be derived from studying the distorted galaxy images. The reconstruction shows far more detail about the galaxy than could have been seen in the absence of lensing. As the image shows, this galaxy contains regions of star formation glowing like bright Christmas tree bulbs. These are much brighter than any star-formation regions in our Milky Way Galaxy. (credit: modification of work by NASA, ESA, and Z. Levay (STScI))

We should note that the visible mass in a galaxy is not the only possible gravitational lens. Dark matter can also reveal itself by producing this effect. Astronomers are using lensed images from all over the sky to learn more about where dark matter is located and how much of it exists.

LINK TO LEARNING

You can use the Gravitational Lensing Simulator (https://openstax.org/l/30gravitlensing) to explore how the distance and mass of a cluster of galaxies affects the offset of lensed images of a very distant galaxy. Instructions are available by clicking on Help.

Superclusters and Voids

After astronomers discovered clusters of galaxies, they naturally wondered whether there were still larger structures in the universe. Do clusters of galaxies gather together? To answer this question, we must be able to map large parts of the universe in three dimensions. We must know not only the position of each galaxy on the sky (that's two dimensions) but also its distance from us (the third dimension).

This means we must be able to measure the redshift of each galaxy in our map. Taking a spectrum of each individual galaxy to do this is a much more time-consuming task than simply counting galaxies seen in different directions on the sky, as Hubble did. Today, astronomers have found ways to get the spectra of many galaxies in the same field of view (sometimes hundreds or even thousands at a time) to cut down the time it takes to finish their three-dimensional maps. Larger telescopes are also able to measure the redshifts—and therefore the distances—of much more distant galaxies and (again) to do so much more quickly than previously possible.

Another challenge astronomers faced in deciding how to go about constructing a map of the universe is similar to that confronted by the first team of explorers in a huge, uncharted territory on Earth. Since there is only one band of explorers and an enormous amount of land, they have to make choices about where to go first. One strategy might be to strike out in a straight line in order to get a sense of the terrain. They might, for example, cross some mostly empty prairies and then hit a dense forest. As they make their way through the forest, they learn how thick it is in the direction they are traveling, but not its width to their left or right. Then a river crosses their path; as they wade across, they can measure its width but learn nothing about its length. Still, as they go on in their straight line, they begin to get some sense of what the landscape is like and can make at least part of a map. Other explorers, striking out in other directions, will someday help fill in the remaining parts of that map.

Astronomers have traditionally had to make the same sort of choices. We cannot explore the universe in every direction to infinite "depth" or sensitivity: there are far too many galaxies and far too few telescopes to do the job. But we can pick a single direction or a small slice of the sky and start mapping the galaxies. Margaret Geller, the late John Huchra, and their students at the Harvard-Smithsonian Center for Astrophysics pioneered this technique, and several other groups have extended their work to cover larger volumes of space.

VOYAGERS IN ASTRONOMY

Margaret Geller: Cosmic Surveyor

Born in 1947, Margaret Geller is the daughter of a chemist who encouraged her interest in science and helped her visualize the three-dimensional structure of molecules as a child. (It was a skill that would later come in very handy for visualizing the three-dimensional structure of the universe.) She remembers being bored in elementary school, but she was encouraged to read on her own by her parents. Her recollections also include subtle messages from teachers that mathematics (her strong early interest) was not a field for girls, but she did not allow herself to be deterred.

Geller obtained a BA in physics from the University of California at Berkeley and became the second woman to receive a PhD in physics from Princeton. There, while working with James Peebles, one of the world's leading cosmologists, she became interested in problems relating to the large-scale structure of the universe. In 1980, she accepted a research position at the Harvard-Smithsonian Center for Astrophysics, one of the nation's most dynamic institutions for astronomy research. She saw that to make progress in understanding how galaxies and clusters are organized, a far more intensive series of surveys was required. Although it would not bear fruit for many years, Geller and her collaborators began the long, arduous task of mapping the galaxies (Figure 28.20).

Figure 28.20 Margaret Geller. Geller's work mapping and researching galaxies has helped us to better understand the structure of the universe. (credit: modification of work by Massimo Ramella)

Her team was fortunate to be given access to a telescope that could be dedicated to their project, the 60-inch reflector on Mount Hopkins, near Tucson, Arizona, where they and their assistants took spectra to determine galaxy distances. To get a slice of the universe, they pointed their telescope at a predetermined position in the sky and then let the rotation of Earth bring new galaxies into their field of view. In this way, they measured the positions and redshifts of over 18,000 galaxies and made a wide range of interesting maps to display their data. Their surveys now include "slices" in both the Northern and Southern Hemispheres.

As news of her important work spread beyond the community of astronomers, Geller received a MacArthur Foundation Fellowship in 1990. These fellowships, popularly called "genius awards," are designed to recognize truly creative work in a wide range of fields. Geller continues to have a strong interest in visualization and has (with filmmaker Boyd Estus) made several award-winning videos explaining her work to nonscientists (one is titled *So Many Galaxies . . . So Little Time*). She has appeared on a variety of national news and documentary programs, including the *MacNeil/Lehrer NewsHour*, *The Astronomers*, and *The Infinite Voyage*. Energetic and outspoken, she has given talks on her work to many audiences around the country, and works hard to find ways to explain the significance of her pioneering surveys to the public.

"It's exciting to discover something that nobody's seen before. [To be] one of the first three people to ever see that slice of the universe [was] sort of being like Columbus. . . . Nobody expected such a striking pattern!"—Margaret Geller

LINK TO LEARNING

Find out more about Geller and Huchra's work (including interviews with Geller) in this 4-minute NOVA (https://openstax.org/l/30gellhucwork) video. You can also learn more about their conclusions (https://openstax.org/l/30gellhucconc) and additional research it led to.

The largest universe mapping project to date is the Sloan Digital Sky Survey (see the Making Connections feature box Astronomy and Technology: The Sloan Digital Sky Survey at the end of this section). A plot of the distribution of galaxies mapped by the Sloan survey is shown in Figure 28.21. To the surprise of astronomers, maps like the one in the figure showed that clusters of galaxies are not arranged uniformly throughout the universe, but are found in huge filamentary **superclusters** that look like great arcs of inkblots splattered across a page. The Local Group is part of a supercluster we call the Virgo Supercluster because it also includes the giant Virgo cluster of galaxies. The superclusters resemble an irregularly torn sheet of paper or a pancake in shape—they can extend for hundreds of millions of light-years in two dimensions, but are only 10 to 20 million light-years thick in the third dimension. Detailed study of some of these structures shows that their masses are a few times 10^{16} M_{Sun}, which is 10,000 times more massive than the Milky Way Galaxy.

LINK TO LEARNING

Check out this animated visualization (https://openstax.org/l/30anivisslosur) of large-scale structure from the Sloan survey.

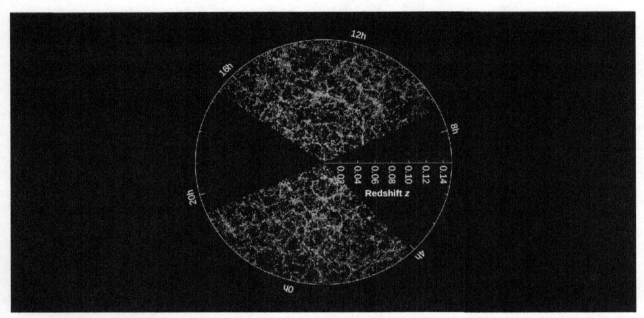

Figure 28.21 Sloan Digital Sky Survey Map of the Large-Scale Structure of the Universe. This image shows slices from the SDSS map. The point at the center corresponds to the Milky Way and might say "You Are Here!" Points on the map moving outward from the center are farther away. The distance to the galaxies is indicated by their redshifts (following Hubble's law), shown on the horizontal line going right from the center. The redshift $z = \Delta\lambda/\lambda$, where $\Delta\lambda$ is the difference between the observed wavelength and the wavelength λ emitted by a nonmoving source in the laboratory. Hour angle on the sky is shown around the circumference of the circular graph. The colors of the galaxies indicate the ages of their stars, with the redder color showing galaxies that are made of older stars. The outer circle is at a distance of two billion light-years from us. Note that red (older stars) galaxies are more strongly clustered than blue galaxies (young stars). The unmapped areas are where our view of the universe is obstructed by dust in our own Galaxy. (credit: modification of work by M. Blanton and the Sloan Digital Sky Survey)

Separating the filaments and sheets in a supercluster are **voids**, which look like huge empty bubbles walled in by the great arcs of galaxies. They have typical diameters of 150 million light-years, with the clusters of galaxies concentrated along their walls. The whole arrangement of filaments and voids reminds us of a sponge, the inside of a honeycomb, or a hunk of Swiss cheese with very large holes. If you take a good slice or cross-section through any of these, you will see something that looks roughly like Figure 28.21.

Before these voids were discovered, most astronomers would probably have predicted that the regions between giant clusters of galaxies were filled with many small groups of galaxies, or even with isolated individual galaxies. Careful searches within these voids have found few galaxies of any kind. Apparently, 90 percent of the galaxies occupy less than 10 percent of the volume of space.

EXAMPLE 28.1

Galaxy Distribution

To determine the distribution of galaxies in three-dimensional space, astronomers have to measure their positions and their redshifts. The larger the volume of space surveyed, the more likely the measurement is a fair sample of the universe as a whole. However, the work involved increases very rapidly as you increase the volume covered by the survey.

Let's do a quick calculation to see why this is so.

Suppose that you have completed a survey of all the galaxies within 30 million light-years and you now want to survey to 60 million light-years. What volume of space is covered by your second survey? How much larger is this volume than the volume of your first survey? Remember that the volume of a sphere, V, is given by the formula $V = 4/3\pi R^3$, where R is the radius of the sphere.

Solution

Since the volume of a sphere depends on R^3 and the second survey reaches twice as far in distance, it will cover a volume that is $2^3 = 8$ times larger. The total volume covered by the second survey will be $(4/3)\pi \times (60$ million light-years$)^3 = 9 \times 10^{23}$ light-years^3.

Check Your Learning

Suppose you now want to expand your survey to 90 million light-years. What volume of space is covered, and how much larger is it than the volume of the second survey?

Answer:

The total volume covered is $(4/3)\pi \times (90$ million light-years$)^3 = 3.05 \times 10^{24}$ light-years^3. The survey reaches 3 times as far in distance, so it will cover a volume that is $3^3 = 27$ times larger.

Even larger, more sensitive telescopes and surveys are currently being designed and built to peer farther and farther out in space and back in time. The new 50-meter Large Millimeter Telescope in Mexico and the Atacama Large Millimeter Array in Chile can detect far-infrared and millimeter-wave radiation from massive starbursting galaxies at redshifts and thus distances more than 90% of the way back to the Big Bang. These cannot be observed with visible light because their star formation regions are wrapped in clouds of thick dust. And in 2021, the 6.5-meter-diameter James Webb Space Telescope is scheduled to launch. It will be the first new major visible light and near-infrared telescope in space since Hubble was launched more than 30 years earlier. One of the major goals of this telescope is to observe directly the light of the first galaxies and even the first stars to shine, less than half a billion years after the Big Bang.

At this point, if you have been thinking about our discussions of the expanding universe in Galaxies, you may be wondering what exactly in Figure 28.21 is expanding. We know that the galaxies and clusters of galaxies are held together by their gravity and do not expand as the universe does. However, the voids do grow larger and the filaments move farther apart as space stretches (see The Big Bang).

MAKING CONNECTIONS

Astronomy and Technology: The Sloan Digital Sky Survey

In Edwin Hubble's day, spectra of galaxies had to be taken one at a time. The faint light of a distant galaxy gathered by a large telescope was put through a slit, and then a spectrometer (also called a spectrograph) was used to separate the colors and record the spectrum. This was a laborious process, ill suited to the demands of making large-scale maps that require the redshifts of many thousands of galaxies.

But new technology has come to the rescue of astronomers who seek three-dimensional maps of the universe of galaxies. One ambitious survey of the sky was produced using a special telescope, camera, and spectrograph atop the Sacramento Mountains of New Mexico. Called the Sloan Digital Sky Survey (SDSS), after the foundation that provided a large part of the funding, the program used a 2.5-meter telescope (about the same aperture as the Hubble) as a wide-angle astronomical camera. During a mapping program lasting more than ten years, astronomers used the SDSS's 30 charge-coupled devices (CCDs)—sensitive electronic light detectors similar to those used in many digital cameras and cell phones—to take images of over 500 million objects and spectra of over 3 million, covering more than one-quarter of the celestial sphere. Like many large projects in modern science, the Sloan Survey involved scientists and engineers from many different institutions, ranging from universities to national laboratories.

Every clear night for more than a decade, astronomers used the instrument to make images recording the position and brightness of celestial objects in long strips of the sky. The information in each strip was digitally recorded and preserved for future generations. When the seeing (recall this term from Astronomical Instruments) was only adequate, the telescope was used for taking spectra of galaxies and quasars—but it did so for up to *640 objects at a time*.

The key to the success of the project was a series of *optical fibers*, thin tubes of flexible glass that can transmit light from a source to the CCD that then records the spectrum. After taking images of a part of the sky and identifying which objects are galaxies, project scientists drilled an aluminum plate with holes for attaching fibers at the location of each galaxy. The telescope was then pointed at the right section of the sky, and the fibers led the light of each galaxy to the spectrometer for individual recording (Figure 28.22).

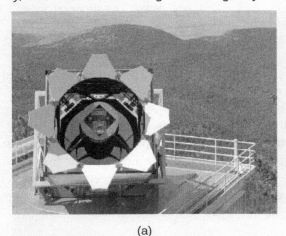
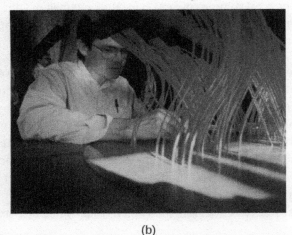

(a)   (b)

Figure 28.22 Sloan Digital Sky Survey. (a) The Sloan Digital Sky Survey telescope is seen here in front of the Sacramento Mountains in New Mexico. (b) Astronomer Richard Kron inserts some of the optical fibers into the pre-drilled plate to enable the instruments to make many spectra of galaxies at the same time. (credit a, b: modification of work by the Sloan Digital Sky Survey)

About an hour was sufficient for each set of spectra, and the pre-drilled aluminum plates could be switched quickly. Thus, it was possible to take as many as 5000 spectra in one night (provided the weather was good enough).

The galaxy survey led to a more comprehensive map of the sky than has ever before been possible, allowing astronomers to test their ideas about large-scale structure and the evolution of galaxies against an impressive array of real data.

The information recorded by the Sloan Survey staggers the imagination. The data came in at 8 megabytes per second (this means 8 million individual numbers or characters every second). Over the course of the project, scientists recorded over 15 terabytes, or 15 thousand billion bytes, which they estimate is comparable to the information contained in the Library of Congress. Organizing and sorting this volume of data and extracting the useful scientific results it contains is a formidable challenge, even in our information age. Like many other fields, astronomy has now entered an era of "Big Data," requiring supercomputers and advanced computer algorithms to sift through all those terabytes of data efficiently.

One very successful solution to the challenge of dealing with such large datasets is to turn to "citizen science," or crowd-sourcing, an approach the SDSS helped pioneer. The human eye is very good at recognizing subtle differences among shapes, such as between two different spiral galaxies, while computers often fail at such tasks. When Sloan project astronomers wanted to catalog the shapes of some of the millions of galaxies in their new images, they launched the "Galaxy Zoo" project: volunteers around the world were given a short training course online, then were provided with a few dozen galaxy images to classify by eye. The project was wildly successful, resulting in over 40 million galaxy classifications by more than 100,000 volunteers and the discovery of whole new types of galaxies.

LINK TO LEARNING

Learn more about how you can be part of the project of classifying galaxies (https://openstax.org/l/30classgalax) in this citizen science effort. This program is part of a whole series of "citizen science" projects (https://openstax.org/l/30citizscien) that enable people in all walks of life to be part of the research that professional astronomers (and scholars in a growing number of fields) need help with.

28.4 | The Challenge of Dark Matter

Learning Objectives

By the end of this section, you will be able to:

> Explain how astronomers know that the solar system contains very little dark matter
> Summarize the evidence for dark matter in most galaxies
> Explain how we know that galaxy clusters are dominated by dark matter
> Relate the presence of dark matter to the average mass-to-light ratio of huge volumes of space containing many galaxies

So far this chapter has focused almost entirely on matter that radiates electromagnetic energy—stars, planets, gas, and dust. But, as we have pointed out in several earlier chapters (especially The Milky Way Galaxy), it is now clear that galaxies contain large amounts of dark matter as well. There is much more dark matter, in fact, than matter we can see—which means it would be foolish to ignore the effect of this unseen material in our theories about the structure of the universe. (As many a ship captain in the polar seas found out too late, the part of the iceberg visible above the ocean's surface was not necessarily the only part he needed to pay attention to.) Dark matter turns out to be extremely important in determining the evolution of galaxies and of the universe as a whole.

The idea that much of the universe is filled with dark matter may seem like a bizarre concept, but we can cite a historical example of "dark matter" much closer to home. In the mid-nineteenth century, measurements showed that the planet Uranus did not follow exactly the orbit predicted from Newton's laws if one added up the gravitational forces of all the known objects in the solar system. Some people worried that Newton's laws may simply not work so far out in our solar system. But the more straightforward interpretation was to attribute Uranus' orbital deviations to the gravitational effects of a new planet that had not yet been seen. Calculations showed where that planet had to be, and Neptune was discovered just about in the predicted location.

In the same way, astronomers now routinely determine the location and amount of dark matter in galaxies by measuring its gravitational effects on objects we can see. And, by measuring the way that galaxies move in clusters, scientists have discovered that dark matter is also distributed among the galaxies in the clusters. Since the environment surrounding a galaxy is important in its development, dark matter must play a central role in galaxy evolution as well. Indeed, it appears that dark matter makes up most of the matter in the universe. But what *is* dark matter? What is it made of? We'll look next at the search for dark matter and the quest to determine its nature.

Dark Matter in the Local Neighborhood

Is there dark matter in our own solar system? Astronomers have examined the orbits of the known planets and of spacecraft as they journey to the outer planets and beyond. No deviations have been found from the orbits predicted on the basis of the masses of objects already discovered in our solar system and the theory of gravity. We therefore conclude that there is no evidence that there are large amounts of dark matter nearby.

Astronomers have also looked for evidence of dark matter in the region of the Milky Way Galaxy that lies within a few hundred light-years of the Sun. In this vicinity, most of the stars are restricted to a thin disk. It is possible to calculate how much mass the disk must contain in order to keep the stars from wandering far above or below it. The total matter that must be in the disk is less than twice the amount of luminous matter. This means that no more than half of the mass in the region near the Sun can be dark matter.

Dark Matter in and around Galaxies

In contrast to our local neighborhood near the Sun and solar system, there is (as we saw in The Milky Way Galaxy) ample evidence strongly suggesting that about 90% of the mass in the entire galaxy is in the form of a halo of dark matter. In other words, there is apparently about nine times more dark matter than visible matter. Astronomers have found some stars in the outer regions of the Milky Way beyond its bright disk, and these stars are revolving very rapidly around its center. The mass contained in all the stars and all the interstellar matter we can detect in the galaxy does not exert enough gravitational force to explain how those fast-moving stars remain in their orbits and do not fly away. Only by having large amounts of unseen matter could the galaxy be holding on to those fast-moving outer stars. The same result is found for other spiral galaxies as well.

Figure 28.23 is an example of the kinds of observations astronomers are making, for the Triangulum galaxy, a member of our Local Group. The observed rotation of spiral galaxies like Andromeda is usually seen in plots, known as *rotation curves,* that show velocity versus distance from the galaxy center. Such plots suggest that the dark matter is found in a large halo surrounding the luminous parts of each galaxy. The radius of the halos around the Milky Way and Andromeda may be as large as 300,000 light-years, much larger than the visible size of these galaxies.

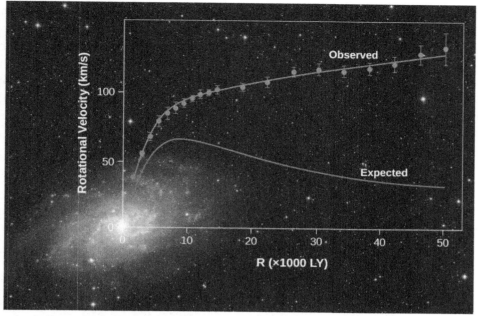

Figure 28.23 Rotation Indicates Dark Matter. We see the Milky Way's neighbor, the Triangulum galaxy, with a graph that shows the velocity at which stars and clouds of gas orbit the galaxy at different distances from the center (red line). As is true of the Milky Way, the rotational velocity (or orbital speed) does not decrease with distance from the center, which is what you would expect if an assembly of objects rotates around a common center. A calculation (blue line) based on the total mass visible as stars, gas, and dust predicts that the velocity should be much lower at larger distances from the center. The discrepancy between the two curves implies the presence of a halo of massive dark matter extending outside the boundary of the luminous matter. This dark matter causes everything in the galaxy to orbit faster than the observed matter alone could explain. (credit background: modification of work by ESO)

Dark Matter in Clusters of Galaxies

Galaxies in clusters also move around: they orbit the cluster's center of mass. It is not possible for us to follow a galaxy around its entire orbit because that typically takes about a billion years. It is possible, however, to measure the velocities with which galaxies in a cluster are moving, and then estimate what the total mass in the cluster must be to keep the individual galaxies from flying out of the cluster. The observations indicate that the mass of the galaxies alone cannot keep the cluster together—some other gravity must again be present. The total amount of dark matter in clusters exceeds by more than ten times the luminous mass contained within the galaxies themselves, indicating that dark matter exists between galaxies as well as inside them.

There is another approach we can take to measuring the amount of dark matter in clusters of galaxies. As we saw, the universe is expanding, but this expansion is not perfectly uniform, thanks to the interfering hand of gravity. Suppose, for example, that a galaxy lies outside but relatively close to a rich cluster of galaxies. The gravitational force of the cluster will tug on that neighboring galaxy and slow down the rate at which it moves away from the cluster due to the expansion of the universe.

Consider the Local Group of galaxies, lying on the outskirts of the Virgo Supercluster. The mass concentrated at the center of the Virgo Cluster exerts a gravitational force on the Local Group. As a result, the Local Group is moving away from the center of the Virgo Cluster at a velocity a few hundred kilometers per second slower than the Hubble law predicts. By measuring such deviations from a smooth expansion, astronomers can estimate the total amount of mass contained in large clusters.

There are two other very useful methods for measuring the amount of dark matter in galaxy clusters, and both of them have produced results in general agreement with the method of measuring galaxy velocities: gravitational lensing and X-ray emission. Let's take a look at both.

As Albert Einstein showed in his theory of general relativity, the presence of mass bends the surrounding fabric of spacetime. Light follows those bends, so very massive objects can bend light significantly. You saw examples of this in the Astronomy Basics feature box Gravitational Lensing in the previous section. Visible galaxies are not the only possible gravitational lenses. Dark matter can also reveal its presence by producing this effect. Figure 28.24 shows a galaxy cluster that is acting like a gravitational lens; the streaks and arcs you see on the picture are lensed images of more distant galaxies. Gravitational lensing is well enough understood that astronomers can use the many ovals and arcs seen in this image to calculate detailed maps of how much matter there is in the cluster and how that mass is distributed. The result from studies of many such gravitational lens clusters shows that, like individual galaxies, galaxy clusters contain more than ten times as much dark matter as luminous matter.

Figure 28.24 Cluster Abell 2218. This view from the Hubble Space Telescope shows the massive galaxy cluster Abell 2218 at a distance of about 2 billion light-years. Most of the yellowish objects are galaxies belonging to the cluster. But notice the numerous long, thin streaks, many of them blue; those are the distorted and magnified images of even more distant background galaxies, gravitationally lensed by the enormous mass of the intervening cluster. By carefully analyzing the lensed images, astronomers can construct a map of the dark matter that dominates the mass of the cluster. (credit: modification of work by NASA, ESA, and Johan Richard (Caltech))

The third method astronomers use to detect and measure dark matter in galaxy clusters is to image them in the light of X-rays. When the first sensitive X-ray telescopes were launched into orbit around Earth in the 1970s and trained on massive galaxy clusters, it was quickly discovered that the clusters emit copious X-ray radiation (see Figure 28.25). Most stars do not emit much X-ray radiation, and neither does most of the gas or dust between the stars inside galaxies. What could be emitting the X-rays seen from virtually all massive galaxy clusters?

It turns out that just as galaxies have gas distributed between their stars, clusters of galaxies have gas distributed between their galaxies. The particles in these huge reservoirs of gas are not just sitting still; rather, they are constantly moving, zooming around under the influence of the cluster's immense gravity like mini planets around a giant sun. As they move and bump against each other, the gas heats up hotter and hotter until, at temperatures as high as 100 million K, it shines brightly at X-ray wavelengths. The more mass the cluster has, the faster the motions, the hotter the gas, and the brighter the X-rays. Astronomers calculate that the mass present to induce those motions must be about ten times the mass they can see in the clusters, including all the galaxies and all the gas. Once again, this is evidence that the galaxy clusters are seen to be dominated by dark matter.

Figure 28.25 X-Ray Image of a Galaxy Cluster. This composite image shows the galaxy cluster Abell 1689 at a distance of 2.3 billion light-years. The finely detailed views of the galaxies, most of them yellow, are in visible and near-infrared light from the Hubble Space Telescope, while the diffuse purple haze shows X-rays as seen by Chandra X-ray Observatory. The abundant X-rays, the gravitationally lensed images (thin curving arcs) of background galaxies, and the measured velocities of galaxies in the clusters all show that the total mass of Abell 1689—most of it dark matter—is about 10^{15} solar masses. (credit: modification of work by NASA/ESA/JPL-Caltech/Yale/CNRS)

Mass-to-Light Ratio

We described the use of the mass-to-light ratio to characterize the matter in galaxies or clusters of galaxies in Properties of Galaxies. For systems containing mostly old stars, the mass-to-light ratio is typically 10 to 20, where mass and light are measured in units of the Sun's mass and luminosity. A mass-to-light ratio of 100 or more is a signal that a substantial amount of dark matter is present. Table 28.1 summarizes the results of measurements of mass-to-light ratios for various classes of objects. Very large mass-to-light ratios are found for all systems of galaxy size and larger, indicating that dark matter is present in all these types of objects. This is why we say that dark matter apparently makes up most of the total mass of the universe.

Mass-To-Light Ratios

Type of Object	Mass-to-Light Ratio
Sun	1
Matter in vicinity of Sun	2
Mass in Milky Way within 80,000 light-years of the center	10
Small groups of galaxies	50–150
Rich clusters of galaxies	250–300

Table 28.1

The clustering of galaxies can be used to derive the total amount of mass in a given region of space, while visible radiation is a good indicator of where the luminous mass is. Studies show that the dark matter and luminous matter are very closely associated. The dark matter halos do extend beyond the luminous boundaries of the galaxies that they surround. However, where there are large clusters of galaxies, you will also find large amounts of dark matter. Voids in the galaxy distribution are also voids in the distribution of dark matter.

What Is the Dark Matter?

How do we go about figuring out what the dark matter consists of? The technique we might use depends on its composition. Let's consider the possibility that some of the dark matter is made up of normal particles: protons, neutrons, and electrons. Suppose these particles were assembled into black holes, brown dwarfs, or white dwarfs. If the black holes had no accretion disks, they would be invisible to us. White and brown dwarfs do emit some radiation but have such low luminosities that they cannot be seen at distances greater than a few thousand light-years.

We can, however, look for such compact objects because they can act as gravitational lenses. (See the Astronomy Basics feature box Gravitational Lensing.) Suppose the dark matter in the halo of the Milky Way were made up of black holes, brown dwarfs, and white dwarfs. These objects have been whimsically dubbed MACHOs (MAssive Compact Halo Objects). If an invisible MACHO passes directly between a distant star and Earth, it acts as a gravitational lens, focusing the light from the distant star. This causes the star to appear to brighten over a time interval of a few hours to several days before returning to its normal brightness. Since we can't predict when any given star might brighten this way, we have to monitor huge numbers of stars to catch one in the act. There are not enough astronomers to keep monitoring so many stars, but today's automated telescopes and computer systems can do it for us.

Research teams making observations of millions of stars in the nearby galaxy called the Large Magellanic Cloud have reported several examples of the type of brightening expected if MACHOs are present in the halo of the Milky Way (Figure 28.26). However, there are not enough MACHOs in the halo of the Milky Way to account for the mass of the dark matter in the halo.

Figure 28.26 Large and Small Magellanic Clouds. Here, the two small galaxies we call the Large Magellanic Cloud and Small

Magellanic Cloud can be seen above the auxiliary telescopes for the Very Large Telescope Array on Cerro Paranal in Chile. You can see from the number of stars that are visible that this is a very dark site for doing astronomy. (credit: ESO/J. Colosimo)

This result, along with a variety of other experiments, leads us to conclude that the types of matter we are familiar with can make up only a tiny portion of the dark matter. Another possibility is that dark matter is composed of some new type of particle—one that researchers are now trying to detect in laboratories here on Earth (see The Big Bang).

The kinds of dark matter particles that astronomers and physicists have proposed generally fall into two main categories: hot and cold dark matter. The terms *hot* and *cold* don't refer to true temperatures, but rather to the average velocities of the particles, analogous to how we might think of particles of air moving in your room right now. In a cold room, the air particles move more slowly on average than in a warm room.

In the early universe, if dark matter particles easily moved fast and far compared to the lumps and bumps of ordinary matter that eventually became galaxies and larger structures, we call those particles **hot dark matter**. In that case, smaller lumps and bumps would be smeared out by the particle motions, meaning fewer small galaxies would get made.

On the other hand, if the dark matter particles moved slowly and covered only small distances compared to the sizes of the lumps in the early universe, we call that **cold dark matter**. Their slow speeds and energy would mean that even the smaller lumps of ordinary matter would survive to grow into small galaxies. By looking at when galaxies formed and how they evolve, we can use observations to distinguish between the two kinds of dark matter. So far, observations seem most consistent with models based on cold dark matter.

Solving the dark matter problem is one of the biggest challenges facing astronomers. After all, we can hardly understand the evolution of galaxies and the long-term history of the universe without understanding what its most massive component is made of. For example, we need to know just what role dark matter played in starting the higher-density "seeds" that led to the formation of galaxies. And since many galaxies have large halos made of dark matter, how does this affect their interactions with one another and the shapes and types of galaxies that their collisions create?

Astronomers armed with various theories are working hard to produce models of galaxy structure and evolution that take dark matter into account in just the right way. Even though we don't know what the dark matter is, we do have some clues about how it affected the formation of the very first galaxies. As we will see in The Big Bang, careful measurements of the microwave radiation left over after the Big Bang have allowed astronomers to set very tight limits on the actual sizes of those early seeds that led to the formation of the large galaxies that we see in today's universe. Astronomers have also measured the relative numbers and distances between galaxies and clusters of different sizes in the universe today. So far, most of the evidence seems to weigh heavily in favor of cold dark matter, and most current models of galaxy and large-scale structure formation use cold dark matter as their main ingredient.

As if the presence of dark matter—a mysterious substance that exerts gravity and outweighs all the known stars and galaxies in the universe but does not emit or absorb light—were not enough, there is an even more baffling and equally important constituent of the universe that has only recently been discovered: we have called it **dark energy** in parallel with dark matter. We will say more about it and explore its effects on the evolution of the universe in The Big Bang. For now, we can complete our inventory of the contents of the universe by noting that it appears that the entire universe contains some mysterious energy that pushes spacetime apart, taking galaxies and the larger structures made of galaxies along with it. Observations show that dark energy becomes more and more important relative to gravity as the universe ages. As a result, the expansion of the universe is accelerating, and this acceleration seems to be happening mostly since the universe was about half its current age.

What we see when we peer out into the universe—the light from trillions of stars in hundreds of billions of galaxies wrapped in intricate veils of gas and dust—is therefore actually only a sprinkling of icing on top of the cake: as we will see in The Big Bang, when we look outside galaxies and clusters of galaxies at the universe as a whole, astronomers find that for every gram of luminous normal matter, such as protons, neutrons, electrons, and atoms in the universe, there are about 4 grams of nonluminous normal matter, mainly intergalactic hydrogen and helium. There are about 27 grams of dark matter, and the energy equivalent (remember Einstein's famous $E = mc^2$) of about 68 grams of dark energy. Dark matter, and (as we will see) even more so dark energy, are dramatic demonstrations of what we have tried to emphasize throughout this book: science is always a "progress report," and we often encounter areas where we have more questions than answers.

Let's next put together all these clues to trace the life history of galaxies and large-scale structure in the universe. What follows is the current consensus, but research in this field is moving rapidly, and some of these ideas will probably be modified as new observations are made.

28.5 | The Formation and Evolution of Galaxies and Structure in the Universe

Learning Objectives

By the end of this section, you will be able to:

> Summarize the main theories attempting to explain how individual galaxies formed
> Explain how tiny "seeds" of dark matter in the early universe grew by gravitational attraction over billions of years into the largest structures observed in the universe: galaxy clusters and superclusters, filaments, and voids

As with most branches of natural science, astronomers and cosmologists always want to know the answer to the question, "How did it get that way?" What made galaxies and galaxy clusters, superclusters, voids, and filaments look the way they do? The existence of such large filaments of galaxies and voids is an interesting puzzle because we have evidence (to be discussed in The Big Bang) that the universe was extremely smooth even a few hundred thousand years after forming. The challenge for theoreticians is to understand how a nearly featureless universe changed into the complex and lumpy one that we see today. Armed with our observations and current understanding of galaxy evolution over cosmic time, dark matter, and large-scale structure, we are now prepared to try to answer that question on some of the largest possible scales in the universe. As we will see, the short answer to how the universe got this way is "dark matter + gravity + time."

How Galaxies Form and Grow

We've already seen that galaxies were more numerous, but smaller, bluer, and clumpier, in the distant past than they are today, and that galaxy mergers play a significant role in their evolution. At the same time, we have observed quasars and galaxies that emitted their light when the universe was less than a billion years old—so we know that large condensations of matter had begun to form at least that early. We also saw in Active Galaxies, Quasars, and Supermassive Black Holes that many quasars are found in the centers of elliptical galaxies. This means that some of the first large concentrations of matter must have evolved into the elliptical galaxies that we see in today's universe. It seems likely that the supermassive black holes in the centers of galaxies and the spherical distribution of ordinary matter around them formed at the same time and through related physical processes.

Dramatic confirmation of that picture arrived only in the last decade, when astronomers discovered a curious empirical relationship: as we saw in Active Galaxies, Quasars, and Supermassive Black Holes, the more massive a galaxy is, the more massive its central black hole is. Somehow, the black hole and the galaxy "know" enough about each other to match their growth rates.

There have been two main types of galaxy formation models to explain all those observations. The first asserts that massive elliptical galaxies formed in a single, rapid collapse of gas and dark matter, during which virtually all the gas was turned quickly into stars. Afterward the galaxies changed only slowly as the stars evolved. This is what astronomers call a "top-down" scenario.

The second model suggests that today's giant ellipticals were formed mostly through mergers of smaller galaxies that had already converted at least some of their gas into stars—a "bottom-up" scenario. In other words, astronomers have debated whether giant ellipticals formed most of their stars in the large galaxy that we see today or in separate small galaxies that subsequently merged.

Since we see some luminous quasars from when the universe was less than a billion years old, it is likely that at least some giant ellipticals began their evolution very early through the collapse of a single cloud. However, the best evidence also seems to show that mature *giant* elliptical galaxies like the ones we see nearby were rare before the universe was about 6 billion years old and that they are much more common today than they were when the universe was young. Observations also indicate that most of the gas in elliptical galaxies was converted to stars by the time the universe was about 3 billion years old, so it appears that elliptical galaxies have not formed many new stars since then. They are often said to be "red and dead"—that is, they mostly contain old, cool, red stars, and there is little or no new star formation going on.

These observations (when considered together) suggest that the giant elliptical galaxies that we see nearby formed from a combination of both top-down and bottom-up mechanisms, with the most massive galaxies forming in the densest clusters where both processes happened very early and quickly in the history of the universe.

The situation with spiral galaxies is apparently very different. The bulges of these galaxies formed early, like the elliptical galaxies (Figure 28.27). However, the disks formed later (remember that the stars in the disk of the Milky Way are younger than the stars in the bulge and the halo) and still contain gas and dust. However, the rate of star formation in spirals today is about ten times lower than it was 8 billion years ago. The number of stars being formed drops as the gas is used up. So spirals seem to form mostly "bottom up" but over a longer time than ellipticals and in a more complex way, with at least two distinct phases.

Rapid Collapse

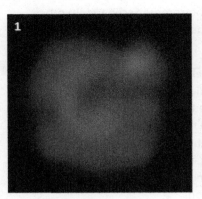

Primordial hydrogen cloud.

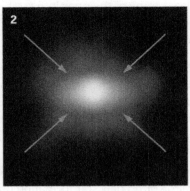

Cloud collapses under gravity.

Large bulge of ancient stars dominates galaxy.

Environmental Effects

Disk galaxy and companion.

Smaller galaxy falls into disk galaxy.

Bulge inflates with addition of young stars and gas.

Figure 28.27 Growth of Spiral Bulges. The nuclear bulges of some spiral galaxies formed through the collapse of a single protogalactic cloud (top row). Others grew over time through mergers with other smaller galaxies (bottom row).

Hubble originally thought that elliptical galaxies were young and would eventually turn into spirals, an idea we now know is not true. In fact, as we saw above, it's more likely the other way around: two spirals that crash together under their mutual gravity can turn into an elliptical.

Despite these advances in our understanding of how galaxies form and evolve, many questions remain. For example, it's even possible, given current evidence, that spiral galaxies could lose their spiral arms and disks in a merger event, making them look more like an elliptical or irregular galaxy, and then regain the disk and arms again later if enough gas remains available. The story of how galaxies assume their final shapes is still being written as we learn more about galaxies and their environment.

Forming Galaxy Clusters, Superclusters, Voids, and Filaments

If individual galaxies seem to grow mostly by assembling smaller pieces together gravitationally over cosmic time, what about the clusters of galaxies and larger structures such as those seen in Figure 28.21? How do we explain the large-scale maps that show galaxies distributed on the walls of huge sponge- or bubble-like structures spanning hundreds of millions of light-years?

As we saw, observations have found increasing evidence for concentrations, filaments, clusters, and superclusters of galaxies when the universe was less than 3 billion years old (Figure 28.28). This means that large concentrations of galaxies had already come together when the universe was less than a quarter as old as it is now.

Figure 28.28 Merging Galaxies in a Distant Cluster. This Hubble image shows the core of one of the most distant galaxy clusters yet discovered, SpARCS 1049+56; we are seeing it as it was nearly 10 billion years ago. The surprise delivered by the image was the "train wreck" of chaotic galaxy shapes and blue tidal tails: apparently there are several galaxies right in the core that are merging together, the probable cause of a massive burst of star formation and bright infrared emission from the cluster. (credit: modification of work by NASA/STScI/ESA/JPL-Caltech/McGill)

Almost all the currently favored models of how large-scale structure formed in the universe tell a story similar to that for individual galaxies: tiny dark matter "seeds" in the hot cosmic soup after the Big Bang grew by gravity into larger and larger structures as cosmic time ticked on (Figure 28.29). The final models we construct will need to be able to explain the size, shape, age, number, and spatial distribution of galaxies, clusters, and filaments—not only today, but also far back in time. Therefore, astronomers are working hard to measure and then to model those features of large-scale structure as accurately as possible. So far, a mixture of 5% normal atoms, 27% cold dark matter, and 68% dark energy seems to be the best way to explain all the evidence currently available (see The Big Bang).

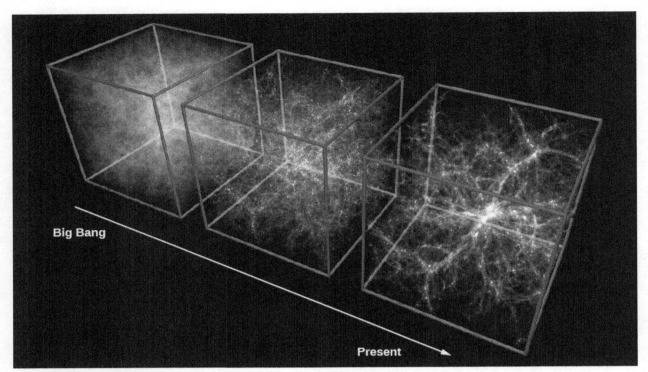

Figure 28.29 Growth of Large-Scale Structure as Calculated by Supercomputers. The boxes show how filaments and superclusters of galaxies grow over time, from a relatively smooth distribution of dark matter and gas, with few galaxies formed in the first 2 billion years after the Big Bang, to the very clumpy strings of galaxies with large voids today. Compare the last image in this sequence with the actual distribution of nearby galaxies shown in Figure 28.21. (credit: modification of work by CXC/MPE/ V.Springel)

The box at left is labeled "Big Bang," the box at center is unlabeled and the box at right is labeled "Present". A white arrow points from left to right representing the direction of time.

Scientists even have a model to explain how a nearly uniform, hot "soup" of particles and energy at the beginning of time acquired the Swiss-cheese-like structure that we now see on the largest scales. As we will see in The Big Bang, when the universe was only a few hundred thousand years old, *everything* was at a temperature of a few thousand degrees. Theorists suggest that at that early time, all the hot gas was vibrating, much as sound waves vibrate the air of a nightclub with an especially loud band. This vibrating could have concentrated matter into high-density peaks and created emptier spaces between them. When the universe cooled, the concentrations of matter were "frozen in," and galaxies ultimately formed from the matter in these high-density regions.

The Big Picture

To finish this chapter, let's put all these ideas together to tell a coherent story of how the universe came to look the way it does. Initially, as we said, the distribution of matter (both luminous and dark) was nearly, but not quite exactly, smooth and uniform. That "not quite" is the key to everything. Here and there were lumps where the density of matter (both luminous and dark) was ever so slightly higher than average.

Initially, each individual lump expanded because the whole universe was expanding. However, as the universe continued to expand, the regions of higher density acquired still more mass because they exerted a slightly larger than average gravitational force on surrounding material. If the inward pull of gravity was high enough, the denser individual regions ultimately stopped expanding. They then began to collapse into irregularly shaped blobs (that's the technical term astronomers use!). In many regions the collapse was more rapid in one direction, so the concentrations of matter were not spherical but came to resemble giant clumps, pancakes, and rope-like filaments—each much larger than individual galaxies.

These elongated clumps existed throughout the early universe, oriented in different directions and collapsing at different rates. The clumps provided the framework for the large-scale filamentary and bubble-like structures that we see preserved in the universe today.

The universe then proceeded to "build itself" from the bottom up. Within the clumps, smaller structures formed first, then merged to build larger ones, like Lego pieces being put together one by one to create a giant Lego metropolis. The first dense concentrations of matter that collapsed were the size of small dwarf galaxies or globular clusters—which helps explain why globular clusters are the oldest things in the Milky Way and most other galaxies. These fragments then gradually assembled to build galaxies, galaxy clusters, and, ultimately, superclusters of galaxies.

According to this picture, small galaxies and large star clusters first formed in the highest density regions of all—the filaments and nodes where the pancakes intersect—when the universe was about two percent of its current age. Some stars may have formed even before the first star clusters and galaxies came into existence. Some galaxy-galaxy collisions triggered massive bursts of star formation, and some of these led to the formation of black holes. In that rich, crowded environment, black holes found constant food and grew in mass. The development of massive black holes then triggered quasars and other active galactic nuclei whose powerful outflows of energy and matter shut off the star formation in their host galaxies. The early universe must have been an exciting place!

Clusters of galaxies then formed as individual galaxies congregated, drawn together by their mutual gravitational attraction (Figure 28.30). First, a few galaxies came together to form groups, much like our own Local Group. Then the groups began combining to form clusters and, eventually, superclusters. This model predicts that clusters and superclusters should still be in the process of gathering together, and observations do in fact suggest that clusters are still gathering up their flocks of galaxies and collecting more gas as it flows in along filaments. In some instances we even see entire clusters of galaxies merging together.

Figure 28.30 Formation of Cluster of Galaxies. This schematic diagram shows how galaxies might have formed if small clouds formed first and then congregated to form galaxies and then clusters of galaxies.

Most giant elliptical galaxies formed through the collision and merger of many smaller fragments. Some spiral galaxies may have formed in relatively isolated regions from a single cloud of gas that collapsed to make a flattened disk, but others acquired additional stars, gas, and dark matter through collisions, and the stars acquired through these collisions now populate their halos and bulges. As we have seen, our Milky Way is still capturing small galaxies and adding them to its halo, and probably also pulling fresh gas from these galaxies into its disk.

🔑 Key Terms

cold dark matter   slow-moving massive particles, not yet identified, that don't absorb, emit, or reflect light or other electromagnetic radiation

cosmological principle   the assumption that, on the large scale, the universe at any given time is the same everywhere—isotropic and homogeneous

dark energy   an energy that is causing the expansion of the universe to accelerate; the source of this energy is not yet understood

evolution (of galaxies)   changes in individual galaxies over cosmic time, inferred by observing snapshots of many different galaxies at different times in their lives

galactic cannibalism   a process by which a larger galaxy strips material from or completely swallows a smaller one

homogeneous   having a consistent and even distribution of matter that is the same everywhere

hot dark matter   massive particles, not yet identified, that don't absorb, emit, or reflect light or other electromagnetic radiation; hot dark matter is faster-moving material than cold dark matter

isotropic   the same in all directions

Local Group   a small cluster of galaxies to which our Galaxy belongs

merger   a collision between galaxies (of roughly comparable size) that combine to form a single new structure

starburst   a galaxy or merger of multiple galaxies that turns gas into stars much faster than usual

supercluster   a large region of space (more than 100 million light-years across) where groups and clusters of galaxies are more concentrated; a cluster of clusters of galaxies

void   a region between clusters and superclusters of galaxies that appears relatively empty of galaxies

📖 Summary

28.1 Observations of Distant Galaxies

When we look at distant galaxies, we are looking back in time. We have now seen galaxies as they were when the universe was about 500 million years old—only about four percent as old as it is now. The universe now is 13.8 billion years old. The color of a galaxy is an indicator of the age of the stars that populate it. Blue galaxies must contain a lot of hot, massive, young stars. Galaxies that contain only old stars tend to be yellowish red. The first generation of stars formed when the universe was only a few hundred million years old. Galaxies observed when the universe was only a few billion years old tend to be smaller than today's galaxies, to have more irregular shapes, and to have more rapid star formation than the galaxies we see nearby in today's universe. This shows that the smaller galaxy fragments assembled themselves into the larger galaxies we see today.

28.2 Galaxy Mergers and Active Galactic Nuclei

When galaxies of comparable size collide and coalesce we call it a merger, but when a small galaxy is swallowed by a much larger one, we use the term galactic cannibalism. Collisions play an important role in the evolution of galaxies. If the collision involves at least one galaxy rich in interstellar matter, the resulting compression of the gas will result in a burst of star formation, leading to a starburst galaxy. Mergers were much more common when the universe was young, and many of the most distant galaxies that we see are starburst galaxies that are involved in collisions. Active galactic nuclei powered by supermassive black holes in the centers of most galaxies can have major effects on the host galaxy, including shutting off star formation.

28.3 The Distribution of Galaxies in Space

Counts of galaxies in various directions establish that the universe on the large scale is homogeneous and isotropic (the same everywhere and the same in all directions, apart from evolutionary changes with time). The sameness of the universe everywhere is referred to as the cosmological principle. Galaxies are grouped together in clusters. The Milky Way Galaxy is a member of the Local Group, which contains at least 54 member galaxies. Rich clusters (such as Virgo and Coma) contain thousands or tens of thousands of galaxies. Galaxy clusters often group together with other clusters to form large-scale structures called superclusters, which can extend over distances of several hundred million light-years. Clusters and superclusters are found in filamentary structures that are huge but fill only a small fraction of space. Most of space consists of large voids between superclusters, with nearly all galaxies confined to less than 10% of the total volume.

28.4 The Challenge of Dark Matter

Stars move much faster in their orbits around the centers of galaxies, and galaxies around centers of galaxy clusters, than they should according to the gravity of all the luminous matter (stars, gas, and dust) astronomers can detect. This discrepancy implies that galaxies and galaxy clusters are dominated by dark matter rather than normal luminous matter. Gravitational lensing and X-ray radiation from massive galaxy clusters confirm the presence of dark matter. Galaxies and clusters of galaxies contain about 10 times more dark matter than luminous matter. While some of the dark matter may be made up of ordinary matter (protons, neutrons, and electrons), perhaps in the form of very faint stars or black holes, most of it probably consists of some totally new type of particle not yet detected on Earth. Observations of gravitational lensing effects on distant objects have been used to look in the outer region of our Galaxy for any dark matter in the form of compact, dim stars or star remnants, but not enough such objects have been found to account for all the dark matter.

28.5 The Formation and Evolution of Galaxies and Structure in the Universe

Initially, luminous and dark matter in the universe was distributed almost—but not quite—uniformly. The challenge for galaxy formation theories is to show how this "not quite" smooth distribution of matter developed the structures—galaxies and galaxy clusters—that we see today. It is likely that the filamentary distribution of galaxies and voids was built in near the beginning, before stars and galaxies began to form. The first condensations of matter were about the mass of a large star cluster or a small galaxy. These smaller structures then merged over cosmic time to form large galaxies, clusters of galaxies, and superclusters of galaxies. Superclusters today are still gathering up more galaxies, gas, and dark matter. And spiral galaxies like the Milky Way are still acquiring material by capturing small galaxies near them.

 For Further Exploration

Articles

Andrews, B. "What Are Galaxies Trying to Tell Us?" *Astronomy* (February 2011): 24. Introduction to our understanding of the shapes and evolution of different types of galaxies.

Barger, A. "The Midlife Crisis of the Cosmos." *Scientific American* (January 2005): 46. On how our time differs from the early universe in terms of what galaxies are doing, and what role supermassive black holes play.

Berman, B. "The Missing Universe." *Astronomy* (April 2014): 24. Brief review of dark matter, what it could be, and modified theories of gravity that can also explain it.

Faber, S., et al. "Staring Back to Cosmic Dawn." *Sky & Telescope* (June 2014): 18. Program to see the most distant and earliest galaxies with the Hubble.

Geller, M., & Huchra, J. "Mapping the Universe." *Sky & Telescope* (August 1991): 134. On their project mapping the location of galaxies in three dimensions.

Hooper, D. "Dark Matter in the Discovery Age." *Sky & Telescope* (January 2013): 26. On experiments looking for the nature of dark matter.

James, C. R. "The Hubble Deep Field: The Picture Worth a Trillion Stars." *Astronomy* (November 2015): 44. Detailed history and results, plus the Hubble Ultra-Deep Field.

Kaufmann, G., & van den Bosch, F. "The Life Cycle of Galaxies." *Scientific American* (June 2002): 46. On the evolution of galaxies and how the different shapes of galaxies develop.

Knapp, G. "Mining the Heavens: The Sloan Digital Sky Survey." *Sky & Telescope* (August 1997): 40.

Kron, R., & Butler, S. "Stars and Strips Forever." *Astronomy* (February 1999): 48. On the Sloan Digital Survey.

Kruesi, L. "What Do We Really Know about Dark Matter?" *Astronomy* (November 2009): 28. Focuses on what dark matter could be and experiments to find out.

Larson, R., & Bromm, V. "The First Stars in the Universe." *Scientific American* (December 2001): 64. On the dark ages and the birth of the first stars.

Nadis, S. "Exploring the Galaxy-Black Hole Connection." *Astronomy* (May 2010): 28. About the role of massive black holes in the evolution of galaxies.

Nadis, S. "Astronomers Reveal the Universe's Hidden Structure." *Astronomy* (September 2013): 44. How dark matter is the scaffolding on which the visible universe rests.

Schilling, G. "Hubble Goes the Distance." *Sky & Telescope* (January 2015): 20. Using gravitational lensing with HST to see the most distant galaxies.

Strauss, M. "Reading the Blueprints of Creation." *Scientific American* (February 2004): 54. On large-scale surveys of galaxies and what they tell us about the organization of the early universe.

Tytell, D. "A Wide Deep Field: Getting the Big Picture." *Sky & Telescope* (September 2001): 42. On the NOAO survey of deep sky objects.

Villard, R. "How Gravity's Grand Illusion Reveals the Universe." *Astronomy* (January 2013): 44. On gravitational lensing and what it teaches us.

Websites

Assembly of Galaxies: http://jwst.nasa.gov/galaxies.html (http://jwst.nasa.gov/galaxies.html). Introductory background information about galaxies: what we know and what we want to learn.

Brief History of Gravitational Lensing: http://www.einstein-online.info/spotlights/grav_lensing_history (http://www.einstein-online.info/spotlights/grav_lensing_history). From Einstein OnLine.

Cosmic Structures: http://skyserver.sdss.org/dr1/en/astro/structures/structures.asp (http://skyserver.sdss.org/dr1/en/astro/structures/structures.asp). Brief review page on how galaxies are organized, from the Sloan Survey.

Gravitational Lensing Discoveries from the Hubble Space Telescope: https://esahubble.org/news/?search=gravitational+lens (https://esahubble.org/news/?search=gravitational+lens). A chronological list of news releases and images.

Local Group of Galaxies: http://www.atlasoftheuniverse.com/localgr.html (http://www.atlasoftheuniverse.com/localgr.html). Clickable map from the Atlas of the Universe project. See also their Virgo Cluster page: http://www.atlasoftheuniverse.com/galgrps/vir.html (http://www.atlasoftheuniverse.com/galgrps/vir.html).

Sloan Digital Sky Survey Website: https://www.youtube.com/watch?v=1RXpHiCNsKU (https://www.youtube.com/watch?v=1RXpHiCNsKU). Includes nontechnical and technical parts.

Spyglasses into the Universe: http://www.spacetelescope.org/science/gravitational_lensing/ (http://www.spacetelescope.org/science/gravitational_lensing/). Hubble page on gravitational lensing; includes links to videos.

Virgo Cluster of Galaxies: http://messier.seds.org/more/virgo.html (http://messier.seds.org/more/virgo.html).

972   28 • Collaborative Group Activities

A page with brief information and links to maps, images, etc.

Videos

Cosmic Simulations: http://www.tapir.caltech.edu/~phopkins/Site/Movies_cosmo.html (http://www.tapir.caltech.edu/~phopkins/Site/Movies_cosmo.html). Beautiful videos with computer simulations of how galaxies form, from the FIRE group.

Cosmology of the Local Universe: http://irfu.cea.fr/cosmography (http://irfu.cea.fr/cosmography). Narrated flythrough of maps of galaxies showing the closer regions of the universe (17:35).

Gravitational Lensing: https://www.youtube.com/watch?v=4Z71RtwoOas (https://www.youtube.com/watch?v=4Z71RtwoOas). Video from Fermilab, with Dr. Don Lincoln (7:14).

How Galaxies Were Cooked from the Primordial Soup: https://www.youtube.com/watch?v=wqNNCm7SNyw (https://www.youtube.com/watch?v=wqNNCm7SNyw). A 2013 public talk by Dr. Sandra Faber of Lick Observatory about the evolution of galaxies; part of the Silicon Valley Astronomy Lecture Series (1:19:33).

Hubble Extreme Deep Field Pushes Back Frontiers of Time and Space: https://www.youtube.com/watch?v=gu_VhzhlqGw (https://www.youtube.com/watch?v=gu_VhzhlqGw). Brief 2012 video (2:42).

Looking Deeply into the Universe in 3-D: https://www.eso.org/public/videos/eso1507a/ (https://www.eso.org/public/videos/eso1507a/). 2015 ESOCast video on how the Very Large Telescopes are used to explore the Hubble Ultra-Deep Field and learn more about the faintest and most distant galaxies (5:12).

Millennium Simulation: http://wwwmpa.mpa-garching.mpg.de/galform/virgo/millennium (http://wwwmpa.mpa-garching.mpg.de/galform/virgo/millennium). A supercomputer in Germany follows the evolution of a representative large box as the universe evolves.

Movies of flying through the large-scale local structure: http://www.ifa.hawaii.edu/~tully/ (http://www.ifa.hawaii.edu/~tully/). By Brent Tully.

Shedding Light on Dark Matter: https://www.youtube.com/watch?v=bZW_B9CC-gI (https://www.youtube.com/watch?v=bZW_B9CC-gI). 2008 TED talk on galaxies and dark matter by physicist Patricia Burchat (17:08).

Sloan Digital Sky Survey overview movies: https://www.youtube.com/watch?v=9vfOqVHyohw (https://www.youtube.com/watch?v=9vfOqVHyohw).

Virtual Universe: https://www.youtube.com/watch?v=SY0bKE10ZDM (https://www.youtube.com/watch?v=SY0bKE10ZDM). An MIT model of a section of universe evolving, with dark matter included (4:11).

When Two Galaxies Collide: http://www.openculture.com/2009/04/when_galaxies_collide.html (http://www.openculture.com/2009/04/when_galaxies_collide.html). Computer simulation, which stops at various points and shows a Hubble image of just such a system in nature (1:37).

Collaborative Group Activities

A. Suppose you developed a theory to account for the evolution of New York City. Have your group discuss whether it would resemble the development of structure in the universe (as we have described it in this chapter). What elements of your model for NYC resemble the astronomers' model for the growth of structure in the universe? Which elements do not match?

B. Most astronomers believe that dark matter exists and is a large fraction of the total matter in the universe. At the same time, most astronomers do not believe that UFOs are evidence that we are being visited by aliens from another world. Yet astronomers have never actually seen either dark matter or a UFO. Why do you think one idea is widely accepted by scientists and the other is not? Which idea do you think is more believable? Give your reasoning.

Access for free at openstax.org

C. Someone in your group describes the redshift surveys of galaxies to a friend, who says he's never heard of a bigger waste of effort. Who cares, he asks, about the large-scale structure of the universe? What is your group's reaction, and what reasons could you come up with for putting money into figuring out how the universe is organized?

D. The leader of a small but very wealthy country is obsessed by maps. She has put together a fabulous collection of Earth maps, purchased all the maps of other planets that astronomers have assembled, and now wants to commission the best possible map of the entire universe. Your group is selected to advise her. What sort of instruments and surveys should she invest in to produce a good map of the cosmos? Be as specific as you can.

E. Download a high-resolution image of a rich galaxy cluster from the Hubble Space Telescope (see the list of gravitational lens news stories in the "For Further Exploration" section). See if your group can work together to identify gravitational arcs, the images of distant background galaxies distorted by the mass of the cluster. How many can you find? Can you identify any multiple images of the same background galaxy? (If anyone in the group gets really interested, there is a Citizen Science project called Spacewarps, where you can help astronomers identify gravitational lenses on their images: https://spacewarps.org.)

F. You get so excited about gravitational lensing that you begin to talk about it with an intelligent friend who has not yet taken an astronomy course. After hearing you out, this friend starts to worry. He says, "If gravitational lenses can distort quasar images, sometimes creating multiple, or ghost, images of the same object, then how can we trust any point of light in the sky to be real? Maybe many of the stars we see are just ghost images or lensed images too!" Have your group discuss how to respond. (Hint: Think about the path that the light of a quasar took on its way to us and the path the light of a typical star takes.)

G. The 8.4-meter Large Synoptic Survey Telescope (LSST), currently under construction atop Cerro Pachón, a mountain in northern Chile, will survey the entire sky with its 3.2-gigapixel camera every few days, looking for transient, or temporary, objects that make a brief appearance in the sky before fading from view, including asteroids and Kuiper belt objects in our solar system, and supernovae and other explosive high-energy events in the distant universe. When it's fully operating sometime after 2021, the LSST will produce up to 30 terabytes of data *every night.* (A terabyte is 1000 gigabytes, which is the unit you probably use to rate your computer or memory stick capacity.) With your group, consider what you think might be some challenges of dealing with that quantity of data every night in a scientifically productive but efficient way. Can you propose any solutions to those challenges?

H. Quasars are rare now but were much more numerous when the universe was about one-quarter of its current age. The total star formation taking place in galaxies across the universe peaked at about the same redshift. Does your group think this is a coincidence? Why or why not?

I. One way to see how well the ideas in astronomy (like those in this chapter) have penetrated popular culture is to see whether you can find astronomical words in the marketplace. A short web search for the term "dark matter" turns up both a brand of coffee and a brand of "muscle growth accelerator" with that name. How many other terms used in this chapter can your group find in the world of products? (What's a really popular type of Android cell phone, for example?)

J. What's your complete address in the universe? Group members should write out their full address, based on the information in this chapter (and the rest of the book). After your postal code and country, you may want to add continent, planet, planetary system, galaxy, etc. Then each group member should explain this address to a family member or student not taking astronomy.

📖 Exercises

Review Questions

1. How are distant (young) galaxies different from the galaxies that we see in the universe today?

2. What is the evidence that star formation began when the universe was only a few hundred million years old?

3. Describe the evolution of an elliptical galaxy. How does the evolution of a spiral galaxy differ from that of an elliptical?

4. Explain what we mean when we call the universe homogeneous and isotropic. Would you say that the distribution of elephants on Earth is homogeneous and isotropic? Why?

5. Describe the organization of galaxies into groupings, from the Local Group to superclusters.

6. What is the evidence that a large fraction of the matter in the universe is invisible?

7. When astronomers make maps of the structure of the universe on the largest scales, how do they find the superclusters of galaxies to be arranged?

8. How does the presence of an active galactic nucleus in a starburst galaxy affect the starburst process?

Thought Questions

9. Describe how you might use the color of a galaxy to determine something about what kinds of stars it contains.

10. Suppose a galaxy formed stars for a few million years and then stopped (and no other galaxy merged or collided with it). What would be the most massive stars on the main sequence after 500 million years? After 10 billion years? How would the color of the galaxy change over this time span? (Refer to Evolution from the Main Sequence to Red Giants.)

11. Given the ideas presented here about how galaxies form, would you expect to find a giant elliptical galaxy in the Local Group? Why or why not? Is there in fact a giant elliptical in the Local Group?

12. Can an elliptical galaxy evolve into a spiral? Explain your answer. Can a spiral turn into an elliptical? How?

13. If we see a double image of a quasar produced by a gravitational lens and can obtain a spectrum of the galaxy that is acting as the gravitational lens, we can then put limits on the distance to the quasar. Explain how.

14. The left panel of Figure 27.1 shows a cluster of yellow galaxies that produces several images of blue galaxies through gravitational lensing. Which are more distant—the blue galaxies or the yellow galaxies? The light in the galaxies comes from stars. How do the temperatures of the stars that dominate the light of the cluster galaxies differ from the temperatures of the stars that dominate the light of the blue-lensed galaxy? Which galaxy's light is dominated by young stars?

15. Suppose you are standing in the center of a large, densely populated city that is exactly circular, surrounded by a ring of suburbs with lower-density population, surrounded in turn by a ring of farmland. From this specific location, would you say the population distribution is isotropic? Homogeneous?

16. Astronomers have been making maps by observing a slice of the universe and seeing where the galaxies lie within that slice. If the universe is isotropic and homogeneous, why do they need more than one slice? Suppose they now want to make each slice extend farther into the universe. What do they need to do?

17. Human civilization is about 10,000 years old as measured by the development of agriculture. If your telescope collects starlight tonight that has been traveling for 10,000 years, is that star inside or outside our Milky Way Galaxy? Is it likely that the star has changed much during that time?

18. Given that only about 5% of the galaxies visible in the Hubble Deep Field are bright enough for astronomers to study spectroscopically, they need to make the most of the other 95%. One technique is to use their colors and apparent brightnesses to try to roughly estimate their redshift. How do you think the inaccuracy of this redshift estimation technique (compared to actually measuring the redshift from a spectrum) might affect our ability to make maps of large-scale structures such as the filaments and voids shown in Figure 28.21?

Figuring for Yourself

19. Using the information from Example 28.1, how much fainter an object will you have to be able to measure in order to include the same kinds of galaxies in your second survey? Remember that the brightness of an object varies as the inverse square of the distance.

20. Using the information from Example 28.1, if galaxies are distributed homogeneously, how many times more of them would you expect to count on your second survey?

21. Using the information from Example 28.1, how much longer will it take you to do your second survey?

22. Galaxies are found in the "walls" of huge voids; very few galaxies are found in the voids themselves. The text says that the structure of filaments and voids has been present in the universe since shortly after the expansion began 13.8 billion years ago. In science, we always have to check to see whether some conclusion is contradicted by any other information we have. In this case, we can ask whether the voids would have filled up with galaxies in roughly 14 billion years. Observations show that in addition to the motion associated with the expansion of the universe, the galaxies in the walls of the voids are moving in random directions at typical speeds of 300 km/s. At least some of them will be moving into the voids. How far into the void will a galaxy move in 14 billion years? Is it a reasonable hypothesis that the voids have existed for 14 billion years?

23. Calculate the velocity, the distance, and the look-back time of the most distant galaxies in Figure 28.21 using the Hubble constant given in this text and the redshift given in the diagram. Remember the Doppler formula for velocity $\left(v = c \times \frac{\Delta \lambda}{\lambda} \right)$ and the Hubble law ($v = H \times d$, where d is the distance to a galaxy). For these low velocities, you can neglect relativistic effects.

24. Assume that dark matter is uniformly distributed throughout the Milky Way, not just in the outer halo but also throughout the bulge and in the disk, where the solar system lives. How much dark matter would you expect there to be inside the solar system? Would you expect that to be easily detectable? Hint: For the radius of the Milky Way's dark matter halo, use $R = 300{,}000$ light-years; for the solar system's radius, use 100 AU; and start by calculating the ratio of the two volumes.

25. The simulated box of galaxy filaments and superclusters shown in Figure 28.29 stretches across 1 billion light-years. If you were to make a scale model where that box covered the core of a university campus, say 1 km, then how big would the Milky Way Galaxy be? How far away would the Andromeda galaxy be in the scale model?

26. The first objects to collapse gravitationally after the Big Bang might have been globular cluster-size galaxy pieces, with masses around 10^6 solar masses. Suppose you merge two of those together, then merge two larger pieces together, and so on, Lego-style, until you reach a Milky Way mass, about 10^{12} solar masses. How many merger generations would that take, and how many original pieces? (Hint: Think in powers of 2.)

29

The Big Bang

Figure 29.1 Space Telescope of the Future. This drawing shows the James Webb Space Telescope, which is currently planned for launch in 2021. The silver sunshade shadows the primary mirror and science instruments. The primary mirror is 6.5 meters (21 feet) in diameter. Before and during launch, the mirror will be folded up. After the telescope is placed in its orbit, ground controllers will command it to unfold the mirror petals. To see distant galaxies whose light has been shifted to long wavelengths, the telescope will carry several instruments for taking infrared images and spectra. (credit: modification of work by NASA)

Chapter Outline

29.1 The Age of the Universe
29.2 A Model of the Universe
29.3 The Beginning of the Universe
29.4 The Cosmic Microwave Background
29.5 What Is the Universe Really Made Of?
29.6 The Inflationary Universe
29.7 The Anthropic Principle

Thinking Ahead

In previous chapters, we explored the contents of the universe—planets, stars, and galaxies—and learned about how these objects change with time. But what about the universe as a whole? How old is it? What did it look like in the beginning? How has it changed since then? What will be its fate?

Cosmology is the study of the universe as a whole and is the subject of this chapter. The story of observational cosmology really begins in 1929 when Edwin Hubble published observations of redshifts and distances for a small sample of galaxies and showed the then-revolutionary result that we live in an expanding universe—one which in the past was denser, hotter, and smoother. From this early discovery, astronomers developed many predictions about the origin and evolution of the universe and then tested those predictions with observations. In this chapter, we will describe what we already know about the history of our dynamic universe and highlight some of the mysteries that remain.

29.1 | The Age of the Universe

Learning Objectives

By the end of this section, you will be able to:

> Describe how we estimate the age of the universe
> Explain how changes in the rate of expansion over time affect estimates of the age of the universe
> Describe the evidence that dark energy exists and that the rate of expansion is currently accelerating
> Describe some independent evidence for the age of the universe that is consistent with the age estimate based on the rate of expansion

To explore the history of the universe, we will follow the same path that astronomers followed historically—beginning with studies of the nearby universe and then probing ever-more-distant objects and looking further back in time.

The realization that the universe changes with time came in the 1920s and 1930s when measurements of the redshifts of a large sample of galaxies became available. With hindsight, it is surprising that scientists were so shocked to discover that the universe is expanding. In fact, our theories of gravity demand that the universe must be either expanding or contracting. To show what we mean, let's begin with a universe of finite size—say a giant ball of a thousand galaxies. All these galaxies attract each other because of their gravity. If they were initially stationary, they would inevitably begin to move closer together and eventually collide. They could avoid this collapse only if for some reason they happened to be moving away from each other at high speeds. In just the same way, only if a rocket is launched at high enough speed can it avoid falling back to Earth.

The problem of what happens in an infinite universe is harder to solve, but Einstein (and others) used his theory of general relativity (which we described in Black Holes and Curved Spacetime) to show that even infinite universes cannot be static. Since astronomers at that time did not yet know the universe was expanding (and Einstein himself was philosophically unwilling to accept a universe in motion), he changed his equations by introducing an arbitrary new term (we might call it a fudge factor) called the **cosmological constant**. This constant represented a hypothetical force of repulsion that could balance gravitational attraction on the largest scales and permit galaxies to remain at fixed distances from one another. That way, the universe could remain still.

(a)   (b)

Figure 29.2 Einstein and Hubble. (a) Albert Einstein is shown in a 1921 photograph. (b) Edwin Hubble at work in the Mt. Wilson Observatory.

About a decade later, Hubble, and his coworkers reported that the universe is expanding, so that no mysterious balancing force is needed. (We discussed this in the chapter on Galaxies.) Einstein is reported to have said that the introduction of the cosmological constant was "the biggest blunder of my life." As we shall see later in this chapter, however, relatively recent observations indicate that the expansion is *accelerating*. Observations are now being carried out to determine whether this acceleration is consistent with a cosmological constant. In a way, it may turn out that Einstein was right after all.

LINK TO LEARNING

View this web exhibit (https://openstax.org/l/30exhcosmAIPCHP) on the history of our thinking about cosmology, with images and biographies, from the American Institute of Physics Center for the History of Physics.

The Hubble Time

If we had a movie of the expanding universe and ran the film *backward*, what would we see? The galaxies, instead of moving apart, would move *together* in our movie—getting closer and closer all the time. Eventually, we would find that all the matter we can see today was once concentrated in an infinitesimally small volume. Astronomers identify this time with the *beginning of the universe*. The explosion of that concentrated universe at the beginning of time is called the **Big Bang** (not a bad term, since you can't have a bigger bang than one that creates the entire universe). But when did this bang occur?

We can make a reasonable estimate of the time since the universal expansion began. To see how astronomers do this, let's begin with an analogy. Suppose your astronomy class decides to have a party (a kind of "Big Bang") at someone's home to celebrate the end of the semester. Unfortunately, everyone is celebrating with so much enthusiasm that the neighbors call the police, who arrive and send everyone away at the same moment. You get home at 2 a.m., still somewhat upset about the way the party ended, and realize you forgot to look at your watch to see what time the police got there. But you use a map to measure that the distance between the party and your house is 40 kilometers. And you also remember that you drove the whole trip at a steady speed of 80 kilometers/hour (since you were worried about the police cars following you). Therefore, the trip must have taken:

$$\text{time} = \frac{\text{distance}}{\text{velocity}} = \frac{40 \text{ kilometers}}{80 \text{ kilometers/hour}} = 0.5 \text{ hours}$$

So the party must have broken up at 1:30 a.m.

No humans were around to look at their watches when the universe began, but we can use the same technique to estimate when the galaxies began moving away from each other. (Remember that, in reality, it is space that is expanding, not the galaxies that are moving through static space.) If we can measure how far apart the galaxies are now, and how fast they are moving, we can figure out how long a trip it's been.

Let's call the age of the universe measured in this way T_0. Let's first do a simple case by assuming that the expansion has been at a constant rate ever since the expansion of the universe began. In this case, the time it has taken a galaxy to move a distance, d, away from the Milky Way (remember that at the beginning the galaxies were all together in a very tiny volume) is (as in our example)

$$T_0 = d/v$$

where v is the velocity of the galaxy. If we can measure the speed with which galaxies are moving away, and also the distances between them, we can establish how long ago the expansion began.

Making such measurements should sound very familiar. This is just what Hubble and many astronomers after him needed to do in order to establish the Hubble law and the Hubble constant. We learned in Galaxies that a galaxy's distance and its velocity in the expanding universe are related by

$$V = H \times d$$

where *H* is the Hubble constant. Combining these two expressions gives us

$$T_0 = \frac{d}{v} = \frac{d}{(H \times d)} = \frac{1}{H}$$

We see, then, that the work of calculating this time was already done for us when astronomers measured the Hubble constant. The age of the universe estimated in this way turns out to be just the *reciprocal of the Hubble constant* (that is, 1/*H*). This age estimate is sometimes called the Hubble time. For a Hubble constant of 20 kilometers/second per million light-years, the Hubble time is about 15 billion years. (By the way, the unit used by astronomers for the Hubble constant is kilometers/second per million parsecs. In these units, the Hubble constant is equal to about 70 kilometers/second per million parsecs, again with an uncertainty of about 5%.)

To make numbers easier to remember, we have done some rounding here. Estimates for the Hubble constant are actually closer to 21 or 22 kilometers/second per million light-years, which would make the age closer to 14 billion years. But there is still about a 5% uncertainty in the Hubble constant, which means the age of the universe estimated in this way is also uncertain by about 5%.

To put these uncertainties in perspective, however, you should know that 50 years ago, the uncertainty was a factor of 2. Remarkable progress toward pinning down the Hubble constant has been made in the last couple of decades.

The Role of Deceleration

The Hubble time is the right age for the universe only if the expansion rate has been constant throughout the time since the expansion of the universe began. Continuing with our end-of-the-semester-party analogy, this is equivalent to assuming that you traveled home from the party at a constant rate, when in fact this may not have been the case. At first, mad about having to leave, you may have driven fast, but then as you calmed down—and thought about police cars on the highway—you may have begun to slow down until you were driving at a more socially acceptable speed (such as 80 kilometers/hour). In this case, given that you were driving faster at the beginning, the trip home would have taken less than a half-hour.

In the same way, in calculating the Hubble time, we have assumed that the expansion rate has been constant throughout all of time. It turns out that this is not a good assumption. Earlier in their thinking about this, astronomers expected that the rate of expansion should be slowing down. We know that matter creates gravity, whereby all objects pull on all other objects. The mutual attraction between galaxies was expected to slow the expansion as time passed. This means that, if gravity were the only force acting (a big *if*, as we shall see in the next section), then the rate of expansion must have been faster in the past than it is today. In this case, we would say the universe has been *decelerating* since the beginning.

How much it has decelerated depends on the importance of gravity in slowing the expansion. If the universe were nearly empty, the role of gravity would be minor. Then the deceleration would be close to zero, and the universe would have been expanding at a constant rate. But in a universe with any significant density of matter, the pull of gravity means that the rate of expansion should be slower now than it used to be. If we use the current rate of expansion to estimate how long it took the galaxies to reach their current separations, we will overestimate the age of the universe—just as we may have overestimated the time it took for you to get home from the party.

A Universal Acceleration

Astronomers spent several decades looking for evidence that the expansion was decelerating, but they were not successful. What they needed were 1) larger telescopes so that they could measure the redshifts of more distant galaxies and 2) a very luminous *standard bulb* (or standard candle), that is, some astronomical object with known luminosity that produces an enormous amount of energy and can be observed at distances of a billion light-years or more.

Recall that we discussed standard bulbs in the chapter on Galaxies. If we compare how luminous a standard bulb is supposed to be and how dim it actually looks in our telescopes, the difference allows us to calculate its distance. The redshift of the galaxy such a bulb is in can tell us how fast it is moving in the universe. So we can measure its distance and motion independently.

These two requirements were finally met in the 1990s. Astronomers showed that supernovae of type Ia (see The Death of Stars), with some corrections based on the shapes of their light curves, are standard bulbs. This type of supernova occurs when a white dwarf accretes enough material from a companion star to exceed the Chandrasekhar limit and then collapses and explodes. At the time of maximum brightness, these dramatic supernovae can briefly outshine the galaxies that host them, and hence, they can be observed at very large distances. Large 8- to 10-meter telescopes can be used to obtain the spectra needed to measure the redshifts of the host galaxies (Figure 29.3).

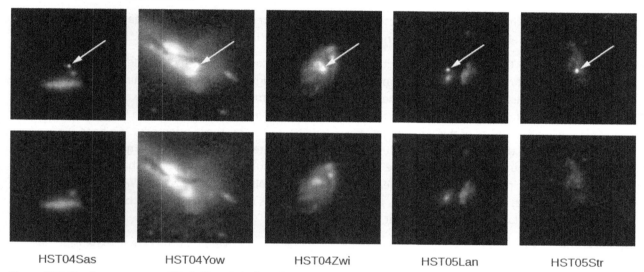

Figure 29.3 Five Supernovae and Their Host Galaxies. The top row shows each galaxy and its supernova (arrow). The bottom row shows the same galaxies either before or after the supernovae exploded. (credit: modification of work by NASA, ESA, and A. Riess (STScI))

The result of painstaking, careful study of these supernovae in a range of galaxies, carried out by two groups of researchers, was published in 1998. It was shocking—and so revolutionary that their discovery received the 2011 Nobel Prize in Physics. What the researchers found was that these type Ia supernovae in distant galaxies were fainter than expected from Hubble's law, given the measured redshifts of their host galaxies. In other words, distances estimated from the supernovae used as standard bulbs disagreed with the distances measured from the redshifts.

If the universe were decelerating, we would expect the far-away supernovae to be *brighter* than expected. The slowing down would have kept them closer to us. Instead, they were *fainter*, which at first seemed to make no sense.

Before accepting this shocking development, astronomers first explored the possibility that the supernovae might not really be as useful as standard bulbs as they thought. Perhaps the supernovae appeared too faint because dust along our line of sight to them absorbed some of their light. Or perhaps the supernovae at large distances were for some reason intrinsically less luminous than nearby supernovae of type Ia.

A host of more detailed observations ruled out these possibilities. Scientists then had to consider the alternative that the distance estimated from the redshift was incorrect. Distances derived from redshifts assume that the Hubble constant has been truly constant for all time. We saw that one way it might not be constant is that the expansion is slowing down. But suppose neither assumption is right (steady speed or slowing down.)

Suppose, instead, that the universe is *accelerating*. If the universe is expanding faster now than it was billions of years ago, our motion away from the distant supernovae has sped up since the explosion occurred, sweeping us farther away from them. The light of the explosion has to travel a greater distance to reach us than if the expansion rate were constant. The farther the light travels, the fainter it appears. This conclusion would explain the supernova observations in a natural way, and this has now been substantiated by many additional observations over the last couple of decades. It really seems that *the expansion of the universe is accelerating*, a notion so unexpected that astronomers at first resisted considering it.

How can the expansion of the universe be speeding up? If you want to accelerate your car, you must supply energy by stepping on the gas. Similarly, energy must be supplied to accelerate the expansion of the universe. The discovery of the acceleration was shocking because scientists still have no idea what the source of the energy is. Scientists call whatever it is **dark energy**, which is a clear sign of how little we understand it.

Note that this new component of the universe is not the dark matter we talked about in earlier chapters. Dark energy is something else that we have also not yet detected in our laboratories on Earth.

What is dark energy? One possibility is that it is the cosmological constant, which is an energy associated with the vacuum of "empty" space itself. Quantum mechanics (the intriguing theory of how things behave at the atomic and subatomic levels) tells us that the source of this vacuum energy might be tiny elementary particles that flicker in and out of existence everywhere throughout the universe. Various attempts have been made to calculate how big the effects of this vacuum energy should be, but so far these attempts have been unsuccessful. In fact, the order of magnitude of theoretical estimates of the vacuum energy based on the quantum mechanics of matter and the value required to account for the acceleration of the expansion of the universe differ by an incredible factor of at least 10^{120} (that is a 1 followed by 120 zeros)! Various other theories have been suggested, but the bottom line is that, although there is compelling evidence that dark energy exists, we do not yet know the source of that energy.

Whatever the dark energy turns out to be, we should note that the discovery that the rate of expansion has not been constant since the beginning of the universe complicates the calculation of the age of the universe. Interestingly, the acceleration seems not to have started with the Big Bang. During the first several billion years after the Big Bang, when galaxies were close together, gravity was strong enough to slow the expansion. As galaxies moved farther apart, the effect of gravity weakened. Several billion years after the Big Bang, dark energy took over, and the expansion began to accelerate (Figure 29.4).

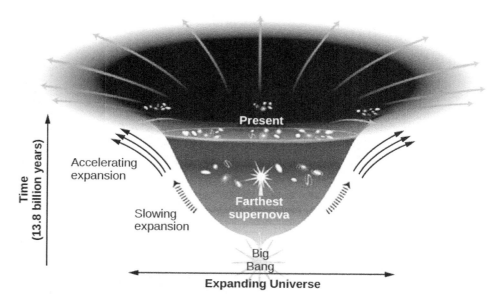

Figure 29.4 Changes in the Rate of Expansion of the Universe Since Its Beginning 13.8 Billion Years Ago. The more the diagram spreads out horizontally, the faster the change in the velocity of expansion. After a period of very rapid expansion at the beginning, which scientists call inflation and which we will discuss later in this chapter, the expansion began to decelerate. Galaxies were then close together, and their mutual gravitational attraction slowed the expansion. After a few billion years, when galaxies were farther apart, the influence of gravity began to weaken. Dark energy then took over and caused the expansion to accelerate. (credit: modification of work by Ann Feild (STScI))

Deceleration works to make the age of the universe estimated by the simple relation $T_0 = 1/H$ seem older than it really is, whereas acceleration works to make it seem younger. By happy coincidence, our best estimates of how much deceleration and acceleration occurred lead to an answer for the age very close to $T_0 = 1/H$. The best current estimate is that the universe is 13.8 billion years old with an uncertainty of only about 100 million years.

Throughout this chapter, we have referred to the Hubble *constant*. We now know that the Hubble constant does change with time. It is, however, constant everywhere in the universe at any given time. When we say the Hubble constant is about 70 kilometers/second/million parsecs, we mean that this is the value of the Hubble constant at the current time.

Comparing Ages

We now have one estimate for the age of the universe from its expansion. Is this estimate consistent with other observations? For example, are the oldest stars or other astronomical objects younger than 13.8 billion years? After all, the universe has to be at least as old as the oldest objects in it.

In our Galaxy and others, the oldest stars are found in the globular clusters (Figure 29.5), which can be dated using the models of stellar evolution described in the chapter Stars from Adolescence to Old Age.

Figure 29.5 Globular Cluster 47 Tucanae. This NASA/ESA Hubble Space Telescope image shows a globular cluster known as 47 Tucanae, since it is in the constellation of Tucana (The Toucan) in the southern sky. The second-brightest globular cluster in the night sky, it includes hundreds of thousands of stars. Globular clusters are among the oldest objects in our Galaxy and can be used to estimate its age. (credit: NASA, ESA, and the Hubble Heritage (STScI/AURA)-ESA/Hubble Collaboration)

The accuracy of the age estimates of the globular clusters has improved markedly in recent years for two reasons. First, models of interiors of globular cluster stars have been improved, mainly through better information about how atoms absorb radiation as they make their way from the center of a star out into space. Second, observations from satellites have improved the accuracy of our measurements of the distances to these clusters. The conclusion is that the oldest stars formed about 12–13 billion years ago.

This age estimate has recently been confirmed by the study of the spectrum of uranium in the stars. The isotope uranium-238 is radioactive and decays (changes into another element) over time. (Uranium-238 gets its designation because it has 92 protons and 146 neutrons.) We know (from how stars and supernovae make elements) how much uranium-238 is generally made compared to other elements. Suppose we measure the amount of uranium relative to nonradioactive elements in a very old star and in our own Sun, and compare the abundances. With those pieces of information, we can estimate how much longer the uranium has been decaying in the very old star because we know from our own Sun how much uranium decays in 4.5 billion years.

The line of uranium is very weak and hard to make out even in the Sun, but it has now been measured in one extremely old star using the European Very Large Telescope (Figure 29.6). Comparing the abundance with that in the solar system, whose age we know, astronomers estimate the star is 12.5 billion years old, with an uncertainty of about 3 billion years. While the uncertainty is large, this work is important confirmation of the ages estimated by studies of the globular cluster stars. Note that the uranium age estimate is completely independent; it does not depend on either the measurement of distances or on models of the interiors of stars.

Figure 29.6 European Extremely Large Telescope, European Very Large Telescope, and the Colosseum. The European Extremely Large Telescope (E-ELT) is currently under construction in Chile. This image compares the size of the E-ELT (left) with the four 8-meter telescopes of the European Very Large Telescope (center) and with the Colosseum in Rome (right). The mirror of the E-ELT will be 39 meters in diameter. Astronomers are building a new generation of giant telescopes in order to observe very distant galaxies and understand what they were like when they were newly formed and the universe was young. (credit: modification of work by ESO)

As we shall see later in this chapter, the globular cluster stars probably did not form until the expansion of the universe had been underway for at least a few hundred million years. Accordingly, their ages are consistent with the 13.8 billion-year age estimated from the expansion rate.

29.2 | A Model of the Universe

Learning Objectives

By the end of this section, you will be able to:

> Explain how the rate of expansion of the universe affects its evolution
> Describe four possibilities for the evolution of the universe
> Explain what is expanding when we say that the universe is expanding
> Define critical density and the evidence that matter alone in the universe is much smaller than the critical density
> Describe what the observations say about the likely long-term future of the universe

Let's now use the results about the expansion of the universe to look at how these ideas might be applied to develop a model for the evolution of the universe as a whole. With this model, astronomers can make predictions about how the universe has evolved so far and what will happen to it in the future.

The Expanding Universe

Every model of the universe must include the expansion we observe. Another key element of the models is that the cosmological principle (which we discussed in The Evolution and Distribution of Galaxies) is valid: on the large scale, the universe at any given time is the same everywhere (homogeneous and isotropic). As a result, the expansion rate must be the same everywhere during any epoch of cosmic time. If so, we don't need to think about the entire universe when we think about the expansion, we can just look at any sufficiently large portion of it. (Some models for dark energy would allow the expansion rate to be different in different directions, and scientists are designing experiments to test this idea. However, until such evidence is found, we will assume that the cosmological principle applies throughout the universe.)

In Galaxies, we hinted that when we think of the expansion of the universe, it is more correct to think of space itself stretching rather than of galaxies moving through static space. Nevertheless, we have since been discussing the redshifts of galaxies as if they resulted from the motion of the galaxies themselves.

Now, however, it is time to finally put such simplistic notions behind us and take a more sophisticated look at the cosmic expansion. Recall from our discussion of Einstein's theory of general relativity (in the chapter on Black Holes and Curved Spacetime) that space—or, more precisely, spacetime—is not a mere backdrop to the action of the universe, as Newton thought. Rather, it is an active participant—affected by and in turn affecting the matter and energy in the universe.

Since the expansion of the universe is the stretching of all spacetime, all points in the universe are stretching together. Thus, the expansion began *everywhere at once*. Unfortunately for tourist agencies of the future, there is no location you can visit where the stretching of space began or where we can say that the Big Bang happened.

To describe just how space stretches, we say the cosmic expansion causes the universe to undergo a uniform change in *scale* over time. By scale we mean, for example, the distance between two clusters of galaxies. It is customary to represent the scale by the factor R; if R doubles, then the distance between the clusters has doubled. Since the universe is expanding at the same rate everywhere, the change in R tells us how much it has expanded (or contracted) at any given time. For a static universe, R would be constant as time passes. In an expanding universe, R increases with time.

If it is space that is stretching rather than galaxies moving through space, then why do the galaxies show redshifts in their spectra? When you were young and naïve—a few chapters ago—it was fine to discuss the redshifts of distant galaxies as if they resulted from their motion away from us. But now that you are an older and wiser student of cosmology, this view will simply not do.

A more accurate view of the redshifts of galaxies is that the light waves are stretched by the stretching of the space they travel through. Think about the light from a remote galaxy. As it moves away from its source, the light has to travel through space. If space is stretching during all the time the light is traveling, the light waves will be stretched as well. A redshift is a stretching of waves—the wavelength of each wave increases (Figure 29.7). Light from more distant galaxies travels for more time than light from closer ones. This means that the light has stretched more than light from closer ones and thus shows a greater redshift.

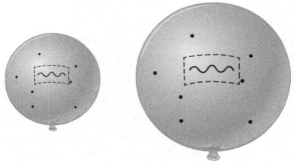

Figure 29.7 Expansion and Redshift. As an elastic surface expands, a wave on its surface stretches. For light waves, the increase in wavelength would be seen as a redshift.

Thus, what the measured redshift of light from an object is telling us is how much the universe has expanded since the light left the object. If the universe has expanded by a factor of 2, then the wavelength of the light (and all electromagnetic waves from the same source) will have doubled.

Models of the Expansion

Before astronomers knew about dark energy or had a good measurement of how much matter exists in the universe, they made speculative models about how the universe might evolve over time. The four possible scenarios are shown in Figure 29.8. In this diagram, time moves forward from the bottom upward, and the scale of space increases by the horizontal circles becoming wider.

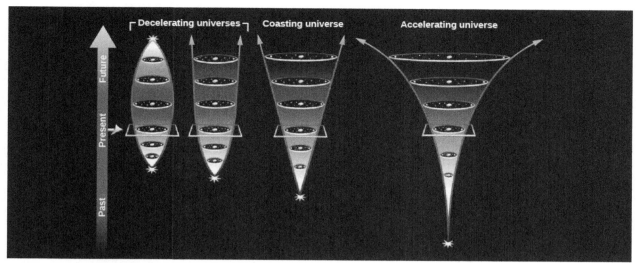

Figure 29.8 Four Possible Models of the Universe. The yellow square marks the present in all four cases, and for all four, the Hubble constant is equal to the same value at the present time. Time is measured in the vertical direction. The first two universes on the left are ones in which the rate of expansion slows over time. The one on the left will eventually slow, come to a stop and reverse, ending up in a "big crunch," while the one next to it will continue to expand forever, but ever-more slowly as time passes. The "coasting" universe is one that expands at a constant rate given by the Hubble constant throughout all of cosmic time. The accelerating universe on the right will continue to expand faster and faster forever. (credit: modification of work by NASA/ESA)

The simplest scenario of an expanding universe would be one in which R increases with time at a constant rate. But you already know that life is not so simple. The universe contains a great deal of mass and its gravity decelerates the expansion—by a large amount if the universe contains a lot of matter, or by a negligible amount if the universe is nearly empty. Then there is the observed acceleration, which astronomers blame on a kind of dark energy.

Let's first explore the range of possibilities with models for different amounts of mass in the universe and for different contributions by dark energy. In some models—as we shall see—the universe expands forever. In others, it stops expanding and starts to contract. After looking at the extreme possibilities, we will look at recent observations that allow us to choose the most likely scenario.

We should perhaps pause for a minute to note how remarkable it is that we can do this at all. Our understanding of the principles that underlie how the universe works on the large scale and our observations of how the objects in the universe change with time allow us to model the evolution of the entire cosmos these days. It is one of the loftiest achievements of the human mind.

What astronomers look at in practice, to determine the kind of universe we live in, is the *average density* of the universe. This is the mass of matter (including the equivalent mass of energy)[1] that would be contained in each unit of volume (say, 1 cubic centimeter) if all the stars, galaxies, and other objects were taken apart, atom by atom, and if all those particles, along with the light and other energy, were distributed throughout all of space with absolute uniformity. If the average density is low, there is less mass and less gravity, and the universe will not decelerate very much. It can therefore expand forever. Higher average density, on the other hand, means there is more mass and more gravity and that the stretching of space might slow down enough that the expansion will eventually stop. An extremely high density might even cause the universe to collapse again.

For a given rate of expansion, there is a **critical density**—the mass per unit volume that will be just enough to slow the expansion to zero at some time infinitely far in the future. If the actual density is higher than this critical density, then the expansion will ultimately reverse and the universe will begin to contract. If the actual density is lower, then the universe will expand forever.

These various possibilities are illustrated in Figure 29.9. In this graph, one of the most comprehensive in all of

---

1 By equivalent mass we mean that which would result if the energy were turned into mass using Einstein's formula, $E = mc^2$.

science, we chart the development of the scale of space in the cosmos against the passage of time. Time increases to the right, and the scale of the universe, R, increases upward in the figure. Today, at the point marked "present" along the time axis, R is increasing in each model. We know that the galaxies are currently expanding away from each other, no matter which model is right. (The same situation holds for a baseball thrown high into the air. While it may eventually fall back down, near the beginning of the throw it moves upward most rapidly.)

The various lines moving across the graph correspond to different models of the universe. The straight dashed line corresponds to the empty universe with no deceleration; it intercepts the time axis at a time, T_0 (the Hubble time), in the past. This is not a realistic model but gives us a measure to compare other models to. The curves below the dashed line represent models with no dark energy and with varying amounts of deceleration, starting from the Big Bang at shorter times in the past. The curve above the dashed line shows what happens if the expansion is accelerating. Let's take a closer look at the future according to the different models.

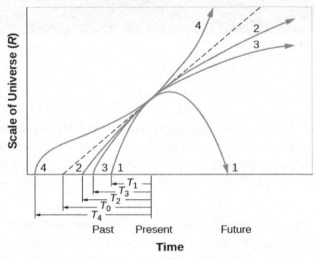

Figure 29.9 Models of the Universe. This graph plots R, the scale of the universe, against time for various cosmological models. Curve 1 represents a universe where the density is greater than the critical value; this model predicts that the universe will eventually collapse. Curve 2 represents a universe with a density lower than critical; the universe will continue to expand but at an ever-slower rate. Curve 3 is a critical-density universe; in this universe, the expansion will gradually slow to a stop infinitely far in the future. Curve 4 represents a universe that is accelerating because of the effects of dark energy. The dashed line is for an empty universe, one in which the expansion is not slowed by gravity or accelerated by dark energy. Time is very compressed on this graph.

Let's start with curve 1 in Figure 29.9. In this case, the actual density of the universe is higher than the critical density and there is no dark energy. This universe will stop expanding at some time in the future and begin contracting. This model is called a **closed universe** and corresponds to the universe on the left in Figure 29.8. Eventually, the scale drops to zero, which means that space will have shrunk to an infinitely small size. The noted physicist John Wheeler called this the "big crunch," because matter, energy, space, and time would all be crushed out of existence. Note that the "big crunch" is the opposite of the Big Bang—it is an *implosion*. The universe is not expanding but rather collapsing in upon itself.

Some scientists speculated that another Big Bang might follow the crunch, giving rise to a new expansion phase, and then another contraction—perhaps oscillating between successive Big Bangs and big crunches indefinitely in the past and future. Such speculation was sometimes referred to as the oscillating theory of the universe. The challenge for theorists was how to describe the transition from collapse (when space and time themselves disappear into the big crunch) to expansion. With the discovery of dark energy, however, it does not appear that the universe will experience a big crunch, so we can put worrying about it on the back burner.

If the density of the universe is less than the critical density (curve 2 in Figure 29.9 and the universe second from the left in Figure 29.8), gravity is never important enough to stop the expansion, and so the universe expands forever. Such a universe is infinite and this model is called an **open universe**. Time and space begin with the Big Bang, but they have no end; the universe simply continues expanding, always a bit more slowly as time goes on. Groups of galaxies eventually get so far apart that it would be difficult for observers in any of them to see the others. (See the feature box on What Might the Universe Be Like in the Distant Future? for more about the distant future in the closed and open universe models.)

At the critical density (curve 3), the universe can just barely expand forever. The critical-density universe has an age of exactly two-thirds T_0, where T_0 is the age of the empty universe. Universes that will someday begin to contract have ages less than two-thirds T_0.

In an empty universe (the dashed line Figure 29.9 and the coasting universe in Figure 29.8), neither gravity nor dark energy is important enough to affect the expansion rate, which is therefore constant throughout all time.

In a universe with dark energy, the rate of the expansion will increase with time, and the expansion will continue at an ever-faster rate. Curve 4 in Figure 29.9, which represents this universe, has a complicated shape. In the beginning, when the matter is all very close together, the rate of expansion is most influenced by gravity. Dark energy appears to act only over large scales and thus becomes more important as the universe grows larger and the matter begins to thin out. In this model, at first the universe slows down, but as space stretches, the acceleration plays a greater role and the expansion speeds up.

The Cosmic Tug of War

We might summarize our discussion so far by saying that a "tug of war" is going on in the universe between the forces that push everything apart and the gravitational attraction of matter, which pulls everything together. If we can determine who will win this tug of war, we will learn the ultimate fate of the universe.

The first thing we need to know is the density of the universe. Is it greater than, less than, or equal to the critical density? The critical density today depends on the value of the expansion rate today, H_0. If the Hubble constant is around 20 kilometers/second per million light-years, the critical density is about 10^{-26} kg/m^3. Let's see how this value compares with the actual density of the universe.

EXAMPLE 29.1

Critical Density of the Universe

As we discussed, the critical density is that combination of matter and energy that brings the universe coasting to a stop at time infinity. Einstein's equations lead to the following expression for the critical density (ρ_{crit}):

$$\rho_{crit} = \frac{3H^2}{8\pi G}$$

where H is the Hubble constant and G is the universal constant of gravity (6.67×10^{-11} Nm^2/kg^2).

Solution

Let's substitute our values and see what we get. Take an $H = 22$ km/s per million light-years. We need to convert both km and light-years into meters for consistency. A million light-years $= 10^6 \times 9.5 \times 10^{15}$ m $= 9.5 \times 10^{21}$ m. And 22 km/s $= 2.2 \times 10^4$ m/s. That makes $H = 2.3 \times 10^{-18}$ /s and $H^2 = 5.36 \times 10^{-36}$ /s^2. So,

$$\rho_{crit} = \frac{3 \times 5.36 \times 10^{-36}}{8 \times 3.14 \times 6.67 \times 10^{-11}} = 9.6 \times 10^{-27} \text{ kg/m}^3$$

which we can round off to the 10^{-26} kg/m^3. (To make the units work out, you have to know that N, the unit of force, is the same as kg $\times$ m/s^2.)

Now we can compare densities we measure in the universe to this critical value. Note that density is mass per unit volume, but energy has an equivalent mass of $m = E/c^2$ (from Einstein's equation $E = mc^2$).

Check Your Learning

a. A single grain of dust has a mass of about 1.1 × 10⁻¹³ kg. If the average mass-energy density of space is equal to the critical density on average, how much space would be required to produce a total mass-energy equal to a dust grain?

b. If the Hubble constant were twice what it actually is, how much would the critical density be?

Answer:

a. In this case, the average mass-energy in a volume V of space is $E = \rho_{crit} V$. Thus, for space with critical density, we require that

$$V = \frac{E_{grain}}{\rho_{crit}} = \frac{1.1 \times 10^{-13}\,\text{kg}}{9.6 \times 10^{-26}\,\text{kg/m}^3} = 1.15 \times 10^{12}\,\text{m}^3 = (10{,}500\,\text{m})^3 \cong (10.5\,\text{km})^3$$

Thus, the sides of a cube of space with mass-energy density averaging that of the critical density would need to be slightly greater than 10 km to contain the total energy equal to a single grain of dust!

b. Since the critical density goes as the square of the Hubble constant, by doubling the Hubble parameter, the critical density would increase by a factor a four. So if the Hubble constant was 44 km/s per million light-years instead of 22 km/s per million light-years, the critical density would be $\rho_{crit} = 4 \times 9.6 \times 10^{-27}\,\text{kg/m}^3 = 3.8 \times 10^{-26}\,\text{kg/m}^3$.

We can start our survey of how dense the cosmos is by ignoring the dark energy and just estimating the density of all matter in the universe, including ordinary matter and dark matter. Here is where the cosmological principle really comes in handy. Since the universe is the same all over (at least on large scales), we only need to measure how much matter exists in a (large) representative sample of it. This is similar to the way a representative survey of a few thousand people can tell us whom the millions of residents of the US prefer for president.

There are several methods by which we can try to determine the average density of matter in space. One way is to count all the galaxies out to a given distance and use estimates of their masses, including dark matter, to calculate the average density. Such estimates indicate a density of about 1 to 2 × 10⁻²⁷ kg/m³ (10 to 20% of critical), which by itself is too small to stop the expansion.

A lot of the dark matter lies outside the boundaries of galaxies, so this inventory is not yet complete. But even if we add an estimate of the dark matter outside galaxies, our total won't rise beyond about 30% of the critical density. We'll pin these numbers down more precisely later in this chapter, where we will also include the effects of dark energy.

In any case, even if we ignore dark energy, the evidence is that the universe will continue to expand forever. The discovery of dark energy that is causing the rate of expansion to speed up only strengthens this conclusion. Things definitely do not look good for fans of the closed universe (big crunch) model.

MAKING CONNECTIONS

What Might the Universe Be Like in the Distant Future?

Some say the world will end in fire,
Some say in ice.
From what I've tasted of desire
I hold with those who favor fire.

—From the poem "Fire and Ice" by Robert Frost (1923)

Given the destructive power of impacting asteroids, expanding red giants, and nearby supernovae, our species may not be around in the remote future. Nevertheless, you might enjoy speculating about what it would be like to live in a much, much older universe.

The observed acceleration makes it likely that we will have continued expansion into the indefinite future. If the universe expands forever (R increases without limit), the clusters of galaxies will spread ever farther apart with time. As eons pass, the universe will get thinner, colder, and darker.

Within each galaxy, stars will continue to go through their lives, eventually becoming white dwarfs, neutron stars, and black holes. Low-mass stars might take a long time to finish their evolution, but in this model, we would literally have all the time in the world. Ultimately, even the white dwarfs will cool down to be black dwarfs, any neutron stars that reveal themselves as pulsars will slowly stop spinning, and black holes with accretion disks will one day complete their "meals." The remains of stars will all be dark and difficult to observe.

This means that the light that now reveals galaxies to us will eventually go out. Even if a small pocket of raw material were left in one unsung corner of a galaxy, ready to be turned into a fresh cluster of stars, we will only have to wait until the time that their evolution is also complete. And time is one thing this model of the universe has plenty of. There will surely come a time when all the stars are out, galaxies are as dark as space, and no source of heat remains to help living things survive. Then the lifeless galaxies will just continue to move apart in their lightless realm.

If this view of the future seems discouraging (from a human perspective), keep in mind that we fundamentally do not understand why the expansion rate is currently accelerating. Thus, our speculations about the future are just that: speculations. You might take heart in the knowledge that science is always a progress report. The most advanced ideas about the universe from a hundred years ago now strike us as rather primitive. It may well be that our best models of today will in a hundred or a thousand years also seem rather simplistic and that there are other factors determining the ultimate fate of the universe of which we are still completely unaware.

Ages of Distant Galaxies

In the chapter on Galaxies, we discussed how we can use Hubble's law to measure the distance to a galaxy. But that simple method only works with galaxies that are not too far away. Once we get to large distances, we are looking so far into the past that we must take into account changes in the rate of the expansion of the universe. Since we cannot measure these changes directly, we must assume one of the models of the universe to be able to convert large redshifts into distances.

This is why astronomers squirm when reporters and students ask them exactly how far away some newly discovered distant quasar or galaxy is. We really can't give an answer without first explaining the model of the universe we are assuming in calculating it (by which time a reporter or student is long gone or asleep). Specifically, we must use a model that includes the change in the expansion rate with time. The key ingredients of the model are the amounts of matter, including dark matter, and the equivalent mass (according to $E = mc^2$) of the dark energy along with the Hubble constant.

Elsewhere in this book, we have estimated the mass density of ordinary matter plus dark matter as roughly 0.3 times the critical density, and the mass equivalent of dark energy as roughly 0.7 times the critical density. We will refer to these values as the "standard model of the universe." The latest (slightly improved) estimates for these values and the evidence for them will be given later in this chapter. Calculations also require the current value of the Hubble constant. For Table 29.1, we have adopted a Hubble constant of 67.3 kilometers/second/million parsecs (rather than rounding it to 70 kilometers/second/million parsecs), which is consistent with the

13.8 billion-year age of the universe estimated by the latest observations.

Once we assume a model, we can use it to calculate the age of the universe at the time an object emitted the light we see. As an example, Table 29.1 lists the times that light was emitted by objects at different redshifts as fractions of the current age of the universe. The times are given for two very different models so you can get a feeling for the fact that the calculated ages are fairly similar. The first model assumes that the universe has a critical density of matter and no dark energy. The second model is the standard model described in the preceding paragraph. The first column in the table is the redshift, which is given by the equation $z = \Delta\lambda/\lambda_0$ and is a measure of how much the wavelength of light has been stretched by the expansion of the universe on its long journey to us.

Ages of the Universe at Different Redshifts

Redshift	Percent of Current Age of Universe When the Light Was Emitted (mass = critical density)	Percent of Current Age of Universe When the Light Was Emitted (mass = 0.3 critical density; dark energy = 0.7 critical density)
0	100 (now)	100 (now)
0.5	54	63
1.0	35	43
2.0	19	24
3.0	13	16
4.0	9	11
5.0	7	9
8.0	4	5
11.9	2.1	2.7
Infinite	0	0

Table 29.1

Notice that as we find objects with higher and higher redshifts, we are looking back to smaller and smaller fractions of the age of the universe. The highest observed redshifts as this book is being written are close to 12 (Figure 29.10). As Table 29.1 shows, we are seeing these galaxies as they were when the universe was only about 3% as old as it is now. They were already formed only about 700 million years after the Big Bang.

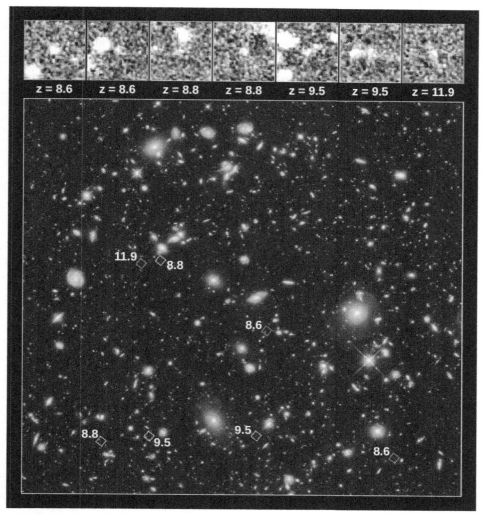

Figure 29.10 Hubble Ultra-Deep Field. This image, called the Hubble Ultra Deep Field, shows faint galaxies, seen very far away and therefore very far back in time. The colored squares in the main image outline the locations of the galaxies. Enlarged views of each galaxy are shown in the black-and-white images. The red lines mark each galaxy's location. The "redshift" of each galaxy is indicated below each box, denoted by the symbol "z." The redshift measures how much a galaxy's ultraviolet and visible light has been stretched to infrared wavelengths by the universe's expansion. The larger the redshift, the more distant the galaxy, and therefore the further astronomers are seeing back in time. One of the seven galaxies may be a distance breaker, observed at a redshift of 11.9. If this redshift is confirmed by additional measurements, the galaxy is seen as it appeared only 380 million years after the Big Bang, when the universe was less than 3% of its present age. (credit: modification of work by NASA, ESA, R. Ellis (Caltech), and the UDF 2012 Team)

29.3 | The Beginning of the Universe

Learning Objectives

By the end of this section, you will be able to:

> Describe what the universe was like during the first few minutes after it began to expand
> Explain how the first new elements were formed during the first few minutes after the Big Bang
> Describe how the contents of the universe change as the temperature of the universe decreases

The best evidence we have today indicates that the first galaxies did not begin to form until a few hundred million years after the Big Bang. What were things like before there were galaxies and space had not yet stretched very significantly? Amazingly, scientists have been able to calculate in some detail what was happening in the universe in the first few minutes after the Big Bang.

The History of the Idea

It is one thing to say the universe had a beginning (as the equations of general relativity imply) and quite another to describe that beginning. The Belgian priest and cosmologist Georges Lemaître was probably the first to propose a specific model for the Big Bang itself (Figure 29.11). He envisioned all the matter of the universe starting in one great bulk he called the *primeval atom*, which then broke into tremendous numbers of pieces. Each of these pieces continued to fragment further until they became the present atoms of the universe, created in a vast nuclear fission. In a popular account of his theory, Lemaître wrote, "The evolution of the world could be compared to a display of fireworks just ended—some few red wisps, ashes, and smoke. Standing on a well-cooled cinder, we see the slow fading of the suns and we try to recall the vanished brilliance of the origin of the worlds."

Figure 29.11 Abbé Georges Lemaître (1894–1966). This Belgian cosmologist studied theology at Mechelen and mathematics and physics at the University of Leuven. It was there that he began to explore the expansion of the universe and postulated its explosive beginning. He actually predicted Hubble's law 2 years before its verification, and he was the first to consider seriously the physical processes by which the universe began.

LINK TO LEARNING

View a short video (https://openstax.org/l/30Lemaitrevid) about the work of Lemaître, considered by some to be the father of the Big Bang theory.

Physicists today know much more about nuclear physics than was known in the 1920s, and they have shown that the primeval fission model cannot be correct. Yet Lemaître's vision was in some respects quite prophetic. We still believe that everything was together at the beginning; it was just not in the form of matter we now know. Basic physical principles tell us that when the universe was much denser, it was also much hotter, and that it cools as it expands, much as gas cools when sprayed from an aerosol can.

By the 1940s, scientists knew that fusion of hydrogen into helium was the source of the Sun's energy. Fusion requires high temperatures, and the early universe must have been hot. Based on these ideas, American physicist George Gamow (Figure 29.12) suggested a universe with a different kind of beginning that involved nuclear **fusion** instead of fission. Ralph Alpher worked out the details for his PhD thesis, and the results were published in 1948. (Gamow, who had a quirky sense of humor, decided at the last minute to add the name of physicist Hans Bethe to their paper, so that the coauthors on this paper about the beginning of things would be Alpher, Bethe, and Gamow, a pun on the first three letters of the Greek alphabet: alpha, beta, and gamma.)

Gamow's universe started with fundamental particles that built up the heavy elements by fusion in the Big Bang.

Figure 29.12 George Gamow and Collaborators. This composite image shows George Gamow emerging like a genie from a bottle of ylem, a Greek term for the original substance from which the world formed. Gamow revived the term to describe the material of the hot Big Bang. Flanking him are Robert Herman (left) and Ralph Alpher (right), with whom he collaborated in working out the physics of the Big Bang. (The modern composer Karlheinz Stockhausen was inspired by Gamow's ideas to write a piece of music called *Ylem*, in which the players actually move away from the stage as they perform, simulating the expansion of the universe.)

Gamow's ideas were close to our modern view, except we now know that the early universe remained hot enough for fusion for only a short while. Thus, only the three lightest elements—hydrogen, helium, and a small amount of lithium—were formed in appreciable abundances at the beginning. The heavier elements formed later in stars. Since the 1940s, many astronomers and physicists have worked on a detailed theory of what happened in the early stages of the universe.

The First Few Minutes

Let's start with the first few minutes following the Big Bang. Three basic ideas hold the key to tracing the changes that occurred during the time just after the universe began. The first, as we have already mentioned, is that the universe cools as it expands. Figure 29.13 shows how the temperature changes with the passage of time. Note that a huge span of time, from a tiny fraction of a second to billions of years, is summarized in this diagram. In the first fraction of a second, the universe was unimaginably hot. By the time 0.01 second had elapsed, the temperature had dropped to 100 billion (10^{11}) K. After about 3 minutes, it had fallen to about 1 billion (10^9) K, still some 70 times hotter than the interior of the Sun. After a few hundred thousand years, the temperature was down to a mere 3000 K, and the universe has continued to cool since that time.

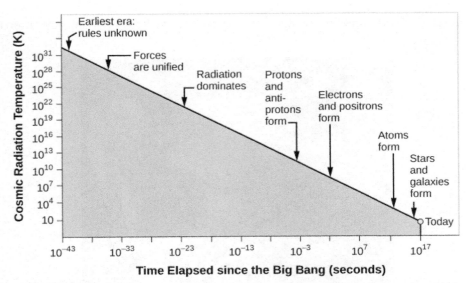

Figure 29.13 Temperature of the Universe. This graph shows how the temperature of the universe varies with time as predicted by the standard model of the Big Bang. Note that both the temperature (vertical axis) and the time in seconds (horizontal axis) change over vast scales on this compressed diagram.

All of these temperatures but the last are derived from theoretical calculations since (obviously) no one was there to measure them directly. As we shall see in the next section, however, we have actually detected the feeble glow of radiation emitted at a time when the universe was a few hundred thousand years old. We can measure the characteristics of that radiation to learn what things were like long ago. Indeed, the fact that we have found this ancient glow is one of the strongest arguments in favor of the Big Bang model.

The second step in understanding the evolution of the universe is to realize that at very early times, it was so hot that it contained mostly radiation (and not the matter that we see today). The photons that filled the universe could collide and produce material particles; that is, under the conditions just after the Big Bang, energy could turn into matter (and matter could turn into energy). We can calculate how much mass is produced from a given amount of energy by using Einstein's formula $E = mc^2$ (see the chapter on The Sun: A Nuclear Powerhouse).

The idea that energy could turn into matter in the universe at large is a new one for many students, since it is not part of our everyday experience. That's because, when we compare the universe today to what it was like right after the Big Bang, we live in cold, hard times. The photons in the universe today typically have far-less energy than the amount required to make new matter. In the discussion on the source of the Sun's energy in The Sun: A Nuclear Powerhouse, we briefly mentioned that when subatomic particles of matter and *antimatter* collide, they turn into pure energy. But the reverse, energy turning into matter and antimatter, is equally possible. This process has been observed in particle accelerators around the world. If we have enough energy, under the right circumstances, new particles of matter (and antimatter) are indeed created —and the conditions were right during the first few minutes after the expansion of the universe began.

Our third key point is that the hotter the universe was, the more energetic were the photons available to make matter and antimatter (see Figure 29.13). To take a specific example, at a temperature of 6 billion (6×10^9) K, the collision of two typical photons can create an electron and its antimatter counterpart, a positron. If the temperature exceeds 10^{14} K, much more massive protons and antiprotons can be created.

The Evolution of the Early Universe

Keeping these three ideas in mind, we can trace the evolution of the universe from the time it was about 0.01 second old and had a temperature of about 100 billion K. Why not begin at the very beginning? There are as yet no theories that allow us penetrate to a time before about 10^{-43} second (this number is a decimal point followed by 42 zeros and then a one). It is so small that we cannot relate it to anything in our everyday experience. When the universe was that young, its density was so high that the theory of general relativity is not adequate to describe it, and even the concept of time breaks down.

Scientists, by the way, have been somewhat more successful in describing the universe when it was older than 10^{-43} second but still less than about 0.01 second old. We will take a look at some of these ideas later in this chapter, but for now, we want to start with somewhat more familiar situations.

By the time the universe was 0.01 second old, it consisted of a soup of matter and radiation; the matter included protons and neutrons, leftovers from an even younger and hotter universe. Each particle collided rapidly with other particles. The temperature was no longer high enough to allow colliding photons to produce neutrons or protons, but it was sufficient for the production of electrons and positrons (Figure 29.14). There was probably also a sea of exotic subatomic particles that would later play a role as dark matter. All the particles jiggled about on their own; it was still much too hot for protons and neutrons to combine to form the nuclei of atoms.

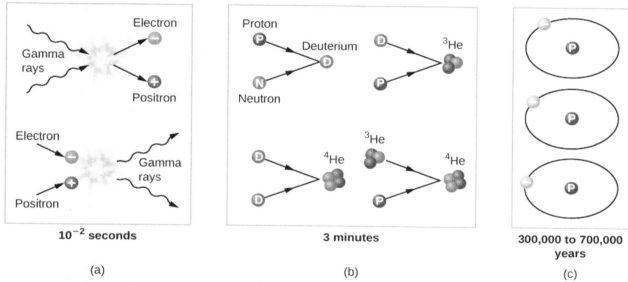

Figure 29.14 Particle Interactions in the Early Universe. (a) In the first fractions of a second, when the universe was very hot, energy was converted into particles and antiparticles. The reverse reaction also happened: a particle and antiparticle could collide and produce energy. (b) As the temperature of the universe decreased, the energy of typical photons became too low to create matter. Instead, existing particles fused to create such nuclei as deuterium and helium. (c) Later, it became cool enough for electrons to settle down with nuclei and make neutral atoms. Most of the universe was still hydrogen.

Think of the universe at this time as a seething cauldron, with photons colliding and interchanging energy, and sometimes being destroyed to create a pair of particles. The particles also collided with one another. Frequently, a matter particle and an antimatter particle met and turned each other into a burst of gamma-ray radiation.

Among the particles created in the early phases of the universe was the ghostly neutrino (see The Sun: A Nuclear Powerhouse), which today interacts only very rarely with ordinary matter. In the crowded conditions of the very early universe, however, neutrinos ran into so many electrons and positrons that they experienced frequent interactions despite their "antisocial" natures.

By the time the universe was a little more than 1 second old, the density had dropped to the point where neutrinos no longer interacted with matter but simply traveled freely through space. In fact, these neutrinos should now be all around us. Since they have been traveling through space unimpeded (and hence unchanged) since the universe was 1 second old, measurements of their properties would offer one of the best tests of the Big Bang model. Unfortunately, the very characteristic that makes them so useful—the fact that they interact so weakly with matter that they have survived unaltered for all but the first second of time—also renders them unable to be measured, at least with present techniques. Perhaps someday someone will devise a way to capture these elusive messengers from the past.

Atomic Nuclei Form

When the universe was about 3 minutes old and its temperature was down to about 900 million K, protons and neutrons could combine. At higher temperatures, these atomic nuclei had immediately been blasted apart by interactions with high-energy photons and thus could not survive. But at the temperatures and densities reached between 3 and 4 minutes after the beginning, **deuterium** (a proton and neutron) lasted long enough that collisions could convert some of it into helium, ([Figure 29.14](#)). In essence, the entire universe was acting the way centers of stars do today—fusing new elements from simpler components. In addition, a little bit of element 3, **lithium**, could also form.

This burst of cosmic fusion was only a brief interlude, however. By 4 minutes after the Big Bang, more helium was having trouble forming. The universe was still expanding and cooling down. After the formation of helium and some lithium, the temperature had dropped so low that the fusion of helium nuclei into still-heavier elements could not occur. No elements beyond lithium could form in the first few minutes. That 4-minute period was the end of the time when the entire universe was a fusion factory. In the cool universe we know today, the fusion of new elements is limited to the centers of stars and the explosions of supernovae.

Still, the fact that the Big Bang model allows the creation of a good deal of helium is the answer to a long-standing mystery in astronomy. Put simply, there is just too much helium in the universe to be explained by what happens inside stars. All the generations of stars that have produced helium since the Big Bang cannot account for the quantity of helium we observe. Furthermore, even the oldest stars and the most distant galaxies show significant amounts of helium. These observations find a natural explanation in the synthesis of helium by the Big Bang itself during the first few minutes of time. We estimate that *10 times more helium* was manufactured in the first 4 minutes of the universe than in all the generations of stars during the succeeding 10 to 15 billion years.

Learning from Deuterium

We can learn many things from the way the early universe made atomic nuclei. It turns out that all of the deuterium (a hydrogen nucleus with a neutron in it) in the universe was formed during the first 4 minutes. In stars, any region hot enough to fuse two protons to form a deuterium nucleus is also hot enough to change it further—either by destroying it through a collision with an energetic photon or by converting it into helium through nuclear reactions.

The amount of deuterium that can be produced in the first 4 minutes of creation depends on the density of the universe at the time deuterium was formed. If the density were relatively high, nearly all the deuterium would have been converted into helium through interactions with protons, just as it is in stars. If the density were relatively low, then the universe would have expanded and thinned out rapidly enough that some deuterium would have survived. The amount of deuterium we see today thus gives us a clue to the density of the universe when it was about 4 minutes old. Theoretical models can relate the density then to the density now; thus, measurements of the abundance of deuterium today can give us an estimate of the current density of the universe.

The measurements of deuterium indicate that the present-day density of ordinary matter—protons and neutrons—is about 5×10^{-28} kg/m^3. Deuterium can only provide an estimate of the density of ordinary matter because the abundance of deuterium is determined by the particles that interact to form it, namely protons and neutrons alone. From the abundance of deuterium, we know that not enough protons and neutrons are present, by a factor of about 20, to produce a critical-density universe.

We do know, however, that there are dark matter particles that add to the overall matter density of the universe, which is then higher than what is calculated for ordinary matter alone. Because dark matter particles do not affect the production of deuterium, measurement of the deuterium abundance cannot tell us how much dark matter exists. Dark matter is made of some exotic kind of particle, not yet detected in any earthbound laboratory. It is definitely not made of protons and neutrons like the readers of this book.

29.4 | The Cosmic Microwave Background

Learning Objectives

By the end of this section, you will be able to:

> Explain why we can observe the afterglow of the hot, early universe
> Discuss the properties of this afterglow as we see it today, including its average temperature and the size of its temperature fluctuations
> Describe open, flat, and curved universes and explain which type of universe is supported by observations
> Summarize our current knowledge of the basic properties of the universe including its age and contents

The description of the first few minutes of the universe is based on theoretical calculations. It is crucial, however, that a scientific theory should be testable. What predictions does it make? And do observations show those predictions to be accurate? One success of the theory of the first few minutes of the universe is the correct prediction of the amount of helium in the universe.

Another prediction is that a significant milestone in the history of the universe occurred about 380,000 years after the Big Bang. Scientists have directly observed what the universe was like at this early stage, and these observations offer some of the strongest support for the Big Bang theory. To find out what this milestone was, let's look at what theory tells us about what happened during the first few hundred thousand years after the Big Bang.

The fusion of helium and lithium was completed when the universe was about 4 minutes old. The universe then continued to resemble the interior of a star in some ways for a few hundred thousand years more. It remained hot and opaque, with radiation being scattered from one particle to another. It was still too hot for electrons to "settle down" and become associated with a particular nucleus; such free electrons are especially effective at scattering photons, thus ensuring that no radiation ever got very far in the early universe without having its path changed. In a way, the universe was like an enormous crowd right after a popular concert; if you get separated from a friend, even if he is wearing a flashing button, it is impossible to see through the dense crowd to spot him. Only after the crowd clears is there a path for the light from his button to reach you.

The Universe Becomes Transparent

Not until a few hundred thousand years after the Big Bang, when the temperature had dropped to about 3000 K and the density of atomic nuclei to about 1000 per cubic centimeter, did the electrons and nuclei manage to combine to form stable atoms of hydrogen and helium (Figure 29.14). With no free electrons to scatter photons, the universe became transparent for the first time in cosmic history. From this point on, matter and radiation interacted much less frequently; we say that they *decoupled* from each other and evolved separately. Suddenly, electromagnetic radiation could really travel, and it has been traveling through the universe ever since.

Discovery of the Cosmic Background Radiation

If the model of the universe described in the previous section is correct, then—as we look far outward in the universe and thus far back in time—the first "afterglow" of the hot, early universe should still be detectable. Observations of it would be very strong evidence that our theoretical calculations about how the universe evolved are correct. As we shall see, we have indeed detected the radiation emitted at this **photon decoupling time**, when radiation began to stream freely through the universe without interacting with matter (Figure 29.15).

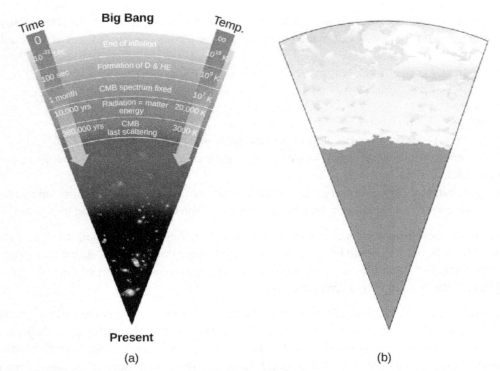

Big Bang

Time | Temp.

0 | ∞

End of inflation

10^{-32} sec | 10^{19} K

Formation of D & HE

100 sec | 10^9 K

CMB spectrum fixed

1 month | 10^7 K

Radiation = matter energy

10,000 yrs | 20,000 K

CMB last scattering

380,000 yrs | 3000 K

Present

(a)   (b)

Figure 29.15 Cosmic Microwave Background and Clouds Compared. (a) Early in the universe, photons (electromagnetic energy) were scattering off the crowded, hot, charged particles and could not get very far without colliding with another particle. But after electrons and photons settled into neutral atoms, there was far less scattering, and photons could travel over vast distances. The universe became transparent. As we look out in space and back in time, we can't see back beyond this time. (b) This is similar to what happens when we see clouds in Earth's atmosphere. Water droplets in a cloud scatter light very efficiently, but clear air lets light travel over long distances. So as we look up into the atmosphere, our vision is blocked by the cloud layers and we can't see beyond them. (credit: modification of work by NASA)

The detection of this afterglow was initially an accident. In the late 1940s, Ralph Alpher and Robert Herman, working with George Gamow, realized that just before the universe became transparent, it must have been radiating like a blackbody at a temperature of about 3000 K—the temperature at which hydrogen atoms could begin to form. If we could have seen that radiation just after neutral atoms formed, it would have resembled radiation from a reddish star. It was as if a giant fireball filled the whole universe.

But that was nearly 14 billion years ago, and, in the meantime, the scale of the universe has increased a thousand fold. This expansion has increased the wavelength of the radiation by a factor of 1000 (see Figure 29.7). According to Wien's law, which relates wavelength and temperature, the expansion has correspondingly lowered the temperature by a factor of 1000 (see the chapter on Radiation and Spectra). The cosmic background behaves like a blackbody and should therefore have a spectrum that obeys Wien's Law.

Alpher and Herman predicted that the glow from the fireball should now be at radio wavelengths and should resemble the radiation from a blackbody at a temperature only a few degrees above absolute zero. Since the fireball was everywhere throughout the universe, the radiation left over from it should also be everywhere. If our eyes were sensitive to radio wavelengths, the whole sky would appear to glow very faintly. However, our eyes can't see at these wavelengths, and at the time Alpher and Herman made their prediction, there were no instruments that could detect the glow. Over the years, their prediction was forgotten.

In the mid-1960s, in Holmdel, New Jersey, Arno Penzias and Robert Wilson of AT&T's Bell Laboratories had built a delicate microwave antenna (Figure 29.16) to measure astronomical sources, including supernova remnants like Cassiopeia A (see the chapter on The Death of Stars). They were plagued with some unexpected background noise, just like faint static on a radio, which they could not get rid of. The puzzling thing about this radiation was that it seemed to be coming from all directions at once. This is very unusual in astronomy: after all, most radiation has a specific direction where it is strongest—the direction of the Sun, or a supernova remnant, or the disk of the Milky Way, for example.

Figure 29.16 Robert Wilson (left) and Arno Penzias (right). These two scientists are standing in front of the horn-shaped antenna with which they discovered the cosmic background radiation. The photo was taken in 1978, just after they received the Nobel Prize in physics.

Penzias and Wilson at first thought that any radiation appearing to come from all directions must originate from inside their telescope, so they took everything apart to look for the source of the noise. They even found that some pigeons had roosted inside the big horn-shaped antenna and had left (as Penzias delicately put it) "a layer of white, sticky, dielectric substance coating the inside of the antenna." However, nothing the scientists did could reduce the background radiation to zero, and they reluctantly came to accept that it must be real, and it must be coming from space.

Penzias and Wilson were not cosmologists, but as they began to discuss their puzzling discovery with other scientists, they were quickly put in touch with a group of astronomers and physicists at Princeton University (a short drive away). These astronomers had—as it happened—been redoing the calculations of Alpher and Herman from the 1940s and also realized that the radiation from the decoupling time should be detectable as a faint afterglow of radio waves. The different calculations of what the observed temperature would be for this **cosmic microwave background (CMB)**[2] were uncertain, but all predicted less than 40 K.

Penzias and Wilson found the distribution of intensity at different radio wavelengths to correspond to a temperature of 3.5 K. This is very cold—closer to absolute zero than most other astronomical measurements—and a testament to how much space (and the waves within it) has stretched. Their measurements have been repeated with better instruments, which give us a reading of 2.73 K. So Penzias and Wilson came very close. Rounding this value, scientists often refer to "the 3-degree microwave background."

Many other experiments on Earth and in space soon confirmed the discovery by Penzias and Wilson: The radiation was indeed coming from all directions (it was isotropic) and matched the predictions of the Big Bang theory with remarkable precision. Penzias and Wilson had inadvertently observed the glow from the primeval fireball. They received the Nobel Prize for their work in 1978. And just before his death in 1966, Lemaître learned that his "vanished brilliance" had been discovered and confirmed.

LINK TO LEARNING

You may enjoy watching *Three Degrees*, a 26-minute video (https://openstax.org/l/30threedegvid) from Bell Labs about Penzias and Wilson's discovery of the cosmic background radiation (with interesting historical footage).

---

2   Recall that microwaves are in the radio region of the electromagnetic spectrum.

Properties of the Cosmic Microwave Background

One issue that worried astronomers is that Penzias and Wilson were measuring the background radiation filling space through Earth's atmosphere. What if that atmosphere is a source of radio waves or somehow affected their measurements? It would be better to measure something this important from space.

The first accurate measurements of the CMB were made with a satellite orbiting Earth. Named the Cosmic Background Explorer (COBE), it was launched by NASA in November 1989. The data it received quickly showed that the CMB closely matches that expected from a blackbody with a temperature of 2.73 K (Figure 29.17). This is exactly the result expected if the CMB was indeed redshifted radiation emitted by a hot gas that filled all of space shortly after the universe began.

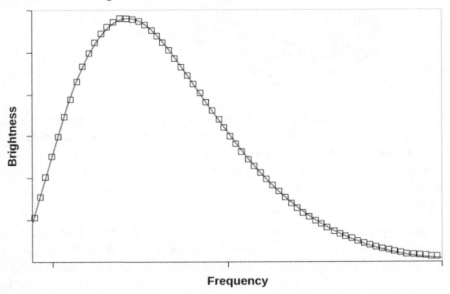

Figure 29.17 Cosmic Background Radiation. The solid line shows how the intensity of radiation should change with wavelength for a blackbody with a temperature of 2.73 K. The boxes show the intensity of the cosmic background radiation as measured at various wavelengths by COBE's instruments. The fit is perfect. When this graph was first shown at a meeting of astronomers, they gave it a standing ovation.

The first important conclusion from measurements of the CMB, therefore, is that the universe we have today has indeed evolved from a hot, uniform state. This observation also provides direct support for the general idea that we live in an evolving universe, since the universe is cooler today than it was in the beginning.

Small Differences in the CMB

It was known even before the launch of COBE that the CMB is extremely *isotropic*. In fact, its uniformity in every direction is one of the best confirmations of the cosmological principle— that the universe is homogenous and isotropic.

According to our theories, however, the temperature could not have been *perfectly* uniform when the CMB was emitted. After all, the CMB is radiation that was scattered from the particles in the universe at the time of decoupling. If the radiation were completely smooth, then all those particles must have been distributed through space absolutely evenly. Yet it is those particles that have become all the galaxies and stars (and astronomy students) that now inhabit the cosmos. Had the particles been completely smoothly distributed, they could not have formed all the large-scale structures now present in the universe—the clusters and superclusters of galaxies discussed in the last few chapters.

The early universe must have had tiny density fluctuations from which such structures could evolve. Regions of higher-than-average density would have attracted additional matter and eventually grown into the galaxies and clusters that we see today. It turns out that these denser regions would appear to us to be colder spots, that is, they would have lower-than-average temperatures.

The reason that temperature and density are related can be explained this way. At the time of decoupling, photons in a slightly denser portion of space had to expend some of their energy to escape the gravitational force exerted by the surrounding gas. In losing energy, the photons became slightly colder than the overall average temperature at the time of decoupling. Vice versa, photons that were located in a slightly less dense portion of space lost less energy upon leaving it than other photons, thus appearing slightly hotter than average. Therefore, if the seeds of present-day galaxies existed at the time that the CMB was emitted, we should see some slight variations in the CMB temperature as we look in different directions in the sky.

Scientists working with the data from the COBE satellite did indeed detect very subtle temperature differences—about 1 part in 100,000—in the CMB. The regions of lower-than-average temperature come in a variety of sizes, but even the smallest of the colder areas detected by COBE is far too large to be the precursor of an individual galaxy, or even a supercluster of galaxies. This is because the COBE instrument had "blurry vision" (poor resolution) and could only measure large patches of the sky. We needed instruments with "sharper vision."

The most detailed measurements of the CMB have been obtained by two satellites launched more recently than COBE. The results from the first of these satellites, the Wilkinson Microwave Anisotropy Probe (WMAP) spacecraft, were published in 2003. In 2015, measurements from the Planck satellite extended the WMAP measurements to even-higher spatial resolution and lower noise (Figure 29.18).

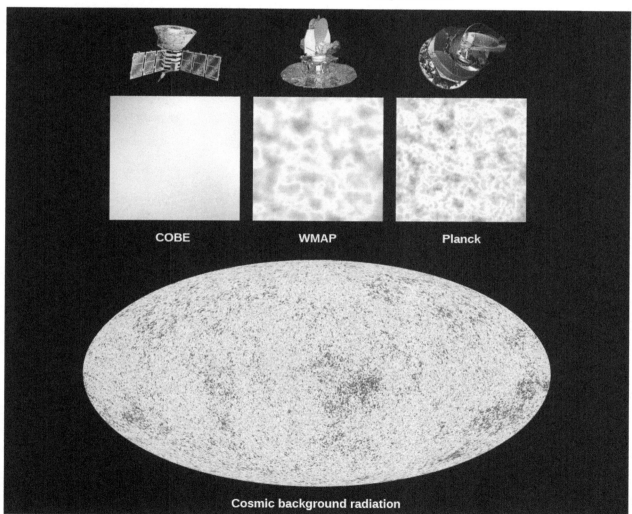

COBE   WMAP   Planck

Cosmic background radiation

Figure 29.18 CMB Observations. This comparison shows how much detail can be seen in the observations of three satellites used to measure the CMB. The CMB is a snapshot of the oldest light in our universe, imprinted on the sky when the universe was just about 380,000 years old. The first spacecraft, launched in 1989, is NASA's Cosmic Background Explorer, or COBE. WMAP was launched in

2001, and Planck was launched in 2009. The three panels show 10-square-degree patches of all-sky maps. This cosmic background radiation image (bottom) is an all-sky map of the CMB as observed by the Planck mission. The colors in the map represent different temperatures: red for warmer and blue for cooler. These tiny temperature fluctuations correspond to regions of slightly different densities, representing the seeds of all future structures: the stars, galaxies, and galaxy clusters of today. (credit top: modification of work by NASA/JPL-Caltech/ESA; credit bottom: modification of work by ESA and the Planck Collaboration)

Theoretical calculations show that the sizes of the hot and cold spots in the CMB depend on the geometry of the universe and hence on its total density. (It's not at all obvious that it should do so, and it takes some pretty fancy calculations—way beyond the level of our text—to make the connection, but having such a dependence is very useful.) The total density we are discussing here includes both the amount of mass in the universe and the mass equivalent of the dark energy. That is, we must add together mass and energy: ordinary matter, dark matter, and the dark energy that is speeding up the expansion.

To see why this works, remember (from the chapter on Black Holes and Curved Spacetime) that with his theory of general relativity, Einstein showed that matter can curve space and that the amount of curvature depends on the amount of matter present. Therefore, the total amount of matter in the universe (including dark matter and the equivalent matter contribution by dark energy), determines the overall geometry of space. Just like the geometry of space around a black hole has a curvature to it, so the entire universe may have a curvature. Let's take a look at the possibilities (Figure 29.19).

If the density of matter is higher than the critical density, the universe will eventually collapse. In such a closed universe, two initially parallel rays of light will eventually meet. This kind of geometry is referred to as spherical geometry. If the density of matter is less than critical, the universe will expand forever. Two initially parallel rays of light will diverge, and this is referred to as hyperbolic geometry. In a critical-density universe, two parallel light rays never meet, and the expansion comes to a halt only at some time infinitely far in the future. We refer to this as a **flat universe**, and the kind of Euclidean geometry you learned in high school applies in this type of universe.

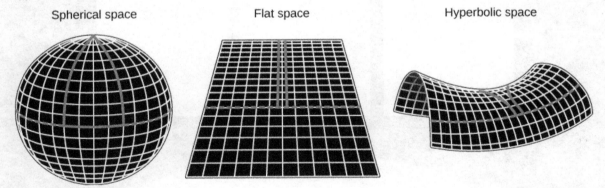

Figure 29.19 Picturing Space Curvature for the Entire Universe. The density of matter and energy determines the overall geometry of space. If the density of the universe is greater than the critical density, then the universe will ultimately collapse and space is said to be *closed* like the surface of a sphere. If the density exactly equals the critical density, then space is *flat* like a sheet of paper; the universe will expand forever, with the rate of expansion coming to a halt infinitely far in the future. If the density is less than critical, then the expansion will continue forever and space is said to be *open* and negatively curved like the surface of a saddle (where more space than you expect opens up as you move farther away). Note that the red lines in each diagram show what happens in each kind of space—they are initially parallel but follow different paths depending on the curvature of space. Remember that these drawings are trying to show how space for the entire universe is "warped"—this can't be seen locally in the small amount of space that we humans occupy.

If the density of the universe is equal to the critical density, then the hot and cold spots in the CMB should typically be about a degree in size. If the density is greater than critical, then the typical sizes will be larger than one degree. If the universe has a density less than critical, then the structures will appear smaller. In Figure 29.20, you can see the differences easily. WMAP and Planck observations of the CMB confirmed earlier experiments that we do indeed live in a flat, critical-density universe.

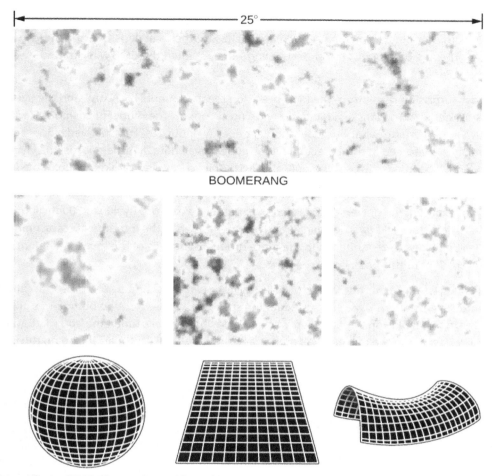

Figure 29.20 Comparison of CMB Observations with Possible Models of the Universe. Cosmological simulations predict that if our universe has critical density, then the CMB images will be dominated by hot and cold spots of around one degree in size (bottom center). If, on the other hand, the density is higher than critical (and the universe will ultimately collapse), then the images' hot and cold spots will appear larger than one degree (bottom left). If the density of the universe is less than critical (and the expansion will continue forever), then the structures will appear smaller (bottom right). As the measurements show, the universe is at critical density. The measurements shown were made by a balloon-borne instrument called BOOMERanG (Balloon Observations of Millimetric Extragalactic Radiation and Geophysics), which was flown in Antarctica. Subsequent satellite observations by WMAP and Planck confirm the BOOMERanG result. (credit: modification of work by NASA)

Key numbers from an analysis of the Planck data give us the best values currently available for some of the basic properties of the universe:

- Age of universe: 13.799 ± 0.038 billion years (Note: That means we know the age of the universe to within 38 million years. Amazing!)
- Hubble constant: 67.31 ± 0.96 kilometers/second/million parsecs (in the units we've been using, 20.65 kilometers/second/million light-years)
- Fraction of universe's content that is "dark energy": 68.5% ± 1.3%
- Fraction of the universe's content that is matter: 31.5% ± 1.3%

Note that this value for the Hubble constant is slightly smaller than the value of 70 kilometers/second/million parsecs that we have adopted in this book. In fact, the value derived from measurements of redshifts is 73 kilometers/second/million parsecs. So precise is modern cosmology these days that scientists are working hard to resolve this discrepancy. The fact that the difference between these two independent measurements is so small is actually a remarkable achievement. Only a few decades ago, astronomers were arguing about whether the Hubble constant was around 50 kilometers/second/million parsecs or 100 kilometers/second/million parsecs.

Analysis of Planck data also shows that ordinary matter (mainly protons and neutrons) makes up 4.9% of the total density. Dark matter plus normal matter add up to 31.5% of the total density. Dark energy contributes the remaining 68.5%. The age of the universe at decoupling—that is, when the CMB was emitted—was 380,000 years.

Perhaps the most surprising result from the high-precision measurements by WMAP and the even higher-precision measurements from Planck is that there were no surprises. The model of cosmology with ordinary matter at about 5%, dark matter at about 25%, and dark energy about 70% has survived since the late 1990s when cosmologists were forced in that direction by the supernovae data. In other words, the very strange universe that we have been describing, with only about 5% of its contents being made up of the kinds of matter we are familiar with here on Earth, really seems to be the universe we live in.

After the CMB was emitted, the universe continued to expand and cool off. By 400 to 500 million years after the Big Bang, the very first stars and galaxies had already formed. Deep in the interiors of stars, matter was reheated, nuclear reactions were ignited, and the more gradual synthesis of the heavier elements that we have discussed throughout this book began.

We conclude this quick tour of our model of the early universe with a reminder. You must not think of the Big Bang as a *localized* explosion *in space*, like an exploding superstar. There were no boundaries and there was no single site where the explosion happened. It was an explosion *of space* (and time and matter and energy) that happened everywhere in the universe. All matter and energy that exist today, including the particles of which you are made, came from the Big Bang. We were, and still are, in the midst of a Big Bang; it is all around us.

29.5 | What Is the Universe Really Made Of?

Learning Objectives

By the end of this section, you will be able to:

> Specify what fraction of the density of the universe is contributed by stars and galaxies and how much ordinary matter (such as hydrogen, helium, and other elements we are familiar with here on Earth) makes up the overall density

> Describe how ideas about the contents of the universe have changed over the last 50 years

> Explain why it is so difficult to determine what dark matter really is

> Explain why dark matter helped galaxies form quickly in the early universe

> Summarize the evolution of the universe from the time the CMB was emitted to the present day

The model of the universe we described in the previous section is the simplest model that explains the observations. It assumes that general relativity is the correct theory of gravity throughout the universe. With this assumption, the model then accounts for the existence and structure of the CMB; the abundances of the light elements deuterium, helium, and lithium; and the acceleration of the expansion of the universe. All of the observations to date support the validity of the model, which is referred to as the standard (or concordance) model of cosmology.

Figure 29.21 and Table 29.2 summarize the current best estimates of the contents of the universe. Luminous matter in stars and galaxies and neutrinos contributes about 1% of the mass required to reach critical density. Another 4% is mainly in the form of hydrogen and helium in the space between stars and in intergalactic space. Dark matter accounts for about an additional 27% of the critical density. The mass equivalent of dark energy (according to $E = mc^2$) then supplies the remaining 68% of the critical density.

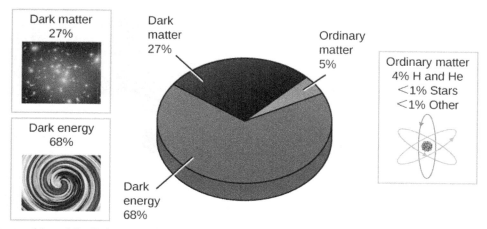

Figure 29.21 Composition of the Universe. Only about 5% of all the mass and energy in the universe is matter with which we are familiar here on Earth. Most ordinary matter consists of hydrogen and helium located in interstellar and intergalactic space. Only about one-half of 1% of the critical density of the universe is found in stars. Dark matter and dark energy, which have not yet been detected in earthbound laboratories, account for 95% of the contents of the universe.

What Different Kinds of Objects Contribute to the Density of the Universe

Object	Density as a Percent of Critical Density
Luminous matter (stars, etc.)	<1
Hydrogen and helium in interstellar and intergalactic space	4
Dark matter	27
Equivalent mass density of the dark energy	68

Table 29.2

This table should shock you. What we are saying is that 95% of the stuff of the universe is either dark matter or dark energy—neither of which has ever been detected in a laboratory here on Earth. This whole textbook, which has focused on objects that emit electromagnetic radiation, has generally been ignoring 95% of what is out there. Who says there aren't big mysteries yet to solve in science!

Figure 29.22 shows how our ideas of the composition of the universe have changed over just the past three decades. The fraction of the universe that we think is made of the same particles as astronomy students has been decreasing steadily.

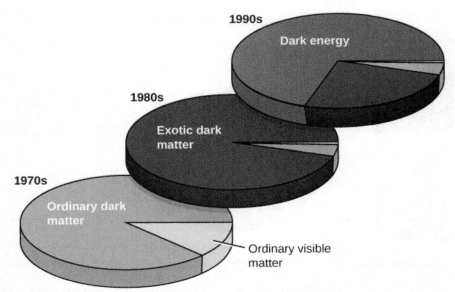

Figure 29.22 Changing Estimates of the Content of the Universe. This diagram shows the changes in our understanding of the contents of the universe over the past three decades. In the 1970s, we suspected that most of the matter in the universe was invisible, but we thought that this matter might be ordinary matter (protons, neutrons, etc.) that was simply not producing electromagnetic radiation. By the 1980s, it was becoming likely that most of the dark matter was made of something we had not yet detected on Earth. By the late 1990s, a variety of experiments had shown that we live in a critical-density universe and that dark energy contributes about 70% of what is required to reach critical density. Note how the estimate of the relative importance of ordinary luminous matter (shown in yellow) has diminished over time.

What Is Dark Matter?

Many astronomers find the situation we have described very satisfying. Several independent experiments now agree on the type of universe we live in and on the inventory of what it contains. We seem to be very close to having a cosmological model that explains nearly everything. Others are not yet ready to jump on the bandwagon. They say, "show me the 96% of the universe we can't detect directly—for example, find me some dark matter!"

At first, astronomers thought that **dark matter** might be hidden in objects that appear dark because they emit no light (e.g., black holes) or that are too faint to be observed at large distances (e.g., planets or white dwarfs). However, these objects would be made of ordinary matter, and the deuterium abundance tells us that no more than 5% of the critical density consists of ordinary matter.

Another possible form that dark matter can take is some type of elementary particle that we have not yet detected here on Earth—a particle that has mass and exists in sufficient abundance to contribute 23% of the critical density. Some physics theories predict the existence of such particles. One class of these particles has been given the name WIMPs, which stands for **weakly interacting massive particles**. Since these particles do not participate in nuclear reactions leading to the production of deuterium, the deuterium abundance puts no limits on how many WIMPs might be in the universe. (A number of other exotic particles have also been suggested as prime constituents of dark matter, but we will confine our discussion to WIMPs as a useful example.)

If large numbers of WIMPs do exist, then some of them should be passing through our physics laboratories right now. The trick is to catch them. Since by definition they interact only weakly (infrequently) with other matter, the chances that they will have a measurable effect are small. We don't know the mass of these particles, but various theories suggest that it might be a few to a few hundred times the mass of a proton. If WIMPs are 60 times the mass of a proton, there would be about 10 million of them passing through your outstretched hand every second—with absolutely no effect on you. If that seems too mind-boggling, bear in mind that neutrinos interact weakly with ordinary matter, and yet we were able to "catch" them eventually.

Despite the challenges, more than 30 experiments designed to detect WIMPS are in operation or in the planning stages. Predictions of how many times WIMPs might actually collide with the nucleus of an atom in the instrument designed to detect them are in the range of 1 event per year to 1 event per 1000 years per kilogram of detector. The detector must therefore be large. It must be shielded from radioactivity or other types of particles, such as neutrons, passing through it, and hence these detectors are placed in deep mines. The energy imparted to an atomic nucleus in the detector by collision with a WIMP will be small, and so the detector must be cooled to a very low temperature.

The WIMP detectors are made out of crystals of germanium, silicon, or xenon. The detectors are cooled to a few thousandths of a degree—very close to absolute zero. That means that the atoms in the detector are so cold that they are scarcely vibrating at all. If a dark matter particle collides with one of the atoms, it will cause the whole crystal to vibrate and the temperature therefore to increase ever so slightly. Some other interactions may generate a detectable flash of light.

A different kind of search for WIMPs is being conducted at the Large Hadron Collider (LHC) at CERN, Europe's particle physics lab near Geneva, Switzerland. In this experiment, protons collide with enough energy potentially to produce WIMPs. The LHC detectors cannot detect the WIMPs directly, but if WIMPs are produced, they will pass through the detectors, carrying energy away with them. Experimenters will then add up all the energy that they detect as a result of the collisions of protons to determine if any energy is missing.

So far, none of these experiments has detected WIMPs. Will the newer experiments pay off? Or will scientists have to search for some other explanation for dark matter? Only time will tell (Figure 29.23).

Figure 29.23 Dark Matter. This cartoon from NASA takes a humorous look at how little we yet understand about dark matter. (credit: NASA)

Dark Matter and the Formation of Galaxies

As elusive as dark matter may be in the current-day universe, galaxies could not have formed quickly without it. Galaxies grew from density fluctuations in the early universe, and some had already formed only about 400–500 million years after the Big Bang. The observations with WMAP, Planck, and other experiments give us information on the size of those density fluctuations. It turns out that the density variations we observe are too small to have formed galaxies so soon after the Big Bang. In the hot, early universe, energetic photons collided with hydrogen and helium, and kept them moving so rapidly that gravity was still not strong enough to cause the atoms to come together to form galaxies. How can we reconcile this with the fact that galaxies *did* form and are all around us?

Our instruments that measure the CMB give us information about density fluctuations only for *ordinary matter*, which interacts with radiation. Dark matter, as its name indicates, does not interact with photons at all. Dark matter could have had much greater variations in density and been able to come together to form gravitational "traps" that could then have begun to attract ordinary matter immediately after the universe became transparent. As ordinary matter became increasingly concentrated, it could have turned into galaxies quickly thanks to these dark matter traps.

For an analogy, imagine a boulevard with traffic lights every half mile or so. Suppose you are part of a motorcade of cars accompanied by police who lead you past each light, even if it is red. So, too, when the early universe was opaque, radiation interacted with ordinary matter, imparting energy to it and carrying it along, sweeping past the concentrations of dark matter. Now suppose the police leave the motorcade, which then encounters some red lights. The lights act as traffic traps; approaching cars now have to stop, and so they bunch up. Likewise, after the early universe became transparent, ordinary matter interacted with radiation only occasionally and so could fall into the dark matter traps.

The Universe in a Nutshell

In the previous sections of this chapter, we traced the evolution of the universe progressively further back in time. Astronomical discovery has followed this path historically, as new instruments and new techniques have allowed us to probe ever closer to the beginning of time. The rate of expansion of the universe was determined from measurements of nearby galaxies. Determinations of the abundances of deuterium, helium, and lithium based on nearby stars and galaxies were used to put limits on how much ordinary matter is in the universe. The motions of stars in galaxies and of galaxies within clusters of galaxies could only be explained if there were large quantities of dark matter. Measurements of supernovae that exploded when the universe was about half as old as it is now indicated that the rate of expansion of the universe has sped up since those explosions occurred. Observations of extremely faint galaxies show that galaxies had begun to form when the universe was only 400–500 million years old. And observations of the CMB confirmed early theories that the universe was initially very hot.

But all this moving further and further backward in time might have left you a bit dizzy. So now let's instead show how the universe evolves as time moves forward.

Figure 29.24 summarizes the entire history of the observable universe from the beginning in a single diagram. The universe was very hot when it began to expand. We have fossil remnants of the very early universe in the form of neutrons, protons, electrons, and neutrinos, and the atomic nuclei that formed when the universe was 3–4 minutes old: deuterium, helium, and a small amount of lithium. Dark matter also remains, but we do not yet know what form it is in.

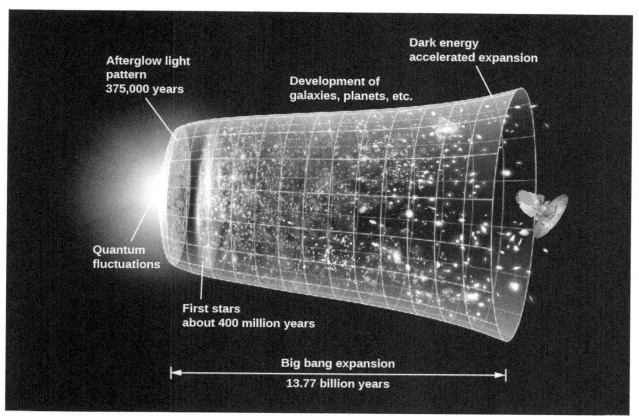

Figure 29.24 History of the Universe. This image summarizes the changes that have occurred in the universe during the last 13.8 billion years. Protons, deuterium, helium, and some lithium were produced in the initial fireball. About 380,000 years after the Big Bang, the universe became transparent to electromagnetic radiation for the first time. COBE, WMAP, Planck, and other instruments have been used to study the radiation that was emitted at that time and that is still visible today (the CMB). The universe was then dark (except for this background radiation) until the first stars and galaxies began to form only a few hundred million years after the Big Bang. Existing space and ground-based telescopes have made substantial progress in studying the subsequent evolution of galaxies. (credit: modification of work by NASA/WMAP Science Team)

The universe gradually cooled; when it was about 380,000 years old, and at a temperature of about 3000 K, electrons combined with protons to form hydrogen atoms. At this point, as we saw, the universe became transparent to light, and astronomers have detected the CMB emitted at this time. The universe still contained no stars or galaxies, and so it entered what astronomers call "the dark ages" (since stars were not lighting up the darkness). During the next several hundred million years, small fluctuations in the density of the dark matter grew, forming gravitational traps that concentrated the ordinary matter, which began to form galaxies about 400–500 million years after the Big Bang.

By the time the universe was about a billion years old, it had entered its own renaissance: it was again blazing with radiation, but this time from newly formed stars, star clusters, and small galaxies. Over the next several billion years, small galaxies merged to form the giants we see today. Clusters and superclusters of galaxies began to grow, and the universe eventually began to resemble what we see nearby.

During the next 20 years, astronomers plan to build giant new telescopes both in space and on the ground to explore even further back in time. In 2021, the James Webb Space Telescope, a 6.5-meter telescope that is the successor to the Hubble Space Telescope, will be launched and assembled in space. The predictions are that with this powerful instrument (see <u>Figure 29.1</u>) we should be able to look back far enough to analyze in detail the formation of the first galaxies.

29.6 | The Inflationary Universe

Learning Objectives

By the end of this section, you will be able to:

> Describe two important properties of the universe that the simple Big Bang model cannot explain
> Explain why these two characteristics of the universe can be accounted for if there was a period of rapid expansion (inflation) of the universe just after the Big Bang
> Name the four forces that control all physical processes in the universe

The hot Big Bang model that we have been describing is remarkably successful. It accounts for the expansion of the universe, explains the observations of the CMB, and correctly predicts the abundances of the light elements. As it turns out, this model also predicts that there should be exactly three types of neutrinos in nature, and this prediction has been confirmed by experiments with high-energy accelerators. We can't relax just yet, however. This standard model of the universe doesn't explain *all* the observations we have made about the universe as a whole.

Problems with the Standard Big Bang Model

There are a number of characteristics of the universe that can only be explained by considering further what might have happened before the emission of the CMB. One problem with the standard Big Bang model is that it does not explain why the density of the universe is equal to the critical density. The mass density could have been, after all, so low and the effects of dark energy so high that the expansion would have been too rapid to form any galaxies at all. Alternatively, there could have been so much matter that the universe would have already begun to contract long before now. Why is the universe balanced so precisely on the knife edge of the critical density?

Another puzzle is the remarkable *uniformity* of the universe. The temperature of the CMB is the same to about 1 part in 100,000 everywhere we look. This sameness might be expected if all the parts of the visible universe were in contact at some point in time and had the time to come to the same temperature. In the same way, if we put some ice into a glass of lukewarm water and wait a while, the ice will melt and the water will cool down until they are the same temperature.

However, if we accept the standard Big Bang model, all parts of the visible universe were *not* in contact at any time. The fastest that information can go from one point to another is the speed of light. There is a maximum distance that light can have traveled from any point since the time the universe began—that's the distance light could have covered since then. This distance is called that point's *horizon distance* because anything farther away is "below its horizon"—unable to make contact with it. One region of space separated by more than the horizon distance from another has been completely isolated from it through the entire history of the universe.

If we measure the CMB in two opposite directions in the sky, we are observing regions that were significantly beyond each other's horizon distance at the time the CMB was emitted. We can see both regions, but *they* can never have seen each other. Why, then, are their temperatures so precisely the same? According to the standard Big Bang model, they have never been able to exchange information, and there is no reason they should have identical temperatures. (It's a little like seeing the clothes that all the students wear at two schools in different parts of the world become identical, without the students ever having been in contact.) The only explanation we could suggest was simply that the universe somehow *started out* being absolutely uniform (which is like saying all students were born liking the same clothes). Scientists are always uncomfortable when they must appeal to a special set of initial conditions to account for what they see.

The Inflationary Hypothesis

Some physicists suggested that these fundamental characteristics of the cosmos—its flatness and uniformity—can be explained if shortly after the Big Bang (and before the emission of the CMB), the universe experienced a sudden increase in size. A model universe in which this rapid, early expansion occurs is called an **inflationary universe**. The inflationary universe is identical to the Big Bang universe for all time after the first 10^{-30} second. Prior to that, the model suggests that there was a brief period of extraordinarily rapid expansion or inflation, during which the scale of the universe increased by a factor of about 10^{50} times more than predicted by standard Big Bang models (Figure 29.25).

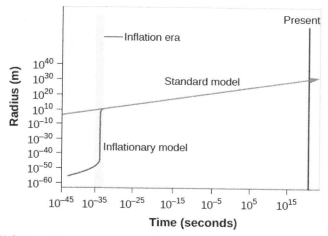

Figure 29.25 Expansion of the Universe. This graph shows how the scale factor of the observable universe changes with time for the standard Big Bang model (red line) and for the inflationary model (blue line). (Note that the time scale at the bottom is extremely compressed.) During inflation, regions that were very small and in contact with each other are suddenly blown up to be much larger and outside each other's horizon distance. The two models are the same for all times after 10^{-30} second.

Prior to (and during) inflation, all the parts of the universe that we can now see were so small and close to each other that they *could* exchange information, that is, the horizon distance included all of the universe that we can now observe. Before (and during) inflation, there was adequate time for the observable universe to homogenize itself and come to the same temperature. Then, inflation expanded those regions tremendously, so that many parts of the universe are now beyond each other's horizon.

Another appeal of the inflationary model is its prediction that the density of the universe should be exactly equal to the critical density. To see why this is so, remember that curvature of spacetime is intimately linked to the density of matter. If the universe began with some curvature of its spacetime, one analogy for it might be the skin of a balloon. The period of inflation was equivalent to blowing up the balloon to a tremendous size. The universe became so big that from our vantage point, no curvature should be visible (Figure 29.26). In the same way, Earth's surface is so big that it looks flat to us no matter where we are. Calculations show that a universe with no curvature is one that is at critical density. Universes with densities either higher or lower than the critical density would show marked curvature. But we saw that the observations of the CMB in Figure 29.18, which show that the universe has critical density, rule out the possibility that space is significantly curved.

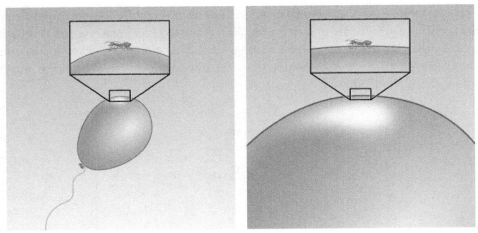

Figure 29.26 Analogy for Inflation. During a period of rapid inflation, a curved balloon grows so large that to any local observer it looks flat. The inset shows the geometry from the ant's point of view.

Grand Unified Theories

While inflation is an intriguing idea and widely accepted by researchers, we cannot directly observe events so early in the universe. The conditions at the time of inflation were so extreme that we cannot reproduce them in our laboratories or high-energy accelerators, but scientists have some ideas about what the universe might have been like. These ideas are called **grand unified theories** or GUTs.

In GUT models, the forces that we are familiar with here on Earth, including gravity and electromagnetism, behaved very differently in the extreme conditions of the early universe than they do today. In physical science, the term *force* is used to describe anything that can change the motion of a particle or body. One of the remarkable discoveries of modern science is that all known physical processes can be described through the action of just four forces: gravity, electromagnetism, the strong nuclear force, and the weak nuclear force (Table 29.3).

The Forces of Nature

Force	Relative Strength Today	Range of Action	Important Applications
Gravity	1	Whole universe	Motions of planets, stars, galaxies
Electromagnetism	10^{36}	Whole universe	Atoms, molecules, electricity, magnetic fields
Weak nuclear force	10^{33}	10^{-17} meters	Radioactive decay
Strong nuclear force	10^{38}	10^{-15} meters	The existence of atomic nuclei

Table 29.3

Gravity is perhaps the most familiar force, and certainly appears strong if you jump off a tall building. However, the force of gravity between two elementary particles—say two protons—is by far the weakest of the four forces. Electromagnetism—which includes both magnetic and electrical forces, holds atoms together, and produces the electromagnetic radiation that we use to study the universe—is much stronger, as you can see in Table 29.3. The weak nuclear force is only weak in comparison to its strong "cousin," but it is in fact much stronger than gravity.

Both the weak and strong nuclear forces differ from the first two forces in that they act only over very small distances—those comparable to the size of an atomic nucleus or less. The weak force is involved in radioactive decay and in reactions that result in the production of neutrinos. The strong force holds protons and neutrons together in an atomic nucleus.

Physicists have wondered why there are four forces in the universe—why not 300 or, preferably, just one? An important hint comes from the name *electromagnetic force*. For a long time, scientists thought that the forces of electricity and magnetism were separate, but James Clerk Maxwell (see the chapter on Radiation and Spectra) was able to *unify* these forces—to show that they are aspects of the same phenomenon. In the same way, many scientists (including Einstein) have wondered if the four forces we now know could also be unified. Physicists have actually developed GUTs that unify three of the four forces (but not gravity).

In these theories, the strong, weak, and electromagnetic forces are not three independent forces but instead are different manifestations or aspects of what is, in fact, a single force. The theories predict that at high enough temperatures, there would be only one force. At lower temperatures (like the ones in the universe today), however, this single force has changed into three different forces (Figure 29.27). Just as different gases or liquids freeze at different temperatures, we can say that the different forces "froze out" of the unified force at different temperatures. Unfortunately, the temperatures at which the three forces acted as one force are so high that they cannot be reached in any laboratory on Earth. Only the early universe, at times prior to 10^{-35} second, was hot enough to unify these forces.

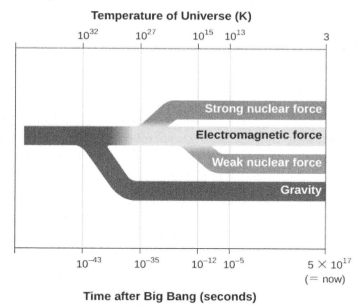

Figure 29.27 Four Forces That Govern the Universe. The behavior of the four forces depends on the temperature of the universe. This diagram (inspired by some grand unified theories) shows that at very early times when the temperature of the universe was very high, all four forces resembled one another and were indistinguishable. As the universe cooled, the forces took on separate and distinctive characteristics.

Many physicists think that gravity was also unified with the three other forces at still higher temperatures, and scientists have tried to develop a theory that combines all four forces. For example, in string theory, the point-like particles of matter that we have discussed in this book are replaced by one-dimensional objects called strings. In this theory, infinitesimal strings, which have length but not height or width, are the building blocks used to construct all the forms of matter and energy in the universe. These strings exist in 11-dimensional space (not the 4-dimensional spacetime with which we are familiar). The strings vibrate in the various dimensions, and depending on how they vibrate, they are seen in our world as matter or gravity or light. As you can imagine, the mathematics of string theory is very complex, and the theory remains untested by experiments. Even the largest particle accelerators on Earth do not achieve high enough energy to show whether string theory applies to the real world.

String theory is interesting to scientists because it is currently the only approach that seems to have the potential of combining all four forces to produce what physicists have termed the Theory of Everything.[3] Theories of the earliest phases of the universe must take both quantum mechanics and gravity into account, but at the simplest level, gravity and quantum mechanics are incompatible. General relativity, our best theory of gravity, says that the motions of objects can be predicted exactly. Quantum mechanics says you can only calculate the probability (chance) that an object will do something. String theory is an attempt to resolve this paradox. The mathematics that underpins string theory is elegant and beautiful, but it remains to be seen whether it will make predictions that can be tested by observations in yet-to-be-developed, high-energy accelerators on Earth or by observations of the early universe.

The earliest period in the history of the universe from time zero to 10^{-43} second is called the Planck time. The universe was unimaginably hot and dense, and theorists believe that at this time, quantum effects of gravity dominated physical interactions—and, as we have just discussed, we have no tested theory of quantum gravity. Inflation is hypothesized to have occurred somewhat later, when the universe was between perhaps 10^{-35} and 10^{-33} second old and the temperature was 10^{27} to 10^{28} K. This rapid expansion took place when three forces (electromagnetic, strong, and weak) are thought to have been unified, and this is when GUTs are applicable.

After inflation, the universe continued to expand (but more slowly) and to cool. An important milestone was reached when the temperature was down to 10^{15} K and the universe was 10^{-10} second old. Under these conditions, all four forces were separate and distinct. High-energy particle accelerators can achieve similar conditions, and so theories of the history of the universe from this point on have a sound basis in experiments.

As yet, we have no direct evidence of what the conditions were during the inflationary epoch, and the ideas presented here are speculative. Researchers are trying to devise some experimental tests. For example, the quantum fluctuations in the very early universe would have caused variations in density and produced gravitational waves that may have left a detectable imprint on the CMB. Detection of such an imprint will require observations with equipment whose sensitivity is improved from what we have today. Ultimately, however, it may provide confirmation that we live in a universe that once experienced an epoch of rapid inflation.

If you are typical of the students who read this book, you may have found this brief discussion of dark matter, inflation, and cosmology a bit frustrating. We have offered glimpses of theories and observations, but have raised more questions than we have answered. What is dark matter? What is dark energy? Inflation explains the observations of flatness and uniformity of the university, but did it actually happen? These ideas are at the forefront of modern science, where progress almost always leads to new puzzles, and much more work is needed before we can see clearly. Bear in mind that less than a century has passed since Hubble demonstrated the existence of other galaxies. The quest to understand just how the universe of galaxies came to be will keep astronomers busy for a long time to come.

29.7 | The Anthropic Principle

Learning Objectives

By the end of this section, you will be able to:

> Name some properties of the universe that, if different, would have precluded the development of humans

---

3 This name became the title of a film about physicist Stephen Hawking in 2014.

Despite our uncertainties, we must admit that the picture we have developed about the evolution of our universe is a remarkable one. With new telescopes, we have begun to collect enough observational evidence that we can describe how the universe evolved from a mere fraction of a second after the expansion began. Although this is an impressive achievement, there are still some characteristics of the universe that we cannot explain. And yet, it turns out that if these characteristics were any different, we would not be here to ask about them. Let's look at some of these "lucky accidents," beginning with the observations of the cosmic microwave background (CMB).

Lucky Accidents

As we described in this chapter, the CMB is radiation that was emitted when the universe was a few hundred thousand years old. Observations show that the temperature of the radiation varies from one region to another, typically by about 10 millionths of a degree, and these temperature differences signal small differences in density. But suppose the tiny, early fluctuations in density had been much smaller. Then calculations show that the pull of gravity near them would have been so small that no galaxies would ever have formed.

What if the fluctuations in density had been much larger? Then it is possible that very dense regions would have condensed, and these would simply have collapsed directly to black holes without ever forming galaxies and stars. Even if galaxies had been able to form in such a universe, space would have been filled with intense X-rays and gamma rays, and it would have been difficult for life forms to develop and survive. The density of stars within galaxies would be so high that interactions and collisions among them would be frequent. In such a universe, any planetary systems could rarely survive long enough for life to develop.

So for us to be here, the density fluctuations need to be "just right"—not too big and not too small.

Another lucky accident is that the universe is finely balanced between expansion and contraction. It is expanding, but very slowly. If the expansion had been at a much higher rate, all of the matter would have thinned out before galaxies could form. If everything were expanding at a much slower rate, then gravity would have "won." The expansion would have reversed and all of the matter would have recollapsed, probably into a black hole—again, no stars, no planets, no life.

The development of life on Earth depends on still-luckier coincidences. Had matter and antimatter been present initially in exactly equal proportions, then all matter would have been annihilated and turned into pure energy. We owe our existence to the fact that there was slightly more matter than antimatter. (After most of the matter made contact with an equal amount of antimatter, turning into energy, a small amount of additional matter must have been present. We are all descendants of that bit of "unbalanced" matter.)

If nuclear fusion reactions occurred at a somewhat faster rate than they actually do, then at the time of the initial fireball, all of the matter would have been converted from hydrogen into helium into carbon and all the way into iron (the most stable nucleus). That would mean that no stars would have formed, since the existence of stars depends on there being light elements that can undergo fusion in the main-sequence stage and make the stars shine. In addition, the structure of atomic nuclei had to be just right to make it possible for three helium atoms to come together easily to fuse carbon, which is the basis of life. If the *triple-alpha process* we discussed in the chapter on Stars from Adolescence to Old Age were too unlikely, not enough carbon would have formed to lead to biology as we know it. At the same time, it had to be hard enough to fuse carbon into oxygen that a large amount of carbon survived for billions of years.

There are additional factors that have contributed to life like us being possible. Neutrinos have to interact with matter at just the right, albeit infrequent, rate. Supernova explosions occur when neutrinos escape from the cores of collapsing stars, deposit some of their energy in the surrounding stellar envelope, and cause it to blow out and away into space. The heavy elements that are ejected in such explosions are essential ingredients of life here on Earth. If neutrinos did not interact with matter at all, they would escape from the cores of collapsing stars without causing the explosion. If neutrinos interacted strongly with matter, they would remain trapped in the stellar core. In either case, the heavy elements would remain locked up inside the collapsing star.

If gravity were a much stronger force than it is, stars could form with much smaller masses, and their lifetimes would be measured in years rather than billions of years. Chemical processes, on the other hand, would not be sped up if gravity were a stronger force, and so there would be no time for life to develop while stars were so short-lived. Even if life did develop in a stronger-gravity universe, life forms would have to be tiny or they could not stand up or move around.

What Had to Be, Had to Be

In summary, we see that a specific set of rules and conditions in the universe has allowed complexity and life on Earth to develop. As yet, we have no theory to explain why this "right" set of conditions occurred. For this reason, many scientists are beginning to accept an idea we call the **anthropic principle**—namely, that the physical laws we observe must be what they are precisely because these are the only laws that allow for the existence of humans.

Some scientists speculate that our universe is but one of countless universes, each with a different set of physical laws—an idea that is sometimes referred to as the **multiverse**. Some of those universes might be stillborn, collapsing before any structure forms. Others may expand so quickly that they remain essentially featureless with no stars and galaxies. In other words, there may be a much larger multiverse that contains our own universe and many others. This multiverse (existing perhaps in more dimensions that we can become aware of) is infinite and eternal; it generates many, many inflating regions, each of which evolves into a separate universe, which may be completely unlike any of the other separate universes. Our universe is then the way it is because it is the only way it could be and have humans like ourselves in it to discover its properties and ask such questions.

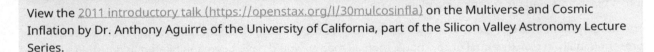

LINK TO LEARNING

View the 2011 introductory talk (https://openstax.org/l/30mulcosinfla) on the Multiverse and Cosmic Inflation by Dr. Anthony Aguirre of the University of California, part of the Silicon Valley Astronomy Lecture Series.

It is difficult to know how to test these ideas since we can never make contact with any other universe. For most scientists, our discussion in this section borders on the philosophical and metaphysical. Perhaps in the future our understanding of physics will develop to the point that we can know why the gravitational constant is as strong as it is, why the universe is expanding at exactly the rate it is, and why all of the other "lucky accidents" happened—why they were inevitable and could be no other way. Then this anthropic idea would no longer be necessary. No one knows, however, whether we will ever have an explanation for why this universe works the way it does.

We have come a long way in our voyage through the universe. We have learned a remarkable amount about *how* and *when* the cosmos came to be, but the question of *why* the universe is the way it is remains as elusive as ever.

🔑 Key Terms

anthropic principle   idea that physical laws must be the way they are because otherwise we could not be here to measure them

Big Bang   the theory of cosmology in which the expansion of the universe began with a primeval explosion (of space, time, matter, and energy)

closed universe   a model in which the universe expands from a Big Bang, stops, and then contracts to a big crunch

cosmic microwave background (CMB)   microwave radiation coming from all directions that is the redshifted afterglow of the Big Bang

cosmological constant   the term in the equations of general relativity that represents a repulsive force in the universe

cosmology   the study of the organization and evolution of the universe

critical density   in cosmology, the density that is just sufficient to bring the expansion of the universe to a stop after infinite time

dark energy   the energy that is causing the expansion of the universe to accelerate; its existence is inferred from observations of distant supernovae

dark matter   nonluminous material, whose nature we don't yet understand, but whose presence can be inferred because of its gravitational influence on luminous matter

deuterium   a form of hydrogen in which the nucleus of each atom consists of one proton and one neutron

flat universe   a model of the universe that has a critical density and in which the geometry of the universe is flat, like a sheet of paper

fusion   the building of heavier atomic nuclei from lighter ones

grand unified theories   (GUTs) physical theories that attempt to describe the four forces of nature as different manifestations of a single force

inflationary universe   a theory of cosmology in which the universe is assumed to have undergone a phase of very rapid expansion when the universe was about 10^{-35} second old; after this period of rapid expansion, the standard Big Bang and inflationary models are identical

lithium   the third element in the periodic table; lithium nuclei with three protons and four neutrons were manufactured during the first few minutes of the expansion of the universe

multiverse   the speculative idea that our universe is just one of many universes, each with its own set of physical laws

open universe   a model in which the density of the universe is not high enough to bring the expansion of the universe to a halt

photon decoupling time   when radiation began to stream freely through the universe without interacting with matter

weakly interacting massive particles   (WIMPs) weakly interacting massive particles are one of the candidates for the composition of dark matter

📄 Summary

29.1 The Age of the Universe

Cosmology is the study of the organization and evolution of the universe. The universe is expanding, and this is one of the key observational starting points for modern cosmological theories. Modern observations show that the rate of expansion has not been constant throughout the life of the universe. Initially, when galaxies were close together, the effects of gravity were stronger than the effects of dark energy, and the expansion rate gradually slowed. As galaxies moved farther apart, the influence of gravity on the expansion rate weakened. Measurements of distant supernovae show that when the universe was about half its current age, dark energy began to dominate the rate of expansion and caused it to speed up. In order to estimate the age of the universe, we must allow for changes in the rate of expansion. After allowing for these effects, astronomers estimate that all of the matter within the observable universe was concentrated in an extremely small volume 13.8 billion years ago, a time we call the Big Bang.

29.2 A Model of the Universe

For describing the large-scale properties of the universe, a model that is isotropic and homogeneous (same everywhere) is a pretty good approximation of reality. The universe is expanding, which means that the universe undergoes a change in scale with time; space stretches and distances grow larger by the same factor everywhere at a given time. Observations show that the mass density of the universe is less than the critical density. In other words, there is not enough matter in the universe to stop the expansion. With the discovery of dark energy, which is accelerating the rate of expansion, the observational evidence is strong that the universe will expand forever. Observations tell us that the expansion started about 13.8 billion years ago.

29.3 The Beginning of the Universe

Lemaître, Alpher, and Gamow first worked out the ideas that are today called the Big Bang theory. The universe cools as it expands. The energy of photons is determined by their temperature, and calculations show that in the hot, early universe, photons had so much energy that when they collided with one another, they could produce material particles. As the universe expanded and cooled, protons and neutrons formed first, then came electrons and positrons. Next, fusion reactions produced deuterium, helium, and lithium nuclei. Measurements of the deuterium abundance in today's universe show that the total amount of ordinary matter in the universe is only about 5% of the critical density.

29.4 The Cosmic Microwave Background

When the universe became cool enough to form neutral hydrogen atoms, the universe became transparent to radiation. Scientists have detected the cosmic microwave background (CMB) radiation from this time during the hot, early universe. Measurements with the COBE satellite show that the CMB acts like a blackbody with a temperature of 2.73 K. Tiny fluctuations in the CMB show us the seeds of large-scale structures in the universe. Detailed measurements of these fluctuations show that we live in a critical-density universe and that the critical density is composed of 31% matter, including dark matter, and 69% dark energy. Ordinary matter—the kinds of elementary particles we find on Earth—make up only about 5% of the critical density. CMB measurements also indicate that the universe is 13.8 billion years old.

29.5 What Is the Universe Really Made Of?

Twenty-seven percent of the critical density of the universe is composed of dark matter. To explain so much dark matter, some physics theories predict that additional types of particles should exist. One type has been given the name of WIMPs (weakly interacting massive particles), and scientists are now conducting experiments to try to detect them in the laboratory. Dark matter plays an essential role in forming galaxies. Since, by definition, these particles interact only very weakly (if at all) with radiation, they could have congregated while the universe was still very hot and filled with radiation. They would thus have formed gravitational traps that quickly attracted and concentrated ordinary matter after the universe became transparent, and matter and radiation decoupled. This rapid concentration of matter enabled galaxies to form by the time the universe was only 400–500 million years old.

29.6 The Inflationary Universe

The Big Bang model does not explain why the CMB has the same temperature in all directions. Neither does it explain why the density of the universe is so close to critical density. These observations can be explained if the universe experienced a period of rapid expansion, which scientists call inflation, about 10^{-35} second after the Big Bang. New grand unified theories (GUTs) are being developed to describe physical processes in the universe before and at the time that inflation occurred.

29.7 The Anthropic Principle

Recently, many cosmologists have noted that the existence of humans depends on the fact that many properties of the universe—the size of density fluctuations in the early universe, the strength of gravity, the structure of atoms—were just right. The idea that physical laws must be the way they are because otherwise we could not be here to measure them is called the anthropic principle. Some scientists speculate that there may be a multiverse of universes, in which ours is just one.

 For Further Exploration

Articles

The January 2021 issue of *Astronomy* magazine is devoted to our modern understanding of the beginning and end of the universe.

Kruesi, L. "Cosmology: 5 Things You Need to Know." *Astronomy* (May 2007): 28. Five questions students often ask, and how modern cosmologists answer them.

Kruesi, L. "How Planck Has Redefined the Universe." *Astronomy* (October 2013): 28. Good review of what this space mission has told us about the CMB and the universe.

Lineweaver, C. & Davis, T. "Misconceptions about the Big Bang." *Scientific American* (March 2005): 36. Some basic ideas about modern cosmology clarified, using general relativity.

Nadis, S. "Sizing Up Inflation." *Sky & Telescope* (November 2005): 32. Nice review of the origin and modern variants on the inflationary idea.

Nadis, S. "How We Could See Another Universe." *Astronomy* (June 2009): 24. On modern ideas about multiverses and how such bubbles of space-time might collide.

Nadis, S. "Dark Energy's New Face: How Exploding Stars Are Changing our View." *Astronomy* (July 2012): 45. About our improving understanding of the complexities of type Ia supernovae.

Naze, Y. "The Priest, the Universe, and the Big Bang." *Astronomy* (November 2007): 40. On the life and work of Georges Lemaître.

Panek, R. "Going Over to the Dark Side." *Sky & Telescope* (February 2009): 22. A history of the observations and theories about dark energy.

Pendrick, D. "Is the Big Bang in Trouble?" *Astronomy* (April 2009): 48. This sensationally titled article is really more of a quick review of how modern ideas and observations are fleshing out the Big Bang hypothesis (and raising questions.)

Reddy, F. "How the Universe Will End." *Astronomy* (September 2014): 38. Brief discussion of local and general future scenarios.

Riess, A. and Turner, M. "The Expanding Universe: From Slowdown to Speedup." *Scientific American* (September 2008): 62.

Turner, M. "The Origin of the Universe." *Scientific American* (September 2009): 36. An introduction to modern cosmology.

Websites

Cosmology Primer: https://preposterousuniverse.com/cosmologyprimer/ (https://preposterousuniverse.com/cosmologyprimer/). Caltech Astrophysicist Sean Carroll offers a non-technical site with brief overviews of many key topics in modern cosmology.

How Big Is the Universe?: http://www.pbs.org/wgbh/nova/space/how-big-universe.html (http://www.pbs.org/

wgbh/nova/space/how-big-universe.html). A clear essay by a noted astronomer Brent Tully summarizes some key ideas in cosmology and introduces the notion of the acceleration of the universe.

Universe 101: WMAP Mission Introduction to the Universe: http://map.gsfc.nasa.gov/universe/ (http://map.gsfc.nasa.gov/universe/). Concise NASA primer on cosmological ideas from the WMAP mission team.

Cosmic Times Project: http://cosmictimes.gsfc.nasa.gov/ (http://cosmictimes.gsfc.nasa.gov/). James Lochner and Barbara Mattson have compiled a rich resource of twentieth-century cosmology history in the form of news reports on key events, from NASA's Goddard Space Flight Center.

Videos

The Day We Found the Universe: https://www.youtube.com/watch?v=HV23qWIieBw (https://www.youtube.com/watch?v=HV23qWIieBw). Distinguished science writer Marcia Bartusiak discusses Hubble's work and the discovery of the expansion of the cosmos.

Images of the Infant Universe: https://www.youtube.com/watch?v=x0AqCwElyUk (https://www.youtube.com/ watch?v=x0AqCwElyUk). Lloyd Knox's public talk on the latest discoveries about the CMB and what they mean for cosmology (1:16:00).

Runaway Universe: https://www.youtube.com/watch?v=kNYVFrnmcOU (https://www.youtube.com/ watch?v=kNYVFrnmcOU). Roger Blandford (Stanford Linear Accelerator Center) public lecture on the discovery and meaning of cosmic acceleration and dark energy (1:08:08).

From the Big Bang to the Nobel Prize and on to the James Webb Space Telescope and the Discovery of Alien Life: http://svs.gsfc.nasa.gov/vis/a010000/a010300/a010370/index.html (http://svs.gsfc.nasa.gov/vis/a010000/ a010300/a010370/index.html). John Mather, NASA Goddard (1:01:02). His Nobel Prize talk from Dec. 8, 2006 can be found at http://www.nobelprize.org/mediaplayer/index.php?id=74&view=1 (http://www.nobelprize.org/ mediaplayer/index.php?id=74&view=1).

Supernovae and the Discovery of the Accelerating Universe: https://www.youtube.com/watch?v=_D6cwrl0CxA (https://www.youtube.com/watch?v=_D6cwrl0CxA). A public lecture by Nobel Laureate Adam Riess in 2019 (58 min).

Collaborative Group Activities

A. This chapter deals with some pretty big questions and ideas. Some belief systems teach us that there are questions to which "we were not meant to know" the answers. Other people feel that if our minds and instruments are capable of exploring a question, then it becomes part of our birthright as thinking human beings. Have your group discuss your personal reactions to discussing questions like the beginning of time and space, and the ultimate fate of the universe. Does it make you nervous to hear about scientists discussing these issues? Or is it exciting to know that we can now gather scientific evidence about the origin and fate of the cosmos? (In discussing this, you may find that members of your group strongly disagree; try to be respectful of others' points of view.)

B. A popular model of the universe in the 1950s and 1960s was the so-called steady-state cosmology. In this model, the universe was not only the same everywhere and in all directions (homogeneous and isotropic), but also the same *at all times*. We know the universe is expanding and the galaxies are thinning out, and so this model hypothesized that new matter was continually coming into existence to fill in the space between galaxies as they moved farther apart. If so, the infinite universe did not have to have a sudden beginning, but could simply exist forever in a steady state. Have your group discuss your reaction to this model. Do you find it more appealing philosophically than the Big Bang model? Can you cite some evidence that indicates that the universe was not the same billions of years ago as it is now—that it is not in a steady state?

C. One of the lucky accidents that characterizes our universe is the fact that the time scale for the development of intelligent life on Earth and the lifetime of the Sun are comparable. Have your group discuss what would happen if the two time scales were very different. Suppose, for example, that the time for intelligent life to evolve was 10 times greater than the main-sequence lifetime of the Sun. Would our civilization have ever developed? Now suppose the time for intelligent life to evolve is ten times shorter than the main-sequence lifetime of the Sun. Would we be around? (This latter discussion requires considerable thought, including such ideas as what the early stages in the Sun's life were like and how much the early Earth was bombarded by asteroids and comets.)

D. The grand ideas discussed in this chapter have a powerful effect on the human imagination, not just for scientists, but also for artists, composers, dramatists, and writers. Here we list just a few of these responses to cosmology. Each member of your group can select one of these, learn more about it, and then report back, either to the group or to the whole class.

- The California poet Robinson Jeffers was the brother of an astronomer who worked at the Lick Observatory. His poem "Margrave" is a meditation on cosmology and on the kidnap and murder of a child: http://www.poemhunter.com/best-poems/robinson-jeffers/margrave/.
- In the science fiction story "The Gravity Mine" by Stephen Baxter, the energy of evaporating supermassive black holes is the last hope of living beings in the far future in an ever-expanding universe. The story has poetic description of the ultimate fate of matter and life and is available online at: http://www.infinityplus.co.uk/stories/gravitymine.htm.
- The musical piece *YLEM* by Karlheinz Stockhausen takes its title from the ancient Greek term for primeval material revived by George Gamow. It tries to portray the oscillating universe in musical terms. Players actually expand through the concert hall, just as the universe does, and then return and expand again. See: http://www.karlheinzstockhausen.org/ylem_english.htm.
- The musical piece *Supernova Sonata* http://www.astro.uvic.ca/~alexhp/new/supernova_sonata.html by Alex Parker and Melissa Graham is based on the characteristics of 241 type Ia supernova explosions, the ones that have helped astronomers discover the acceleration of the expanding universe.
- Gregory Benford's short story "The Final Now" envisions the end of an accelerating open universe, and blends religious and scientific imagery in a very poetic way. Available free online at: http://www.tor.com/stories/2010/03/the-final-now.

E. When Einstein learned about Hubble's work showing that the universe of galaxies is expanding, he called his introduction of the cosmological constant into his general theory of relativity his "biggest blunder." Can your group think of other "big blunders" from the history of astronomy, where the thinking of astronomers was too conservative and the universe turned out to be more complicated or required more "outside-the-box" thinking?

🗐 Exercises

Review Questions

1. What are the basic observations about the universe that any theory of cosmology must explain?

2. Describe some possible futures for the universe that scientists have come up with. What property of the universe determines which of these possibilities is the correct one?

3. What does the term Hubble time mean in cosmology, and what is the current best calculation for the Hubble time?

4. Which formed first: hydrogen nuclei or hydrogen atoms? Explain the sequence of events that led to each.

5. Describe at least two characteristics of the universe that are explained by the standard Big Bang model.

6. Describe two properties of the universe that are not explained by the standard Big Bang model (without inflation). How does inflation explain these two properties?

7. Why do astronomers believe there must be dark matter that is not in the form of atoms with protons and neutrons?

8. What is dark energy and what evidence do astronomers have that it is an important component of the universe?

9. Thinking about the ideas of space and time in Einstein's general theory of relativity, how do we explain the fact that all galaxies outside our Local Group show a redshift?

10. Astronomers have found that there is more helium in the universe than stars could have made in the 13.8 billion years that the universe has been in existence. How does the Big Bang scenario solve this problem?

11. Describe the anthropic principle. What are some properties of the universe that make it "ready" to have life forms like you in it?

12. Describe the evidence that the expansion of the universe is accelerating.

Thought Questions

13. What is the most useful probe of the early evolution of the universe: a giant elliptical galaxy or an irregular galaxy such as the Large Magellanic Cloud? Why?

14. What are the advantages and disadvantages of using quasars to probe the early history of the universe?

15. Would acceleration of the universe occur if it were composed entirely of matter (that is, if there were no dark energy)?

16. Suppose the universe expands forever. Describe what will become of the radiation from the primeval fireball. What will the future evolution of galaxies be like? Could life as we know it survive forever in such a universe? Why?

17. Some theorists expected that observations would show that the density of matter in the universe is just equal to the critical density. Do the current observations support this hypothesis?

18. There are a variety of ways of estimating the ages of various objects in the universe. Describe two of these ways, and indicate how well they agree with one another and with the age of the universe itself as estimated by its expansion.

19. Since the time of Copernicus, each revolution in astronomy has moved humans farther from the center of the universe. Now it appears that we may not even be made of the most common form of matter. Trace the changes in scientific thought about the central nature of Earth, the Sun, and our Galaxy on a cosmic scale. Explain how the notion that most of the universe is made of dark matter continues this "Copernican tradition."

20. The anthropic principle suggests that in some sense we are observing a special kind of universe; if the universe were different, we could never have come to exist. Comment on how this fits with the Copernican tradition described in Exercise 29.19.

21. Penzias and Wilson's discovery of the Cosmic Microwave Background (CMB) is a nice example of scientific *serendipity*—something that is found by chance but turns out to have a positive outcome. What were they looking for and what did they discover?

22. Construct a timeline for the universe and indicate when various significant events occurred, from the beginning of the expansion to the formation of the Sun to the appearance of humans on Earth.

Figuring for Yourself

23. Suppose the Hubble constant were not 22 but 33 km/s per million light-years. Then what would the critical density be?

24. Assume that the average galaxy contains 10^{11} M_{Sun} and that the average distance between galaxies is 10 million light-years. Calculate the average density of matter (mass per unit volume) in galaxies. What fraction is this of the critical density we calculated in the chapter?

25. The CMB contains roughly 400 million photons per m^3. The energy of each photon depends on its wavelength. Calculate the typical wavelength of a CMB photon. Hint: The CMB is blackbody radiation at a temperature of 2.73 K. According to Wien's law, the peak wave length in nanometers is given by $\lambda_{max} = \frac{3 \times 10^6}{T}$. Calculate the wavelength at which the CMB is a maximum and, to make the units consistent, convert this wavelength from nanometers to meters.

26. Following up on Exercise 29.25 calculate the energy of a typical photon. Assume for this approximate calculation that each photon has the wavelength calculated in Exercise 29.25. The energy of a photon is given by $E = \frac{hc}{\lambda}$, where h is Planck's constant and is equal to 6.626 × 10^{-34} J × s, c is the speed of light in m/s, and λ is the wavelength in m.

27. Continuing the thinking in Exercise 29.25 and Exercise 29.26, calculate the energy in a cubic meter of space, multiply the energy per photon calculated in Exercise 29.26 by the number of photons per cubic meter given above.

28. Continuing the thinking in the last three exercises, convert this energy to an equivalent in mass, use Einstein's equation $E = mc^2$. Hint: Divide the energy per m^3 calculated in Exercise 29.27 by the speed of light squared. Check your units; you should have an answer in kg/m^3. Now compare this answer with the critical density. Your answer should be several powers of 10 smaller than the critical density. In other words, you have found for yourself that the contribution of the CMB photons to the overall density of the universe is much, much smaller than the contribution made by stars and galaxies.

29. There is still some uncertainty in the Hubble constant. (a) Current estimates range from about 19.9 km/s per million light-years to 23 km/s per million light-years. Assume that the Hubble constant has been constant since the Big Bang. What is the possible range in the ages of the universe? Use the equation in the text, $T_0 = \frac{1}{H}$, and make sure you use consistent units. (b) Twenty years ago, estimates for the Hubble constant ranged from 50 to 100 km/s per Mps. What are the possible ages for the universe from those values? Can you rule out some of these possibilities on the basis of other evidence?

30. It is possible to derive the age of the universe given the value of the Hubble constant and the distance to a galaxy, again with the assumption that the value of the Hubble constant has not changed since the Big Bang. Consider a galaxy at a distance of 400 million light-years receding from us at a velocity, v. If the Hubble constant is 20 km/s per million light-years, what is its velocity? How long ago was that galaxy right next door to our own Galaxy if it has always been receding at its present rate? Express your answer in years. Since the universe began when all galaxies were very close together, this number is a rough estimate for the age of the universe.

Figure 30.1 Astrobiology: The Road to Life in the Universe. In this fanciful montage produced by a NASA artist, we see one roadmap for discovering life in the universe. Learning more about the origin, evolution, and properties of life on Earth aids us in searching for evidence of life beyond our planet. Our neighbor world, Mars, had warmer, wetter conditions billions of years ago that might have helped life there begin. Farther out, Jupiter's moon Europa represents the icy moons of the outer solar system. Beneath their shells of solid ice may lie vast oceans of liquid water that could support biology. Beyond our solar system are stars that host their own planets, some of which might be similar to Earth in the ability to support liquid water—and a thriving biosphere—at the planet's surface. Research is pushing actively in all these directions with the goal of proving a scientific answer to the question, "Are we alone?" (credit: modification of work by NASA)

Chapter Outline

30.1 The Cosmic Context for Life
30.2 Astrobiology
30.3 Searching for Life beyond Earth
30.4 The Search for Extraterrestrial Intelligence

Thinking Ahead

As we have learned more about the universe, we have naturally wondered whether there might be other forms of life out there. The ancient question, "Are we alone in the universe?" connects us to generations of humans before us. While in the past, this question was in the realm of philosophy or science fiction, today we have the means to seek an answer through scientific inquiry. In this chapter, we will consider how life began on Earth, whether the same processes could have led to life on other worlds, and how we might seek evidence of life elsewhere. This is the science of astrobiology.

The search for life on other planets is not the same as the search for *intelligent* life, which (if it exists) is surely much rarer. Learning more about the origin, evolution, and properties of life on Earth aids us in searching for evidence of all kinds of life beyond that on our planet.

30.1 | The Cosmic Context for Life

Learning Objectives

By the end of this section, you will be able to:

> Describe the chemical and environmental conditions that make Earth hospitable to life
> Discuss the assumption underlying the Copernican principle and outline its implications for modern-day astronomers
> Understand the questions underlying the Fermi paradox

We saw that the universe was born in the Big Bang about 14 billion years ago. After the initial hot, dense fireball of creation cooled sufficiently for atoms to exist, all matter consisted of hydrogen and helium (with a very small amount of lithium). As the universe aged, processes within stars created the other elements, including those that make up Earth (such as iron, silicon, magnesium, and oxygen) and those required for life as we know it, such as carbon, oxygen, and nitrogen. These and other elements combined in space to produce a wide variety of compounds that form the basis of life on Earth. In particular, life on Earth is based on the presence of a key unit known as an **organic molecule**, a molecule that contains carbon. Especially important are the hydrocarbons, chemical compounds made up entirely of hydrogen and carbon, which serve as the basis for our biological chemistry, or *biochemistry*. While we do not understand the details of how life on Earth began, it is clear that to make creatures like us possible, events like the ones we described must have occurred, resulting in what is called the *chemical evolution* of the universe.

What Made Earth Hospitable to Life?

About 5 billion years ago, a cloud of gas and dust in this cosmic neighborhood began to collapse under its own weight. Out of this cloud formed the Sun and its planets, together with all the smaller bodies, such as comets, that also orbit the Sun (Figure 30.2). The third planet from the Sun, as it cooled, eventually allowed the formation of large quantities of liquid water on its surface.

Figure 30.2 Comet Hyakutake. This image was captured in 1996 by NASA photographer Bill Ingalls. Comet impacts can deliver both water and a variety of interesting chemicals, including some organic chemicals, to Earth. (credit: NASA/Bill Ingalls)

The chemical variety and moderate conditions on Earth eventually led to the formation of molecules that could make copies of themselves (reproduce), which is essential for beginning life. Over the billions of years of Earth history, life evolved and became more complex. The course of evolution was punctuated by occasional planet-wide changes caused by collisions with some of the smaller bodies that did not make it into the Sun or one of its accompanying worlds. As we saw in the chapter on Earth as a Planet, mammals may owe their domination of Earth's surface to just such a collision 65 million years ago, which led to the extinction of the dinosaurs (along with the majority of other living things). The details of such mass extinctions are currently the focus of a great deal of scientific interest.

Through many twisting turns, the course of evolution on Earth produced a creature with self-consciousness, able to ask questions about its own origins and place in the cosmos (Figure 30.3). Like most of Earth, this creature is composed of atoms that were forged in earlier generations of stars—in this case, assembled into both its body and brain. We might say that through the thoughts of human beings, the matter in the universe can become aware of itself.

Figure 30.3 Young Human. Human beings have the intellect to wonder about their planet and what lies beyond it. Through them (and perhaps other intelligent life), the universe becomes aware of itself. (credit: Andrew Fraknoi)

Think about those atoms in your body for a minute. They are merely on loan to you from the lending library of atoms that make up our local corner of the universe. Atoms of many kinds circulate through your body and then leave it—with each breath you inhale and exhale and the food you eat and excrete. Even the atoms that take up more permanent residence in your tissues will not be part of you much longer than you are alive. Ultimately, you will return your atoms to the vast reservoir of Earth, where they will be incorporated into other structures and even other living things in the millennia to come.

This picture of *cosmic evolution*, of our descent from the stars, has been obtained through the efforts of scientists in many fields over many decades. Some of its details are still tentative and incomplete, but we feel reasonably confident in its broad outlines. It is remarkable how much we have been able to learn in the short time we have had the instruments to probe the physical nature of the universe.

The Copernican Principle

Our study of astronomy has taught us that we have always been wrong in the past whenever we have claimed that Earth is somehow unique. Galileo, using the newly invented technology of the telescope, showed us that Earth is not the center of the solar system, but merely one of a number of objects orbiting the Sun. Our study of the stars has demonstrated that the Sun itself is a rather undistinguished star, halfway through its long main-sequence stage like so many billions of others. There seems nothing special about our position in the Milky Way Galaxy either, and nothing surprising about our Galaxy's position in either its own group or its supercluster.

The discovery of planets around other stars confirms our idea that the formation of planets is a natural consequence of the formation of stars. We have identified thousands of exoplanets—planets orbiting around other stars, from huge ones orbiting close to their stars (informally called "hot Jupiters") down to planets smaller than Earth. A steady stream of exoplanet discoveries is leading to the conclusion that earthlike planets occur frequently—enough that there are likely many billions of "exo-Earths" in our own Milky Way Galaxy alone. From a planetary perspective, Earth-size planets are not unusual.

Philosophers of science sometimes call the idea that there is nothing special about our place in the universe the *Copernican principle.* Given all of the above, most scientists would be surprised if life were limited to our planet and had started nowhere else. There are billions of stars in our Galaxy old enough for life to have developed on a planet around them, and there are billions of other galaxies as well. Astronomers and biologists have long conjectured that a series of events similar to those on the early Earth probably led to living organisms on many planets around other stars, and possibly even on other planets in our solar system, such as Mars.

The real scientific issue (which we do not currently know the answer to) is whether organic biochemistry is likely or unlikely in the universe at large. Are we a fortunate and exceedingly rare outcome of chemical evolution, or is organic biochemistry a regular part of the chemical evolution of the cosmos? We do not yet know the answer to this question, but data, even an exceedingly small amount (like finding "unrelated to us" living systems on a world like Europa), will help us arrive at it.

So Where Are They?

If the Copernican principle is applied to life, then biology may be rather common among planets. Taken to its logical limit, the Copernican principle also suggests that intelligent life like us might be common. Intelligence like ours has some very special properties, including an ability to make progress through the application of technology. Organic life around other (older) stars may have started a billion years earlier than we did on Earth, so they may have had a lot more time to develop advanced technology such as sending information, probes, or even life-forms between stars.

Faced with such a prospect, physicist Enrico Fermi asked a question several decades ago that is now called the *Fermi paradox*: where are they? If life and intelligence are common and have such tremendous capacity for growth, why is there not a network of galactic civilizations whose presence extends even into a "latecomer" planetary system like ours?

Several solutions have been suggested to the Fermi paradox. Perhaps life is common but intelligence (or at least technological civilization) is rare. Perhaps such a network will come about in the future but has not yet had the time to develop. Maybe there are invisible streams of data flowing past us all the time that we are not advanced enough or sensitive enough to detect. Maybe advanced species make it a practice not to interfere with immature, developing consciousness such as our own. Or perhaps civilizations that reach a certain level of technology then self-destruct, meaning there are no other civilizations now existing in our Galaxy. We do not yet know whether any advanced life is out there and, if it is, why we are not aware of it. Still, you might want to keep these issues in mind as you read the rest of this chapter.

LINK TO LEARNING

Is there a network of galactic civilizations beyond our solar system? If so, why can't we see them? Explore the possibilities in the cartoon video (https://openstax.org/l/30fermparadox) "The Fermi Paradox—Where Are All the Aliens?"

30.2 | Astrobiology

Learning Objectives

By the end of this section, you will be able to:

> Describe the chemical building blocks required for life
> Describe the molecular systems and processes driving the origin and evolution of life
> Describe the characteristics of a habitable environment
> Describe some of the extreme conditions on Earth, and explain how certain organisms have adapted to these conditions

Scientists today take a multidisciplinary approach to studying the origin, evolution, distribution, and ultimate fate of life in the universe; this field of study is known as **astrobiology**. You may also sometimes hear this field referred to as *exobiology* or *bioastronomy*. Astrobiology brings together astronomers, planetary scientists, chemists, geologists, and biologists (among others) to work on the same problems from their various perspectives.

Among the issues that astrobiologists explore are the conditions in which life arose on Earth and the reasons for the extraordinary adaptability of life on our planet. They are also involved in identifying habitable worlds beyond Earth and in trying to understand in practical terms how to look for life on those worlds. Let's look at some of these issues in more detail.

The Building Blocks of Life

While no unambiguous evidence for life has yet been found anywhere beyond Earth, life's chemical building blocks have been detected in a wide range of extraterrestrial environments. Meteorites (which you learned about in Cosmic Samples and the Origin of the Solar System) have been found to contain two kinds of substances whose chemical structures mark them as having an extraterrestrial origin—amino acids and sugars. **Amino acids** are **organic compounds** that are the molecular building blocks of proteins. **Proteins** are key biological molecules that provide the structure and function of the body's tissues and organs and essentially carry out the "work" of the cell. When we examine the gas and dust around comets, we also find a number of organic molecules—compounds that on Earth are associated with the chemistry of life.

Expanding beyond our solar system, one of the most interesting results of modern radio astronomy has been the discovery of organic molecules in giant clouds of gas and dust between stars. More than 100 different molecules have been identified in these reservoirs of cosmic raw material, including formaldehyde, alcohol, and others we know as important stepping stones in the development of life on Earth. Using radio telescopes and radio spectrometers, astronomers can measure the abundances of various chemicals in these clouds. We find organic molecules most readily in regions where the interstellar dust is most abundant, and it turns out these are precisely the regions where star formation (and probably planet formation) happen most easily (Figure 30.4).

Figure 30.4 Cloud of Gas and Dust. This cloud of gas and dust in the constellation of Scorpius is the sort of region where complex molecules are found. It is also the sort of cloud where new stars form from the reservoir of gas and dust in the cloud. Radiation from a group of hot stars (off the picture to the bottom left) called the Scorpius OB Association is "eating into" the cloud, sweeping it into an elongated shape and causing the reddish glow seen at its tip. (credit: Dr. Robert Gendler)

Clearly the early Earth itself produced some of the molecular building blocks of life. Since the early 1950s, scientists have tried to duplicate in their laboratories the chemical pathways that led to life on our planet. In a series of experiments known as the *Miller-Urey experiments*, pioneered by Stanley Miller and Harold Urey at the University of Chicago, biochemists have simulated conditions on early Earth and have been able to produce some of the fundamental building blocks of life, including those that form proteins and other large biological molecules known as nucleic acids (which we will discuss shortly).

Although these experiments produced encouraging results, there are some problems with them. The most interesting chemistry from a biological perspective takes place with hydrogen-rich or *reducing* gases, such as ammonia and methane. However, the early atmosphere of Earth was probably dominated by carbon dioxide (as Venus' and Mars' atmospheres still are today) and may not have contained an abundance of reducing gases comparable to that used in Miller-Urey type experiments. Hydrothermal vents—seafloor systems in which ocean water is superheated and circulated through crustal or mantle rocks before reemerging into the ocean—have also been suggested as potential contributors of organic compounds on the early Earth, and such sources would not require Earth to have an early reducing atmosphere.

Both earthly and extraterrestrial sources may have contributed to Earth's early supply of organic molecules, although we have more direct evidence for the latter. It is even conceivable that life itself originated elsewhere and was seeded onto our planet—although this, of course, does not solve the problem of how that life originated to begin with.

LINK TO LEARNING

Hydrothermal vents are beginning to seem more likely as early contributors to the organic compounds found on Earth. Read about hydrothermal vents and watch videos and slideshows on these and other deep-sea wonders at the Woods Hole Oceanographic Institution (https://openstax.org/l/30wohooceins) website.

Try an interactive simulation of hydrothermal-vent circulation at the Dive and Discover (https://openstax.org/l/30divediscover) website.

The Origin and Early Evolution of Life

The carbon compounds that form the chemical basis of life may be common in the universe, but it is still a giant step from these building blocks to a living cell. Even the simplest molecules of the **genes** (the basic functional units that carry the genetic, or hereditary, material in a cell) contain millions of molecular units, each arranged in a precise sequence. Furthermore, even the most primitive life required two special capabilities: a means of extracting energy from its environment, and a means of encoding and replicating information in order to make faithful copies of itself. Biologists today can see ways that either of these capabilities might have formed in a natural environment, but we are still a long way from knowing how the two came together in the first life-forms.

We have no solid evidence for the pathway that led to the origin of life on our planet except for whatever early history may be retained in the biochemistry of modern life. Indeed, we have very little direct evidence of what Earth itself was like during its earliest history—our planet is so effective at resurfacing itself through plate tectonics (see the chapter on Earth as a Planet) that very few rocks remain from this early period. In the earlier chapter on Cratered Worlds, you learned that Earth was subjected to a heavy bombardment—a period of large impact events—some 3.8 to 4.1 billion years ago. Large impacts would have been energetic enough to heat-sterilize the surface layers of Earth, so that even if life had begun by this time, it might well have been wiped out.

When the large impacts ceased, the scene was set for a more peaceful environment on our planet. If the oceans of Earth contained accumulated organic material from any of the sources already mentioned, the ingredients were available to make living organisms. We do not understand in any detail the sequence of events that led from molecules to biology, but there is fossil evidence of microbial life in 3.5-billion-year-old rocks, and possible (debated) evidence for life as far back as 3.8 billion years.

Life as we know it employs two main molecular systems: the functional molecules known as proteins, which carry out the chemical work of the cell, and information-containing molecules of **DNA (deoxyribonucleic acid)** that store information about how to create the cell and its chemical and structural components. The origin of life is sometimes considered a "chicken and egg problem" because, in modern biology, neither of these systems works without the other. It is our proteins that assemble DNA strands in the precise order required to store information, but the proteins are created based on information stored in DNA. Which came first? Some origin of life researchers believe that prebiotic chemistry was based on molecules that could both store information and do the chemical work of the cell. It has been suggested that **RNA (ribonucleic acid)**, a molecule that aids in the flow of genetic information from DNA to proteins, might have served such a purpose. The idea of an early "RNA world" has become increasingly accepted, but a great deal remains to be understood about the origin of life.

Perhaps the most important innovation in the history of biology, apart from the origin of life itself, was the discovery of the process of **photosynthesis**, the complex sequence of chemical reactions through which some living things can use sunlight to manufacture products that store energy (such as carbohydrates), releasing oxygen as one by-product. Previously, life had to make do with sources of chemical energy available on Earth or delivered from space. But the abundant energy available in sunlight could support a larger and more productive biosphere, as well as some biochemical reactions not previously possible for life. One of these was the production of oxygen (as a waste product) from carbon dioxide, and the increase in atmospheric levels of oxygen about 2.4 billion years ago means that oxygen-producing photosynthesis must have emerged and become globally important by this time. In fact, it is likely that oxygen-producing photosynthesis emerged considerably earlier.

Some forms of chemical evidence contained in ancient rocks, such as the solid, layered rock formations known as **stromatolites**, are thought to be the fossils of oxygen-producing photosynthetic bacteria in rocks that are almost 3.5 billion years old (Figure 30.5). It is generally thought that a simpler form of photosynthesis that does not produce oxygen (and is still used by some bacteria today) probably preceded oxygen-producing photosynthesis, and there is strong fossil evidence that one or the other type of photosynthesis was functioning on Earth at least as far back as 3.4 billion years ago.

(a)   (b)

Figure 30.5 Stromatolites Preserve the Earliest Physical Representation of Life on Earth. In their reach for sunlight, the single-celled microbes formed mats that trapped sediments in the water above them. Such trapped sediments fell and formed layers on top of the mats. The microbes then climbed atop the sediment layers and trapped more sediment. What is found in the rock record are (a) the solidified, curved sedimentary layers that are signatures of biological activity. The earliest known stromatolite is 3.47 billion years old and is found in Western Australia. (b) This more recent example is in Lake Thetis, also in Western Australia. (credit a: modification of work by James St. John; credit b: modification of work by Ruth Ellison)

The free oxygen produced by photosynthesis began accumulating in our atmosphere about 2.4 billion years ago. The interaction of sunlight with oxygen can produce ozone (which has three atoms of oxygen per molecule, as compared to the two atoms per molecule in the oxygen we breathe), which accumulated in a layer high in Earth's atmosphere. As it does on Earth today, this ozone provided protection from the Sun's damaging ultraviolet radiation. This allowed life to colonize the landmasses of our planet instead of remaining only in the ocean.

The rise in oxygen levels was deadly to some microbes because, as a highly reactive chemical, it can irreversibly damage some of the biomolecules that early life had developed in the absence of oxygen. For other microbes, it was a boon: combining oxygen with organic matter or other reduced chemicals generates a lot of energy—you can see this when a log burns, for example—and many forms of life adopted this way of living. This new energy source made possible a great proliferation of organisms, which continued to evolve in an oxygen-rich environment.

The details of that evolution are properly the subject of biology courses, but the process of evolution by natural selection (survival of the fittest) provides a clear explanation for the development of Earth's remarkable variety of life-forms. It does not, however, directly solve the mystery of life's earliest beginnings. We hypothesize that life will arise whenever conditions are appropriate, but this hypothesis is just another form of the Copernican principle. We now have the potential to address this hypothesis with observations. If a second example of life is found in our solar system or a nearby star, it would imply that life emerges commonly enough that the universe is likely filled with biology. To make such observations, however, we must first decide where to focus our search.

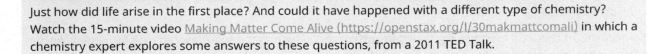

LINK TO LEARNING

Just how did life arise in the first place? And could it have happened with a different type of chemistry? Watch the 15-minute video Making Matter Come Alive (https://openstax.org/l/30makmattcomali) in which a chemistry expert explores some answers to these questions, from a 2011 TED Talk.

Habitable Environments

Among the staggering number of objects in our solar system, Galaxy, and universe, some may have conditions suitable for life, while others do not. Understanding what conditions and features make a **habitable environment**—an environment capable of hosting life—is important both for understanding how widespread habitable environments may be in the universe and for focusing a search for life beyond Earth. Here, we discuss habitability from the perspective of the life we know. We will explore the basic requirements of life and, in the following section, consider the full range of environmental conditions on Earth where life is found. While we can't entirely rule out the possibility that other life-forms might have biochemistry based on alternatives to carbon and liquid water, such life "as we don't know it" is still completely speculative. In our discussion here, we are focusing on habitability for life that is chemically similar to that on Earth.

Life requires a solvent (a liquid in which chemicals can dissolve) that enables the construction of biomolecules and the interactions between them. For life as we know it, that solvent is water, which has a variety of properties that are critical to how our biochemistry works. Water is abundant in the universe, but life requires that water be in liquid form (rather than ice or gas) in order to properly fill its role in biochemistry. That is the case only within a certain range of temperatures and pressures—too high or too low in either variable, and water takes the form of a solid or a gas. Identifying environments where water is present within the appropriate range of temperature and pressure is thus an important first step in identifying habitable environments. Indeed, a "follow the water" strategy has been, and continues to be, a key driver in the exploration of planets both within and beyond our solar system.

Our biochemistry is based on molecules made of carbon, hydrogen, nitrogen, oxygen, phosphorus, and sulfur. Carbon is at the core of organic chemistry. Its ability to form four bonds, both with itself and with the other elements of life, allows for the formation of a vast number of potential molecules on which to base biochemistry. The remaining elements contribute structure and chemical reactivity to our biomolecules, and form the basis of many of the interactions among them. These "biogenic elements," sometimes referred to with the acronym CHNOPS (carbon, hydrogen, nitrogen, oxygen, phosphorus, and sulfur), are the raw materials from which life is assembled, and an accessible supply of them is a second requirement of habitability.

As we learned in previous chapters on nuclear fusion and the life story of the stars, carbon, nitrogen, oxygen, phosphorus, and sulfur are all formed by fusion within stars and then distributed out into their galaxy as those stars die. But how they are distributed among the planets that form within a new star system, in what form, and how chemical, physical, and geological processes on those planets cycle the elements into structures that are accessible to biology, can have significant impacts on the distribution of life. In Earth's oceans, for example, the abundance of phytoplankton (simple organisms that are the base of the ocean food chain) in surface waters can vary by a thousand-fold because the supply of nitrogen differs from place to place (Figure 30.6). Understanding what processes control the accessibility of elements at all scales is thus a critical part of identifying habitable environments.

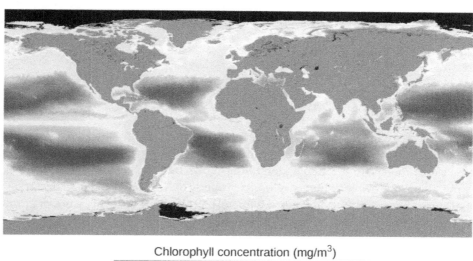

Chlorophyll concentration (mg/m^3)

0.01   0.03   0.1   0.3   1   3   10

Figure 30.6 Chlorophyll Abundance. The abundance of chlorophyll (an indicator of photosynthetic bacteria and algae) varies by almost a thousand-fold across the ocean basins. That variation is almost entirely due to the availability of nitrogen—one of the major "biogenic elements" in forms that can be used by life. (credit: modification of work by NASA, Gene C. Feldman)

With these first two requirements, we have the elemental raw materials of life and a solvent in which to assemble them into the complicated molecules that drive our biochemistry. But carrying out that assembly and maintaining the complicated biochemical machinery of life takes energy. You fulfill your own requirement for energy every time you eat food or take a breath, and you would not live for long if you failed to do either on a regular basis. Life on Earth makes use of two main types of energy: for you, these are the oxygen in the air you breathe and the organic molecules in your food. But life overall can use a much wider array of chemicals and, while all animals require oxygen, many bacteria do not. One of the earliest known life processes, which still operates in some modern microorganisms, combines hydrogen and carbon dioxide to make methane, releasing energy in the process. There are microorganisms that "breathe" metals that would be toxic to us, and even some that breathe in sulfur and breathe out sulfuric acid. Plants and photosynthetic microorganisms have also evolved mechanisms to use the energy in light directly.

Water in the liquid phase, the biogenic elements, and energy are the fundamental requirements for habitability. But are there additional environmental constraints? We consider this in the next section.

Figure 30.7 Grand Prismatic Spring in Yellowstone National Park. This hot spring, where water emerges from the bluish center at temperatures near the local boiling point (about 92 °C), supports a thriving array of microbial life. The green, yellow, and orange colors around the edges come from thick "mats" of photosynthetic bacteria. In fact, their coloration in part demonstrates their use of light energy—some wavelengths of incoming sunlight are selectively captured for energy; the rest are reflected back. Since it lacks the captured wavelengths, this light is now different in color than the sunlight that illuminates it. The blue part of the spring has temperatures too high to allow photosynthetic life (hence the lack of color except that supplied by water itself), but life is still present. Here, at nearly boiling temperatures, bacteria use the chemical energy supplied by the combination of hydrogen and other chemicals with oxygen. (credit: modification of work by Domenico Salvagnin)

Life in Extreme Conditions

At a chemical level, life consists of many types of molecules that interact with one another to carry out the processes of life. In addition to water, elemental raw materials, and energy, life also needs an environment in which those complicated molecules are stable (don't break down before they can do their jobs) and their interactions are possible. Your own biochemistry works properly only within a very narrow range of about 10 °C in body temperature and two-tenths of a unit in blood pH (pH is a numerical measure of acidity, or the amount of free hydrogen ions). Beyond those limits, you are in serious danger.

Life overall must also have limits to the conditions in which it can properly work but, as we will see, they are much broader than human limits. The resources that fuel life are distributed across a very wide range of conditions. For example, there is abundant chemical energy to be had in hot springs that are essentially boiling acid (see Figure 30.7). This provides ample incentive for evolution to fill as much of that range with life as is biochemically possible. An organism (usually a microbe) that tolerates or even thrives under conditions that most of the life around us would consider hostile, such as very high or low temperature or acidity, is known as an **extremophile** (where the suffix -*phile* means "lover of"). Let's have a look at some of the conditions that can challenge life and the organisms that have managed to carve out a niche at the far reaches of possibility.

Both high and low temperatures can cause a problem for life. As a large organism, you are able to maintain an almost constant body temperature whether it is colder or warmer in the environment around you. But this is not possible at the tiny size of microorganisms; whatever the temperature in the outside world is also the temperature of the microbe, and its biochemistry must be able to function at that temperature. High temperatures are the enemy of complexity—increasing thermal energy tends to break apart big molecules into smaller and smaller bits, and life needs to stabilize the molecules with stronger bonds and special proteins. But this approach has its limits.

Nevertheless, as noted earlier, high-temperature environments like hot springs and hydrothermal vents often offer abundant sources of chemical energy and therefore drive the evolution of organisms that can tolerate high temperatures (see Figure 30.8); such an organism is called a **thermophile**. Currently, the high temperature record holder is a methane-producing microorganism that can grow at 122 °C, where the pressure also is so high that water still does not boil. That's amazing when you think about it. We cook our food—meaning, we alter the chemistry and structure of its biomolecules—by boiling it at a temperature of 100 °C. In fact, food begins to cook at much lower temperatures than this. And yet, there are organisms whose biochemistry remains intact and operates just fine at temperatures 20 degrees higher.

Figure 30.8 Hydrothermal Vent on the Sea Floor. What appears to be black smoke is actually superheated water filled with minerals of metal sulfide. Hydrothermal vent fluid can represent a rich source of chemical energy, and therefore a driver for the evolution of microorganisms that can tolerate high temperatures. Bacteria feeding on this chemical energy form the base of a food chain that can support thriving communities of animals—in this case, a dense patch of red and white tubeworms growing around the base of the vent. (credit: modification of work by the University of Washington; NOAA/OAR/OER)

Cold can also be a problem, in part because it slows down metabolism to very low levels, but also because it can cause physical changes in biomolecules. Cell membranes—the molecular envelopes that surround cells and allow their exchange of chemicals with the world outside—are basically made of fatlike molecules. And just as fat congeals when it cools, membranes crystallize, changing how they function in the exchange of materials in and out of the cell. Some cold-adapted cells (called *psychrophiles*) have changed the chemical composition of their membranes in order to cope with this problem; but again, there are limits. Thus far, the coldest temperature at which any microbe has been shown to reproduce is about –25 °C.

Conditions that are very acidic or alkaline can also be problematic for life because many of our important molecules, like proteins and DNA, are broken down under such conditions. For example, household drain cleaner, which does its job by breaking down the chemical structure of things like hair clogs, is a very alkaline solution. The most acid-tolerant organisms (*acidophiles*) are capable of living at pH values near zero—about ten million times more acidic than your blood (Figure 30.9). At the other extreme, some *alkaliphiles* can grow at pH levels of about 13, which is comparable to the pH of household bleach and almost a million times more alkaline than your blood.

Figure 30.9 Spain's Rio Tinto. With a pH close to 2, Rio Tinto is literally a river of acid. Acid-loving microorganisms (acidophiles) not only thrive in these waters, their metabolic activities help generate the acid in the first place. The rusty red color that gives the river its name comes from high levels of iron dissolved in the waters.

High levels of salts in the environment can also cause a problem for life because the salt blocks some cellular functions. Humans recognized this centuries ago and began to salt-cure food to keep it from spoiling—meaning, to keep it from being colonized by microorganisms. Yet some microbes have evolved to grow in water that is saturated in sodium chloride (table salt)—about ten times as salty as seawater (Figure 30.10).

Figure 30.10 Salt Ponds. The waters of an evaporative salt works near San Francisco are colored pink by thriving communities of photosynthetic organisms. These waters are about ten times as salty as seawater—enough for sodium chloride to begin to crystallize out—yet some organisms can survive and thrive in these conditions. (credit: modification of work by NASA)

Very high pressures can literally squeeze life's biomolecules, causing them to adopt more compact forms that do not work very well. But we still find life—not just microbial, but even animal life—at the bottoms of our ocean trenches, where pressures are more than 1000 times atmospheric pressure. Many other adaptions to environmental "extremes" are also known. There is even an organism, *Deinococcus radiodurans,* that can tolerate ionizing radiation (such as that released by radioactive elements) a thousand times more intense than you would be able to withstand. It is also very good at surviving extreme desiccation (drying out) and a variety of metals that would be toxic to humans.

From many such examples, we can conclude that life is capable of tolerating a wide range of environmental extremes—so much so that we have to work hard to identify places where life can't exist. A few such places are known—for example, the waters of hydrothermal vents at over 300 °C appear too hot to support any life—and finding these places helps define the possibility for life elsewhere. The study of extremophiles over the last few decades has expanded our sense of the range of conditions life can survive and, in doing so, has made many scientists more optimistic about the possibility that life might exist beyond Earth.

30.3 | Searching for Life beyond Earth

Learning Objectives

By the end of this section, you will be able to:
> Outline what we have learned from exploration of the environment on Mars
> Identify where in the solar system life is most likely sustainable and why
> Describe some key missions and their findings in our search for life beyond our solar system
> Explain the use of biomarkers in the search for evidence of life beyond our solar system

Astronomers and planetary scientists continue to search for life in the solar system and the universe at large. In this section, we discuss two kinds of searches. First is the direct exploration of planets within our own solar system, especially Mars and some of the icy moons of the outer solar system. Second is the even more difficult task of searching for evidence of life—a **biomarker**—on planets circling other stars. In the next section, we will examine SETI, the *search for extraterrestrial intelligence*. As you will see, the approaches taken in these three cases are very different, even though the goal of each is the same: to determine if life on Earth is unique in the universe.

Life on Mars

The possibility that Mars hosts, or has hosted, life has a rich history dating back to the "canals" that some people claimed to see on the martian surface toward the end of the nineteenth century and the beginning of the twentieth. With the dawn of the space age came the possibility to address this question up close through a progression of missions to Mars that began with the first successful flyby of a robotic spacecraft in 1964 and have led to the deployment of capable rovers, like *Curiosity* and *Perseverance*, with instruments to look for organic chemistry.

The earliest missions to Mars provided some hints that liquid water—one of life's primary requirements—may once have flowed on the surface, and later missions have strengthened this conclusion. The NASA Viking landers, whose purpose was to search directly for evidence of life on Mars, arrived on Mars in 1976. Viking's onboard instruments found no organic molecules (the stuff of which life is made), and no evidence of biological activity in the martian soils it analyzed.

This result is not particularly surprising because, despite the evidence of flowing liquid water in the past, liquid water on the surface of Mars is generally not stable today. Over much of Mars, temperatures and pressures at the surface are so low that pure water would either freeze or boil away (under very low pressures, water will boil at a much lower temperature than usual). To make matters worse, unlike Earth, Mars does not have a magnetic field and ozone layer to protect the surface from harmful solar ultraviolet radiation and energetic particles. However, Viking's analyses of the soil said nothing about whether life may have existed in Mars' distant past, when liquid water was more abundant. We do know that water in the form of ice exists in abundance on Mars, not so deep beneath its surface. Water vapor is also a constituent of the atmosphere of Mars.

Since the visit of Viking, our understanding of Mars has deepened spectacularly. Orbiting spacecraft have provided ever-more detailed images of the surface and detected the presence of minerals that could have formed only in the presence of liquid water. Two bold surface missions, the Mars Exploration Rovers *Spirit* and

Opportunity (2004), followed by the much larger *Curiosity* Rover (2012), confirmed these remote-sensing data. All three rovers found abundant evidence for a past history of liquid water, revealed not only from the mineralogy of rocks they analyzed, but also from the unique layering of rock formations.

Curiosity has gone a step beyond evidence for water and confirmed the existence of habitable environments on ancient Mars. "Habitable" means not only that liquid water was present, but that life's requirements for energy and elemental raw materials could also have been met. The strongest evidence of an ancient habitable environment came from analyzing a very fine-grained rock called a mudstone—a rock type that is widespread on Earth but was unknown on Mars until *Curiosity* found it (see Figure 30.11). The mudstone can tell us a great deal about the wet environments in which they formed. The *Perseverance* rover is collecting samples of sedimentary rock in a former lakebed, to later return to Earth for laboratory analysis.

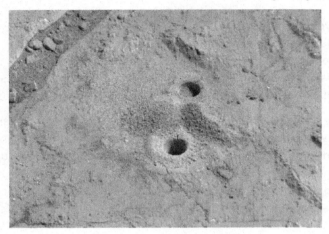

Figure 30.11 Mudstone. Shown are the first holes drilled by NASA's *Curiosity* Mars rover into a mudstone, with "fresh" drill-pilings around the holes. Notice the difference in color between the red ancient martian surface and the gray newly exposed rock powder that came from the drill holes. Each drill hole is about 0.6 inch (1.6 cm) in diameter. (credit: modification of work by NASA/JPL-Caltech/MSSS)

Five decades of robotic exploration have allowed us to develop a picture of how Mars evolved through time. Early Mars had epochs of warmer and wetter conditions that would have been conducive to life at the surface. However, Mars eventually lost much of its early atmosphere and the surface water began to dry up. As that happened, the ever-shrinking reservoirs of liquid water on the martian surface became saltier and more acidic, until the surface finally had no significant liquid water and was bathed in harsh solar radiation. The surface thus became uninhabitable, but this might not be the case for the planet overall.

Reservoirs of ice and liquid water could still exist underground, where pressure and temperature conditions make it stable. There is recent evidence to suggest that liquid water (probably very salty water) can occasionally (and briefly) flow on the surface even today. Thus, Mars might even have habitable conditions in the present day, but of a much different sort than we normally think of on Earth.

Our study of Mars reveals a planet with a fascinating history—one that saw its ability to host surface life dwindle billions of years ago, but perhaps allowing life to adapt and survive in favorable environmental niches. Even if life did not survive, we expect that we might find evidence of life if it ever took hold on Mars. If it is there, it is hidden in the crust, and we are still learning how best to decipher that evidence.

Life in the Outer Solar System

The massive gas and ice giant planets of the outer solar system—Jupiter, Saturn, Uranus, and Neptune—are almost certainly not habitable for life as we know it, but some of their moons might be (see Figure 30.12). Although these worlds in the outer solar system contain abundant water, they receive so little warming sunlight in their distant orbits that it was long believed they would be "geologically dead" balls of hard-frozen ice and rock. But, as we saw in the chapter on Rings, Moons, and Pluto, missions to the outer solar system have found something much more interesting.

Jupiter's moon Europa revealed itself to the Voyager and Galileo missions as an active world whose icy surface apparently conceals an ocean with a depth of tens to perhaps a hundred kilometers. As the moon orbits Jupiter, the planet's massive gravity creates tides on Europa—just as our own Moon's gravity creates our ocean tides—and the friction of all that pushing and pulling generates enough heat to keep the water in liquid form (Figure 30.13). Similar tides act upon other moons if they orbit close to the planet. Scientists now think that six or more of the outer solar system's icy moons may harbor liquid water oceans for the same reason. Among these, Europa and Enceladus, a moon of Saturn, have thus far been of greatest interest to astrobiologists.

Figure 30.12 Jupiter's Moons. The Galilean moons of Jupiter are shown to relative scale and arranged in order of their orbital distance from Jupiter. At far left, Io orbits closest to Jupiter and so experiences the strongest tidal heating by Jupiter's massive gravity. This effect is so strong that Io is thought to be the most volcanically active body in our solar system. At far right, Callisto shows a surface scarred by billions of years' worth of craters—an indication that the moon's surface is old and that Callisto may be far less active than its sibling moons. Between these hot and cold extremes, Europa, second from left, orbits at a distance where Jupiter's tidal heating may be "just right" to sustain a liquid water ocean beneath its icy crust. (credit: modification of work by NASA/JPL/DLR)

Europa has probably had an ocean for most or all of its history, but habitability requires more than just liquid water. Life also requires energy, and because sunlight does not penetrate below the kilometers-thick ice crust of Europa, this would have to be chemical energy. One of Europa's key attributes from an astrobiology perspective is that its ocean is most likely in direct contact with an underlying rocky mantle, and the interaction of water and rocks—especially at high temperatures, as within Earth's hydrothermal vent systems—yields a *reducing chemistry* (where molecules tend to give up electrons readily) that is like one half of a chemical battery. To complete the battery and provide energy that could be used by life requires that an *oxidizing chemistry* (where molecules tend to accept electrons readily) also be available. On Earth, when chemically reducing vent fluids meet oxygen-containing seawater, the energy that becomes available often supports thriving communities of microorganisms and animals on the sea floor, far from the light of the Sun.

The Galileo mission found that Europa's icy surface does contain an abundance of oxidizing chemicals. This means that availability of energy to support life depends very much on whether the chemistry of the surface and the ocean can mix, despite the kilometers of ice in between. That Europa's ice crust appears geologically "young" (only tens of millions of years old, on average) and that it is active makes it tantalizing to think that such mixing might indeed occur. Understanding whether and how much exchange occurs between the surface and ocean of Europa will be a key science objective of future missions to Europa, and a major step forward in understanding whether this moon could be a cradle of life.

Figure 30.13 Jupiter's Moon Europa, as Imaged by NASA's Galileo Mission. The relative scarcity of craters on Europa suggests a surface that is "geologically young," and the network of colored ridges and cracks suggests constant activity and motion. Galileo's instruments also strongly suggested the presence of a massive ocean of salty liquid water beneath the icy crust. (credit: modification of work by NASA/JPL-Caltech/SETI Institute)

In 2005, the Cassini mission performed a close flyby of a small (500-kilometer diameter) moon of Saturn, Enceladus (Figure 30.14), and made a remarkable discovery. Plumes of gas and icy material were venting from the moon's south polar region at a collective rate of about 250 kilograms of material per second. Several observations, including the discovery of salts associated with the icy material, suggest that their source is a liquid water ocean beneath tens of kilometers of ice. Although it remains to be shown definitively whether the ocean is local or global, transient or long-lived, it does appear to be in contact, and to have reacted, with a rocky interior. As on Europa, this is probably a necessary—though not sufficient—condition for habitability. What makes Enceladus so enticing to planetary scientists, though, are those plumes of material that seem to come directly from its ocean: samples of the interior are there for the taking by any spacecraft sent flying through. For a future mission, such samples could yield evidence not only of whether Enceladus is habitable but, indeed, of whether it is home to life.

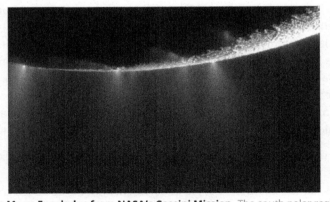

Figure 30.14 Image of Saturn's Moon Enceladus from NASA's Cassini Mission. The south polar region was found to have multiple plumes of ice and gas that, combined, are venting about 250 kilograms of material per second into space. Such features suggest that Enceladus, like Europa, has a sub-ice ocean. (credit: NASA/JPL/ SSI)

Saturn's big moon Titan is very different from both Enceladus and Europa (see Figure 30.15). Although it may host a liquid water layer deep within its interior, it is the surface of Titan and its unusual chemistry that makes this moon such an interesting place. Titan's thick atmosphere—the only one among moons in the solar system—is composed mostly of nitrogen but also of about 5% methane. In the upper atmosphere, the Sun's ultraviolet light breaks apart and recombines these molecules into more complex organic compounds that are collectively known as *tholins*. The tholins shroud Titan in an orange haze, and imagery from Cassini and from the Huygens probe that descended to Titan's surface show that heavier particles appear to accumulate on the surface, even forming "dunes" that are cut and sculpted by flows of liquid hydrocarbons (such as liquid methane). Some scientists see this organic chemical factory as a natural laboratory that may yield some clues about the solar system's early chemistry—perhaps even chemistry that could support the origin of life.

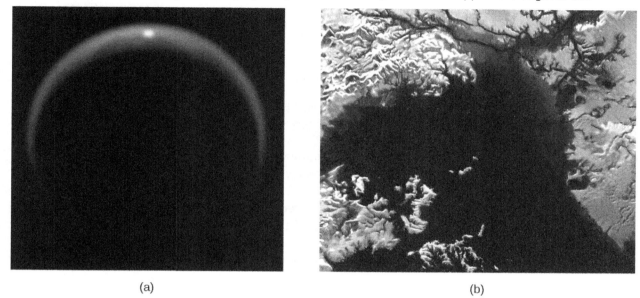

(a)   (b)

Figure 30.15 Image of Saturn's Moon Titan from NASA's Cassini Mission. (a) The hazy orange glow comes from Titan's thick atmosphere (the only one known among the moons of the solar system). That atmosphere is mostly nitrogen but also contains methane and potentially a variety of complex organic compounds. The bright spot near the top of the image is sunlight reflected from a very flat surface—almost certainly a liquid. We see this effect, called "glint," when sunlight reflects off the surface of a lake or ocean. (b) Cassini radar imagery shows what look very much like landforms and lakes on the surface of Titan. But the surface lakes and oceans of Titan are not water; they are probably made of liquid hydrocarbons like methane and ethane. (credit a: modification of work by NASA/JPL/University of Arizona/DLR; credit b: modification of work by NASA/JPL-Caltech/ASI)

LINK TO LEARNING

In January 2005, the Huygens probe descended to the surface of Titan and relayed data, including imagery of the landing site, for about 90 minutes. You can watch a video (https://openstax.org/l/30huytatsurf) about the descent of Huygens to Titan's surface.

Habitable Planets Orbiting Other Stars

One of the most exciting developments in astronomy during the last two decades is the ability to detect exoplanets—planets orbiting other stars. As we saw in the chapter on the formation of stars and planets, since the discovery of the first exoplanet in 1995, there have been thousands of confirmed detections, and many more candidates that are not yet confirmed. These include several dozen possibly habitable exoplanets. Such numbers finally allow us to make some predictions about exoplanets and their life-hosting potential. The majority of stars with mass similar to the Sun appear to host at least one planet, with multi-planet systems like our own not unusual. How many of these planets might be habitable, and how could we search for life there?

LINK TO LEARNING

The NASA Exoplanet Archive (https://openstax.org/l/30NASAexoarc) is an up-to-date searchable online source of data and tools on everything to do with exoplanets. Explore stellar and exoplanet parameters and characteristics, find the latest news on exoplanet discoveries, plot your own data interactively, and link to other related resources.

In evaluating the prospect for life in distant planetary systems, astrobiologists have developed the idea of a **habitable zone**—a region around a star where suitable conditions might exist for life. This concept focuses on life's requirement for liquid water, and the habitable zone is generally thought of as the range of distances from the central star in which water could be present in liquid form at a planet's surface. In our own solar system, for example, Venus has surface temperatures far above the boiling point of water and Mars has surface temperatures that are almost always below the freezing point of water. Earth, which orbits between the two, has a surface temperature that is "just right" to keep much of our surface water in liquid form.

Whether surface temperatures are suitable for maintaining liquid water depends on a planet's "radiation budget" —how much starlight energy it absorbs and retains—and whether or how processes like winds and ocean circulation distribute that energy around the planet. How much stellar energy a planet receives, in turn, depends on how much and what sort of light the star emits and how far the planet is from that star,[1] how much it reflects back to space, and how effectively the planet's atmosphere can retain heat through the greenhouse effect (see Earth as a Planet). All of these can vary substantially, and all matter a lot. For example, Venus receives about twice as much starlight per square meter as Earth but, because of its dense cloud cover, also reflects about twice as much of that light back to space as Earth does. Mars receives only about half as much starlight as Earth, but also reflects only about half as much. Thus, despite their differing orbital distances, the three planets actually absorb comparable amounts of sunlight energy. Why, then, are they so dramatically different?

As we learned in several chapters about the planets, some of the gases that make up planetary atmospheres are very effective at trapping infrared light—the very range of wavelengths at which planets radiate thermal energy back out to space—and this can raise the planet's surface temperature quite a bit more than would otherwise be the case. This is the same "greenhouse effect" that is of such concern for global warming on our planet. Earth's natural greenhouse effect, which comes mostly from water vapor and carbon dioxide in the atmosphere, raises our average surface temperature by about 33 °C over the value it would have if there were no greenhouse gases in the atmosphere. Mars has a very thin atmosphere and thus very little greenhouse warming (about 2 °C worth), while Venus has a massive carbon dioxide atmosphere that creates very strong greenhouse warming (about 510 °C worth). These worlds are much colder and much hotter, respectively, than Earth would be if moved into their orbits. Thus, we must consider the nature of any atmosphere as well as the distance from the star in evaluating the range of habitability.

Of course, as we have learned, stars also vary widely in the intensity and spectrum (the wavelengths of light) they emit. Some are much brighter and hotter (bluer), while others are significantly dimmer and cooler (redder), and the distance of the habitable zone varies accordingly. For example, the habitable zone around M-dwarf stars is 3 to 30 times closer in than for G-type (Sun-like) stars. There is a lot of interest in whether such systems could be habitable because—although they have some potential downsides for supporting life—M-dwarf stars are by far the most numerous and long-lived in our Galaxy.

The luminosity of stars like the Sun also increases over their main-sequence lifetime, and this means that the habitable zone migrates outward as a star system ages. Calculations indicate that the power output of the Sun, for example, has increased by at least 30% over the past 4 billion years. Thus, Venus was once within the habitable zone, while Earth received a level of solar energy insufficient to keep the modern Earth (with its present atmosphere) from freezing over. In spite of this, there is plenty of geological evidence that liquid water was present on Earth's surface billions of years ago. The phenomenon of increasing stellar output and an outwardly migrating habitable zone has led to another concept: the *continuously* habitable zone is defined by the range of orbits that would remain within the habitable zone during the entire lifetime of the star system. As you might imagine, the continuously habitable zone is quite a bit narrower than the habitable zone

---

1   The amount of starlight received per unit area of a planet's surface (per square meter, for example) decreases with the square of the distance from the star. Thus, when the orbital distance doubles, the illumination decreases by 4 times (2^2), and when the orbital distance increases tenfold, the illumination decreases by 100 times (10^2). Venus and Mars orbit the sun at about 72% and 152% of Earth's orbital distance, respectively, so Venus receives about $1/(0.72)^2 = 1.92$ (about twice) and Mars about $1/(1.52)^2 = 0.43$ (about half) as much light per square meter of planet surface as Earth does.

is at any one time in a star's history. The nearest star to the Sun, Proxima Centauri, is an M star that has a planet with a mass of at least 1.3 Earth masses, taking about 11 days to orbit. At the distance for such a quick orbit (0.05 AU), the planet may be in the habitable zone of its star, although whether conditions on such a planet near such a star are hospitable for life is a matter of great scientific debate.

Even when planets orbit within the habitable zone of their star, it is no guarantee that they are habitable. For example, Venus today has virtually no water, so even if it were suddenly moved to a "just right" orbit within the habitable zone, a critical requirement for life would still be lacking.

Scientists are working to understand all the factors that define the habitable zone and the habitability of planets orbiting within that zone because this will be our primary guide in targeting exoplanets on which to seek evidence of life. As technology for detecting exoplanets has advanced, so too has our potential to find Earth-size worlds within the habitable zones of their parent stars. Of the confirmed or candidate exoplanets known at the time of writing, nearly 300 are considered to be orbiting within the habitable zone and more than 10% of those are roughly Earth-size.

LINK TO LEARNING

Explore the habitable universe at the online Planetary Habitability Laboratory (https://openstax.org/l/30planhabitlab) created by the University of Puerto Rico at Arecibo. See the potentially habitable exoplanets and other interesting places in the universe, watch video clips, and link to numerous related resources on astrobiology.

Biomarkers

Our observations suggest increasingly that Earth-size planets orbiting within the habitable zone may be common in the Galaxy—current estimates suggest that more than 40% of stars have at least one. But are any of them inhabited? With no ability to send probes there to sample, we will have to derive the answer from the light and other radiation that come to us from these faraway systems (Figure 30.16). What types of observations might constitute good evidence for life?

Figure 30.16 Earth, as Seen by NASA's Voyager 1. In this image, taken from 4 billion miles away, Earth appears as a "pale blue dot"

To be sure, we need to look for robust biospheres (atmospheres, surfaces, and/or oceans) capable of creating planet-scale change. Earth hosts such a biosphere: the composition of our atmosphere and the spectrum of light reflected from our planet differ considerably from what would be expected in the absence of life. Presently, Earth is the only body in our solar system for which this is true, despite the possibility that habitable conditions might prevail in the subsurface of Mars or inside the icy moons of the outer solar system. Even if life exists on these worlds, it is very unlikely that it could yield planet-scale changes that are both telescopically observable and clearly biological in origin.

What makes Earth "special" among the potentially habitable worlds in our solar system is that it has a photosynthetic biosphere. This requires the presence of liquid water at the planet's surface, where organisms have direct access to sunlight. The habitable zone concept focuses on this requirement for surface liquid water—even though we know that subsurface habitable conditions could prevail at more distant orbits—exactly because these worlds would have biospheres detectable at a distance.

Indeed, plants and photosynthetic microorganisms are so abundant at Earth's surface that they affect the color of the light that our planet reflects out into space—we appear greener in visible wavelengths and reflect more near-infrared light than we otherwise would. Moreover, photosynthesis has changed Earth's atmosphere at a large scale—more than 20% of our atmosphere comes from the photosynthetic waste product, oxygen. Such high levels would be very difficult to explain in the absence of life. Other gases, such as nitrous oxide and methane, when found simultaneously with oxygen, have also been suggested as possible indicators of life. When sufficiently abundant in an atmosphere, such gases could be detected by their effect on the spectrum of light that a planet emits or reflects. (As we saw in the chapter on exoplanets, astronomers today are beginning to have the capability of detecting the spectrum of the atmospheres of some planets orbiting other stars.)

Astronomers have thus concluded that, at least initially, a search for life outside our solar system should focus on exoplanets that are as much like Earth as possible—roughly Earth-size planets orbiting in the habitable zone—and look for the presence of gases in the atmosphere or colors in the visible spectrum that are hard to explain except by the presence of biology. Simple, right? In reality, the search for exoplanet life poses many challenges.

As you might imagine, this task is more challenging for planetary systems that are farther away and, in practical terms, this will limit our search to the habitable worlds closest to our own. Should we become limited to a very small number of nearby targets, it will also become important to consider the habitability of planets orbiting the M-dwarfs we discussed above.

If we manage to separate out a clean signal from the planet and find some features in the light spectrum that might be indicative of life, we will need to work hard to think of any nonbiological process that might account for them. "Life is the hypothesis of last resort," noted astronomer Carl Sagan—meaning that we must exhaust all other explanations for what we see before claiming to have found evidence of extraterrestrial biology. This requires some understanding of what processes might operate on worlds that we will know relatively little about; what we find on Earth can serve as a guide but also has potential to lead us astray (Figure 30.17).

Recall, for example, that it would be extremely difficult to account for the abundance of oxygen in Earth's atmosphere except by the presence of biology. But it has been hypothesized that oxygen could build up to substantial levels on planets orbiting M-dwarf stars through the action of ultraviolet radiation on the atmosphere—with no need for biology. It will be critical to understand where such "false positives" might exist in carrying out our search.

We need to understand that we might not be able to detect biospheres even if they exist. Life has flourished on Earth for perhaps 3.5 billion years, but the atmospheric "biosignatures" that, today, would supply good evidence for life to distant astronomers have not been present for all of that time. Oxygen, for example, accumulated to detectable levels in our atmosphere only a little over 2 billion years ago. Could life on Earth have been detected before that time? Scientists are working actively to understand what additional features might have provided evidence of life on Earth during that early history, and thereby help our chances of finding life beyond.

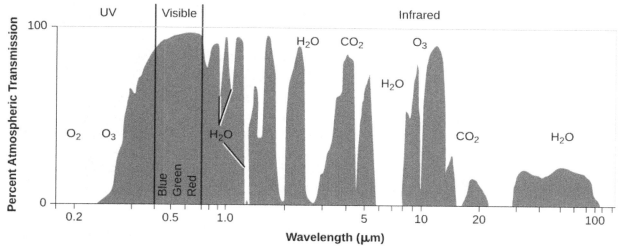

Figure 30.17 Spectrum of Light Transmitted through Earth's Atmosphere. This graph shows wavelengths ranging from ultraviolet (far left) to infrared. The many downward "spikes" come from absorption of particular wavelengths by molecules in Earth's atmosphere. Some of these compounds, like water and the combination oxygen/ozone and methane, might reveal Earth as both habitable and inhabited. We will have to rely on this sort of information to seek life on exoplanets, but our spectra will be of much poorer quality than this one, in part because we will receive so little light from the planet. (credit: modification of work by NASA)

30.4 | The Search for Extraterrestrial Intelligence

Learning Objectives

By the end of this section, you will be able to:

> Explain why spaceships from extraterrestrial civilizations are unlikely to have visited us
> List efforts by humankind to communicate with other civilizations via messages on spacecraft
> Understand the various SETI programs scientists are undertaking

Given all the developments discussed in this chapter, it seems likely that life could have developed on many planets around other stars. Even if that life is microbial, we saw that we may soon have ways to search for chemical biosignatures. This search is of fundamental importance for understanding biology, but it does not answer the question, "Are we alone?" that we raised at the beginning of this chapter. When we ask this question, many people think of other intelligent creatures, perhaps beings that have developed technology similar to our own. If any intelligent, technical civilizations have arisen, as has happened on Earth in the most recent blink of cosmic time, how could we make contact with them?

This problem is similar to making contact with people who live in a remote part of Earth. If students in the United States want to converse with students in Australia, for example, they have two choices. Either one group gets on an airplane and travels to meet the other, or they communicate by sending a message remotely. Given how expensive airline tickets are, most students would probably select the message route.

In the same way, if we want to get in touch with intelligent life around other stars, we can travel, or we can try to exchange messages. Because of the great distances involved, interstellar space travel would be very slow and prohibitively expensive. The fastest spacecraft the human species has built so far would take almost 80,000 years to get to the nearest star. While we could certainly design a faster craft, the more quickly we require it to travel, the greater the energy cost involved. To reach neighboring stars in less than a human life span, we would have to travel close to the speed of light. In that case, however, the expense would become truly astronomical.

Interstellar Travel

Bernard Oliver, an engineer with an abiding interest in life elsewhere, made a revealing calculation about the costs of rapid interstellar space travel. Since we do not know what sort of technology we (or other civilizations) might someday develop, Oliver considered a trip to the nearest star (and back again) in a spaceship with a "perfect engine"—one that would convert its fuel into energy with 100% efficiency. Even with a perfect engine, the energy cost of a single round-trip journey at 70% the speed of light turns out to be equivalent to several hundred thousand years' worth of total U.S. electrical energy consumption. The cost of such travel is literally out of this world.

This is one reason astronomers are so skeptical about claims that UFOs are spaceships from extraterrestrial civilizations. Given the distance and energy expense involved, it seems unlikely that the dozens of UFOs (and even UFO abductions) claimed each year could be visitors from other stars so fascinated by Earth civilization that they are willing to expend fantastically large amounts of energy or time to reach us. Nor does it seem credible that these visitors have made this long and expensive journey and then systematically avoided contacting our governments or political and intellectual leaders.

Not every UFO report has been explained (in many cases, the observations are sketchy or contradictory). But investigation almost always converts them to IFOs (identified flying objects) or NFOs (not-at-all flying objects). While some are hoaxes, others are natural phenomena, such as bright planets, ball lightning, fireballs (bright meteors), or even flocks of birds that landed in an oil slick to make their bellies reflective. Still others are human craft, such as private planes with some lights missing, or secret military aircraft. It is also interesting that the group of people who most avidly look at the night sky, the amateur astronomers, have never reported UFO sightings. Further, not a single UFO has ever left behind any physical evidence that can be tested in a laboratory and shown to be of nonterrestrial origin.

Another common aspect of belief that aliens are visiting Earth comes from people who have difficulty accepting human accomplishments. There are many books and TV shows, for example, that assert that humans could not have built the great pyramids of Egypt, and therefore they must have been built by aliens. The huge statues (called Moai) on Easter Island are also sometimes claimed to have been built by aliens. Some people even think that the accomplishments of space exploration today are based on alien technology.

However, the evidence from archaeology and history is clear: ancient monuments were built by ancient *people*, whose brains and ingenuity were every bit as capable as ours are today, even if they didn't have electronic textbooks like you do.

Messages on Spacecraft

While space travel by living creatures seems very difficult, robot probes can travel over long distances and over long periods of time. Five spacecraft—two Pioneers, two Voyagers, and New Horizons—are now leaving the solar system. At their coasting speeds, they will take hundreds of thousands or millions of years to get anywhere close to another star. On the other hand, they were the first products of human technology to go beyond our home system, so we wanted to put messages on board to show where they came from.

Each Pioneer carries a plaque with a pictorial message engraved on a gold-anodized aluminum plate (Figure 30.18). The Voyagers, launched in 1977, have audio and video records attached, which allowed the inclusion of over 100 photographs and a selection of music from around the world. Given the enormous space between stars in our section of the Galaxy, it is very unlikely that these messages will ever be received by anyone. They are more like a note in a bottle thrown into the sea by a shipwrecked sailor, with no realistic expectation of its being found soon but a slim hope that perhaps someday, somehow, someone will know of the sender's fate.

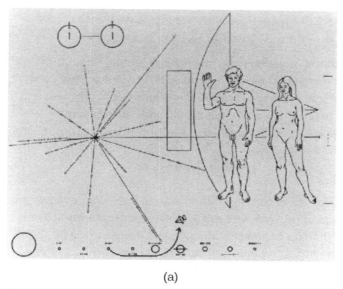 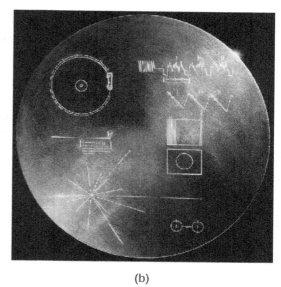

(a)   (b)

Figure 30.18 Interstellar Messages. (a) This is the image engraved on the plaques aboard the Pioneer 10 and 11 spacecraft. The human figures are drawn in proportion to the spacecraft, which is shown behind them. The Sun and planets in the solar system can be seen at the bottom, with the trajectory that the spacecraft followed. The lines and markings in the left center show the positions and pulse periods for a number of pulsars, which might help locate the spacecraft's origins in space and time. (b) Encoded onto a gold-coated copper disk, the Voyager record contains 118 photographs, 90 minutes of music from around the world, greetings in almost 60 languages, and other audio material. It is a summary of the sights and sounds of Earth. (credit a, b: modification of work by NASA)

MAKING CONNECTIONS

The Voyager Message

An Excerpt from the Voyager Record:

"We cast this message into the cosmos. It is likely to survive a billion years into our future, when our civilization is profoundly altered. . . . If [another] civilization intercepts Voyager and can understand these recorded contents, here is our message:

This is a present from a small, distant world, a token of our sounds, our science, our images, our music, our thoughts, and our feelings. We are attempting to survive our time so we may live into yours. We hope, someday, having solved the problems we face, to join a community of galactic civilizations. This record represents our hope and our determination, and our goodwill in a vast and awesome universe."

—Jimmy Carter, President of the United States of America, June 16, 1977

Communicating with the Stars

If direct visits among stars are unlikely, we must turn to the alternative for making contact: exchanging messages. Here the news is a lot better. We already use a messenger—light or, more generally, electromagnetic waves—that moves through space at the fastest speed in the universe. Traveling at 300,000 kilometers per second, light reaches the nearest star in only 4 years and does so at a tiny fraction of the cost of sending material objects. These advantages are so clear and obvious that we assume they will occur to any other species of intelligent beings that develop technology.

However, we have access to a wide spectrum of electromagnetic radiation, ranging from the longest-wavelength radio waves to the shortest-wavelength gamma rays. Which would be the best for interstellar communication? It would not be smart to select a wavelength that is easily absorbed by interstellar gas and dust, or one that is unlikely to penetrate the atmosphere of a planet like ours. Nor would we want to pick a wavelength that has lots of competition for attention in our neighborhood.

One final criterion makes the selection easier: we want the radiation to be inexpensive enough to produce in large quantities. When we consider all these requirements, radio waves turn out to be the best answer. Being the lowest-frequency (and lowest-energy) band of the spectrum, they are not very expensive to produce, and we already use them extensively for communications on Earth. They are not significantly absorbed by interstellar dust and gas. With some exceptions, they easily pass through Earth's atmosphere and through the atmospheres of the other planets we are acquainted with.

The Cosmic Haystack

Having made the decision that radio is the most likely means of communication among intelligent civilizations, we still have many questions and a daunting task ahead of us. Shall we *send* a message, or try to *receive* one? Obviously, if every civilization decides to receive only, then no one will be sending, and everyone will be disappointed. On the other hand, it may be appropriate for us to *begin* by listening, since we are likely to be among the most primitive civilizations in the Galaxy who are interested in exchanging messages.

We do not make this statement to insult the human species (which, with certain exceptions, we are rather fond of). Instead, we base it on the fact that humans have had the ability to receive (or send) a radio message across interstellar distances for only a few decades. Compared to the ages of the stars and the Galaxy, this is a mere instant. If there are civilizations out there that are ahead of us in development by even a short time (in the cosmic sense), they are likely to have a technology head start of many, many years.

In other words, we, who have just started, may well be the "youngest" species in the Galaxy with this capability (see the discussion in Example 30.1). Just as the youngest members of a community are often told to be quiet and listen to their elders for a while before they say something foolish, so may we want to begin our exercise in extraterrestrial communication by listening.

Even restricting our activities to listening, however, leaves us with an array of challenging questions. For example, if an extraterrestrial civilization's signal is too weak to be detected by our present-day radio telescopes, we will not detect them. In addition, it would be very expensive for an extraterrestrial civilization to broadcast on a huge number of channels. Most likely, they select one or a few channels for their particular message. Communicating on a narrow band of channels also helps distinguish an artificial message from the radio static that comes from natural cosmic processes. But the radio band contains an astronomically large number of possible channels. How can we know in advance which one they have selected, and how they have coded their message into the signal?

Table 30.1 summarizes these and other factors that scientists must grapple with when trying to tune in to radio messages from distant civilizations. Because their success depends on either guessing right about so many factors or searching through all the possibilities for each factor, some scientists have compared their quest to looking for a needle in a haystack. Thus, they like to say that the list of factors in Table 30.1 defines the *cosmic haystack problem*.

The Cosmic Haystack Problem: Some Questions about an Extraterrestrial Message

Factors
From which direction (which star) is the message coming?
On what channels (or frequencies) is the message being broadcast?

Table 30.1

Factors
How wide in frequency is the channel?
How strong is the signal (can our radio telescopes detect it)?
Is the signal continuous, or does it shut off at times (as, for example, a lighthouse beam does when it turns away from us)?
Does the signal drift (change) in frequency because of the changing relative motion of the source and the receiver?
How is the message encoded in the signal (how do we decipher it)?
Can we even recognize a message from a completely alien species? Might it take a form we don't at all expect?

Table 30.1

Radio Searches

Although the cosmic haystack problem seems daunting, many other research problems in astronomy also require a large investment of time, equipment, and patient effort. And, of course, if we don't search, we're sure not to find anything.

The very first search was conducted by astronomer Frank Drake in 1960, using the 85-foot antenna at the National Radio Astronomy Observatory (Figure 30.19). Called Project Ozma, after the queen of the exotic Land of Oz in the children's stories of L. Frank Baum, his experiment involved looking at about 7200 channels and two nearby stars over a period of 200 hours. Although he found nothing, Drake demonstrated that we had the technology to do such a search, and set the stage for the more sophisticated projects that followed.

(a)   (b)

Figure 30.19 Project Ozma and the Allen Telescope Array. (a) This 25th anniversary photo shows some members of the Project Ozma team standing in front of the 85-foot radio telescope with which the 1960 search for extraterrestrial messages was performed. Frank Drake is in the back row, second from the right. (b) The Allen Telescope Array in California is made up of 42 small antennas linked together. This system allows simultaneous observations of multiple sources with millions of separate frequency channels. (credit a: modification of work by NRAO; credit b: modification of work by Colby Gutierrez-Kraybill)

Receivers are constantly improving, and the sensitivity of SETI programs—**SETI** stands for the search for extraterrestrial intelligence—is advancing rapidly. Equally important, modern electronics and software allow simultaneous searches on millions of frequencies (channels). If we can thus cover a broad frequency range, the cosmic haystack problem of guessing the right frequency largely goes away. One powerful telescope array (funded with an initial contribution from Microsoft founder Paul Allen) that is built for SETI searches is the Allen Telescope in Northern California. Other radio telescopes being used for such searches have included the giant Arecibo radio dish in Puerto Rico (which, sadly, collapsed due to storm damage in late 2020), the recently completed, and even larger, FAST dish in China, and the Green Bank Telescope in West Virginia, which is the largest steerable radio telescope in the world.

In 2015, Silicon Valley investor and philanthropist Yuri Milner donated 100 million dollars to the University of California, Berkeley, to enhance the SETI endeavor for ten years. The project, called B*reakthrough: Listen*, is enlisting more telescopes around the world in the search and is developing sophisticated artificial intelligence programs to scour the incoming signals for intelligent messages. The aim is to search more stars and more channels than any SETI project so far.

What kind of signals do we hope to pick up? We on Earth are inadvertently sending out a flood of radio signals, dominated by military radar systems. This is a kind of leakage signal, similar to the wasted light energy that is beamed upward by poorly designed streetlights and advertising signs. Could we detect a similar leakage of radio signals from another civilization? The answer is just barely, but only for the nearest stars. For the most part, therefore, current radio SETI searches are looking for beacons, assuming that civilizations might be intentionally drawing attention to themselves or perhaps sending a message to another world or outpost that lies in our direction. Our prospects for success depend on how often civilizations arise, how long they last, and how patient they are about broadcasting their locations to the cosmos.

VOYAGERS IN ASTRONOMY

Jill Tarter: Trying to Make Contact

1997 was quite a year for Jill Cornell Tarter (Figure 30.20), one of the world's leading scientists in the SETI field. The SETI Institute announced that she would be the recipient of its first endowed chair (the equivalent of an endowed research professorship) named in honor of Bernard Oliver. The National Science Foundation approved a proposal by a group of scientists and educators she headed to develop an innovative hands-on high school curriculum based on the ideas of cosmic evolution (the topics of this chapter). And, at roughly the same time, she was being besieged with requests for media interviews as news reports identified her as the model for Ellie Arroway, the protagonist of *Contact*, Carl Sagan's best-selling novel about SETI. The book had been made into a high-budget science fiction film, starring Jodie Foster, who had talked with Tarter before taking the role.

Figure 30.20 Jill Tarter (credit: Christian Schidlowski)

Tarter is quick to point out, "Carl Sagan wrote a book about a woman who does what I do, not about me." Still, as the only woman in such a senior position in the small field of SETI, she was the center of a great deal of public attention. (However, colleagues and reporters pointed out that this was nothing compared to what would happen if her search for radio signals from other civilizations recorded a success.)

Being the only woman in a group is not a new situation to Tarter, who often found herself the only woman in her advanced science or math classes. Her father had encouraged her, both in her interest in science and her "tinkering." As an undergraduate at Cornell University, she majored in engineering physics. That training became key to putting together and maintaining the complex systems that automatically scan for signals from other civilizations.

Switching to astrophysics for her graduate studies, she wrote a PhD thesis that, among other topics, considered the formation of failed stars—those whose mass was not sufficient to ignite the nuclear reactions that power more massive stars like our own Sun. Tarter coined the term "brown dwarf" for these small, dim objects, and it has remained the name astronomers use ever since.

It was while she was still in graduate school that Stuart Bowyer, one of her professors at the University of California, Berkeley, asked her if she wanted to be involved in a small experiment to siphon off a bit of radiation from a radio telescope as astronomers used it year in and year out and see if there was any hint of an intelligently coded radio message buried in the radio noise. Her engineering and computer programming skills became essential to the project, and soon she was hooked on the search for life elsewhere.

Thus began an illustrious career working full time searching for what she calls the technosignatures (signs of technology) from extraterrestrial civilizations. Tarter has received many awards, including being elected fellow of the American Association for the Advancement of Science in 2002, the Adler Planetarium Women in Space Science Award in 2003, and a 2009 TED Prize.

LINK TO LEARNING

Watch the TED talk (https://openstax.org/l/30TarterSETI) Jill Tarter gave on the fascination of the search for intelligence.

EXAMPLE 30.1

The Drake Equation

At the first scientific meeting devoted to SETI, Frank Drake wrote an equation on the blackboard that took the difficult question of estimating the number of civilizations in the Galaxy and broke it down into a series of smaller, more manageable questions. Ever since then, both astronomers and students have used this **Drake equation** as a means of approaching the most challenging question: How likely is it that we are alone? Since this is at present an unanswerable question, astronomer Jill Tarter has called the Drake equation a "way of organizing our ignorance." (See Figure 30.21.)

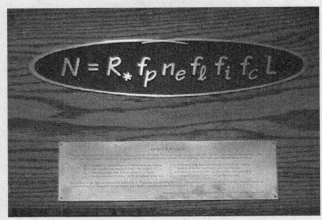

Figure 30.21 Drake Equation. A plaque at the National Radio Astronomy Observatory commemorates the conference where the equation was first discussed. (credit: NRAO/NSF/AUI)

The form of the Drake equation is very simple. To estimate the number of communicating civilizations that currently exist in the Galaxy (we will define these terms more carefully in a moment), we multiply the rate of formation of such civilizations (number per year) by their average lifetime (in years). In symbols,

$$N = R_{\text{total}} \times L$$

To make this formula easier to use (and more interesting), however, Drake separated the rate of formation R_{total} into a series of probabilities:

$$R_{\text{total}} = R_{\text{star}} \times f_{\text{p}} \times f_{\text{e}} \times f_{\text{l}} \times f_{\text{i}} \times f_{\text{c}}$$

R_{star} is the rate of formation of stars like the Sun in our Galaxy, which is, very roughly, around 10 stars per year. Each of the other terms is a fraction or probability (less than or equal to 1.0), and the product of all these probabilities is itself the total probability that each star will have an intelligent, technological, communicating civilization that we might want to talk to. We have:

- f_{p} = the fraction of these stars with planets
- f_{e} = the fraction of the planetary systems that include habitable planets
- f_{l} = the fraction of habitable planets that actually support life
- f_{i} = the fraction of inhabited planets that develop advanced intelligence
- f_{c} = the fraction of these intelligent civilizations that develop science and the technology to build radio telescopes and transmitters

Each of these factors can be discussed and perhaps evaluated, but we must guess at many of the values. In particular, we don't know how to calculate the probability of something that happened once on Earth but has not been observed elsewhere—and these include the development of life, of intelligent life, and of technological life (the last three factors in the equation). One important advance in estimating the terms of the Drake equation comes from the recent discovery of exoplanets. When the Drake equation was first written, no one had any idea whether planets and planetary systems were common. Now we know they are—another example of the Copernican principle.

Solution

Even if we don't know the answers, we can make some guesses and calculate the resulting number N. Let's start with the optimism implicit in the Copernican principle and set the last three terms equal to 1.0. If R is 10 stars/year and if we measure the average lifetime of a technological civilization in years, the units of years cancel. If we also assume that f_p is 0.1, and f_e is 1.0, the equation becomes

$$N = R_{total} \times L = L$$

Now we see the importance of the term L, the lifetime of a communicating civilization (measured in years). We have had this capability (to communicate at the distances of the stars) for only a few decades.

Check Your Learning

Suppose we assume that this stage in our history lasts only one century.

Answer:

With our optimistic assumptions about the other factors, $L = 100$ years and $N = 100$ such civilizations in the entire Galaxy. In that case, there are so few other civilizations like ours that we are unlikely to detect any signals in a SETI search. But suppose the average lifetime is a million years; in that case, there are a million such civilizations in the Galaxy, and some of them may be within range for radio communication.

The most important conclusion from this calculation is that even if we are extremely optimistic about the probabilities, the only way we can expect success from SETI is if other civilizations are much older (and hence probably much more advanced) than ours.

LINK TO LEARNING

Read Frank Drake's own account (https://openstax.org/l/30drakeequat) of how he came up with his "equation." And here is a recent interview (https://openstax.org/l/30frandrakinter) with Frank Drake by one of the authors of this textbook.

SETI outside the Radio Realm

For the reasons discussed above, most SETI programs search for signals at radio wavelengths. But in science, if there are other approaches to answering an unsolved question, we don't want to neglect them. So astronomers have been thinking about other ways we could pick up evidence for the existence of technologically advanced civilizations.

Recently, technology has allowed astronomers to expand the search into the domain of visible light. You might think that it would be hopeless to try to detect a flash of visible light from a planet given the brilliance of the star it orbits. This is why we usually cannot measure the reflected light of planets around other stars. The feeble light of the planet is simply swamped by the "big light" in the neighborhood. So another civilization would need a mighty strong beacon to compete with their star.

However, in recent years, human engineers have learned how to make flashes of light brighter than the Sun. The trick is to "turn on" the light for a very brief time, so that the costs are manageable. But ultra-bright, ultra-short laser pulses (operating for periods of a billionth of a second) can pack a lot of energy and can be coded to carry a message. We also have the technology to detect such short pulses—not with human senses, but with special detectors that can be "tuned" to hunt automatically for such short bursts of light from nearby stars.

Why would any civilization try to outshine its own star in this way? It turns out that the cost of sending an ultra-short laser pulse in the direction of a few promising stars can be less than the cost of sweeping a continuous radio message across the whole sky. Or perhaps they, too, have a special fondness for light messages because one of their senses evolved using light. Several programs are now experimenting with "optical SETI" searches, which can be done with only a modest telescope. (The term *optical* here means using visible light.)

If we let our imaginations expand, we might think of other for *technosignatures*—signs of technology from other civilizations. For example, what if a truly advanced civilization should decide to (or need to) renovate its planetary system to maximize the area for life? It could do so by breaking apart some planets or moons and building a ring of solid material that surrounds or encloses the star and intercepts some or all of its light. This huge artificial ring or sphere might glow very brightly at infrared wavelengths, as the starlight it receives is eventually converted to heat and re-radiated into space. That infrared radiation could be detected by our instruments, and searches for such infrared sources are also underway (Figure 30.22).

Figure 30.22 Wide-Field Infrared Survey Explorer (WISE). Astronomers have used this infrared satellite to search for infrared signatures of enormous construction projects by very advanced civilizations, but their first survey did not reveal any. (credit: modification of work by NASA/JPL-Caltech)

Should We Transmit in Addition to Listening?

Our planet has some leakage of radio waves into space, from FM radio, television, military radars, and communication between Earth and our orbiting spacecraft. However, such leakage radiation is still quite weak, and therefore difficult to detect at the distances of the stars, at least with the radio technology we have. So at the present time our attempts to communicate with other civilizations that may be out there mostly involve trying to receive messages, but not sending any ourselves.

Some scientists, however, think that it is inconsistent to search for beacons from other civilizations without announcing our presence in a similar way. (We discussed earlier the problem that if every other civilization confined itself to listening, no one would ever get in touch.) So, should we be making regular attempts at sending easily decoded messages into space? Some scientists warn that our civilization is too immature and defenseless to announce ourselves at this early point in our development. The decision whether to transmit or not turns out to be an interesting reflection of how we feel about ourselves and our place in the universe.

Discussions of transmission also raise the question of who should speak for planet Earth. Today, anyone and everyone can broadcast radio signals, and many businesses, religious groups, and governments do. It would be a modest step for the same organizations to use or build large radio telescopes and begin intentional transmissions that are much stronger than the signals that leak from Earth today. And if we intercept a signal from an alien civilization, then the issue arises whether to reply.

Who should make the decision about whether, when, and how humanity announces itself to the cosmos? Is there freedom of speech when it comes to sending radio messages to other civilizations? Do all the nations of Earth have to agree before we send a signal strong enough that it has a serious chance of being received at the distances of the stars? How our species reaches a decision about these kinds of questions may well be a test of whether or not there is intelligent life on Earth.

Conclusion

Whether or not we ultimately turn out to be the only intelligent species in our part of the Galaxy, our exploration of the cosmos will surely continue. An important part of that exploration will still be the search for biomarkers from inhabited planets that have not produced technological creatures that send out radio signals. After all, creatures like butterflies and dolphins may never build radio antennas, but we are happy to share our planet with them and would be delighted to find their counterparts on other worlds.

Whether or not life exists elsewhere is just one of the unsolved problems in astronomy that we have discussed in this book. A humble acknowledgment of how much we have left to learn about the universe is one of the fundamental hallmarks of science. This should not, however, prevent us from feeling exhilarated about how much we have already managed to discover, and feeling curious about what else we might find out in the years to come.

Our progress report on the ideas of astronomy ends here, but we hope that your interest in the universe does not. We hope you will keep up with developments in astronomy through media and online, or by going to an occasional public lecture by a local scientist. Who, after all, can even guess all the amazing things that future research projects will reveal about both the universe and our connection with it?

🔑 Key Terms

amino acids   organic compounds that are the molecular building blocks of proteins

astrobiology   the multidisciplinary study of life in the universe: its origin, evolution, distribution, and fate; similar terms are *exobiology* and *bioastronomy*

biomarker   evidence of the presence of life, especially a global indication of life on a planet that could be detected remotely (such as an unusual atmospheric composition)

DNA (deoxyribonucleic acid)   a molecule that stores information about how to replicate a cell and its chemical and structural components

Drake equation   a formula for estimating the number of intelligent, technological civilizations in our Galaxy, first suggested by Frank Drake

extremophile   an organism (usually a microbe) that tolerates or even thrives under conditions that most of the life around us would consider hostile, such as very high or low temperature or acidity

gene   the basic functional unit that carries the genetic (hereditary) material contained in a cell

habitable environment   an environment capable of hosting life

habitable zone   the region around a star in which liquid water could exist on the surface of terrestrial-sized planets, hence the most probable place to look for life in a star's planetary system

organic compound   a compound containing carbon, especially a complex carbon compound; not necessarily produced by life

organic molecule   a combination of carbon and other atoms—primarily hydrogen, oxygen, nitrogen, phosphorus, and sulfur—some of which serve as the basis for our biochemistry

photosynthesis   a complex sequence of chemical reactions through which some living things can use sunlight to manufacture products that store energy (such as carbohydrates), releasing oxygen as one by-product

protein   a key biological molecule that provides the structure and function of the body's tissues and organs, and essentially carries out the chemical work of the cell

RNA (ribonucleic acid)   a molecule that aids in the flow of genetic information from DNA to proteins

SETI   the search for extraterrestrial intelligence; usually applied to searches for radio signals from other civilizations

stromatolites   solid, layered rock formations that are thought to be the fossils of oxygen-producing photosynthetic bacteria in rocks that are 3.5 billion years old

thermophile   an organism that can tolerate high temperatures

📄 Summary

30.1 The Cosmic Context for Life

Life on Earth is based on the presence of a key unit known as an organic molecule, a molecule that contains carbon, especially complex hydrocarbons. Our solar system formed about 5 billion years ago from a cloud of gas and dust enriched by several generations of heavier element production in stars. Life is made up of chemical combinations of these elements made by stars. The Copernican principle, which suggests that there is nothing special about our place in the universe, implies that if life could develop on Earth, it should be able to develop in other places as well. The Fermi paradox asks why, if life is common, more advanced life-forms have not contacted us.

30.2 Astrobiology

The study of life in the universe, including its origin on Earth, is called astrobiology. Life as we know it requires water, certain elemental raw materials (carbon, hydrogen, nitrogen, oxygen, phosphorus, and sulfur), energy, and an environment in which the complex chemistry of life is stable. Carbon-based (or organic) molecules are abundant in space and may also have been produced by processes on Earth. Life appears to have spread around our planet within 400 million years after the end of heavy bombardment, if not sooner. The actual origin of life—the processes leading from chemistry to biology—is not completely understood. Once life took hold, it evolved to use many energy sources, including first a range of different chemistries and later light, and diversified across a range of environmental conditions that humans consider "extreme." This proliferation of life into so many environmental niches, so relatively soon after our planet became habitable, has served to make many scientists optimistic about the chances that life could exist elsewhere.

30.3 Searching for Life beyond Earth

The search for life beyond Earth offers several intriguing targets. Mars appears to have been more similar to Earth during its early history than it is now, with evidence for liquid water on its ancient surface and perhaps even now below ground. The accessibility of the martian surface to our spacecraft offers the exciting potential to directly examine ancient and modern samples for evidence of life. In the outer solar system, the moons Europa and Enceladus likely host vast sub-ice oceans that may directly contact the underlying rocks—a good start in providing habitable conditions—while Titan offers a fascinating laboratory for understanding the sorts of organic chemistry that might ultimately provide materials for life. And the last decade of research on exoplanets leads us to believe that there may be billions of habitable planets in the Milky Way Galaxy. Study of these worlds offers the potential to find biomarkers indicating the presence of life.

30.4 The Search for Extraterrestrial Intelligence

Some astronomers are engaged in the search for extraterrestrial intelligent life (SETI). Because other planetary systems are so far away, traveling to the stars is either very slow or extremely expensive (in terms of energy required). Despite many UFO reports and tremendous media publicity, there is no evidence that any of these are related to extraterrestrial visits. Scientists have determined that the best way to communicate with any intelligent civilizations out there is by using electromagnetic waves, and radio waves seem best suited to the task. So far, they have only begun to comb the many different possible stars, frequencies, signal types, and other factors that make up what we call the cosmic haystack problem. Some astronomers are also undertaking searches for brief, bright pulses of visible light and infrared signatures of huge construction projects by advanced civilizations. If we do find a signal someday, deciding whether to answer and what to answer may be two of the greatest challenges humanity will face.

 For Further Exploration

Articles

Astrobiology

Chyba, C. "The New Search for Life in the Universe." *Astronomy* (May 2010): 34. An overview of astrobiology and the search for life out there in general, with a brief discussion of the search for intelligence.

Dorminey, B. "A New Way to Search for Life in Space." *Astronomy* (June 2014): 44. Finding evidence of photosynthesis on other worlds.

McKay, C., & Garcia, V. "How to Search for Life on Mars." *Scientific American* (June 2014): 44–49. Experiments future probes could perform.

Reed, N. "Why We Haven't Found Another Earth Yet." *Astronomy* (February 2016): 25. On the search for smaller earthlike planets in their star's habitable zones, and where we stand.

Shapiro, R. "A Simpler Origin of Life." *Scientific American* (June 2007): 46. New ideas about what kind of molecules formed first so life could begin.

Simpson, S. "Questioning the Oldest Signs of Life." *Scientific American* (April 2003): 70. On the difficulty of interpreting biosignatures in rocks and the implications for the search for life on other worlds.

SETI

Chandler, D. "The New Search for Alien Intelligence." *Astronomy* (September 2013): 28. Review of various ways of finding other civilizations out there, not just radio wave searches.

Crawford, I. "Where Are They?" *Scientific American* (July 2000): 38. On the Fermi paradox and its resolutions, and on galactic colonization models.

Folger, T. "Contact: The Day After." *Scientific American* (January 2011): 40–45. Journalist reports on efforts to prepare for ET signals; protocols and plans for interpreting messages; and discussions of active SETI.

Kuhn, J., et al. "How to Find ET with Infrared Light." *Astronomy* (June 2013): 30. On tracking alien civilizations by the heat they put out.

Lubick, N. "An Ear to the Stars." *Scientific American* (November 2002): 42. Profile of SETI researcher Jill Tarter.

Nadis, S. "How Many Civilizations Lurk in the Cosmos?" *Astronomy* (April 2010): 24. New estimates for the terms in the Drake equation.

Shostak, S. "Closing in on E.T." *Sky & Telescope* (November 2010): 22. Nice summary of current and proposed efforts to search for intelligent life out there.

Websites

Astrobiology

Astrobiology Web: http://astrobiology.com/ (http://astrobiology.com/). A news site with good information and lots of material.

Exploring Life's Origins: http://exploringorigins.org/index.html (http://exploringorigins.org/index.html). A website for the Exploring Origins Project, part of the multimedia exhibit of the Boston Museum of Science. Explore the origin of life on Earth with an interactive timeline, gain a deeper knowledge of the role of RNA, "build" a cell, and explore links to learn more about astrobiology and other related information.

History of Astrobiology: https://astrobiology.nasa.gov/about/history-of-astrobiology/ (https://astrobiology.nasa.gov/about/history-of-astrobiology/). By Marc Kaufman, on the NASA Astrobiology site.

Life, Here and Beyond: https://astrobiology.nasa.gov/about/ (https://astrobiology.nasa.gov/about/). By Marc Kaufman, on the NASA Astrobiology site.

SETI

Berkeley SETI Research Center: https://seti.berkeley.edu/ (https://seti.berkeley.edu/). The University of California group has received a $100 million grant from a Russian-American billionaire to begin the Breakthrough: Listen project, a major step forward in the number of stars and number of radio channels being searched.

Fermi Paradox: https://www.seti.org/fermi-paradox-0 (https://www.seti.org/fermi-paradox-0). If the Galaxy is teeming with advanced intelligent life, physicist Fermi asked, why is no one visiting us? This is an introduction; the much longer Wikipedia page on this topic (https://en.wikipedia.org/wiki/Fermi_paradox) has more detailed ideas.

Planetary Society: http://www.planetary.org/explore/projects/seti/ (http://www.planetary.org/explore/projects/seti/). This advocacy group for exploration has several pages devoted to the search for life.

Reading list: http://bit.ly/ufoskeptic (http://bit.ly/ufoskeptic). If you would like to see the perspective of scientists and other skeptical investigators on UFO controversies, check out this reading list.

SETI Institute: http://www.seti.org (http://www.seti.org). A key organization in the search for life in the universe; the institute's website is full of information and videos about both astrobiology and SETI.

SETI: http://www.skyandtelescope.com/tag/seti/ (http://www.skyandtelescope.com/tag/seti/). *Sky & Telescope* magazine offers good articles on this topic.

Videos

Astrobiology

Copernicus Complex: Are We Special in the Cosmos?: https://www.youtube.com/watch?v=ERp0AHYRm_Q (https://www.youtube.com/watch?v=ERp0AHYRm_Q). A video of a popular-level talk by Caleb Scharf of Columbia University (1:18:54).

Life at the Edge: Life in Extreme Environments on Earth and the Search for Life in the Universe: https://www.youtube.com/watch?v=91JQmTn0SF0 (https://www.youtube.com/watch?v=91JQmTn0SF0). A video of a 2009 nontechnical lecture by Lynn Rothschild of NASA Ames Research Center (1:31:21).

Saturn's Moon Titan: A World with Rivers, Lakes, and Possibly Even Life: https://www.youtube.com/watch?v=bbkTJeHoOKY (https://www.youtube.com/watch?v=bbkTJeHoOKY). A video of a 2011 talk by Chris McKay of NASA Ames Research Center (1:23:33).

SETI

Allen Telescope Array: The Newest Pitchfork for Exploring the Cosmic Haystack: https://www.youtube.com/watch?v=aqsI1HZCgUM (https://www.youtube.com/watch?v=aqsI1HZCgUM). A 2013 popular-level lecture by Jill Tarter of the SETI Institute (1:45:55).

Search for Extra-Terrestrial Intelligence: Necessarily a Long-Term Strategy: http://www.longnow.org/seminars/02004/jul/09/the-search-for-extra-terrestrial-intelligence-necessarily-a-long-term-strategy/ (http://www.longnow.org/seminars/02004/jul/09/the-search-for-extra-terrestrial-intelligence-necessarily-a-long-term-strategy/). 2004 talk by Jill Tarter at the Long Now Foundation (1:21:13).

The Search for Extra-Terrestrial Life with New Generation Technology: https://www.youtube.com/watch?v=R1ndk08vZqM (https://www.youtube.com/watch?v=R1ndk08vZqM). A briefing in 2020 with Jill Tarter, Andrew Siemion, and Victoria Meadows on how scientists are currently searching for life, including intelligent life.

Search for Intelligent Life Among the Stars: New Strategies: https://www.youtube.com/watch?v=m9WxW2ktcKU (https://www.youtube.com/watch?v=m9WxW2ktcKU). A 2010 nontechnical talk by Seth Shostak of the SETI Institute (1:29:58).

The Breakthrough: Listen Initiative—Our Boldest Effort to Answer our Oldest Question: https://www.youtube.com/watch?v=vQ2sKwwhRgI (https://www.youtube.com/watch?v=vQ2sKwwhRgI). A talk by Andrew Siemion at Harvard, with some nontechnical and some more technical sections.

Collaborative Group Activities

A. If one of the rocks from Mars examined by a future mission to the red planet does turn out to have unambiguous signs of ancient life that formed on Mars, what does your group think would be the implications of such a discovery for science and for our view of life elsewhere? Would such a discovery have any long-term effects on your own thinking?

B. Suppose we receive a message from an intelligent civilization around another star. What does your group think the implications of this discovery would be? How would your own thinking or personal philosophy be affected by such a discovery?

C. A radio message has been received from a civilization around a star 40 light-years away, which contains (in pictures) quite a bit of information about the beings that sent the message. The president of the United States has appointed your group a high-level commission to advise whether humanity should answer the message (which was not particularly directed at us, but comes from a beacon that, like a lighthouse, sweeps out a circle in space). How would you advise the president? Does your group agree on your answer or do you also have a minority view to present?

D. If there is no evidence that UFOs are extraterrestrial visitors, why does your group think that television shows, newspapers, and movies spend so much time and effort publicizing the point of view that UFOs are craft from other worlds? Make a list of reasons. Who stands to gain by exaggerating stories of unknown lights in the sky or simply fabricating stories that alien visitors are already here?

E. Does your group think scientists should simply ignore all the media publicity about UFOs or should they try to respond? If so, how should they respond? Does everyone in the group agree?

F. Suppose your group is the team planning to select the most important sights and sounds of Earth to record and put on board the next interstellar spacecraft. What pictures (or videos) and sounds would you include to represent our planet to another civilization?

G. Let's suppose Earth civilization has decided to broadcast a message announcing our existence to other possible civilizations among the stars. Your group is part of a large task force of scientists, communications specialists, and people from the humanities charged with deciding the form and content of our message. What would you recommend? Make a list of ideas.

H. Think of examples of contact with aliens you have seen in movies and on TV. Discuss with your group how realistic these have been, given what you have learned in this class. Was the contact in person (through traveling) or using messages? Why do you think Hollywood does so many shows and films that are not based on our scientific understanding of the universe?

I. Go through the Drake equation with your group and decide on values for each factor in the estimate. (If you disagree on what a factor should be within the group, you can have a "minority report.") Based on the factors, how many intelligent, communicating civilizations do you estimate to be thriving in our Galaxy right now?

Exercises

Review Questions

1. What is the Copernican principle? Make a list of scientific discoveries that confirm it.

2. Where in the solar system (and beyond) have scientists found evidence of organic molecules?

3. Give a short history of the atoms that are now in your little finger, going back to the beginning of the universe.

4. What is a biomarker? Give some possible examples of biomarkers we might look for beyond the solar system.

5. Why are Mars and Europa the top targets for the study of astrobiology?

6. Why is traveling between the stars (by creatures like us) difficult?

7. What are the advantages to using radio waves for communication between civilizations that live around different stars? List as many as you can.

8. What is the "cosmic haystack problem"? List as many of its components as you can think of.

9. What is a habitable zone?

10. Why is the simultaneous detection of methane and oxygen in an atmosphere a good indication of the existence of a biosphere on that planet?

11. What are two characteristic properties of life that distinguish it from nonliving things?

12. What are the three requirements that scientists believe an environment needs to supply life with in order to be considered habitable?

13. Can you name five environmental conditions that, in their extremes, microbial life has been challenged by and has learned to survive on Earth?

Thought Questions

14. Would a human have been possible during the first generation of stars that formed right after the Big Bang? Why or why not?

15. If we do find life on Mars, what might be some ways to check whether it formed separately from Earth life, or whether exchanges of material between the two planets meant that the two forms of life have a common origin?

16. What kind of evidence do you think would convince astronomers that an extraterrestrial spacecraft has landed on Earth?

17. What are some reasons that more advanced civilizations might want to send out messages to other star systems?

18. What are some answers to the Fermi paradox? Can you think of some that are not discussed in this chapter?

19. Why is there so little evidence of Earth's earliest history and therefore the period when life first began on our planet?

20. Why was the development of photosynthesis a major milestone in the evolution of life?

21. Does all life on Earth require sunshine?

22. Why is life unlikely to be found on the surface of Mars today?

23. In this chapter, we identify these characteristic properties of life: life extracts energy from its environment, and has a means of encoding and replicating information in order to make faithful copies of itself. Does this definition fully capture what we think of as "life"? How might our definition be biased by our terrestrial environment?

24. Given that no sunlight can penetrate Europa's ice shell, what would be the type of energy that could make some form of europan life possible?

25. Why is Saturn's moon Enceladus such an exciting place to send a mission?

26. In addition to an atmosphere dominated by nitrogen, how else is Saturn's moon Titan similar to Earth?

27. How can a planet's atmosphere affect the width of the habitable zone in its planetary system?

28. Why are we limited to finding life on planets orbiting other stars to situations where the biosphere has created planet-scale changes?

Figuring for Yourself

29. Suppose astronomers discover a radio message from a civilization whose planet orbits a star 35 light-years away. Their message encourages us to send a radio answer, which we decide to do. Suppose our governing bodies take 2 years to decide whether and how to answer. When our answer arrives there, their governing bodies also take two of our years to frame an answer to us. How long after we get their first message can we hope to get their reply to ours? (A question for further thinking: Once communication gets going, should we continue to wait for a reply before we send the next message?)

30. The light a planet receives from the Sun (per square meter of planet surface) decreases with the square of the distance from the Sun. So a planet that is twice as far from the Sun as Earth receives $(1/2)^2 = 0.25$ times (25%) as much light and a planet that is three times as far from the Sun receives $(1/3)^2 = 0.11$ times (11%) as much light. How much light is received by the moons of Jupiter and Saturn (compared to Earth), worlds which orbit 5.2 and 9.5 times farther from the Sun than Earth?

31. Think of our Milky Way Galaxy as a flat disk of diameter 100,000 light-years. Suppose we are one of 1000 civilizations, randomly distributed through the disk, interested in communicating via radio waves. How far away would the nearest such civilization be from us (on average)?

A   How to Study for an Introductory Astronomy Class

In this brief appendix, we want to give you some hints for the effective study of astronomy. These suggestions are based on ideas from good teachers and good students around the United States. Your professor will probably have other, more specific suggestions for doing well in your class.

Astronomy, the study of the universe beyond the borders of our planet, is one of the most exciting and rapidly changing branches of science. Even scientists from other fields often confess to having had a lifelong interest in astronomy, though they may now be doing something earthbound—like biology, chemistry, engineering, or writing software.

But some of the things that make astronomy so interesting also make it a challenge for the beginning student. The universe is a big place, full of objects and processes that do not have familiar counterparts here on Earth. Like a visitor to a new country, it will take you a while to feel familiar with the territory or the local customs. Astronomy, like other sciences, also has its own special vocabulary, some of which you will have to learn to communicate well with your professor and classmates.

Still, hundreds of thousands of non-science majors take an introductory astronomy course every year, and surveys show that students from a wide range of backgrounds have succeeded in (and even enjoyed) these classes. Astronomy is for everyone, not just those who are "science oriented."

So, here are some suggestions to help you increase your chances of doing well in your astronomy class.

1. The best advice we can give you is to be sure to leave enough time in your schedule to study the material in this class regularly. It sounds obvious, but it is not very easy to catch up with a subject like astronomy by trying to do everything just before an exam. (As astronomers like to put it, you can't learn the whole universe in one night!) Try to put aside some part of each day, or every other day, when you can have uninterrupted time for reading and studying astronomy.

2. In class, put your phone away and focus on the class activities. If you have to use a laptop or tablet in class, make a pact with yourself that you will *not* check email, get on social media, or play games during class. A number of careful studies of student behavior and grades have shown that students are not as good at such multi-tasking as they think they are, and that students who do *not* use screens during class get significantly better grades in the end.

3. Try to take careful notes during class. Many students start college without good note-taking habits. If you are not a good note-taker, try to get some help. Many colleges and universities have student learning centers that offer short courses, workbooks, tutors, or videos on developing good study habits. Good note-taking skills will also be useful for many jobs or activities you are likely get involved with after college.

4. Try to read each assignment in the textbook twice, once before it is discussed in class, and once afterward. You can highlight text or make notes in your copy of this electronic book: just move your cursor over some text and a box with highlight colors and room for notes will become visible.

5. Form a small astronomy study group with people in your class. Get together with them regularly and discuss what you have been learning. Also, focus on the topics that may be giving group members trouble. Make up sample exam questions and make sure everyone in the group can answer them confidently. If you have always studied alone, you may at first resist this idea, but don't be too hasty to say no. Study groups are a very effective way to digest a large amount of new information.

6. Before each exam, create a concise outline of the main ideas discussed in class and presented in your text. Compare your outline with those of other students as a check on your own study habits.

7. If your professor suggests doing web-based sample quizzes, or looking at online apps, animations, or study guides, take advantage of these resources to enhance your studying.

8. At the end of each chapter in this textbook you, will find four kinds of questions. The Collaborative Group Activities are designed to encourage you follow up on the material in the chapter as a group, rather than individually. Review Questions help you see if you have learned the material in the chapter. Thought

Questions test deeper understanding by asking you to apply your knowledge to new situations. And Figuring for Yourself exercises test and extend some of the mathematical examples in the chapter. (Not all professors will use the math sections; if they don't, you may not have homework from this section.)

9. If you find a topic in the text or in class especially difficult or interesting, talk to your professor or teaching assistant. Many students are scared to show their ignorance in front of their teacher, but we can assure you that most professors and TA's *like* it when students come to office hours and show that they care enough about the course to ask for help.

10. Don't stay up all night before a test and then expect your mind to respond well. For the same reason, don't eat a big meal just before a test, since we all get a little sleepy and don't think as clearly after a big meal. Take many deep breaths and try to relax during the test itself.

11. Don't be too hard on yourself! If astronomy is new to you, many of the ideas and terms in this book may be unfamiliar. Astronomy is like any new language: it may take a while to become a good conversationalist. Practice as much as you can, but also realize that it is natural to feel overwhelmed by the vastness of the universe and the variety of things that are going on in it.

B   Astronomy Websites, Images, and Apps

Throughout the textbook, we suggest useful resources for students on the specific topics in a given chapter. Here, we offer some websites for exploring astronomy in general, plus good sites for viewing and downloading the best astronomy images, and guides to astronomical apps for smartphones and tablets. This is not an exhaustive listing, but merely a series of suggestions to whet the appetite of those wanting to go beyond the textbook.

Websites for Exploring Astronomy in General

Astronomical Organizations

Amateur Astronomy Clubs. In most large cities and a number of rural areas, there are *amateur astronomy clubs*, where those interested in the hobby of astronomy gather to observe the sky, share telescopes, hear speakers, and help educate the public about the night sky. To find an astronomy club near you, you can try the following sites:

- Night Sky Network club finder: http://nightsky.jpl.nasa.gov/club-map.cfm (http://nightsky.jpl.nasa.gov/club-map.cfm).
- *Sky & Telescope* Magazine astronomy clubs and organizations: https://skyandtelescope.org/astronomy-clubs-organizations/ (https://skyandtelescope.org/astronomy-clubs-organizations/).
- *Astronomy* Magazine club finder: http://www.astronomy.com/groups.aspx (http://www.astronomy.com/groups.aspx).
- Astronomical League astronomy clubs and societies: http://www.astroleague.org/societies/all (http://www.astroleague.org/societies/all).
- Go-Astronomy club search: http://www.go-astronomy.com/astro-club-search.htm (http://www.go-astronomy.com/astro-club-search.htm).

American Astronomical Society: http://www.aas.org (http://www.aas.org). Composed mainly of professional astronomers. They have an active education office and various materials for students and the public on the education pages of their website.

Astronomical League: http://www.astroleague.org (http://www.astroleague.org). The league is the umbrella organization of American astronomy clubs. They offer a newsletter, national observing programs, and support for how to form and support a club.

Astronomical Society of the Pacific: http://www.astrosociety.org (http://www.astrosociety.org). Founded in 1889, this international society is devoted to astronomy education and outreach. They have programs, publications, and materials for families, teachers, amateur astronomers, museum guides, and anyone interested in astronomy.

European Space Agency (ESA): http://www.esa.int/ (http://www.esa.int/). Information on European space missions with an excellent gallery of images.

International Astronomical Union (IAU): http://www.iau.org/ (http://www.iau.org/). International organization for professional astronomers; see the menu choice "IAU for the Public" for information on naming astronomical objects and other topics of interest to students.

International Dark-Sky Association: http://www.darksky.org (http://www.darksky.org). Dedicated to combating light pollution, the encroachment of stray light that wastes energy and washes out the glories of the night sky.

NASA: http://www.nasa.gov (http://www.nasa.gov). NASA has a wide range of information on its many websites; the trick is to find what you need. Most space missions and NASA centers have their own sites.

Planetary Society: http://www.planetary.org (http://www.planetary.org). Founded by the late Carl Sagan and others, this group works to encourage planetary exploration and the search for life elsewhere. While much of their work is advocacy, they have some educational outreach too.

Royal Astronomical Society of Canada: http://www.rasc.ca/ (http://www.rasc.ca/). Unites professional and amateur astronomers around Canada; has 28 centers with local activities, plus national magazines and meetings.

Some Astronomical Publications Students Can Read

Astronomy Now: http://www.astronomynow.com/ (http://www.astronomynow.com/). A colorful British monthly, with excellent articles about astronomy, the history of astronomy, and stargazing.

Astronomy: http://www.astronomy.com (http://www.astronomy.com). Has the largest circulation of any magazine devoted to the universe and is designed especially for astronomy hobbyists and armchair astronomers.

Free Astronomy: http://www.astropublishing.com/ (http://www.astropublishing.com/). A new web-based publication, with European roots.

Scientific American: http://www.sciam.com (http://www.sciam.com). Offers one astronomy article about every other issue. These articles, a number of which are reproduced on their website, are at a slightly higher level, but—often being written by the astronomers who have done the work—are authoritative and current.

Sky & Telescope: http://skyandtelescope.com (http://skyandtelescope.com). An older and somewhat higher-level magazine for astronomy hobbyists. Many noted astronomers write for this publication.

Sky News: http://www.skynews.ca/ (http://www.skynews.ca/). A Canadian publication, featuring both astronomy and stargazing information. It also lists Canadian events for hobbyists.

StarDate: https://stardate.org/ (https://stardate.org/). Magazine that accompanies the brief radio program, with a useful website for beginners.

Sites that Cover Astronomy News

Portal to the Universe: http://www.portaltotheuniverse.org/ (http://www.portaltotheuniverse.org/). A site that gathers online astronomy and space news items, blogs, and pictures.

Science@NASA news stories and newscasts: http://science.nasa.gov/science-news/ (http://science.nasa.gov/science-news/). Well-written stories with, of course, a NASA focus.

Space.com: http://www.space.com/news/ (http://www.space.com/news/). A commercial site, but with wide coverage of space and astronomy news.

Universe Today: http://www.universetoday.com/ (http://www.universetoday.com/). Another commercial site, with good articles by science journalists, but a lot of ads.

Sites for Answering Astronomical Questions

Ask an Astrobiologist: http://astrobiology.nasa.gov/ask-an-astrobiologist/ (http://astrobiology.nasa.gov/ask-an-astrobiologist/). On this site from the National Astrobiology Institute at NASA, astronomer David Morrison answered questions about the search for life on other planets, the origin of life on Earth, and many other topics.

Ask an Astronomer at Lick Observatory: http://www.ucolick.org/~mountain/AAA/ (http://www.ucolick.org/~mountain/AAA/). Graduate students and staff members at this California observatory answered selected astronomy questions, particularly from high school students.

Ask an Astrophysicist: http://imagine.gsfc.nasa.gov/docs/ask_astro/ask_an_astronomer.html

(http://imagine.gsfc.nasa.gov/docs/ask_astro/ask_an_astronomer.html). Questions and answers at NASA's Laboratory for High-Energy Astrophysics focus on X-ray and gamma-ray astronomy, and such objects as black holes, quasars, and supernovae.

Ask an Infrared Astronomer: https://coolcosmos.ipac.caltech.edu/asks (https://coolcosmos.ipac.caltech.edu/asks). A site from the California Institute of Technology, with an archive focusing on infrared (heat-ray) astronomy and the discoveries it makes about cool objects in the universe. No longer taking new questions.

Ask the Astronomer: http://www.astronomycafe.net/qadir/qanda.html (http://www.astronomycafe.net/qadir/qanda.html). This site, run by astronomer Sten Odenwald, is no longer active, but lists 3001 answers to questions asked in the mid-1990s. They are nicely organized by topic.

Ask the Experts at PhysLink: http://www.physlink.com/Education/AskExperts/index.cfm (http://www.physlink.com/Education/AskExperts/index.cfm). Lots of physics questions answered, with some astronomy as well, at this physics education site. Most answers are by physics teachers, not astronomers. Still taking new questions.

Ask the Space Scientist: http://image.gsfc.nasa.gov/poetry/ask/askmag.html (http://image.gsfc.nasa.gov/poetry/ask/askmag.html). An archive of questions about the Sun and its interactions with Earth, answered by astronomer Sten Odenwald. Not accepting new questions.

Curious about Astronomy?: http://curious.astro.cornell.edu (http://curious.astro.cornell.edu). An ask-an-astronomer site run by graduate students and professors of astronomy at Cornell University. Has searchable archives and is still answering new questions.

Miscellaneous Sites of Interest

A Guide to Careers in Astronomy: http://aas.org/files/resources/Careers-in-Astronomy.pdf (http://aas.org/files/resources/Careers-in-Astronomy.pdf). From the American Astronomical Society.

Astronomical Pseudo-Science: A Skeptic's Resource List: http://bit.ly/pseudoastro (http://bit.ly/pseudoastro). Readings and websites that analyze such claims as astrology, UFOs, moon-landing denial, creationism, human faces on other worlds, astronomical disasters, and more.

Astronomy for Beginners: http://www.skyandtelescope.com/astronomy-information/ (http://www.skyandtelescope.com/astronomy-information/). A page to find resources for getting into amateur astronomy.

Black Lives in Astronomy: http://bit.ly/blackastro (http://bit.ly/blackastro).

Contributions of Women to Astronomy: http://bit.ly/astronomywomen (http://bit.ly/astronomywomen).

Music Inspired by Astronomy: http://bit.ly/astronomymusic (http://bit.ly/astronomymusic).

Science Fiction Stories with Good Astronomy and Physics: http://bit.ly/astroscifi (http://bit.ly/astroscifi).

Astronomy Calendar: http://bit.ly/astrodates (http://bit.ly/astrodates). Entitled "This Day in Astronomical History," this list provides 158 astronomical anniversaries or events that you can celebrate.

Unheard Voices: The Astronomy of Many Cultures: https://astrosociety.org/education-outreach/resource-guides/multicultural-astronomy.html (https://astrosociety.org/education-outreach/resource-guides/multicultural-astronomy.html). A guide to resources about the astronomy of native, African, Asian, and other non-Western groups.

Videos To Go with Each Chapter of this Textbook: http://bit.ly/shortastronomyvideos (http://bit.ly/shortastronomyvideos).

Selected Websites for Viewing and Downloading Astronomical Images

The Top Image Sites

Astronomy Picture of the Day: http://antwrp.gsfc.nasa.gov/apod/lib/aptree.html (http://antwrp.gsfc.nasa.gov/apod/lib/aptree.html). Two space scientists scour the internet and feature one interesting astronomy image each day.

European Southern Observatory Photo Gallery: http://www.eso.org/public/images/ (http://www.eso.org/public/images/). Magnificent color images from ESO's largest telescopes. See the topical menu at the top.

Hubble Space Telescope Images: http://hubblesite.org/images/gallery (http://hubblesite.org/images/gallery). Starting at this page, you can select from among many hundreds of Hubble pictures by subject. Other ways to approach these images are through the more public-oriented Hubble Gallery (http://hubblesite.org/gallery/ (http://hubblesite.org/gallery/)) or the European ESA site (http://www.spacetelescope.org/images/ (http://www.spacetelescope.org/images/)).

National Optical Astronomy Observatories Image Gallery: http://www.noao.edu/image_gallery/ (http://www.noao.edu/image_gallery/). Growing archive of images from the many telescopes that are at the United States' National Observatories.

Planetary Photojournal: http://photojournal.jpl.nasa.gov/index.html (http://photojournal.jpl.nasa.gov/index.html). Features thousands of images from NASA's extensive set of planetary exploration missions with a good search menu. Does not include most of the missions from other countries.

The World at Night: http://www.twanight.org/newTWAN/index.asp (http://www.twanight.org/newTWAN/index.asp). Dramatic night-sky images by professional photographers who are amateur astronomers. Note that while many of the astronomy sites allow free use of their images, these are copyrighted by photographers who make their living selling them.

Other Useful General Galleries

Anglo-Australian Observatory: https://images.datacentral.org.au/ (https://images.datacentral.org.au/). Great copyrighted color images by leading astro-photographer David Malin and others.

Canada-France-Hawaii Telescope: http://www.cfht.hawaii.edu/HawaiianStarlight/images.html (http://www.cfht.hawaii.edu/HawaiianStarlight/images.html). Remarkable color images from a major telescope on top of the Maunakea peak in Hawaii.

European Space Agency Gallery: http://www.esa.int/spaceinimages/Images (http://www.esa.int/spaceinimages/Images). Access images from such missions as Mars Express, Rosetta, and Herschel.

Gemini Observatory Images: http://www.gemini.edu/index.php?option=com_gallery (http://www.gemini.edu/index.php?option=com_gallery). Images from a pair of large telescopes in the northern and the southern hemispheres.

Isaac Newton Group of Telescopes Image Gallery: http://www.ing.iac.es/PR/images_index.html (http://www.ing.iac.es/PR/images_index.html). Beautiful images from the Herschel, Newton, and Kapteyn telescopes on La Palma.

National Radio Astronomy Observatory Image Gallery: https://public.nrao.edu/gallery/ (https://public.nrao.edu/gallery/). Organized by topic, the images show objects and processes that give off radio waves.

Our Infrared World Gallery: https://coolcosmos.ipac.caltech.edu/infrared_galleries (https://coolcosmos.ipac.caltech.edu/infrared_galleries). Images from a variety of infrared astronomy telescopes and missions. See also their "Cool Cosmos" site for the public: http://coolcosmos.ipac.caltech.edu/ (http://coolcosmos.ipac.caltech.edu/).

Some Galleries on Specific Subjects

Astronaut Photography of Earth: http://eol.jsc.nasa.gov/ (http://eol.jsc.nasa.gov/).

Chandra X-Ray Observatory Images: http://chandra.harvard.edu/photo/category.html (http://chandra.harvard.edu/photo/category.html).

NASA Human Spaceflight Gallery: https://www.flickr.com/photos/nasa2explore (https://www.flickr.com/photos/nasa2explore) or https://spaceflight.nasa.gov/gallery/index.html (https://spaceflight.nasa.gov/gallery/index.html). Astronaut images.

Robert Gendler: http://www.robgendlerastropics.com/ (http://www.robgendlerastropics.com/). One of the amateur astro-photographers who comes closest to being professional.

Sloan Digital Sky Survey Images: http://www.sdss.org/gallery/ (http://www.sdss.org/gallery/).

Solar Dynamics Observatory Gallery: http://sdo.gsfc.nasa.gov/gallery/main (http://sdo.gsfc.nasa.gov/gallery/main). Sun images.

Spitzer Infrared Telescope Images: http://www.spitzer.caltech.edu/images (http://www.spitzer.caltech.edu/images).

Astronomy Apps for Smartphones and Tablets

The Best Astronomy Apps for IOS and Android: https://www.digitaltrends.com/mobile/best-astronomy-apps/ (https://www.digitaltrends.com/mobile/best-astronomy-apps/).

The Best Stargazing Apps for Looking at the Night Sky: https://www.tomsguide.com/round-up/best-stargazing-apps (https://www.tomsguide.com/round-up/best-stargazing-apps).

The 19 Best Astronomy Apps for Stargazing: https://astrobackyard.com/astronomy-apps-for-stargazing/ (https://astrobackyard.com/astronomy-apps-for-stargazing/).

The 10 Best Astronomy Apps for Enjoying the Night Sky: https://www.makeuseof.com/tag/great-android-astronomy-apps/ (https://www.makeuseof.com/tag/great-android-astronomy-apps/).

The Best Astronomy Apps for Smartphones, Tablets, and Computers: https://www.thoughtco.com/best-astronomy-apps-4160999 (https://www.thoughtco.com/best-astronomy-apps-4160999).

C | Scientific Notation

In astronomy (and other sciences), it is often necessary to deal with very large or very small numbers. In fact, when numbers become truly large in everyday life, such as the national debt in the United States, we call them astronomical. Among the ideas astronomers must routinely deal with is that the Earth is 150,000,000,000 meters from the Sun, and the mass of the hydrogen atom is 0.00000000000000000000000000167 kilograms. No one in his or her right mind would want to continue writing so many zeros!

Instead, scientists have agreed on a kind of shorthand notation, which is not only easier to write, but (as we shall see) makes multiplication and division of large and small numbers much less difficult. If you have never used this powers-of-ten notation or scientific notation, it may take a bit of time to get used to it, but you will soon find it much easier than keeping track of all those zeros.

Writing Large Numbers

In scientific notation, we generally agree to have only one number to the left of the decimal point. If a number is not in this format, it must be changed. The number 6 is already in the right format, because for integers, we understand there to be a decimal point to the right of them. So 6 is really 6., and there is indeed only one number to the left of the decimal point. But the number 965 (which is 965.) has three numbers to the left of the decimal point, and is thus ripe for conversion.

To change 965 to proper form, we must make it 9.65 and then keep track of the change we have made. (Think of the number as a weekly salary and suddenly it makes a lot of difference whether we have $965 or $9.65.) We keep track of the number of places we moved the decimal point by expressing it as a power of ten. So 965 becomes 9.65×10^2 or 9.65 multiplied by ten to the second power. The small raised 2 is called an exponent, and it tells us how many times we moved the decimal point to the left.

Note that 10^2 also designates 10 squared, or 10×10, which equals 100. And 9.65×100 is just 965, the number we started with. Another way to look at scientific notation is that we separate out the messy numbers out front, and leave the smooth units of ten for the exponent to denote. So a number like 1,372,568 becomes 1.372568 times a million (10^6) or 1.372568 times 10 multiplied by itself 6 times. We had to move the decimal point six places to the left (from its place after the 8) to get the number into the form where there is only one digit to the left of the decimal point.

The reason we call this powers-of-ten notation is that our counting system is based on increases of ten; each place in our numbering system is ten times greater than the place to the right of it. As you have probably learned, this got started because human beings have ten fingers and we started counting with them. (It is interesting to speculate that if we ever meet intelligent life-forms with only eight fingers, their counting system would probably be a powers-of-eight notation!)

So, in the example we started with, the number of meters from Earth to the Sun is 1.5×10^{11}. Elsewhere in the book, we mention that a string 1 light-year long would fit around Earth's equator 236 million or 236,000,000 times. In scientific notation, this would become 2.36×10^8. Now if you like expressing things in millions, as the annual reports of successful companies do, you might like to write this number as 236×10^6. However, the usual convention is to have only one number to the left of the decimal point.

Writing Small Numbers

Now take a number like 0.00347, which is also not in the standard (agreed-to) form for scientific notation. To put it into that format, we must make the first part of it 3.47 by moving the decimal point three places *to the right*. Note that this motion to the right is the opposite of the motion to the left that we discussed above. To keep track, we call this change negative and put a minus sign in the exponent. Thus 0.00347 becomes 3.47×10^{-3}.

In the example we gave at the beginning, the mass of the hydrogen atom would then be written as 1.67×10^{-27} kg. In this system, one is written as 10^0, a tenth as 10^{-1}, a hundredth as 10^{-2}, and so on. Note that any number, no matter how large or how small, can be expressed in scientific notation.

Multiplication and Division

Scientific notation is not only compact and convenient, it also simplifies arithmetic. To multiply two numbers expressed as powers of ten, you need only multiply the numbers out front and then *add* the exponents. If there are no numbers out front, as in 100 × 100,000, then you just add the exponents (in our notation, $10^2 \times 10^5$ = 10^7). When there are numbers out front, you have to multiply them, but they are much easier to deal with than numbers with many zeros in them.

Here's an example:

$$\left(3 \times 10^5\right) \times \left(2 \times 10^9\right) = 6 \times 10^{14}$$

And here's another example:

$$
\begin{aligned}
0.04 \times 6,000,000 &= \left(4 \times 10^{-2}\right) \times \left(6 \times 10^6\right) \\
&= 24 \times 10^4 \\
&= 2.4 \times 10^5
\end{aligned}
$$

Note in the second example that when we added the exponents, we treated negative exponents as we do in regular arithmetic (–2 plus 6 equals 4). Also, notice that our first result had a 24 in it, which was not in the acceptable form, having two places to the left of the decimal point, and we therefore changed it to 2.4 and changed the exponent accordingly.

To divide, you divide the numbers out front and *subtract* the exponents. Here are several examples:

$$
\begin{aligned}
\frac{1,000,000}{1000} &= \frac{10^6}{10^3} = 10^{(6-3)} = 10^3 \\
\frac{9 \times 10^{12}}{2 \times 10^3} &= 4.5 \times 10^9 \\
\frac{2.8 \times 10^2}{6.2 \times 10^5} &= 0.452 \times 10^{-3} = 4.52 \times 10^{-4}
\end{aligned}
$$

In the last example, our first result was not in the standard form, so we had to change 0.452 into 4.52, and change the exponent accordingly.

If this is the first time that you have met scientific notation, we urge you to practice many examples using it. You might start by solving the exercises below. Like any new language, the notation looks complicated at first but gets easier as you practice it.

Exercises

1. At the end of September, 2015, the New Horizons spacecraft (which encountered Pluto for the first time in July 2015) was 4.898 billion km from Earth. Convert this number to scientific notation. How many astronomical units is this? (An astronomical unit is the distance from Earth to the Sun, or about 150 million km.)
2. During the first six years of its operation, the Hubble Space Telescope circled Earth 37,000 times, for a total of 1,280,000,000 km. Use scientific notation to find the number of km in one orbit.
3. In a large university cafeteria, a soybean-vegetable burger is offered as an alternative to regular hamburgers. If 889,875 burgers were eaten during the course of a school year, and 997 of them were veggie-burgers, what fraction and what percent of the burgers does this represent?
4. In a 2012 Kelton Research poll, 36 percent of adult Americans thought that alien beings have actually landed on Earth. The number of adults in the United States in 2012 was about 222,000,000. Use scientific notation to determine how many adults believe aliens have visited Earth.

5. In the school year 2009–2010, American colleges and universities awarded 2,354,678 degrees. Among these were 48,069 PhD degrees. What fraction of the degrees were PhDs? Express this number as a percent. (Now go and find a job for all those PhDs!)

6. A star 60 light-years away has been found to have a large planet orbiting it. Your uncle wants to know the distance to this planet in old-fashioned miles. Assume light travels 186,000 miles per second, and there are 60 seconds in a minute, 60 minutes in an hour, 24 hours in a day, and 365 days in a year. How many miles away is that star?

Answers

1. 4.898 billion is 4.898×10^9 km. One astronomical unit (AU) is 150 million km = 1.5×10^8 km. Dividing the first number by the second, we get $3.27 \times 10^{(9-8)} = 3.27 \times 10^1$ AU.

2. $\dfrac{1.28 \times 10^9 \text{ km}}{3.7 \times 10^4 \text{ orbits}} = 0.346 \times 10^{(9-4)} = 0.346 \times 10^5 = 3.46 \times 10^4$ km per orbit.

3. $\dfrac{9.97 \times 10^2 \text{ veggie burgers}}{8.90 \times 10^5 \text{ total burgers}} = 1.12 \times 10^{(2-5)} = 1.12 \times 10^{(2-5)} = 1.12 \times 10^{-3}$ (or roughly about one thousandth) of the burgers were vegetarian. Percent means per hundred. So

   $\dfrac{1.12 \times 10^{-3}}{10^{-2}} = 1.12 \times 10^{(-3-(-2))} = 1.12 \times 10^{-1}$ percent (which is roughly one tenth of one percent).

4. 36% is 36 hundredths or 0.36 or 3.6×10^{-1}. Multiply that by 2.22×10^8 and you get about $7.99 \times 10^{(-1+8)} = 7.99 \times 10^7$ or almost 80 million people who believe that aliens have landed on our planet. We need more astronomy courses to educate all those people.

5. $\dfrac{4.81 \times 10^4}{2.35 \times 10^6} = 2.05 \times 10^{(4-6)} = 2.05 \times 10^{-2} =$ about 2%. (Note that in these examples we are rounding off some of the numbers so that we don't have more than 2 places after the decimal point.)

6. One light-year is the distance that light travels in one year. (Usually, we use metric units and not the old British system that the United States is still using, but we are going to humor your uncle and stick with miles.) If light travels 186,000 miles every second, then it will travel 60 times that in a minute, and 60 times that in an hour, and 24 times that in a day, and 365 times that in a year. So we have $1.86 \times 10^5 \times 6.0 \times 10^1 \times 6.0 \times 10^1 \times 2.4 \times 10^1 \times 3.65 \times 10^2$. So we multiply all the numbers out front together and add all the exponents. We get $586.57 \times 10^{10} = 5.86 \times 10^{12}$ miles in a light year (which is roughly 6 trillion miles—a heck of a lot of miles). So if the star is 60 light-years away, its distance in miles is $6 \times 10^1 \times 5.86 \times 10^{12} = 35.16 \times 10^{13} = 3.516 \times 10^{14}$ miles.

D | Units Used in Science

In the American system of measurement (originally developed in England), the fundamental units of length, weight, and time are the foot, pound, and second, respectively. There are also larger and smaller units, which include the ton (2240 lb), the mile (5280 ft), the rod (16 1/2 ft), the yard (3 ft), the inch (1/12 ft), the ounce (1/16 lb), and so on. Such units, whose origins in decisions by British royalty have been forgotten by most people, are quite inconvenient for conversion or doing calculations.

In science, therefore, it is more usual to use the *metric system,* which has been adopted in virtually all countries except the United States. Its great advantage is that every unit increases by a factor of ten, instead of the strange factors in the American system. The fundamental units of the metric system are:

- length: 1 meter (m)
- mass: 1 kilogram (kg)
- time: 1 second (s)

A meter was originally intended to be 1 ten-millionth of the distance from the equator to the North Pole along the surface of Earth. It is about 1.1 yd. A kilogram is the mass that on Earth results in a weight of about 2.2 lb. The second is the same in metric and American units.

Length

The most commonly used quantities of length of the metric system are the following.

Conversions

1 kilometer (km) = 1000 meters = 0.6214 mile

1 meter (m) = 0.001 km = 1.094 yards = 39.37 inches

1 centimeter (cm) = 0.01 meter = 0.3937 inch

1 millimeter (mm) = 0.001 meter = 0.1 cm

1 micrometer (μm) = 0.000001 meter = 0.0001 cm

1 nanometer (nm) = 10^{-9} meter = 10^{-7} cm

Table D1 Length

To convert from the American system, here are a few helpful factors:

- 1 mile = 1.61 km
- 1 inch = 2.54 cm

Mass

Although we don't make the distinction very carefully in everyday life on Earth, strictly speaking the kilogram is a unit of mass (measuring the quantity of matter in a body, roughly how many atoms it has,) while the pound is a unit of weight (measuring how strongly Earth's gravity pulls on a body).

The most commonly used quantities of mass of the metric system are the following.

Conversions

1 metric ton = 10^6 grams = 1000 kg (and it produces a weight of 2.205×10^3 lb on Earth)

1 kg = 1000 grams (and it produces a weight of 2.2046 lb on Earth)

1 gram (g) = 0.0353 oz (and the equivalent weight is 0.002205 lb)

1 milligram (mg) = 0.001 g

Table D2 Mass

A weight of 1 lb is equivalent on Earth to a mass of 0.4536 kg, while a weight of 1 oz is produced by a mass of 28.35 g.

Temperature

Three temperature scales are in general use:

- Fahrenheit (F); water freezes at 32 °F and boils at 212 °F.
- Celsius or centigrade[1] (C); water freezes at 0 °C and boils at 100 °C.
- Kelvin or absolute (K); water freezes at 273 K and boils at 373 K.

All molecular motion ceases at about −459 °F = −273 °C = 0 K, a temperature called *absolute zero*. Kelvin temperature is measured from this lowest possible temperature, and it is the temperature scale most often used in astronomy. Kelvins have the same value as centigrade or Celsius degrees, since the difference between the freezing and boiling points of water is 100 degrees in each. (Note that we just say "kelvins," not kelvin degrees.)

On the Fahrenheit scale, the difference between the freezing and boiling points of water is 180 degrees. Thus, to convert Celsius degrees or kelvins to Fahrenheit degrees, it is necessary to multiply by 180/100 = 9/5. To convert from Fahrenheit degrees to Celsius degrees or kelvins, it is necessary to multiply by 100/180 = 5/9.

The full conversion formulas are:

- K = °C + 273
- °C = 0.555 × (°F − 32)
- °F = (1.8 × °C) + 32

---

1 Celsius is now the name used for centigrade temperature; it has a more modern standardization but differs from the old centigrade scale by less than 0.1°.

E   Some Useful Constants for Astronomy

Physical Constants

Name	Value
speed of light (c)	2.9979×10^8 m/s
gravitational constant (G)	6.674×10^{-11} m^3/(kg s^2)
Planck's constant (h)	6.626×10^{-34} J-s
mass of a hydrogen atom (M_H)	1.673×10^{-27} kg
mass of an electron (M_e)	9.109×10^{-31} kg
Rydberg constant (R_∞)	1.0974×10^7 m^{-1}
Stefan-Boltzmann constant (σ)	5.670×10^{-8} J/(s·m^2 deg^4)[1]
Wien's law constant ($\lambda_{max}T$)	2.898×10^{-3} m K
electron volt (energy) (eV)	1.602×10^{-19} J
energy equivalent of 1 ton TNT	4.2×10^9 J

Table E1

Astronomical Constants

Name	Value
astronomical unit (AU)	1.496×10^{11} m
Light-year (ly)	9.461×10^{15} m
parsec (pc)	3.086×10^{16} m = 3.262 light-years
sidereal year (y)	3.156×10^7 s
mass of Earth (M_{Earth})	5.974×10^{24} kg
equatorial radius of Earth (R_{Earth})	6.378×10^6 m

Table E2

---

1   deg stands for degrees Celsius or kelvins

Name	Value
obliquity of ecliptic	23° 26'
surface gravity of Earth (g)	9.807 m/s^2
escape velocity of Earth (v_{Earth})	1.119 × 10^4 m/s
mass of Sun (M_{Sun})	1.989 × 10^{30} kg
equatorial radius of Sun (R_{Sun})	6.960 × 10^8 m
luminosity of Sun (L_{Sun})	3.85 × 10^{26} W
solar constant (flux of energy received at Earth) (S)	1.368 × 10^3 W/m^2
Hubble constant (H_0)	approximately 20 km/s per million light-years, or approximately 70 km/s per megaparsec

Table E2

F   Physical and Orbital Data for the Planets

Physical Data for the Major Planets

Major Planet	Mean Diameter (km)	Mean Diameter (Earth = 1)	Mass (Earth = 1)	Mean Density (g/cm³)	Rotation Period (d)	Inclination of Equator to Orbit (°)	Surface Gravity (Earth = 1[g])	Velocity of Escape (km/s)
Mercury	4879	0.38	0.055	5.43	58.	0.0	0.38	4.3
Venus	12,104	0.95	0.815	5.24	−243.	177	0.90	10.4
Earth	12,756	1.00	1.00	5.51	1.000	23.4	1.00	11.2
Mars	6779	0.53	0.11	3.93	1.026	25.2	0.38	5.0
Jupiter	140,000	10.9	318	1.33	0.414	3.1	2.53	60.
Saturn	117,000	9.13	95.2	0.69	0.440	26.7	1.07	36.
Uranus	50,700	3.98	14.5	1.27	−0.718	97.9	0.89	21.
Neptune	49,200	3.86	17.2	1.64	0.671	29.6	1.14	23.

Table F1

Physical Data for Well-Studied Dwarf Planets

Well-Studied Dwarf Planet	Diameter (km)	Diameter (Earth = 1)	Mass (Earth = 1)	Mean Density (g/cm³)	Rotation Period (d)	Inclination of Equator to Orbit (°)	Surface Gravity (Earth = 1[g])	Velocity of Escape (km/s)
Ceres	950	0.07	0.0002	2.2	0.378	3	0.03	0.5
Pluto	2470	0.18	0.0024	1.9	−6.387	122	0.06	1.3
Haumea	1700	0.13	0.0007	3	0.163	—	—	0.8
Makemake	1400	0.11	0.0005	2	0.321	—	—	0.8
Eris	2326	0.18	0.0028	2.5	1.25[1]	—	—	1.4

Table F2

---

[1] This measurement is quite uncertain.

Orbital Data for the Major Planets

Major Planet	Semimajor Axis (AU)	Semimajor Axis (10^6 km)	Sidereal Period (y)	Sidereal Period (d)	Mean Orbital Speed (km/s)	Orbital Eccentricity	Inclination of Orbit to Ecliptic (°)
Mercury	0.39	58	0.24	88.0	47.9	0.206	7.0
Venus	0.72	108	0.6	224.7	35.0	0.007	3.4
Earth	1.00	149	1.00	365.2	29.8	0.017	0.0
Mars	1.52	228	1.88	687.0	24.1	0.093	1.9
Jupiter	5.20	778	11.86	—	13.1	0.048	1.3
Saturn	9.54	1427	29.46	—	9.6	0.056	2.5
Uranus	19.19	2871	84.01	—	6.8	0.046	0.8
Neptune	30.06	4497	164.82	—	5.4	0.010	1.8

Table F3

Orbital Data for Well-Studied Dwarf Planets

Well-Studied Dwarf Planet	Semimajor Axis (AU)	Semimajor Axis (10^6 km)	Sidereal Period (y)	Mean Orbital Speed (km/s)	Orbital Eccentricity	Inclination of Orbit to Ecliptic (°)
Ceres	2.77	414.0	4.6	18	0.08	11
Pluto	39.5	5915	248.6	4.7	0.25	17
Haumea	43.1	6452	283.3	4.5	0.19	28
Makemake	45.8	6850	309.9	4.4	0.16	29
Eris	68.0	10,120	560.9	3.4	0.44	44

Table F4

G | Selected Moons of the Planets

Note: As this book goes to press, nearly two hundred moons are now known in the solar system and more are being discovered on a regular basis. Of the major planets, only Mercury and Venus do not have moons. In addition to moons of the planets, there are many moons of asteroids. In this appendix, we list only the largest and most interesting objects that orbit each planet (including dwarf planets). The number given for each planet is discoveries through 2015. For further information see https://solarsystem.nasa.gov/planets/solarsystem/moons (https://solarsystem.nasa.gov/planets/solarsystem/moons) and https://en.wikipedia.org/wiki/List_of_natural_satellites (https://en.wikipedia.org/wiki/List_of_natural_satellites).

Selected Moons of the Planets

Planet (moons)	Satellite Name	Discovery	Semimajor Axis (km × 1000)	Period (d)	Diameter (km)	Mass (10^{20} kg)	Density (g/cm³)
Earth (1)	Moon	—	384	27.32	3476	735	3.3
Mars (2)	Phobos	Hall (1877)	9.4	0.32	23	1×10^{-4}	2.0
	Deimos	Hall (1877)	23.5	1.26	13	2×10^{-5}	1.7
Jupiter (79)	Amalthea	Barnard (1892)	181	0.50	200	—	—
	Thebe	Voyager (1979)	222	0.67	90	—	—
	Io	Galileo (1610)	422	1.77	3630	894	3.6
	Europa	Galileo (1610)	671	3.55	3138	480	3.0
	Ganymede	Galileo (1610)	1070	7.16	5262	1482	1.9
	Callisto	Galileo (1610)	1883	16.69	4800	1077	1.9
	Himalia	Perrine (1904)	11,460	251	170	—	—
Saturn (82)	Pan	Voyager (1985)	133.6	0.58	20	3×10^{-5}	—

Table G1

Planet (moons)	Satellite Name	Discovery	Semimajor Axis (km × 1000)	Period (d)	Diameter (km)	Mass (10^{20} kg)	Density (g/cm^3)
	Atlas	Voyager (1980)	137.7	0.60	40	—	—
	Prometheus	Voyager (1980)	139.4	0.61	80	—	—
	Pandora	Voyager (1980)	141.7	0.63	100	—	—
	Janus	Dollfus (1966)	151.4	0.69	190	—	—
	Epimetheus	Fountain, Larson (1980)	151.4	0.69	120	—	—
	Mimas	Herschel (1789)	186	0.94	394	0.4	1.2
	Enceladus	Herschel (1789)	238	1.37	502	0.8	1.2
	Tethys	Cassini (1684)	295	1.89	1048	7.5	1.3
	Dione	Cassini (1684)	377	2.74	1120	11	1.3
	Rhea	Cassini (1672)	527	4.52	1530	25	1.3
	Titan	Huygens (1655)	1222	15.95	5150	1346	1.9
	Hyperion	Bond, Lassell (1848)	1481	21.3	270	—	—
	Iapetus	Cassini (1671)	3561	79.3	1435	19	1.2
	Phoebe	Pickering (1898)	12,950	550 (R)[1]	220	—	—

Table G1

Planet (moons)	Satellite Name	Discovery	Semimajor Axis (km × 1000)	Period (d)	Diameter (km)	Mass (10^{20} kg)	Density (g/cm³)
Uranus (27)	Puck	Voyager (1985)	86.0	0.76	170	—	—
	Miranda	Kuiper (1948)	130	1.41	485	0.8	1.3
	Ariel	Lassell (1851)	191	2.52	1160	13	1.6
	Umbriel	Lassell (1851)	266	4.14	1190	13	1.4
	Titania	Herschel (1787)	436	8.71	1610	35	1.6
	Oberon	Herschel (1787)	583	13.5	1550	29	1.5
Neptune (14)	Despina	Voyager (1989)	53	0.33	150	—	—
	Galatea	Voyager (1989)	62	0.40	150	—	—
	Larissa	Reitsema, et al (1981)	74	0.55	194	—	—
	Proteus	Voyager (1989)	118	1.12	420	—	—
	Triton	Lassell (1846)	355	5.88 (R)[2]	2720	220	2.1
	Nereid	Kuiper (1949)	5511	360	340	—	—
Pluto (5)	Charon	Christy (1978)	19.7	6.39	1200	—	1.7
	Styx	Showalter et al (2012)	42	20	20	—	—

Table G1

---

1   R stands for retrograde rotation (backward from the direction that most objects in the solar system revolve and rotate).
2   R stands for retrograde rotation (backward from the direction that most objects in the solar system revolve and rotate).

Planet (moons)	Satellite Name	Discovery	Semimajor Axis (km × 1000)	Period (d)	Diameter (km)	Mass (10^{20} kg)	Density (g/cm^3)
	Nix	Weaver et al (2005)	48	24	46	—	2.1
	Kerberos	Showalter et al (2011)	58	24	28	—	1.4
	Hydra	Weaver et al (2005)	65	38	61	—	0.8
Eris (1)	Dysnomea	Brown et al (2005)	38	16	684	—	—
Makemake (1)	(MK2)	Parker et al (2016)	—	—	160	—	—
Haumea (2)	Hi'iaka	Brown et al (2005)	50	49	400	—	—
	Namaka	Brown et al (2005)	39	35	200	—	—

Table G1

H   Future Total Eclipses

Future Total Solar Eclipses

We also include eclipses that are *annular*—where the Moon is directly in front of the Sun, but doesn't fully cover it—leaving a ring of light around the dark Moon's edges)

Future Total Solar Eclipses

Date	Type of Eclipse	Location on Earth[1]
June 10, 2021	Annular	N Canada, Greenland
December 4, 2021	Total	Only in Antarctica
April 20, 2023	Total[2]	Mostly in Indian and Pacific oceans, Indonesia
October 14, 2023	Annular	OR, NV, UT, NM, TX, C America, Colombia, Brazil
April 8, 2024	Total	N Mexico, U.S. (TX to ME), SE Canada and oceans on either side
October 2, 2024	Annular	S Chile, S Argentina, and oceans on either side
February 17, 2026	Annular	Only in Antarctica
August 12, 2026	Total	Greenland, Iceland, Spain
February 6, 2027	Annular	S Pacific, Argentina, Chile, Uruguay, S Atlantic
August 2, 2027	Total	Spain, Morocco, Egypt, Saudi Arabia, Yemen, Arabian Sea
January 26, 2028	Annular	Ecuador, Peru, Brazil, North Atlantic Ocean, Portugal, Spain
July 22, 2028	Total	Indian Ocean, Australia, New Zealand, South Pacific Ocean

Table H1

Future Total Lunar Eclipses

Future Total Lunar Eclipses

Date	Location on Earth
May 26, 2021	E Asia, Australia, Pacific Ocean, W North America, W South America

Table H2

---

1   Remember that a total or annular eclipse is only visible on a narrow track. The same eclipse will be partial over a much larger area, but partial eclipses are not as spectacular as total ones.
2   This is a so-called hybrid eclipse, which is total in some places and annular in others.

Date	Location on Earth
May 16, 2022	N America, S America, Europe, Africa
November 8, 2022	Asia, Australia, Pacific Ocean, N America, S America
March 14, 2025	Pacific Ocean, N America, S America, Atlantic Ocean, W Europe, W Africa
September 7, 2025	Europe, Africa, Asia, Australia, Indian Ocean
March 3, 2026	E Asia, Australia, Pacific Ocean, N America, C America
June 26, 2029	E North America, S America, Atlantic Ocean, W Europe, W Africa
December 20, 2029	E North America, E South America, Atlantic Ocean, Europe, Africa, Asia

Table H2

Additional Resources

For more information and detailed maps about eclipses, see these resources.

- NASA's Eclipse Site: http://eclipse.gsfc.nasa.gov/ (http://eclipse.gsfc.nasa.gov/)
- Mr. Eclipse site for beginners by Dr. Fred Espenak: http://www.mreclipse.com/ (http://www.mreclipse.com/)
- Eclipse Weather and Maps by Meteorologist Jay Anderson: http://eclipsophile.com (http://eclipsophile.com)
- Eclipse Maps by Michael Zeiler: http://www.eclipse-maps.com/Eclipse-Maps/Welcome.html (http://www.eclipse-maps.com/Eclipse-Maps/Welcome.html)
- Eclipse Information and Maps by Xavier Jubier: http://xjubier.free.fr/en/site_pages/eclipses.html (http://xjubier.free.fr/en/site_pages/eclipses.html)

I   The Nearest Stars, Brown Dwarfs, and White Dwarfs

The Nearest Stars, Brown Dwarfs, and White Dwarfs

Star	System	Discovery Name	Distance (light-year)	Spectral Type	Location: RA[1]	Location: Dec[2]	Luminosity (Sun = 1)
		Sun	—	G2 V	—	—	1
1	1	Proxima Centauri	4.2	M5.5 V	14 29	–62 40	5×10^{-5}
2	2	Alpha Centauri A	4.4	G2 V	14 39	–60 50	1.5
3		Alpha Centauri B	4.4	K2 IV	14 39	–60 50	0.5
4	3	Barnard's Star	6.0	M4 V	17 57	+04 42	4.4×10^{-4}
5	4	Luhman 16A	6.5	L8	10 49	–53 19	
6		Luhman 16B	6.5	T1	10 40	–53 19	
7	5	WISE 0855-0714	7.3	Y2	08 55	–07 15	
8	4	Wolf 359	7.8	M6 V	10 56	+07 00	2×10^{-5}
9	5	Lalande 21 185	8.3	M2 V	11 03	+35 58	5.7×10^{-3}
10	6	Sirius A	8.6	A1 V	06 45	–16 42	23.1
11		Sirius B	8.6	DA2[3]	06 45	–16 43	2.5×10^{-3}
12	7	Luyten 726-8 A	8.7	M5.5 V	01 39	–17 57	6×10^{-5}
13		Luyten 726-8 B (UV Ceti)	8.7	M6 V	01 39	–17 57	4×10^{-5}
14	8	Ross 154	9.7	M3.5 V	18 49	–23 50	5×10^{-4}
15	9	Ross 248 (HH Andromedae)	10.3	M5.5 V	23 41	+44 10	1.0×10^{-4}
16	10	Epsilon Eridani	10.5	K2 V	03 32	–09 27	0.29
17	11	Lacaille 9352	10.7	M0.5 V	23 05	–35 51	0.011
18	12	Ross 128 (FI Virginis)	10.9	M4 V	11 47	+00 48	3.4×10^{-4}

Table I1

Star	System	Discovery Name	Distance (light-year)	Spectral Type	Location: RA[1]	Location: Dec[2]	Luminosity (Sun = 1)
19	13	Luyten 789-6 A (EZ Aquarii A)	11.3	M5 V	22 38	−15 17	5×10^{-5}
20		Luyten 789-6 B (EZ Aquarii B)	11.3	M5.5 V	22 38	−15 15	5×10^{-5}
21		Luyten 789-6 C (EZ Aquarii C)	11.3	M6.5 V	22 38	−15 17	2×10^{-5}
22	14	61 Cygni A	11.4	K5 V	21 06	+38 44	0.086
23		61 Cygni B	11.4	K7 V	21 06	+38 44	0.041
24	15	Procyon A	11.4	F5 IV	07 39	+05 13	7.38
25		Procyon B	11.4	wd[4]	07 39	+05 13	5.5×10^{-4}
26	16	Sigma 2398 A	11.5	M3 V	18 42	+59 37	0.003
27		Sigma 2398 B	11.5	M3.5 V	18 42	+59 37	1.4×10^{-3}
28	17	Groombridge 34 A (GX Andromedae)	11.6	M1.5 V	00 18	+44 01	6.4×10^{-3}
29		Groombridge 34 B (GQ Andromedae)	11.6	M3.5 V	00 18	+44 01	4.1×10^{-4}
30	18	Epsilon Indi A	11.8	K5 V	22 03	−56 46	0.150
31		Epsilon Indi Ba	11.7	T1[5]	22 04	−56 46	—
32		Epsilon Indi Bb	11.7	T6[6]	22 04	−56 46	—
33	19	G 51-15 (DX Cancri)	11.8	M6.5 V	08 29	+26 46	1×10^{-5}
34	20	Tau Ceti	11.9	G8.5 V	01 44	−15 56	0.458
35	21	Luyten 372-58	12.0	M5 V	03 35	−44 30	7×10^{-5}
36	22	Luyten 725-32 (YZ Ceti)	12.1	M4.5 V	01 12	−16 59	1.8×10^{-4}

Table I1

1   Location (right ascension) given for Epoch 2000.0
2   Location (declination) given for Epoch 2000.0
3   White dwarf stellar remnant

Star	System	Discovery Name	Distance (light-year)	Spectral Type	Location: RA[1]	Location: Dec[2]	Luminosity (Sun = 1)
37	23	Luyten's Star	12.4	M3.5 V	07 27	+05 13	1.4×10^{-3}
38	24	SCR J184-6357 A	12.6	M8.5 V	18 45	−63 57	1×10^{-6}
39		SCR J184-6357 B	12.7	T6[7]	18 45	−63 57	—
40	25	Teegarden's Star	12.5	M6 V	02 53	+16 52	1×10^{-5}
41	26	Kapteyn's Star	12.8	M1 V	05 11	−45 01	3.8×10^{-3}
42	27	Lacaille 8760 (AX Microscopium)	12.9	K7 V	21 17	−38 52	0.029

Table I1

J   The Brightest Twenty Stars

Note: These are the stars that *appear* the brightest visually, as seen from our vantage point on Earth. They are not necessarily the stars that are intrinsically the most luminous.

The Brightest Twenty Stars										
Name					Proper Motion (arcsec/y)		Right Ascension		Declination	
Traditional	Bayer	Luminosity (Sun = 1)	Distance (light-years)	Spectral Type	RA	Dec	(h)	(m)	(deg)	(min)
Sirius	α Canis Majoris	22.5	8.6	A1 V	−0.5	−1.2	06	45.2	−16	43
Canopus	α Carinae	13,500	309	F0 II	+0.02	+0.02	06	24.0	−52	42
Rigil Kentaurus	α Centauri	1.94	4.32	G2 V + K IV	−3.7	+0.5	14	39.7	−60	50
Arcturus	α Bootis	120	36.72	K1.5 III	−1.1	−2.0	14	15.7	+19	11
Vega	α Lyrae	49	25.04	A0 V	+0.2	+0.3	18	36.9	+38	47
Capella	α Aurigae	140	42.80	G8 III + G0 III	+0.08	−0.4	05	16.7	+46	00
Rigel	β Orionis	50,600	863	B8 I	+0.00	+0.00	05	14.5	−08	12
Procyon	α Canis Minoris	7.31	11.46	F5 IV-V	−0.7	−1.0	07	39.3	+05	14
Achernar	α Eridani	1030	139	B3 V	+0.10	−0.04	01	37.7	−57	14
Betelgeuse	α Orionis	13,200	498	M2 I	+0.02	+0.01	05	55.2	+07	24
Hadar	β Centauri	7050	392	B1 III	−0.03	−0.02	14	03.8	−60	22
Altair	α Aquilae	11.2	16.73	A7 V	+0.5	+0.4	19	50.8	+08	52
Acrux	α Crucis	4090	322	B0.5 IV + B1V	−0.04	−0.01	12	26.6	−63	06
Aldebaran	α Tauri	160	66.64	K5 III	+0.1	−0.2	04	35.9	+16	31
Spica	α Virginis	2030	250	B1 III-IV + B2 V	−0.04	−0.03	13	25.2	−11	10
Antares	α Scorpii	9290	554	M1.5 I + B2.5 V	−0.01	−0.02	16	29.4	−26	26
Pollux	β Geminorum	31.6	33.78	K0 III	−0.6	−0.05	07	45.3	+28	02
Fomalhaut	α Piscis Austrini	17.2	25.13	A3 V	+0.03	−0.2	22	57.6	−29	37
Mimosa	β Crucis	1980	279	B0.5 III	−0.04	−0.02	12	47.7	−59	41
Deneb	α Cygni	50,600	1412	A2 I	+0.00	+0.00	20	41.4	+45	17

Figure J1 The brightest stars typically have names from antiquity. Next to each star's ancient name, we have added a column with its name in the system originated by Bayer (see the Naming Stars feature box.) The distances of the more remote stars are estimated from their spectral types and apparent brightnesses and are only approximate. The luminosities for those stars are approximate to the same degree. Right ascension and declination is given for Epoch 2000.0.

K   The Chemical Elements

The Chemical Elements

Element	Symbol	Atomic Number	Atomic Weight[1]	Percentage of Naturally Occurring Elements in the Universe
Hydrogen	H	1	1.008	75
Helium	He	2	4.003	23
Lithium	Li	3	6.94	6×10^{-7}
Beryllium	Be	4	9.012	1×10^{-7}
Boron	B	5	10.821	1×10^{-7}
Carbon	C	6	12.011	0.5
Nitrogen	N	7	14.007	0.1
Oxygen	O	8	15.999	1
Fluorine	F	9	18.998	4×10^{-5}
Neon	Ne	10	20.180	0.13
Sodium	Na	11	22.990	0.002
Magnesium	Mg	12	24.305	0.06
Aluminum	Al	13	26.982	0.005
Silicon	Si	14	28.085	0.07
Phosphorus	P	15	30.974	7×10^{-4}
Sulfur	S	16	32.06	0.05
Chlorine	Cl	17	35.45	1×10^{-4}
Argon	Ar	18	39.948	0.02
Potassium	K	19	39.098	3×10^{-4}
Calcium	Ca	20	40.078	0.007

Table K1

1   Where mean atomic weights have not been well determined, the atomic mass numbers of the most stable isotopes are given in

Element	Symbol	Atomic Number	Atomic Weight[1]	Percentage of Naturally Occurring Elements in the Universe
Scandium	Sc	21	44.956	3×10^{-6}
Titanium	Ti	22	47.867	3×10^{-4}
Vanadium	V	23	50.942	3×10^{-4}
Chromium	Cr	24	51.996	0.0015
Manganese	Mn	25	54.938	8×10^{-4}
Iron	Fe	26	55.845	0.11
Cobalt	Co	27	58.933	3×10^{-4}
Nickel	Ni	28	58.693	0.006
Copper	Cu	29	63.546	6×10^{-6}
Zinc	Zn	30	65.38	3×10^{-5}
Gallium	Ga	31	69.723	1×10^{-6}
Germanium	Ge	32	72.630	2×10^{-5}
Arsenic	As	33	74.922	8×10^{-7}
Selenium	Se	34	78.971	3×10^{-6}
Bromine	Br	35	79.904	7×10^{-7}
Krypton	Kr	36	83.798	4×10^{-6}
Rubidium	Rb	37	85.468	1×10^{-6}
'Strontium	Sr	38	87.62	4×10^{-6}
Yttrium	Y	39	88.906	7×10^{-7}
Zirconium	Zr	40	91.224	5×10^{-6}
Niobium	Nb	41	92.906	2×10^{-7}
Molybdenum	Mo	42	95.95	5×10^{-7}

Table K1

parentheses.

Element	Symbol	Atomic Number	Atomic Weight[1]	Percentage of Naturally Occurring Elements in the Universe
Technetium	Tc	43	(98)	—
Ruthenium	Ru	44	101.07	4×10^{-7}
Rhodium	Rh	45	102.906	6×10^{-8}
Palladium	Pd	46	106.42	2×10^{-7}
Silver	Ag	47	107.868	6×10^{-8}
Cadmium	Cd	48	112.414	2×10^{-7}
Indium	In	49	114.818	3×10^{-8}
Tin	Sn	50	118.710	4×10^{-7}
Antimony	Sb	51	121.760	4×10^{-8}
Tellurium	Te	52	127.60	9×10^{-7}
Iodine	I	53	126.904	1×10^{-7}
Xenon	Xe	54	131.293	1×10^{-6}
Cesium	Cs	55	132.905	8×10^{-8}
Barium	Ba	56	137.327	1×10^{-6}
Lanthanum	La	57	138.905	2×10^{-7}
Cerium	Ce	58	140.116	1×10^{-6}
Praseodymium	Pr	59	140.907	2×10^{-7}
Neodymium	Nd	60	144.242	1×10^{-6}
Promethium	Pm	61	(145)	—
Samarium	Sm	62	150.36	5×10^{-7}
Europium	Eu	63	151.964	5×10^{-8}
Gadolinium	Gd	64	157.25	2×10^{-7}

Table K1

Element	Symbol	Atomic Number	Atomic Weight[1]	Percentage of Naturally Occurring Elements in the Universe
Terbium	Tb	65	158.925	5×10^{-8}
Dysprosium	Dy	66	162.500	2×10^{-7}
Holmium	Ho	67	164.930	5×10^{-8}
Erbium	Er	68	167.259	2×10^{-7}
Thulium	Tm	69	168.934	1×10^{-8}
Ytterbium	Yb	70	173.054	2×10^{-7}
Lutetium	Lu	71	174.967	1×10^{-8}
Hafnium	Hf	72	178.49	7×10^{-8}
Tantalum	Ta	73	180.948	8×10^{-9}
Tungsten	W	74	183.84	5×10^{-8}
Rhenium	Re	75	186.207	2×10^{-8}
Osmium	Os	76	190.23	3×10^{-7}
Iridium	Ir	77	192.217	2×10^{-7}
Platinum	Pt	78	195.084	5×10^{-7}
Gold	Au	79	196.967	6×10^{-8}
Mercury	Hg	80	200.592	1×10^{-7}
Thallium	TI	81	204.38	5×10^{-8}
Lead	Pb	82	207.2	1×10^{-6}
Bismuth	Bi	83	208.980	7×10^{-8}
Polonium	Po	84	(209)	—
Astatine	At	85	(210)	—
Radon	Rn	86	(222)	—

Table K1

Element	Symbol	Atomic Number	Atomic Weight[1]	Percentage of Naturally Occurring Elements in the Universe
Francium	Fr	87	(223)	—
Radium	Ra	88	(226)	—
Actinium	Ac	89	(227)	—
Thorium	Th	90	232.038	4×10^{-8}
Protactinium	Pa	91	231.036	—
Uranium	U	92	238.029	2×10^{-8}
Neptunium	Np	93	(237)	—
Plutonium	Pu	94	(244)	—
Americium	Am	95	(243)	—
Curium	Cm	96	(247)	—
Berkelium	Bk	97	(247)	—
Californium	Cf	98	(251)	—
Einsteinium	Es	99	(252)	—
Fermium	Fm	100	(257)	—
Mendelevium	Md	101	(258)	—
Nobelium	No	102	(259)	—
Lawrencium	Lr	103	(262)	—
Rutherfordium	Rf	104	(267)	—
Dubnium	Db	105	(268)	—
Seaborgium	Sg	106	(271)	—
Bohrium	Bh	107	(272)	—
Hassium	Hs	108	(270)	—

Table K1

Element	Symbol	Atomic Number	Atomic Weight[1]	Percentage of Naturally Occurring Elements in the Universe
Meitnerium	Mt	109	(276)	—
Darmstadtium	Ds	110	(281)	—
Roentgenium	Rg	111	(280)	—
Copernicium	Cn	112	(285)	—
Nihonium	Nh	113	(284)	—
Flerovium	Fl	114	(289)	—
Moskovium	Mc	115	(288)	—
Livermorium	Lv	116	(293)	—
Tennessine	Ts	117	(294)	—
Oganesson	Og	118	(294)	—

Table K1

L   The Constellations

The Constellations

Constellation (Latin name)	Genitive Case Ending	English Name or Description	Abbreviation	Approximate Position: α (h)	Approximate Position: δ (°)
Andromeda	Andromedae	Princess of Ethiopia	And	1	+40
Antila	Antilae	Air pump	Ant	10	−35
Apus	Apodis	Bird of Paradise	Aps	16	−75
Aquarius	Aquarii	Water bearer	Aqr	23	−15
Aquila	Aquilae	Eagle	Aql	20	+5
Ara	Arae	Altar	Ara	17	−55
Aries	Arietis	Ram	Ari	3	+20
Auriga	Aurigae	Charioteer	Aur	6	+40
Boötes	Boötis	Herdsman	Boo	15	+30
Caelum	Cael	Graving tool	Cae	5	−40
Camelopardus	Camelopardis	Giraffe	Cam	6	+70
Cancer	Cancri	Crab	Cnc	9	+20
Canes Venatici	Canum Venaticorum	Hunting dogs	CVn	13	+40
Canis Major	Canis Majoris	Big dog	CMa	7	−20
Canis Minor	Canis Minoris	Little dog	CMi	8	+5
Capricornus	Capricorni	Sea goat	Cap	21	−20
Carina[1]	Carinae	Keel of Argonauts' ship	Car	9	−60
Cassiopeia	Cassiopeiae	Queen of Ethiopia	Cas	1	+60

Table L1

1   The four constellations Carina, Puppis, Pyxis, and Vela originally formed the single constellation Argo Navis.

Constellation (Latin name)	Genitive Case Ending	English Name or Description	Abbreviation	Approximate Position: α (h)	Approximate Position: δ (°)
Centaurus	Centauri	Centaur	Cen	13	−50
Cepheus	Cephei	King of Ethiopia	Cep	22	+70
Cetus	Ceti	Sea monster (whale)	Cet	2	−10
Chamaeleon	Chamaeleontis	Chameleon	Cha	11	−80
Circinus	Circini	Compasses	Cir	15	−60
Columba	Columbae	Dove	Col	6	−35
Coma Berenices	Comae Berenices	Berenice's hair	Com	13	+20
Corona Australis	Coronae Australis	Southern crown	CrA	19	−40
Corona Borealis	Coronae Borealis	Northern crown	CrB	16	+30
Corvus	Corvi	Crow	Crv	12	−20
Crater	Crateris	Cup	Crt	11	−15
Crux	Crucis	Cross (southern)	Cru	12	−60
Cygnus	Cygni	Swan	Cyg	21	+40
Delphinus	Delphini	Porpoise	Del	21	+10
Dorado	Doradus	Swordfish	Dor	5	−65
Draco	Draconis	Dragon	Dra	17	+65
Equuleus	Equulei	Little horse	Equ	21	+10
Eridanus	Eridani	River	Eri	3	−20
Fornax	Fornacis	Furnace	For	3	−30
Gemini	Geminorum	Twins	Gem	7	+20

Table L1

Constellation (Latin name)	Genitive Case Ending	English Name or Description	Abbreviation	Approximate Position: α (h)	Approximate Position: δ (°)
Grus	Gruis	Crane	Gru	22	−45
Hercules	Herculis	Hercules, son of Zeus	Her	17	+30
Horologium	Horologii	Clock	Hor	3	−60
Hydra	Hydrae	Sea serpent	Hya	10	−20
Hydrus	Hydri	Water snake	Hyi	2	−75
Indus	Indi	Indian	Ind	21	−55
Lacerta	Lacertae	Lizard	Lac	22	+45
Leo	Leonis	Lion	Leo	11	+15
Leo Minor	Leonis Minoris	Little lion	LMi	10	+35
Lepus	Leporis	Hare	Lep	6	−20
Libra	Librae	Balance	Lib	15	−15
Lupus	Lupi	Wolf	Lup	15	−45
Lynx	Lyncis	Lynx	Lyn	8	+45
Lyra	Lyrae	Lyre or harp	Lyr	19	+40
Mensa	Mensae	Table Mountain	Men	5	−80
Microscopium	Microscopii	Microscope	Mic	21	−35
Monoceros	Monocerotis	Unicorn	Mon	7	−5
Musca	Muscae	Fly	Mus	12	−70
Norma	Normae	Carpenter's level	Nor	16	−50
Octans	Octantis	Octant	Oct	22	−85
Ophiuchus	Ophiuchi	Holder of serpent	Oph	17	0

Table L1

Constellation (Latin name)	Genitive Case Ending	English Name or Description	Abbreviation	Approximate Position: α (h)	Approximate Position: δ (°)
Orion	Orionis	Orion, the hunter	Ori	5	+5
Pavo	Pavonis	Peacock	Pav	20	−65
Pegasus	Pegasi	Pegasus, the winged horse	Peg	22	+20
Perseus	Persei	Perseus, hero who saved Andromeda	Per	3	+45
Phoenix	Phoenicis	Phoenix	Phe	1	−50
Pictor	Pictoris	Easel	Pic	6	−55
Pisces	Piscium	Fishes	Psc	1	+15
Piscis Austrinus	Piscis Austrini	Southern fish	PsA	22	−30
Puppis[2]	Puppis	Stern of the Argonauts' ship	Pup	8	−40
Pyxis[3] (=Malus)	Pyxidus	Compass of the Argonauts' ship	Pyx	9	−30
Reticulum	Reticuli	Net	Ret	4	−60
Sagitta	Sagittae	Arrow	Sge	20	+10
Sagittarius	Sagittarii	Archer	Sgr	19	−25
Scorpius	Scorpii	Scorpion	Sco	17	−40
Sculptor	Sculptoris	Sculptor's tools	Scl	0	−30
Scutum	Scuti	Shield	Sct	19	−10
Serpens	Serpentis	Serpent	Ser	17	0
Sextans	Sextantis	Sextant	Sex	10	0
Taurus	Tauri	Bull	Tau	4	+15

Table L1

2   The four constellations Carina, Puppis, Pyxis, and Vela originally formed the single constellation Argo Navis.
3   The four constellations Carina, Puppis, Pyxis, and Vela originally formed the single constellation Argo Navis.

Constellation (Latin name)	Genitive Case Ending	English Name or Description	Abbreviation	Approximate Position: α (h)	Approximate Position: δ (°)
Telescopium	Telescopii	Telescope	Tel	19	−50
Triangulum	Trianguli	Triangle	Tri	2	+30
Triangulum Australe	Trianguli Australis	Southern triangle	TrA	16	−65
Tucana	Tucanae	Toucan	Tuc	0	−65
Ursa Major	Ursae Majoris	Big bear	UMa	11	+50
Ursa Minor	Ursae Minoris	Little bear	UMi	15	+70
Vela[4]	Velorum	Sail of the Argonauts' ship	Vel	9	−50
Virgo	Virginis	Virgin	Vir	13	0
Volans	Volantis	Flying fish	Vol	8	−70
Vulpecula	Vulpeculae	Fox	Vul	20	+25

Table L1

---

4   The four constellations Carina, Puppis, Pyxis, and Vela originally formed the single constellation Argo Navis.

Star Chart and Sky Event Resources

Star Charts

To obtain graphic charts of the sky over your head tonight, there are a number of free online resources.

- One of the easiest is at the Skymaps website: http://www.skymaps.com/downloads.html (http://www.skymaps.com/downloads.html). Here you can print out a free PDF version of the northern sky (for roughly latitude 40°, which is reasonable for much of the United States). Maps of the sky from the equator and the southern hemisphere can also be printed.
- Sky charts and other summaries of astronomical information can also be found at http://www.heavens-above.com/ (http://www.heavens-above.com/).
- A free, open-source computer application that shows the sky at any time from any place is called Stellarium. You can find it at http://www.stellarium.org/ (http://www.stellarium.org/).
- Appendix B provides a section of information for finding astronomy apps for cell phones and tablets. Many of these also provide star charts. If you have a smartphone, you can find a variety of inexpensive apps that allow you to simply hold your phone upward to see what is in the sky behind your phone.
- A *planisphere* is a sky chart that turns inside a round frame and can show you the night sky at your latitude on any date and time of the year. You can buy them at science supply and telescope stores or online. Or you can construct your own from templates at these two websites:
  - Dennis Schatz's AstroAdventures Star Finder: http://dennisschatz.org/activities/Star%20Finder.pdf (http://dennisschatz.org/activities/Star%20Finder.pdf)
  - Uncle Al's Star Wheel: http://www.lawrencehallofscience.org/do_science_now/science_apps_and_activities/starwheels (http://www.lawrencehallofscience.org/do_science_now/science_apps_and_activities/starwheels)

Calendars of Night Sky Events

The following resources offer calendars of night sky events.

- Sea and Sky: http://www.seasky.org/astronomy/astronomy.html (click on the fourth menu button) (http://www.seasky.org/astronomy/astronomy.html)
- *Sky & Telescope's* This Week's Sky at a Glance: http://www.skyandtelescope.com/observing/sky-at-a-glance/ (http://www.skyandtelescope.com/observing/sky-at-a-glance/)
- *Astronomy* Magazine's The Sky This Week: http://www.astronomy.com/observing/sky-this-week (http://www.astronomy.com/observing/sky-this-week)
- Dr. Fred Espenak's Year by Year Tables of Events: http://astropixels.com/ephemeris/ephemeris.html (http://astropixels.com/ephemeris/ephemeris.html)
- Night Sky Network Sky Planner: https://nightsky.jpl.nasa.gov/planner.cfm (https://nightsky.jpl.nasa.gov/planner.cfm)

Index

Symbols

21-cm line   654, 844, 886
3C 273   900, 901

A

absorption spectrum   152
accelerate   54
accretion   474
accretion disk   821, 908, 912, 913, 916
active galactic nuclei (AGN)   906
active galaxies   906
active region   513
Adams   85
adaptive optics   191
afterglow   784
Alcor   590
Algol   597, 598
Allen Telescope Array   1051
Alpha Centauri   141, 588, 623, 625, 627
Alpher   995, 1000
Amino acids   1031
Andromeda galaxy   22, 23, 857, 863, 864, 875, 881, 910, 928, 944
angular momentum   73
Ant Nebula   719
anthropic principle   1018
antimatter   532, 996, 1017
antiparticle   532
apertures   177
aphelion   78
apogee   78
Apollo program   281
apparent brightness   558, 562
apparent magnitudes   43
apparent solar time   106
Arecibo Observatory   198
Aristarchus   40
Aristotle   40
Armstrong   289
association   729
asteroid belt   79, 356, 423
asteroids   224, 399, 422, 422, 423, 424, 425, 432, 469, 470, 483
astrobiology   1030
astrology   46
astronomical unit   620

astronomical unit (AU)   67
astronomy   11
Atacama Large Millimeter/submillimeter Array   198
atmosphere   365, 371
aurora   518
auroras   503
Australia Telescope Compact Array   199

B

B–V index   563
Baade   857, 884
Balmer lines   567
bar   257
Barnard   659
barred spiral galaxy   841
baryon cycle   669
basalt   248
Bayer   626
Bayer letter   627
Bell   773
BeppoSAX   784, 785
Bessel   623
Betelgeuse   38, 562, 566, 601, 626, 626, 682, 723, 724, 725
Bethe   994
Big Bang   979, 993, 1012
big crunch   988
Big Telescope Altazimuth   184
binary star system   779
binary stars   589
binding energy   534
biomarker   1039
black dwarf   758, 791
black hole   761, 772, 782, 787, 789, 813, 814, 816, 817, 819, 820, 822, 823, 850, 851, 853, 853, 908, 910, 916, 920
black widow pulsar   781
blackbody   146
blueshift   165
Bohr   157
Bohr model   159
Borexino experiment   550
Brahe   64, 767
Brookhaven National Laboratory   548

brown dwarfs   568, 569, 593

C

C-type asteroids   425
calendar   108
Callisto   382, 384, 384, 392
Callisto's surface   385
Canberra Deep Space Communication Complex   198
Cannon   564, 567
Carrington Event   514
Cassini   358
celestial equator   32
celestial poles   32
celestial sphere   31
centaurs   446
central bulge   841
cepheid   631
Ceres   80, 80, 220, 220, 399, 422, 423, 425, 430
Chandra X-Ray Observatory   206
Chandrasekhar   756
Chandrasekhar limit   755, 779
CHARA   201
charge-coupled devices (CCDs)   193
Charon   397, 398, 403, 404
Chelyabinsk   268
Cherenkov Telescope Array   208
Chicxulub   269
Chiron   445
CHNOPS   1035
chromatic aberration   180
chromosphere   500
circumpolar zone   33
circumstellar disk   477
climate   260
closed universe   988
CNO cycle   537
COBE   1002
cold dark matter   961
collisions   863
color index   562
Coma cluster   946
comet   436
Comet Churyumov-Gerasimenko   443, 444
comet dust   440

Comet Hale-Bopp   421, 438, 442
Comet Halley   436, 437, 438, 441, 441, 452, 462
Comet LINEAR   451
Comet Shoemaker-Levy 9   449, 450, 451, 452
Comet Shoemaker–Levy 9   232, 232
comets   224, 438, 440, 441, 461, 463, 470, 475, 650, 660
Compton Gamma-Ray Observatory   782
conduction   541
conservation of angular momentum   73
constellation   35, 38
continuous spectrum   152
convection   251, 541, 544
convection currents   371
convective zone   498
coordinates   96
Copernican principle   1030
Copernicus   50, 51, 52, 56, 1024
core   249
corona   500, 502
coronal hole   514
coronal holes   503
coronal mass ejection   515, 547
coronal mass ejection (CME)   512
cosmic haystack problem   1051
cosmic microwave background (CMB)   1001
Cosmic rays   666, 764
cosmological constant   978
cosmological principle   943, 985, 990
cosmology   39
Crab Nebula   10, 767, 773, 774, 776
crater   291, 292
crater counts   290
craters   233, 266
critical density   987
crust   248
Curtis   905
Cygnus X-1   799, 822

D

Dactyl   427, 427

dark energy   961, 982
dark matter   849, 849, 877, 881, 882, 955, 956, 957, 960, 961, 1008
dark matter halo   841
Darwin   118
Davis   548, 549
declination   97
degenerate gas   755
Deimos   427, 428
Delta Cephei   631, 635, 638
density   72, 222
detector   176
deuterium   998
differential galactic rotation   846
differential rotation   505
Differentiation   230
disk   841
dispersion   150
distance   618
DNA (deoxyribonucleic acid)   1033
Doppler   163, 572
Doppler effect   163, 166, 478, 593, 696, 880
Doppler shift   821, 902
Drake   1054
Drake equation   1054
Draper   577
Dreyer   651
dwarf planet   399

E

Eagle Nebula   681
Earth   17, 80, 96, 119, 222, 230, 245, 246, 250, 257, 261, 266, 345, 359, 481, 484
Earth-approaching asteroids   433, 434, 434
Earth's atmosphere   257, 259
Earth's crust   248, 250, 251
Earth's interior   247
Earth's magnetic field   249
Earth's magnetosphere   249
Earth's surface   248
eccentricity   66
eclipsing binary   596
ecliptic   33, 35
Eddington   609
Effelsberg 100-m Telescope   198

Einstein   530, 531, 800, 805, 808, 957, 978
electromagnetic radiation   138
electromagnetic spectrum   143
ellipse   65
Elliptical galaxies   877
emission spectrum   152
Enceladus   383, 408, 409, 410, 485, 1042
energy   530
energy flux   148
energy level   157
epicycle   45
equivalence principle   802
Eratosthenes   41, 42
Eris   219, 220, 399, 446
Eros   428
escape speed   83
escape velocity   813
Eta Carinae   744, 744
Europa   382, 384, 386, 387, 387, 392, 423, 1041, 1042
European Extremely Large Telescope   183, 208
event horizon   816, 816, 910
evolution   929
Ewen   654
excitation   160
exoplanet   478, 695
exoplanets   477
expanding universe   892
extremophile   1036
eyepiece   179

F

F Ring   408, 413, 413
faults   254
Fermi   1030
Fermi Gamma-ray Space Telescope   207
Fermi paradox   1030
Fisher   886
fission   534
Five-hundred-meter Aperture Spherical radio Telescope (FAST)   197
Five-hundred-meter Aperture Spherical Telescope   202
Flamsteed   626
Flamsteed number   627

flat universe   1004
Fleming   564
focus   65, 178
formaldehyde   657
Foucault   98
Fraunhofer   151, 563
frequency   139
fusion   534, 535, 736, 760, 994

G

Gaia   207
galactic cannibalism   938
galactic mass   881
galaxies   872, 875, 880, 928
galaxy clusters   945
galaxy collisions   937
Galileo   53, 54, 55, 56, 64, 95, 177, 505, 589
Galle   85
gamma rays   143
gamma-ray burst   782, 787, 787
gamma-ray bursts   786
Gamow   1000
Ganymede   230, 382, 384, 385, 386, 392
Gaspra   426
Geller   950
Gemini North   184
Gemini South   184
general theory of relativity   800
genes   1032
geocentric   30
Ghez   855
Giacconi   205
giant impact hypothesis   296
Giant Magellan Telescope   183
giant molecular clouds   681
giant planets   221, 354, 359
giant stars   601
giants   570
Gilbert   291
globular cluster   726, 745
Globular clusters   727, 984
Goldstone Deep Space
Communications Complex   198
Goodricke   597, 634
Gran Telescopio Canarias   183
grand unified theories   1014
granite   248
granulation   499

gravitational energy   529
gravitational lens   948, 960
gravitational redshift   812
gravitational wave   824, 824, 825
gravitational waves   790
gravity   74, 811, 818
great circle   96
Great Red Spot   373
Green Bank Telescope   198
greenhouse effect   264, 324
greenhouse gases   264
Gregorian calendar   110
ground state   160

H

H II region   652
H–R diagram   604, 636, 689, 720, 729, 757
habitable environment   1034
habitable zone   1044
Hale   186, 578
Hale Telescope   184
half-life   234
Halley   70, 436, 437
halo   841, 841
Haro   688
Hat Creek Radio Observatory   199
Haumea   220, 446
heat transfer   543
heliocentric   50
helioseismology   546
helium flash   736
Henderson   623
Herbig   688
Herbig-Haro (HH) object   687
Herman   995, 1000
Herschel   85, 589, 651, 836
Hertz   140
Hertzsprung   602
Hewish   773
highlands   287
Hipparchus   42, 558
HL Tau   477, 694
Hobby–Eberly Telescope   183
Holden   660
homogeneous   943
horizon   30
horizon distance   1012
horoscope   47

hot dark matter   961
hot Jupiter   480
hot Jupiters   698, 710
Hubble   188, 634, 872, 873, 889, 977
Hubble constant   890, 980, 989, 1005
Hubble law   903, 980
Hubble Space Telescope   191, 204, 206, 903, 908
Hubble time   980
Hubble's law   890, 891, 991
Huggins   563, 572
Humason   888
hydrostatic equilibrium   541

I

Ida   426, 427, 484
igneous rock   250
impact craters   291, 294
infalling material   912
inflationary universe   1013
infrared   144
Interface Region Imaging
Spectrograph   501
Interference   199
interferometer   199
interferometer array   199
International Astronomical
Union   228
International Date Line   108
International Gamma-Ray
Astrophysics Laboratory   207
International Thermonuclear
Experimental Reactor (ITER)   538
interstellar dust   648, 661, 664
interstellar extinction   663
interstellar gas   651, 653, 655
interstellar mass   649
interstellar medium   668, 668, 668
interstellar medium (ISM)   648
interstellar molecules   657
inverse square law   142
Io   250, 364, 382, 384, 388, 389, 390, 391, 391
ion   162
ionization   162
IRAM   199

irons 468
irregular galaxy 879
isotopes 156
isotropic 943
Itokawa 429

J

James Clerk Maxwell Telescope 199
James Webb Space Telescope 207
Jansky 195
Jansky Very Large Array 198
Janssen 177
jets 912
Jupiter 55, 80, 191, 219, 221, 222, 229, 229, 232, 232, 353, 354, 355, 355, 356, 359, 360, 360, 361, 362, 363, 364, 367, 369, 372, 372, 373, 381, 382, 384, 404, 448, 450, 451, 452, 479, 569, 593
Jupiter's storms 373

K

Kajita 549
Kant 872
Keck 578
Keck I and II 183, 201
Kelvin 528
Kepler 65, 66, 67, 207, 695, 767, 848
Kepler mission 704
Kepler's first law 68
Kepler's second law 68
Kepler's Supernova 767, 780
Kepler's third law 68, 69
Koshiba 549
Kuiper belt 446, 448

L

Large Binocular Telescope 184, 201
Large Magellanic Cloud 22, 29, 632, 633, 768, 771, 879, 960
Large Synoptic Survey Telescope 184
law of conservation of energy 528
Le Verrier 85

Leavitt 632, 872
Lemaître 888, 994
Levy 449
Lick 577
light 141
light curve 631
light pollution 189
light-year 625
LIGO 824, 825
line broadening 881
Lippershey 177
lithium 998
Little Ice Age 518
Local Bubble 670
Local Fluff 670
Local Group 24, 944, 944, 957
long-duration gamma-ray burst 787
Lovell Telescope 198
Lowell 313, 314, 315, 397
luminosity 558
luminosity classes 637
lunar eclipse 119
lunar exploration 280
lunar history 287
lunar missions 284

M

M-type asteroids 425
M87 906, 908, 909, 941
MACHO 960
Magellan Telescopes 184
magnetic field 364
magnetism 137
magnetogram 508
magnetosphere 249, 514, 518
magnitude 559, 559
main sequence 604, 606, 720, 721
main-sequence stars 607
main-sequence turnoff 733
major axis 66
Makemake 220, 399, 446
mantle 248
Mars 10, 44, 66, 80, 217, 222, 230, 258, 263, 311, 313, 316, 317, 325, 328, 329, 331, 334, 334, 337, 343, 345, 387, 428, 481, 484, 1039
martian meteorites 328
mass 72

mass extinction 269
mass loss 740
mass-luminosity relation 594
mass-to-light ratio 882, 959
mass, 72
Maunder 517
Maunder Minimum 517
Maxwell 136, 137, 1015
Maxwell Mountains 482
Mayor 697
McDonald 549
mean solar time 107
Mercury 80, 221, 222, 230, 231, 280, 297, 298, 300, 302, 808, 809
Mercury's composition 297
Mercury's orbit 297
Mercury's rotation 298
Mercury's surface 300
merger 938
meridian 96
MESSENGER 300
Messier 650
Metamorphic rocks 250
meteor 225, 460
Meteor Crater 268, 294
meteor shower 461, 463
meteorite 225, 465
meteorite fall 465
meteorites 468
meter 618
Metius 177
Michell 814
microwave 144
Milky Way Galaxy 18, 20, 29, 195, 628, 649, 655, 662, 667, 668, 710, 836, 841, 843, 864, 881, 912, 944
Miller 1031
Miller-Urey 1032
millisecond pulsars 781
mini-Neptunes 704
Mizar 590
molecular clouds 656, 680
momentum 70
Moon 17, 111, 114, 115, 116, 119, 218, 220, 230, 234, 279, 280, 285, 286, 290, 296, 384, 481, 596, 622
Moon's highlands 287
Moon's interior 285

Moon's origin   296
Moon's properties   280
Moon's surface   286
moons   382, 383
Multi-Element Radio Linked Interferometer Network   199
Multi-Mirror Telescope   184
multiverse   1018

N

near-Earth asteroids (NEAs)   433
near-Earth objects (NEOs)   432
nebula   648, 650
Neptune   81, 85, 221, 222, 229, 353, 355, 356, 357, 359, 360, 361, 363, 363, 365, 370, 371, 373, 374, 383, 395, 397, 405, 446
Neptune's Great Dark Spot   374
Neptune's rings   411, 411
neutrino   532, 536, 548, 549, 762, 772, 997, 997, 1017
neutron star   761, 772, 774, 778, 781, 787, 789
New Horizons   401
Newton   69, 74, 150, 437, 800, 848
Newton's first law   70
Newton's second law   70
Newton's third law   70
Nobeyama Radio Observatory   199
nova   779
nuclear bulge   841
nucleosynthesis   745
nucleus   155, 439

O

Oliver   1048
Olympus Mons   329, 329, 329, 330, 482, 482
Omega Centauri   727, 727
Oort   446
Oort cloud   447
open cluster   726, 759
Open clusters   728
open universe   988
optical SETI   1056
orbit   65
orbital period   67

orbital speed   66
organic compounds   1031
organic molecule   1028
Orion Nebula   14, 238, 476, 594, 651, 682, 683, 845
oscillating theory of the universe   988
Osterbrock   660
ozone   258

P

Pallas   422
parallax   40, 621, 624, 629
Parkes Observatory   198
parsec   624
partial eclipse   123
Pauli   532
Pauli exclusion principle   754
Payne-Gaposchkin   496
perigee   78
perihelion   78
period-luminosity relation   632, 633
perturbations   84
phases   111
Phobos   427, 428
photochemistry   369
photon   140, 158
photon decoupling time   999
photosphere   498
photosynthesis   263, 1033
Piazzi   422
Pickering   590
Pioneer   1048
Pioneer 10   356
Pioneer 11   356
plages   510
Planck   158
Planck time   1016
planetary evolution   345
planetary nebulae   739
planetary system   476
planetesimal   237, 474
planetesimals   238
planets   36
plasma   497, 540
Plate tectonics   251
Pleiades   662, 728
Pluto   219, 220, 220, 221, 231, 397, 398, 399, 401, 401, 402, 402,

446
Polaris   33, 43, 98, 560, 632
population I   858
population II   858
positron   535
precession   43
prime focus   180
primitive meteorites   469
primitive rock   250
prominences   510
proper motion   573
Proteins   1031
protogalactic cloud   860
proton-proton chain   536, 537
protoplanet   474
protostar   685
Proxima Centauri   18, 587, 588, 627, 650, 712, 1045
Ptolemy   44, 46
pulsar   777
pulsars   773
pulsating variable stars   631
Purcell   654
Pythagoras   40

Q

quantum mechanics   982
quasar   900, 907, 932
quasars   901, 915
Queloz   697
quiescent   911

R

Radar   201, 619
radial velocity   164, 572
radiation   136, 541
radiative zone   498
radio astronomy   854
radio observations   845
radio waves   145
radioactive decay   771
radioactivity   234
Reber   196
reddening   663
redshift   165, 888, 986
reflecting telescope   180
refracting telescope   179
Resolution   190
resonance   412
retrograde motion   45

rift zones   253

right ascension   97

ring system   404

ring systems   406

RNA (ribonucleic acid)   1033

ROTSE   785

RR Lyrae   631, 635

runaway greenhouse effect   324

Russell   602, 603, 756

Rutherford   154

S

S-type asteroids   425

Sagan   225

Sagittarius A   851, 852

Sagittarius A*   851, 853, 853

satellite   78, 81

Saturn   80, 221, 222, 224, 224,
229, 353, 355, 355, 358, 359, 360,
361, 362, 363, 368, 369, 370, 372,
383, 392, 404, 475

Saturn's rings   359, 368, 406,
407, 412, 475

scarps   302

Schiaparelli   313

Schmidt   914

Schmitt   282

Schwabe   506

Schwarzschild   816

science   11

Scott   287

seasons   99, 104

Sedimentary rocks   250

seeing   189

seismic waves   247

selection effect   588

semimajor axis   66

SETI   1052, 1052

Seyfert   906

Shapley   837

short-duration gamma-ray
burst   788

sidereal day   105

sidereal month   114

singularity   819

Sirius   37, 106, 558, 559, 560,
566, 572, 574, 587, 593, 595, 600,
608, 609, 609, 626, 627, 758

Slipher   887

Sloan Digital Sky Survey   951,

954

Small Magellanic Cloud   22, 29,
585, 632, 859, 879, 960

SN 1006   766, 780

SN 1054   767, 773, 780

solar atmosphere   495

solar cycle   504, 508, 509, 513,
518

solar day   105

solar dynamo   508

solar eclipse   119

solar flare   511, 514, 547

solar interior   539, 546

solar maximum   508, 515, 515

solar month   114

solar nebula   237, 471, 472

solar neutrino   548

solar wind   502

solstice   101

Southern African Large
Telescope   183

space velocity   574

space weather   514

spacetime   806, 806, 807

spectra   152

spectral class   564, 565

spectral line   507

spectrometer   151

spectroscopic binary   590

spiral arms   846

spiral galaxies   875, 876

Spitzer Space Telescope   207

Spörer   517

Square Kilometre Array   198

standard bulb   883, 885, 981

star cluster   726

star diameter   596

star formation   685

star mass   589

starburst   939

Stefan-Boltzmann law   148

stellar association   726, 729

stellar color   561

stellar evolution   726, 735, 746

stellar spectra   564

stellar temperature   561

stellar wind   686

stones   468

stony-irons   468

stratosphere   257

string theory   1015

stromatolites   1033

strong nuclear force   533

Struve   623

Subaru Telescope   184

subduction   254

Sudbury Neutrino Observatory
549

Sun   18, 99, 119, 494, 504, 510,
513, 528, 535, 539, 546, 562, 566,
619, 670, 723

sunspot   547

sunspot cycle   506, 507

sunspot maxima   506

sunspot minima   506

sunspots   504

super-Earths   704

superclusters   951

supergiant stars   601

supermassive black hole   853,
855, 907, 920, 921

supermassive black holes   851

supernova   667, 762, 762, 766

supernova 1987A   769

Swift   788

synchronous rotation   114

synchrotron radiation   364

T

T Tauri star   686

tail   436

Tarter   1052

tectonic   321

telescope   55, 176

terrestrial planets   220

theory of general relativity   978,
986, 997, 1004

theory of relativity   530

thermophile   1036

Thirty-Meter Telescope   183

Thomson   154

tidal disruption event   918

tidal heating   386, 391

tides   116

time zones   105

Titan   230, 358, 383, 384, 392,
394, 395, 484, 1042

Titan's atmosphere   393

Titan's surface   393

Tombaugh   397, 399

total eclipse   121

trans-Neptunian object   219

transit   699

transit depth   699

transit technique   478

Transiting Exoplanet Survey Satellite (TESS)   207

transition region   500

tree of life   262

triple-alpha process   736

Triton   383, 395, 396, 397, 449

Triton's surface   396

troposphere   257

Tully   886

Tully-Fisher relation   887

Tunguska   267

Tycho's Supernova   767, 780

type Ia supernova   781, 885, 886, 891

type Ia supernovae   780

type Ic supernova   787

type II supernova   762, 781

Tyson   227

U

ultraviolet   144

unbarred spiral galaxy   840

universal law of gravitation   75

Uranus   80, 85, 221, 222, 229, 353, 355, 356, 359, 360, 361, 362, 363, 363, 370, 372, 383, 405, 411, 413

Uranus' rings   410, 411

Urey   1031

V

Valles Marineris   10, 330, 330, 331

Van Allen   366

Van Allen belt   367

variable star   631

variable stars   883

Vega   43, 562, 562, 566, 574, 576, 587, 623, 625, 843

Vela   782

velocity   71

Venus   53, 55, 80, 220, 222, 230, 258, 311, 312, 316, 316, 317, 318, 319, 320, 322, 323, 345, 481, 484, 559

Very Large Telescope   184, 201

Very Long Baseline Array   198

Vesta   80, 231, 234, 422, 423, 426, 426, 430

Virgo Cluster   914, 945, 957

Virgo Supercluster   24

visible light   144

visual binary   589

Vogel   597

voids   952

Volcanoes   256

volume   72

Voyager 2   357

Voyager record   1049

W

watt   528

wave equation   140

wavelength   139

waves   138

weakly interacting massive particles   1008

weather   259

Wegener   252

Westerbork Synthesis Radio Telescope   198

Wheeler   814, 988

Whipple   440

white dwarf   608, 779

white dwarfs   605, 754, 755, 772

Wide-field Infrared Survey Explorer   207

Wien's law   148

WMAP   1003

Wolf   422

X

X-rays   143

XMM-Newton   207

Y

year   33

Yerkes   187, 578

Z

Zeeman effect   507, 507

zenith   30

zero-age main sequence   720

zodiac   37